siRNA and miRNA Gene Silencing

METHODS IN MOLECULAR BIOLOGY™

John M. Walker, SERIES EDITOR

METHODS IN MOLECULAR BIOLOGY™

siRNA and miRNA Gene Silencing

From Bench to Bedside

Edited by

Mouldy Sioud

*Department of Immunology, Institute for Cancer Research,
The Norwegian Radium Hospital, University of Oslo, Norway*

Humana Press

Editor
Mouldy Sioud
Department of Immunology
Institute for Cancer Research
The Norwegian Radium Hospital
University of Oslo
Norway

Series Editor
John M. Walker
School of Life Sciences
University of Hertfordshire
Hatfield, Hertfordshire
AL10 9AB, UK

ISBN: 978-1-60327-546-0
e-ISBN: 978-1-60327-547-7
ISSN: 1064-3745
e-ISSN: 1940-6029
DOI: 10.1007/978-1-60327-547-7

Library of Congress Control Number: 2008939423

Printed on acid-free paper

9 8 7 6 5 4 3 2 1

springer.com

Preface

RNA interference (RNAi) refers to the process by which dsRNA molecules silence a target through the specific destruction of their mRNA molecules. Subsequent to the discovery that small interfering RNAs (siRNAs) mimicking the Dicer cleavage products can silence mammalian genes, RNAi has become the experimental tool of choice to suppress gene expression in a wide variety of organisms. In addition, RNAi has also become a method of choice for key steps in the development of therapeutic agents, from target discovery and validation to the analysis of the mechanisms of action of small molecules. To date, several strategies have been devised to trigger the RNAi pathway, each of which is adapted and optimized for different cell systems. Although the technology has several advantages over other methods, the specificity of gene silencing is not absolute and there is a danger of off-target effects and activation of innate immunity. Also, strategic success of therapeutic siRNAs will depend on the development of a delivery vehicle that can target pathogenic cells and from our understanding of the biogenesis of microRNAs (miRNAs). The purpose of this book is to provide readers with the recent advances in siRNA design, expression, delivery, in vivo imaging, and methods to minimize siRNA unwanted effects and use in patients.

To design an effective siRNA, one must consider the base composition of the chosen site and whether the target site will be accessible. Chap. 1 critically reviews the published design guide rules and presents new statistical and clustering design strategies that are useful for selecting effective siRNA sequences. If the chosen target is an RNA virus that can mutate rapidly, one may consider to target conserved site sequences and/or to combine diverse siRNA sequences.

Recent studies indicated that certain siRNA sequences can activate innate immunity resulting in the production of pro-inflammatory cytokines and type I interferons. Moreover, siRNAs can also silence the expression of unrelated genes, a phenomenon known as off-target effects that is mediated largely by limited target sequence complementary to the seed region of the siRNA guide strand. Unfortunately, the current tools for siRNA design (see Chap. 1) cannot eliminate all the potential unwanted effects of siRNAs. Chap. 2 offers valuable and detailed description of how to eliminate siRNA unwanted effects, including the activation of innate immunity and off-target effects. Also, it describes siRNA-based methods for enhancing tumor immunity.

Notably, some of the main challenges in using siRNAs in vivo are the delivery, tissue targeting, and monitoring of siRNA potency in vivo. In vitro, siRNA duplexes have been delivered to target cells mainly by lipid-mediated transfection or via electroporation. However, these methods are not broadly applicable in patients. Approaches to improve in vivo delivery of siRNAs are currently being pursued using nanoparticles, new lipid formulations, and receptor-mediated targeting. Chaps. 3–8 describe new formulations and strategies with promising applications in vitro and in vivo. While Chap. 5 describes the first multimodal nanoparticles to deliver siRNAs, image siRNA uptake, and monitor gene silencing in tumors, Chap. 6 describes the first detailed protocol for siRNA magnetofection that is applicable in vitro and in vivo.

Being RNA, siRNAs are prone to nuclease-mediated degradation in serum and the cytosol, which has a negative impact on their use in cells and patients. Chemical modifications of ribose (e.g., locked nucleic acids, 2′-deoxy, 2′-fluoro, 2′-*O*-methyl) can enhance nuclease resistance without interfering with siRNA silencing potency. Chap. 9 describes the development of nuclease-resistant siRNAs with the potential to progress into a new class of therapeutic drugs. Chap. 10 describes new vectors for RNAi in which a synthetic siRNA/miRNA is expressed from a synthetic stem-loop precursor based on the miRNA 155 and miRNA 30 precursors. These new vectors offer several advantages over traditional RNAi vectors driven by RNA polymerase III promoters. These include the expression of several artificial miRNAs from a single transcript and tissue-specific expression as discussed in Chap. 3.

By using siRNAs to downregulate gene expression in human cells, a number of therapeutic target genes have been validated both in vitro and in vivo. Several relevant examples are featured in Chaps. 11–15. These include oncogenes, growth factors, immune regulatory factors, urokinase plasminogen activator and its receptor, matrix metalloproteinases, hypoxia-induced factor, and telomerase.

In addition to interfering with endogenous genes, siRNAs have been used to block viral replication. Nevertheless, in vitro and in vivo experiments have revealed potential problems of viral escape mutants. Chap. 16 describes the treatment of respiratory viral diseases with chemically modified new generation of siRNAs. And Chap. 17 describes the recent progress in using siRNAs as treatment for HIV-1 infection and several excellent recommendations are offered.

Notably, the success of siRNAs will depend not only on the development of delivery strategies and chemical modifications, but also on our understanding of miRNA biogenesis. Naturally occurring miRNA are 19–24 nt in length cleaved from 60- to 110-nt hairpin precursors that are produced from large primary transcripts. To date over 1000 miRNAs have been identified in humans. They play critical roles in developmental and physiological processes by regulating target gene expression at the post-transcriptional level. It is therefore not surprising that deregulation of miRNA expression could result in specific disease phenotypes. Recent studies have implicated miRNAs in cancer development. A group of three chapters describe the recent progress in understanding miRNA expression, function, and involvement in diseases. Chaps. 18 and 20 focuses on the recent progress in understanding the components involved in miRNA function, biogenesis, and interference with virus infection, and Chap. 19 demonstrates that intron-derived miRNA can induce RNAi not only in vitro but also in adult mice. Chap. 21 describes the design of effective miRNA sequences and their applications as anti-gene agents. The book ends by describing the first clinical trial in a patient with leukemia using a synthetic siRNA against *Bcl-Abl* fusion transcript (Chap. 22). It was found that the use of siRNAs in humans is safe, thus facilitating the progression of synthetic siRNA-based drugs to clinical trials.

Topics covered in this volume will be of interest to researchers, teachers, students, and biotech companies interested in RNAi, gene regulation, and gene and immunotherapy. It is my hope that the readers will benefit from this collection of excellent chapters dealing with the recent advances of RNAi technology from the bench to bedside.

Finally, I would like to thank the authors for their contributions, Anne Dybwad for excellent editorial assistance, the series editors, John Walkers, and all those involved in the production of the book.

Mouldy Sioud

Contents

Contributors

DIETMAR ABRAHAM • *Laboratory for Cardiovascular Research, Center for Anatomy and Cell Biology, Medical University of Vienna, Vienna, Austria*

SEYEDHOSSEIN AHARINEJAD • *Laboratory for Cardiovascular Research, Center for Anatomy and Cell Biology, Medical University of Vienna, Vienna, Austria*

MARTINA ANTON • *Institute of Experimental Oncology, Technische Universität München, Munich, Germany*

MARIA ARCE-LARRETA • *Department of Neurosurgery, Huntsman Cancer Institute in the Division of Pediatric Hematology/Oncology, University of Utah School of Medicine, Salt Lake City, UT, USA*

FRANK BAAS • *Department of Neurogenetics, AMC, Amsterdam, The Netherlands*

AGNIESZKA BANAS • *Section for Studies on Metastasis, National Cancer Center Research Institute, Tokyo, Japan*

CORINNE BARAT • *Research Center in Infectious Diseases, CHUL Research Center, Quebec, Canada*

SAILEN BARIK • *Department of Microbiology and Immunology, College of Medicine, University of South Alabama, Mobile, AL, USA*

AKUA N. BONSRA • *Department of Pharmacology and the Center for Developmental Genetics, Stony Brook University, Stony Brook, NY, USA*

DAVID Z. CHANG • *Departments of Immunology, The University of Texas M. D. Anderson Cancer Center, Houston, TX, USA*

GUANGWEI DU • *Department of Pharmacology and the Center for Developmental Genetics, Stony Brook University, Stony Brook, NY, USA*

AHMET H. ELMAAGACLI • *Department of Bone Marrow Transplantation, University Hospital of Essen, Essen, Germany*

KEES FLUITER • *Department of Neurogenetics, AMC, Amsterdam, The Netherlands*

JEANNETTE R. FLYNN • *Center for Children, Huntsman Cancer Institute in the Division of Pediatric Hematology/Oncology, University of Utah School of Medicine, Salt Lake City, UT, USA*

MARCO FOLINI • *Dipartimento di Oncologia Sperimentale e Laboratori, Fondazione IRCCS Istituto Nazionale dei Tumori, Milano, Italy*

JOHANNES FRUEHAUF • *Skip Ackerman Center for Molecular Therapeutics, Division of Gastroenterology, Beth Israel Deaconess Medical Center, Harvard Medical School, Boston, MA, USA*

LIFANG GAO • *Prostate Diseases Prevention and Treatment Research Centre, and Department of Pathophysiology, School of Basic Medicine, Jilin University, Changchun, China*

DAVID L. GILLESPIE • *Department of Neurosurgery, Huntsman Cancer Institute in the Division of Pediatric Hematology/Oncology, University of Utah School of Medicine, Salt Lake City, UT, USA*

Christopher S. Gondi • *Department of Cancer Biology and Pharmacology, University of Illinois College of Medicine at Peoria, Peoria, IL, USA*

Baofeng Guo • *Prostate Diseases Prevention and Treatment Research Centre, and Department of Pathophysiology, School of Basic Medicine, Jilin University, Changchun, China*

Hongnian Guo • *Skip Ackerman Center for Molecular Therapeutics, Division of Gastroenterology, Beth Israel Deaconess Medical Center, Harvard Medical School, Boston, MA, USA*

Edelburga Hammerschmid • *Institute of Experimental Oncology, Technische Universität München, Munich, Germany*

Olaf Heidenreich • *Newcastle University, Northern Institute for Cancer Research, Medical School, Newcastle upon Tyne, UK*

Naomi Hokaiwado • *Section for Studies on Metastasis, National Cancer Center Research Institute, Tokyo, Japan*

Kimi Honma • *Koken Bioscience Institute, Tokyo, Japan*

Constantin G. Ioannides • *Departments of Gynecologic Oncology, The University of Texas M. D. Anderson Cancer Center, Houston, TX, USA*

Randy L. Jensen • *Department of Neurosurgery, Huntsman Cancer Institute in the Division of Pediatric Hematology/Oncology, University of Utah School of Medicine, Salt Lake City, UT, USA*

Andrew C. Keates • *Skip Ackerman Center for Molecular Therapeutics, Division of Gastroenterology, Beth Israel Deaconess Medical Center, Harvard Medical School, Boston, MA, USA*

David A. Kelly • *Department of Neurosurgery, Huntsman Cancer Institute in the Division of Pediatric Hematology/Oncology, University of Utah School of Medicine, Salt Lake City, UT, USA*

Michael Koldehoff • *Department of Bone Marrow Transplantation, University Hospital of Essen, Essen, Germany*

Dennis J. Kopecko • *Laboratory of Enteric and Sexually Transmitted Diseases, Center for Biologics Evaluation and Research, Food and Drug Administration, Bethesda, MD, USA*

Chiang J. Li • *Skip Ackerman Center for Molecular Therapeutics, Division of Gastroenterology, Beth Israel Deaconess Medical Center, Harvard Medical School, Boston, MA, USA*

Shi-Lung Lin • *Department of Cell & Neurobiology, Keck School of Medicine, University of Southern California, Los Angeles, CA, USA*

Trevor Lucas • *Laboratory for Cardiovascular Research, Center for Anatomy and Cell Biology, Medical University of Vienna, Vienna, Austria*

Miguel Angel Martínez • *Fundacio irsiCaixa, Universitat Autònoma de Barcelona (UAB), Spain*

Zdravka Medarova • *Molecular Imaging Laboratory, MGH/MIT/HMS Athinoula A. Martinos Center for Biomedical Imaging, Department of Radiology, Massachusetts General Hospital/Harvard Medical School, Charlestown, MA, USA*

Takahi Mine • *Departments of Gynecologic Oncology and Immunology, Kurume University, Kurume, Japan*

Olaf R. F. Mook • *Department of Neurogenetics, AMC, Amsterdam, The Netherlands*

Anna Moore • *Molecular Imaging Laboratory, MGH/MIT/HMS Athinoula A. Martinos Center for Biomedical Imaging, Department of Radiology, Massachusetts General Hospital/Harvard Medical School, Charlestown, MA, USA*

Olga Mykhaylyk • *Institute of Experimental Oncology, Technische Universität München, Munich, Germany*

Thu Nguyen • *Skip Ackerman Center for Molecular Therapeutics, Division of Gastroenterology, Beth Israel Deaconess Medical Center, Harvard Medical School, Boston, MA, USA*

Takahiro Ochiya • *Section for Studies on Metastasis, National Cancer Center Research Institute, Tokyo, Japan*

Dominique L. Ouellet • *Centre de Recherche en Rhumatologie et Immunologie, CHUL Research Center, Quebec, Canada*

Marzia Pennati • *Dipartimento di Oncologia Sperimentale e Laboratori, Fondazione IRCCS Istituto Nazionale dei Tumori, Milano, Italy*

Marjorie P. Perron • *Centre de Recherche en Rhumatologie et Immunologie, CHUL Research Center, Quebec, Canada*

Christian Plank • *Technische Universität München, Munich, Germany*

Isabelle Plante • *Centre de Recherche en Rhumatologie et Immunologie, CHUL Research Center, Quebec, Canada*

Patrick Provost • *Centre de Recherche en Rhumatologie et Immunologie, CHUL Research Center, Quebec, Canada*

Brian T. Ragel • *Department of Neurosurgery, Huntsman Cancer Institute in the Division of Pediatric Hematology/Oncology, University of Utah School of Medicine, Salt Lake City, UT, USA*

Jasti S. Rao • *Department of Cancer Biology and Pharmacology, University of Illinois College of Medicine at Peoria, Peoria, IL, USA*

Joseph Rosenecker • *Forschungszentrum der Kinderklinik und Poliklinik Dr. von Haunersches Kinderspital, Kubus Rückgebäude, München, Germany*

Yueting Shao • *Prostate Diseases Prevention and Treatment Research Centre, and Department of Pathophysiology, School of Basic Medicine, Jilin University, Changchun, China*

Mouldy Sioud • *Department of Immunology, Institute for Cancer Research, The Norwegian Radium Hospital, University of Oslo, Norway*

Shigeru Takasaki • *RIKEN Genomic Sciences Center (GSC),Tsurumi-ku, Yokohama, Kanagawa, Japan*

Fumitaka Takeshita • *Section for Studies on Metastasis, National Cancer Center Research Institute, Tokyo, Japan*

Michel J. Tremblay • *Research Center in Infectious Diseases, CHUL Research Center, Quebec, Canada*

SHERYL R. TRIPP • *ARUP Laboratories, Salt Lake City, UT, USA*
NAOTAKE TSUDA • *Departments of Gynecologic Oncology and Immunology, Kurume University, Kurume, Japan*
JUNZHU WU • *Department of Pharmacology and the Center for Developmental Genetics, Stony Brook University, Stony Brook, NY, USA*
SHUANGLIN XIANG • *Skip Ackerman Center for Molecular Therapeutics, Division of Gastroenterology, Beth Israel Deaconess Medical Center, Harvard Medical School, Boston, MA, USA*
DE-QI XU • *Laboratory of Enteric and Sexually Transmitted Diseases, Center for Biologics Evaluation and Research, Food and Drug Administration, Bethesda, MD, USA*
YOUXIN YANG • *Skip Ackerman Center for Molecular Therapeutics, Division of Gastroenterology, Beth Israel Deaconess Medical Center, Harvard Medical School, Boston, MA, USA*
SHAO-YAO YING • *Department of Cell & Neurobiology, Keck School of Medicine, University of Southern California, Los Angeles, CA, USA*
NADIA ZAFFARONI • *Dipartimento di Oncologia Sperimentale e Laboratori, Fondazione IRCCS Istituto Nazionale dei Tumori, Milano, Italy*
OLIVIER ZELPHATI • *OZ Biosciences, Parc Scientifique et Technologique de Luminy, Marseille, Cedex 9, France*
LING ZHANG • *Prostate Diseases Prevention and Treatment Research Centre, and Department of Pathophysiology, School of Basic Medicine, Jilin University, Changchun, China*
LIJING ZHAO • *Prostate Diseases Prevention and Treatment Research Centre, and Department of Pathophysiology, School of Basic Medicine, Jilin University, Changchun, China*

Chapter 1

Methods for Selecting Effective siRNA Sequences by Using Statistical and Clustering Techniques

Shigeru Takasaki

Abstract

Short interfering RNAs (siRNAs) have been widely used for studying gene functions in mammalian cells but vary markedly in their gene-silencing efficacy. Although many design rules/guidelines for effective siRNAs based on various criteria have been reported recently, there are only a few consistencies among them. This makes it difficult to select effective siRNA sequences targeting mammalian genes. This chapter first reviews the reported siRNA design guidelines and clarifies the problems concerning the current guidelines. It then describes the recently reported new scoring methods for selecting effective siRNA sequences by using statistics and clustering techniques such as the self-organizing map (SOM) and the radial basis function (RBF) network. In the proposed three methods, individual scores are defined as a gene degradation measure based on position-specific statistical significances. The effectiveness of the methods was confirmed by evaluating effective and ineffective siRNAs for recently reported genes and comparison with other reported scoring methods. The sizes (values) of these scores are closely correlated with the degree of gene degradation, and the scores can easily be used for selecting high-potential siRNA candidates. The evaluation results indicate that the proposed new methods are useful for selecting siRNA sequences targeting mammalian mRNA sequences.

Key words: siRNA design, RNA interference, gene silencing, SOM classification, statistical significance, RBF network.

1. Introduction

Although RNA interference (RNAi) has been successfully used for studying gene functions in both plants and invertebrates, many practical obstacles need to be overcome before it becomes an established tool for use in mammalian systems (*1–6*). One of the important problems is designing effective short interfering RNA (siRNA) sequences for target genes. The siRNA responsible for

M. Sioud (ed.), *Methods in Molecular Biology, siRNA and miRNA Gene Silencing, vol. 487*

DOI: 10.1007/978-1-60327-547-7_1

RNA interference varies markedly in its gene-silencing efficacy in mammalian genes, where the gene-silencing effectiveness depends very much on the target sequence positions (sites) selected from the target gene *(7,8)*. Since different siRNAs synthesized for various positions induce different levels of gene silencing, the selection of the target sequence is critical for the effectiveness of the siRNA. We therefore need useful criteria for gene-silencing efficacy when we design siRNA sequences *(9,10)*.

Zamore et al. and Jayasena et al. showed that the 5′ end of the antisense strand might be incorporated into the RNA-induced silencing complex (RISC). Strand incorporation may depend on weaker base pairing and thus an A–T terminus may lead to more strand incorporation than a G–C terminus *(11,12)*. Other factors reported to be related to gene-silencing efficacy are GC content, point-specific nucleotides, specific motif sequences, and secondary structures of mRNA. Several siRNA design rules/guidelines using efficacy-related factors have been reported *(13 –17)*.

Although the positional nucleotide characteristics for siRNA designs seem to be the most important factor determining effective siRNA sequences, there are few consistencies among the reported rules/guidelines *(18–23)*. This implies that these rules/guidelines might result in the generation of many candidates and thus make it difficult to extract a few for synthesizing siRNAs. Furthermore, there is in RNAi a risk of off-target regulation: a possibility that the siRNA will silence other genes whose sequences are similar to that of the target gene. When we use gene silencing for studying gene functions, we have to first somehow select high-potential siRNA candidate sequences and then eliminate possible off-target ones *(24)*.

This chapter first reviews the reported siRNA design guidelines and clarifies the problems concerning the reported guidelines. It then describes the recently reported new scoring methods for selecting effective siRNA sequences by using statistical and clustering techniques *(25–32)*.

In the statistical method, many effective siRNA sequences are examined in the literature *(31)*, because it can be hypothesized that position-specific nucleotides play important roles in gene-silencing efficacy. If specific features of nucleotide frequencies appeared in many effective siRNAs, they mean the positional nucleotide characteristics for siRNA designs. The features of nucleotide frequencies at individual positions are then analyzed by using the statistical significance test. As these features can be considered as new guidelines, a measure (score) for selecting effective siRNA candidates is defined based on the positional features of specific significant nucleotides. The effectiveness of the proposed measure was confirmed by comparing the computed scores with those of the recently reported other selection methods *(28,29,31)*.

The chapter then describes how to extract individual nucleotide features from many effective siRNA sequences by using mathematical clustering techniques – the SOM and the RBF network (*see* later **Sects. 3.2.1** and **3.2.2**) *(25–27)*. In the SOM-based clustering method, siRNA classification from many effective siRNAs is first described. It is then shown how positional nucleotide features are extracted from the classified groups and is demonstrated how the extracted features are integrated as a measure (score). It is finally confirmed that the SOM method is effective by evaluating the relations between the scores and effective/ineffective siRNAs reported in the literature and comparing them with those of other reported scoring methods *(30,33)*.

In the RBF-network-based method, after the siRNA classification is carried out by using the RBF network *(25,26)*, the preferred and unpreferred nucleotides for effective siRNAs at individual positions are determined by significance testing and are used to calculate a score that measures a sequence's potential for gene degradation. The effectiveness of the proposed scoring method was then confirmed by using it to evaluate RNA sequences recently reported to effectively or ineffectively suppress the expression of various genes (see later subsection) and comparing it with other scoring methods *(32,33)*.

As a result of various evaluations, it is found there are good correlations between the sizes (values) of the proposed individual scores and the effectiveness and ineffectiveness of the recently reported siRNA sequences. The evaluation results indicate that the three methods would be useful for selecting siRNA sequences for mammalian genes.

2. RNA Interference and siRNA Sequence Selection Problem

2.1. RNA Interference

RNA interference (RNAi) is a phenomenon that silences gene expression by introducing double-stranded RNA (dsRNA) homologous to the target mRNA *(1)*. After this phenomenon was discovered in the nematode *Caenorhabditis elegans*, it gradually became clear that similar phenomena occur in the cells of plants, fungi, and mammals *(1–6)*. RNAi has been reported to result from the following sequence of events *(2,5,6)*. Long dsRNA is first cleaved into siRNA species by an RNAase III enzyme, Dicer. These siRNAs are then incorporated into an RNA-induced silencing complex (RISC), where the duplex siRNA is unwound so the antisense strand can guide RISC to the target mRNA having the complementary sequence. Finally, the target mRNA is cleaved at a single site in the center of the duplex region between the guide siRNA and the target mRNA *(28)*. Among the events that are still unclear, however, are the mechanism of the target mRNA cleavage and the mechanism by which the center of the duplex region in the RISC is identified. Furthermore, although RNAi has been widely used for studying gene functions, the effectiveness with which the genes in mammalian cells can be silenced this way depends very much on target sequence positions (sites) selected from the target gene. That is, different siRNAs synthesized from various positions induce different levels of gene silencing. This indicates that the selection of the target sequence position (site) is critical for the effectiveness of the siRNA *(3–6)*.

2.2. siRNA Sequence Selection Problem

2.2.1. Related Works Regarding the siRNA Sequence Selection Problem

To use RNAi as a biological tool for mammalian cell experiments, we first need to identify target sequences causing gene degradation. They have so far been identified by using a trail-and-error method *(3,8)*, but siRNAs extracted from different regions of the same gene have varied remarkably in their effectiveness. The difficulty of using the trail-and-error method to select target sequences causing gene silencing increases when the coding regions are long, as they are in mammalian cells. This is because the larger the number of candidates becomes, the more difficult it is to get gene-silencing candidates.

2.2.2. The Reported Guidelines for siRNA Sequence Design

The earliest guidelines for siRNA sequence design were proposed by Elbashir et al. *(4,8,40)*. They suggested that synthesizing siRNA duplexes of 21 nucleotides (nt) length – 19 nt base-paired sequence with 2 nt 3′ overhang at the ends – mediates efficient cleavage of the target mRNA. Their rules are summarized as follows.

(1) Select the target region from the open reading frame (ORF) of a given cDNA sequence preferably 50–100 nt downstream of the start codon. Avoid 5′ or 3′ untranslated regions (UTRs) or regions close to the start codon as these may be richer in regulatory protein-binding sites.

(2) Search for sequences 5′-AA(N19)UU, where N is any nucleotide, in the mRNA sequence and choose those with approximately 50% GC content. Highly G-rich sequences should be avoided because they tend to form G-quartet structures. If there are no 5′-AA(N19)UU motifs present in the target mRNA, search for 5′-AA(N21) or 5′-NA(N21), and synthesize the sense siRNA as 5′-(N19) TT and the antisense siRNA as 5′-(N′19)TT, where N′19 denotes the reverse complement sequence of N19 and T indicates 2′-deoxythymidine.

(3) Blast-search the selected siRNA sequences against EST libraries or mRNA sequences of the respective organism to ensure that only a single gene is targeted.

(4) It may be advisable to synthesize several siRNA duplexes to control for the specificity of the knockdown experiments; those siRNAs duplexes that are effective for silencing should produce exactly the same phenotype. Furthermore, a non-specific siRNA duplex may be needed as control.

(5) If the siRNA does not work, first verify that the target sequence and the cell line used are derived from the same organism. Finally, make sure that the mRNA sequence used for selection of siRNA duplexes is reliable; it could contain sequencing errors, mutations, or polymorphisms.

After that, many siRNA design guidelines/rules were reported as follows. Reynolds et al. analyzed 180 siRNAs systematically, targeting every other position of two 197-base regions of firefly luciferase and human cyclophilin B mRNA (90 siRNAs per gene), and reported the following eight criteria for improving siRNA selection *(18)*.

G1 (Reynolds et al.) eight criteria: (1) G/C content 30–52%, (2) at least 3 As or Ts at positions 15–19, (3) absence of internal repeats, (4) an A at position 19, (5) an A at position 3, (6) a T at position 10, (7) a base other than G or C at position 19, and (8) a base other than G at position 13.

Ui-Tei et al. examined 72 siRNAs targeting six genes and reported four rules for effective siRNA designs *(19)*. They are summarized as follows.

G2 (Ui-Tei et al.) four rules: (1) A or T effective and G or C ineffective at position 19, (2) G or C effective and A or T ineffective at position 1, (3) at least 5 T or A residues from positions 13 to 19, and (4) no GC stretch more than 9 nt long.

Amarzguioui and Prydz analyzed 46 siRNAs targeting four genes and reported the following six rules for effective siRNA designs based on their literature *(20)*.

G3 (Amarzguioui and Prydz) six rules: (1) G or C positive and T negative at position 1, (2) A positive at position 6, (3) T negative at position 10, (4) T positive at position 13, (5) C positive at position 16, and (6) A or T positive and G negative at position 19.

Jagla et al. tested 601 siRNAs targeting one exogenous and three endogenous genes and reported four rules as follows *(22)*.

G4 (Jagla et al.): (1) A or T positive at position 19, (2) A or T positive at position 10, (3) G or C positive at position 1, and (4) more than three A/Ts between positions 13 and 19.

Hsieh et al. examined 138 siRNAs targeting 22 genes and reported the following position-specific characteristics *(21)*.

G5 (Hsieh et al.): (1) T positive and G negative at position 19, (2) C or G positive and A or T negative at position 11, (3) G positive at position 16, (4) A positive at position 13, and (5) C negative at position 6.

The above previous works for positional characteristics in siRNA designs are summarized in Table 1.1a.

Other scoring, screening, and designing methods for functional siRNAs have also been reported recently. Chalk et al. reported seven rules ("Stockholm rules") based on thermodynamic properties. They are (1) total hairpin energy <1, (2) antisense 5′ end binding energy <9, (3) sense 5′ end binding energy in range 5–9 exclusive, (4) GC between 36 and 53%, (5) middle (7–12) binding energy <13, (6) energy difference < 0, and (7) energy difference within – 1 and 0. The score of an siRNA candidate is incremented by one for each rule fulfilled, giving a score range of (0,7) *(13)*.

Huesken et al. reported the screen method of functional siRNAs by using an artificial neural network *(23)*. This network was first trained by 2182 randomly selected siRNAs targeted to 34 genes and was used in the design of a genome-wide siRNA collection with two potent siRNAs per gene.

Teramoto et al. and Ladunga reported functional siRNA selection methods using support vector machine (SVM) *(14,34)*. Teramoto et al. used generalized string kernel (GSK) combined with SVM. siRNA sequences were represented as vectors in a multidimensional feature space according to the numbers of subsequences in each siRNA and classified into effective or ineffective siRNAs *(14)*. Ladunga used SVM with polynomial kernels and constrained optimization models from 572 sequence, thermodynamic, accessibility, and self-hairpin features over 2200

Table 1.1a
Effective and ineffective nucleotides specified in the individual guidelines

	Position	1	3	6	10	11	13	16	19
G1	Preferred		A		T		A/C/T		A/T
G2	Preferred	G/C							A/T
	Unpreferred	A/T							G/C
G3	Preferred	G/C		A			T	C	A/T
	Unpreferred	T			T				G
G4	Preferred	G/C			A/T				A/T
G5	Preferred					C/G	A	G	T
	Unpreferred			C		A/T			G

Position: nucleotide position from 1 to 19 (5′ to 3′, cDNA form). Preferred: effective (positive), unpreferred: ineffective (negative).

published siRNAs *(23,34)*. As the key to SVM success is to collect many useful features of effective siRNA sequences, it may depend on the selected siRNAs.

Holen recently reported siRNA rules based on an apparent overrepresentation or underrepresentation of certain nucleotides in certain positions of the Novartis screen *(23,35)*. The criteria for an siRNA candidate depend on the computed positive and negative scores for each position using the scoring table. The scoring table is generated by the percentage overrepresentation or underrepresentation of individual nucleotides for each position in the Novartis screen *(23)*. Although the method was evaluated by using the other reported siRNA sets, it is unclear what a siRNA candidate is useful for effective gene silencing. In addition, as the original scores in the scoring table are based on the only percentage overrepresentation or underrepresentation of certain nucleotides in certain positions, they may vary drastically depending on what siRNA sets are used. This makes it difficult to evaluate the computed scores for siRNA candidates.

Although secondary structures of siRNA sequences are also considered to be an important factor in predicting siRNA efficacy, there are conflicting results concerning the effects of secondary structures on siRNA functionality. On one hand, some studies suggested that the secondary structure of the siRNA plays a role in determining the efficacy of gene silencing *(37–39)*. On the other hand, other studies did not find any correlation between functionality of the siRNA and second structures of the target mRNA *(7,18,20)*. Therefore this issue still requires further studies.

Features of individual siRNA design rules/algorithms are summarized in **Table 1.1b**.

2.2.3. siRNA Sequence Selection Problems Using the Previous Guidelines

Analyzing **Table 1.1a** and **b**, the problems concerning the reported guidelines will become clear. First is the problem of inconsistencies for nucleotide frequencies of each position. Although there are common preferred and unpreferred nucleotides at both positions 1 and 19 in some guidelines, there are few consistencies at other positions in individual guidelines as listed in **Table 1.1a**. These results indicate that though some rules from the guidelines are suitable for getting effective sequences for specific genes, they might sometimes be unsuitable for selecting sequences for other genes. Since the previous guidelines are based on the analyses of specific genes as listed in **Table 1.1b**, it could be inferred that they are not always effective for many other genes. Therefore if these guidelines were used to select sequence candidates for other mammalian genes, many sequences might be selected as candidates. This is because there are mostly long coding regions in mammalian genes but there are only a few consistencies among the previous guidelines as shown in **Table 1.1a**. As a result, many

Table 1.1b
Features of individual siRNA design rules/algorithms

siRNA design rules	Citation	No. of genes	No. of siRNAs	Description	Technique
Reynolds et al.	(18)	2	197	Sequence features	
Ui-Tei et al.	(19)	6	72	Sequence features	
Amarzguioui et al.	(20)	4	46	Sequence features	
Hsieh et al.	(21)	22	138	Sequence features	
Hesken et al.	(24)	34	2128	Sequence motifs	Neural network
Jagla et al.	(22)	4	601	Sequence features	Decision tree
Holen	(35)	34	400	Sequence features	Percentage
Saetrom	(36)	40	581	Sequence motifs	Genetic Programming
Teramoto et al.	(14)	2	94	Sequence motifs	Support vector machine
Ladunga	(34)	34	2252	Position features	Support vector machine
Calk et al.	(13)	92	398	Binding energy	Regression tree
Takasaki et al.	(30–32)	490	833	Sequence features	Statistics, SOM, RBF

candidate sequences might be selected. That is the second problem. Experimentally evaluating whether the selected sequences provide effective gene degradation is a costly and time-consuming task. This implies that it would be hard to select a few final candidates for synthesizing siRNAs.

3. Statistical and Clustering Techniques for siRNA Sequence Selections

3.1. Significance Test for siRNA Sequence Selections

Since it is possible to test the significance of each nucleotide at individual positions (i.e., 1 to 19) on the basis of the occurrence probabilities of each nucleotide at individual positions and in the entire target sequence population, the significance testing was done with the following equation:

$$Z_s = \frac{|p_{as} - p_b|}{\sqrt{P(1-P)(1/n_{as} + 1/n_b)}}, \quad (1.1)$$

where s is the site (position) 1 to 19, p_{as} is the probability of each nucleotide a occurring at each site s (a = A, G, C, or T), p_b is the occurrence probability of each nucleotide averaged over the entire target sequence population, P is arithmetic mean of p_{as} and p_b, n_{as} is the number of nucleotides at position s and n_b is the total number of nucleotides in all positions *(28)*.

Suppose, for example, that we have 200 effective siRNA sequences. If the occurrence probability of the nucleotide G at position 7 is 0.35 (70/200) and the occurrence probability of G in the entire target sequence population is 0.28 (1064/(200 × 19)), the 95% significance probability of the nucleotide G at position 7 would be indicated by a z value of 2.14.

If Z is equal to or greater than 1.96 and less than 2.56, the significance probability of the nucleotide in the class is 95%. Similarly, if Z is greater than 2.56, the significance probability is 99%. The two-sided statistical test has two types of significance values, higher and lower levels of significance. The higher level (greater than 1.96) means the nucleotides are preferred for effective siRNAs, whereas the lower one (greater than 1.96) means the nucleotides are unpreferred for effective siRNAs.

3.2. Statistical Significant Nucleotide Selection

As the two-sided statistical test has two types of significance values, higher (upper) and lower levels of significance, they are expressed as follows:

Higher-significance nucleotide (HN_p^v) and

Lower-significance nucleotide (LN_p^v),

where H denotes higher, L denotes lower, N is a nucleotide, v: 95 – significance probability is 95% (level of significance = 0.05), v: 99 – significance probability is 99% (level of significance = 0.01), and p: nucleotide position (site) (i.e., 1–19).

3.3. Clustering Techniques for siRNA Sequence Selections

3.3.1. Self-Organizing Map for siRNA Sequence Selections

3.3.1.1. The Self-Organizing Map for siRNA Classification

The SOM is a neural network algorithm introduced by Kohonen *(27)* and has been used to categorize and interpret large, high-dimensional data sets in various research fields. This categorization can be achieved by mapping n-dimensional data points representing similar characteristics onto nearby regions of an r-dimensional space; r is usually much smaller than n and is also usually 2. The most important feature of SOM is its preservation of topology, i.e., the topological relations between high-dimensional data (input vectors or input space) are preserved on the SOM map (two-dimensional grid or output map). In other words, the SOM technique is a tool for mapping similar input patterns onto contiguous locations in the output space. As a result, the SOM can accomplish the clustering or the creation of abstractions of the input space and the visualization of high-dimensional data in two-dimensional displays.

To use the SOM technique for siRNA sequence classifications, it is necessary to transform individual nucleotides (A, T, G and C) into numeric attribute data. The nucleotides A, T, G, and C are therefore transformed into the following digital representations (numerical values):

A = (100) = 1,T = (010) = 2,G = (110) = 3 and C = (001) = 4.

The siRNA sequence GCCAGGTTCTAGAGGATGA, for example, is expressed by the numerical sequence 3441442242131331231. Other numerical data representations for individual nucleotides are, of course, also possible. That is, if individual nucleotides were expressed by other numerical values, e.g., A = 4, T = 2, G = 3, and C = 1, the SOM could accomplish similar classifications.

3.3.1.2. Using the SOM to Obtain siRNA Sequence Classifications

The SOM processes in the case of 200 input vectors are, for example, shown as follows *(30)*.

Preparation

Input vectors: 200 effective siRNA sequences

They are, for example, expressed as follows:

1. GACCCGCGGAGTTTCCCGT=3144434331322244432
2. CTCAATACGCGGCTAACAG=4241121424334211413
3. GATACTCCTTGCCTGATAA=3121423322344231211

----

200. GCAGAGTTCAAAAGCCCTT=3413132241111344422

Output map: an array of nodes (two-dimension nodes, for example, 3 × 3 nodes)

Each node has an associated reference vector of the same size as each input vector.

The reference vectors contained in all nodes are randomly initialized. In case of 3 × 3 nodes, they are initialized as random sequences in the following way:

1-1. TACAGAGTTATTTAGGGTA=2141313221222133321

1-2. CATAGATTCCGTTTCGGGA=4121312244322243331

3-3. GTACAGGGATCCATGCTCA=3214133312441244241

The SOM processes are carried out as follows:

Step 1: Select an input vector from the input data set randomly.

Step 2: Apply individual output node criteria E for the input vector.

Each $\boldsymbol{E}$ is defined as the Euclidean distance between the input vectorx and the reference vector $\boldsymbol{m}$.

$$\boldsymbol{E}_i = \sqrt{\sum_{j=1}^{19}\left(\boldsymbol{x}_j - \boldsymbol{m}_{ij}\right)^2} \qquad (1.2)$$

where

$\boldsymbol{i}$: individual nodes, e.g., $\boldsymbol{i}$ = 1-1, 1-2, … 3-3,

$\boldsymbol{x}_j$: each element in the input vector, $\boldsymbol{j}$ = 1,2, … 19, and

$\boldsymbol{m}_{ij}$: each element in the reference vector, $\boldsymbol{j}$ = 1,2, … 19.

Step 3: Determine the winning node W whose reference vector is the minimum criterion.

$$W = \min_i E_i \tag{1.3}$$

Step 4: Modify the winning and neighbor node reference vector m.

$$m = m + \alpha(x - m) \tag{1.4}$$

where

α is the learning rate ($0 < \alpha < 1$). α is initially close to 1 and diminished with each iteration.

Step 5: Terminate the SOM processes if the change in E is less than a certain level (approximately zero) or the number of iterations is more than that of the specified time-steps; otherwise go to Step 1.

3.3.2. Radial Basis Function Network for siRNA Sequence Selections

An RBF network is a type of artificial network for application to problems of supervised learning, such as regression, classification, and time series prediction. In the supervised learning, the function is learned from the examples (training set) which a teacher supplies. The training set contains elements which consist of paired values of the independent (input) variable and the dependent (output) variable. The RBF network is a new technique for value prediction that demonstrates more robustness and flexibility than traditional regression approaches such as neural networks and polynomial fits. The RBF network works by choosing not just a single nonlinear function, but a weighted sum of a set of nonlinear functions. These weighted functions are the so-called radial basis functions. The RBFs are each fitted to separate regions in the input space. The regions are chosen such that the output is quite similar within a region, so that the RBF is most likely to fit well to the output. For each selected region, an RBF center is created that predicts the average of the region. Data points that fall between regions are predicted by taking a weighted average of the predictions of all centers, where the weight for a center decays rapidly if the center is very far from that data point *(25,26)*. The RBF networks are supervised learning models with a single middle layer of units. They are similar back propagation neural networks but usually faster to train because the RBFs used in the units mean that fewer weight adjustments are needed. Also, RBF networks tend to be more resistant to noisy data than back propagation networks.

3.3.2.1. Preparation

The relations between siRNA sequences and the effectiveness of their gene silencing are shown in **Fig. 1.1a** *(32)*. For simplicity, sense strands of siRNAs (cDNA 5′ to 3′, 19 nucleotides from positions 1 to 19) are described. To use an RBF network for selecting effective siRNA sequences, it is necessary to represent individual nucleotides (A, G, C, and T) as numerical data. The symbols A, G, C, and T are therefore transformed into the following numerical representations: A = 1, G = 2, C = 3, and T = 4. Other numerical data representations for individual nucleotides are, of course, also possible. That is, if individual nucleotides were expressed by other numerical values, e.g., A = 4, G = 3, C = 1, and T = 2, the RBF network could accomplish similar classifications. The original set of siRNAs is thus transformed into a set of numerical values as shown in **Fig. 1.1b**. To make the RBF network for siRNA selections easily understood, siRNA sequences (input vectors) are changed to the vertical style shown in **Fig. 1.1c.** The RBF network can be constructed by adding the hidden and output layers as shown in **Fig. 1.1d**.

3.3.2.2. Effective siRNA Classifications by Using the RBF Network

To carry out effective siRNA classifications by using the RBF network, the data (known effective and ineffective siRNAs) were partitioned into two sets, one of training data and the other of validation data. The processes of the classifications are carried out in two phases: training and validation.

1.Training phase

The training of the RBF network proceeds in two steps. First the hidden layer parameters are determined as a function of the input data (vectors) and then the weights between the hidden and output layers are determined by comparing the target data and the output of the hidden layer. The hidden layer parameters to be determined are the parameters of hyperellipsoids that partition the input data (vectors) into discrete clusters or regions. The parameters locate the center (i.e., the mean) of each ellipsoid's (region or cluster) basis function and describe the extent or spread of the region (i.e., the variance or standard deviation).

The centers of individual clusters are determined as follows:

1. Randomly choose *m* vectors from the input data set to be the centers of *m* basis functions.
2. For each vector *i* in the input dataset compute the Euclidean distance $D_{i,m}$ to each of the *m* basis function centers.

$$D_{i,m} = \left| X_i - M_m \right| = \sqrt{\sum_{j=1}^{19} \left(x_{i,j} - \mu_{m,j} \right)^2}, \tag{1.5}$$

where i is input vector number, e.g., i = 1, 2, ..., TN (the maximum number of vectors in the set of training data), X_i is i-th input vector, $X_i = (x_{i,1}, x_{i,2}, \ldots, x_{i,19})$, and M_m is the location vector or center of the basis function for hidden node m,

$$M_m = \left(\mu_{m,1}, \mu_{m,2}, \ldots, \mu_{m,19} \right).$$

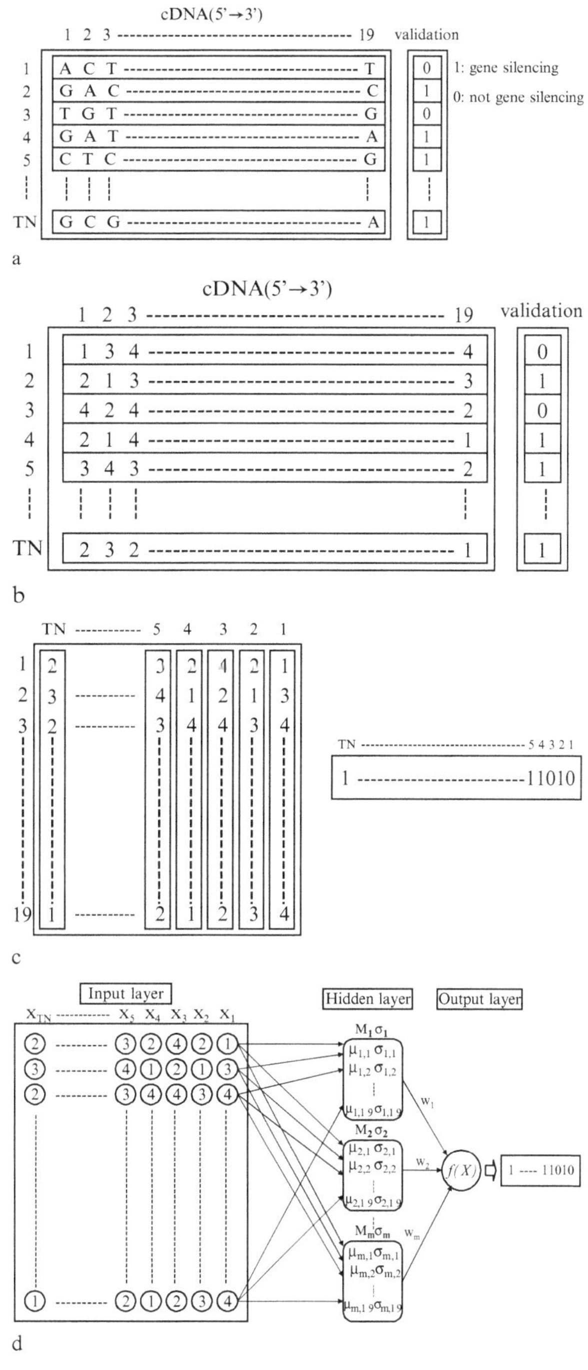

Fig. 1.1. RBF network representation of the relations between effective and ineffective siRNA sequences. (**a**) Relations between siRNA sequences and the effectiveness of their gene silencing. (**b**) siRNA numerical representations made by converting A, G, C, and T to 1, 2, 3, and 4. (**c**) Vertical representations of numerical siRNAs. (**d**) RBF network transformation of the siRNA selection problem. The input layer is the set of numerically represented siRNA sequences. The hidden layer classifies the input vectors into several clusters depending on the similarities or closeness of individual input vectors. The output layer indicates the effectiveness of individual siRNA sequences (1: effective, 0: ineffective).

3. Determine for each input data vector the closest basis function center:

$$\mu_{bf,i} = \boldsymbol{Min}\left\{\boldsymbol{D}_{i,1}, \boldsymbol{D}_{i,2}, \ldots, \boldsymbol{D}_{i,m}\right\} \text{ for } i = 1,2\ldots,TN, \tag{1.6}$$

where $\mu_{bf,i}$ is the closest basis function for the input vector i.

4. For all the input vectors grouped around the basis functions, compute the mean $\boldsymbol{\mu}_m$

$$\mu_m = \frac{\sum BF_i^m}{N_m} \text{ for all } m, \tag{1.7}$$

where $\boldsymbol{BF}_i^m$ is the input vector i of the closest basis function m and N_m is the number of input vectors grouped around the basis function m.

5. Use these grounded means as the new mean values for the m basis functions.

6. While repeating Steps 2–5, compute the change $\boldsymbol{E}$ between the present and previous processes:

$$E = \left|\mu_{m_r} - \mu_{m_{r-1}}\right| \text{ for all } m, \tag{1.8}$$

where r is the number of repeats and the initial value $\mu_{m_0} = 0$.

The termination is carried out if the change is less than a certain level (approximately zero).

The number m of basis functions starts as a small value, e.g., $m = 4$, and increases as the validation data is being evaluated. The variances of the individual basis functions $\left(\sigma_1^2, \sigma_2^2, \ldots, \sigma_m^2\right)$ are computed after the individual basis functions are determined.

$$\sigma_m^2 = \frac{\sum\left(\mu_{bf,i} - \mu_m\right)^2}{N_m} \text{ for all } m \tag{1.9}$$

The radial basis function $GR(i,m)$ for the hidden unit m output of the input vector i is defined as a Gaussian function in the following way:

$$GR(i,m) = e^{-\left(D_{i,m}\right)^2 / 2\sigma_m^2}, \tag{1.10}$$

where σ_m^2 is a measure of the size of the cluster m (i.e., the variance or the square of the standard deviation).

Then all that remains is to find the linear combination of weights that produces the desired output (target) values for each input vector. Since this is a linear problem, convergence is guaranteed and computation proceeds rapidly. This task can be accomplished with an iterative technique based on the perceptron training rule or with various other numerical techniques. Technically, the problem is a matrix inversion problem:

$$T = BW, \tag{1.11}$$

where T is the target vector, W is the to-be-determined weighting vector, and B is the matrix of output values from each hidden unit in response to the input data (calculated from the basis functions, e.g., Eq. 1.10). The matrix is usually not square, so a pseudo inverse may be used to give a minimum least-squares solution.

In the case of the supervised learning, we have already obtained gene-silencing results for all input vectors, e.g., $i = TN$.

$$i = 1,\quad f(X_1) = \sum_{l=1}^{m} w_l GR(1,l) = \sum_{l=1}^{m} w_l e^{-D_{1,l}{}^2/2\sigma_l^2} = 1$$

$$i = 2,\quad f(X_2) = \sum_{l=1}^{m} w_l GR(2,l) = \sum_{l=1}^{m} w_l e^{-D_{2,l}{}^2/2\sigma_l^2} = 0$$

$$i = 3,\quad f(X_3) = \sum_{l=1}^{m} w_l GR(3,l) = \sum_{l=1}^{m} w_l e^{-D_{3,l}{}^2/2\sigma_l^2} = 0$$

$$i = TN,\quad f(X_{TN}) = \sum_{l=1}^{m} w_l GR(TN,l) = \sum_{l=1}^{m} w_l e^{-D_{TN,l}{}^2/2\sigma_l^2} = 0 \tag{1.12}$$

Therefore, $w_1, w_2, \ldots, w_m$ are determined by solving the above linear equations.

After determining the weighting variables, the percentages of effective and ineffective siRNAs can be computed in the individual clusters.

2. **Validation phase**

To evaluate whether the RBF network carried out appropriate (not overtraining) classifications, individual clusters in the classifications were verified by using the validation data. The differences between the percentages of effective and ineffective siRNAs for the training and validation data are computed for individual clusters. If there are few differences between the percentages of effective and ineffective siRNAs for the training and validation data in some classification, it can be inferred that the classification was carried out appropriately. If, on the other hand, there are large differences between them, we must conclude that the classification was not appropriate. The differences therefore indicate the effectiveness of individual classifications by the RBF network. The summation of the differences – the entire error of this partition (cluster) number *m* – is used to compare the error of this partition with other errors of other partitions (clusters).

3. **Determination of the number m of clusters**
The number *m* of basis functions corresponds to the number of partitions (clusters) and is determined on the basis of the minimum error of the individual clusters by using the validation data. That is, after carrying out several classifications while changing the number *m* of clusters, the errors of individual clusters are checked and the number of clusters yielding the minimum error is the desired number, i.e., the optimal classification.

3.3. Evaluation and Training Data

The recently reported effective and ineffective siRNAs were used as the evaluation data. A total of 833 effective siRNAs from 490 genes were used for obtaining frequency ratios of individual nucleotides at individual positions and the training data. Another 847 ineffective siRNAs from 34 genes were also used as the training data for the RBF network *(23)*. The evaluation data were not included in the training data.

3.3.1. Recently Reported Effective and Ineffective siRNA Sequences

Reynolds et al. analyzed 90 siRNAs systematically, targeting every other position of 197-base regions of human *cyclophilin B* mRNA (GenBank accession no. M60875) *(18)*. For simplicity, human *cyclophilin B* is symbolized throughout the present chapter as MG1. From the 90 analyzed siRNA sequences we selected as effective ones the 25 top-ranked sequences for which the MG1 target gene silencing was greater than 80% and selected as ineffective ones the 25 worst-ranked sequences for which MG1 target gene silencing was less than 50% with standard deviation 10%.

Ui-Tei et al. reported 38 effective and 24 ineffective sequences for six genes: *firefly luciferase (PRL-TK)*, *vimentin*, *Oct 4*, *EGFP*, *ECFP*, *and DsRed (19)*. For simplicity, in the rest of this chapter all six of these genes are symbolized as MG2.

Amarzguioui et al. reported 21 effective and 25 ineffective siRNA sequences for four genes: *hTF* (accession no. M16553), *mTF* (accession no. M26071), *PSK* (accession no. J272212), and *CSK* (accession no. NM_004383) *(20)*. For simplicity, in the rest of this chapter these four genes are symbolized as MG3.

Takasaki et al. reported seven effective and seven ineffective siRNA sequences for the *cyclin B1* (accession no. NM_031966)) *(28)*. For simplicity, in the rest of this chapter this gene is symbolized as MG4.

Huesken et al. reported 37 siRNAs for TC10 (accession no. BD135193), UBE2I (accession no. NM_003345), and CDC34 (accession no. NM_004359) *(23)*. The 12 top-ranked effective and 12 worst-ranked ineffective siRNA sequences were selected for these genes. For simplicity, they are symbolized in the rest of this chapter as MG5.

3.3.2. Effective siRNA Sequences

833 known effective siRNA sequences were used for obtaining frequency ratios of individual nucleotides and the training data. They were more than 80% gene silencing at the protein level and were collected from 490 different cDNAs in the published references of the PubMed database *(5,6,29)*.

3.3.3. Ineffective siRNA Sequences

Since it is difficult to select many known ineffective siRNAs from many different genes, we used the 847 worst-ranked ineffective siRNAs (less than 0.612 (normalized inhibitory activity)) from Huesken et al. *(23)* for the training data of the RBF network.

4. Methods for Selecting Effective siRNA Sequences

4.1. siRNA Scoring Method Using Statistical Techniques

4.1.1. Nucleotide Frequency Analysis at Individual Positions

Since it could be inferred that the previous guideline problems were based on specific gene analyses, Takasaki et al. examined whether there are nucleotide occurrence specificities for many effective siRNA sequences reported in the literature (*see* **Sect. 3.3.2**) *(31)*. Nucleotide frequencies at individual positions were listed in **Table 1.2**.

These sequences were analyzed by using Eq. 1.1 and many higher significance and lower significance nucleotides were obtained as listed in **Table 1.3**. They are mostly different from the previously reported nucleotides.

4.1.2. Gene Silencing Measure (Score) Definition

A target siRNA sequence including many higher-significance nucleotides and a few lower-significance nucleotides could be inferred to become a highly effective gene silencer. A measure (priority score *S*) based on this idea is therefore defined in the following way.

$$S = a\left(\sum HN_p^v\right) - b\left(\sum LN_p^v\right) \quad (1.13)$$

where a and b are weighting factors for higher and lower scores, normally $a = b = 1$.

HN_p^v and LN_p^v are higher-significance and lower-significance nucleotides, respectively (*see* **Sect. 3.1**).

Equation 1.13 shows that the larger S becomes, the greater the likelihood that the sequence is effective for gene silencing.

4.1.3. siRNA Sequence Selection Method Based on Statistical Techniques

The method using the score of the statistical significance to select siRNA sequence for the target gene is summarized as follows:

Step 1: Select as candidate sequences those 21-nt sequences with "AA" on the upstream end.

Table 1.2
Nucleotide frequencies at individual positions

Position	1	2	3	4	5	6	7	8	9	10	11	12	13	14	15	16	17	18	19
A	100	250	265	182	226	245	206	248	226	191	208	236	216	235	225	193	240	260	257
G	464	225	191	242	218	199	253	211	187	214	232	206	212	230	202	239	200	199	171
C	173	202	173	245	207	173	219	191	203	229	213	195	206	181	168	209	196	148	192
T	96	156	204	164	182	216	155	183	217	199	180	196	199	187	238	192	197	224	204

Table 1.3
Higher-significance and lower-significance nucleotides and their values

Position	1	2	3	4	5	6	7	8	9	10	11	12	13	14	15	16	17	18	19
Sig. (z)	G (17.7)	A (2.2)	A (3.3)	C (3.8)		T (2)		A (2)	T (2.1)	C (2.4)					T (3.6)			A,T (3,2.5)	A (2.9)
Sig. (z)	A,T (9.4,7.6)	T (2.7)	G (2.8)	A,T (3.1,2.1)		G (2.2)	T (2.8)		G (3.2)	A (2.4)					G,C (2.2,2.5)	A (2.2)	G (2.3)	G,C (2.4,4.1)	G (4.4)

Step 2: Use Eq. 1.13 to calculate the priority score for each candidate sequence.

Step 3: Rank all candidate sequences according to their priority scores and assign them new sequence numbers.

Step 4: Select several of the highest-ranked sequences as target sequences for dsRNA syntheses by considering how close to 50% their GC ratios are and also considering off-target genes.

4.1.4. Evaluation of the Statistical Method

The purpose of the score setting is to indicate the sequence priority for selecting new siRNA candidates. This is because it is necessary to select several of the highest-ranked sequences as target sequences for dsRNA syntheses. From this point of view, the effectiveness of the proposed score for effective and ineffective siRNA sequences (MG1–MG5; *see* **Sect. 3.3.1**) was evaluated by using Eq. 1.13 and **Table 1.3**. Since there were ups and downs in the computed scores of the individual sequence classes, the averages for them were calculated. The average scores of the effective sequences for MG1–MG5 were 7.9, 11.5, 4.5, 23.5, and 3.7, respectively, whereas those of the ineffective sequences were -11.6, –16.2, –5.3, –16.6, and –12.8. These scores therefore reveal that the proposed method might be useful for selections of effective siRNA candidates. The average scores obtained by the previously reported score methods of Reynolds et al., Ui-Tei et al., Amarzguioui and Prydz, and Hsieh et al. were also examined *(33)*. The relative relations between scores of the previous methods and those of the proposed method are shown in **Fig. 1.2**. The results indicate that the previous methods are not always clear correspondences between the scores and the effective and ineffective siRNA sequences. The methods of Reynolds et al. and Hsieh et al., for example, show positive values for both effective and ineffective siRNAs of MG1, MG3, MG4, and MG5, and do not indicate distinct score differences between the effective and ineffective siRNAs. In addition, although the methods of Ui-Tei et al. and Amarzguioui and Prydz indicate the correspondences between the individual average scores and the effective and ineffective siRNAs for MG1–MG5, the relative score differences between the effective and ineffective siRNAs are not so big as shown in **Fig. 1.2**. In addition, the maximum and minimum score ranges of their methods are respectively from 2 to -2 and from 8 to –2 *(33)*. These score ranges imply that the restricted discrete scores might be assigned to the candidate sequences. For example, the method of Ui-Tei et al. may assign five kinds of scores, i.e., (2, 1, 0, –1, –2) to the candidate sequences. This indicates that there might be many same score sequences and the difficulty of selecting several candidates. Therefore, the scores of the previous methods imply that it is difficult to assign the priority of new siRNA candidates according to the obtained scores. On the

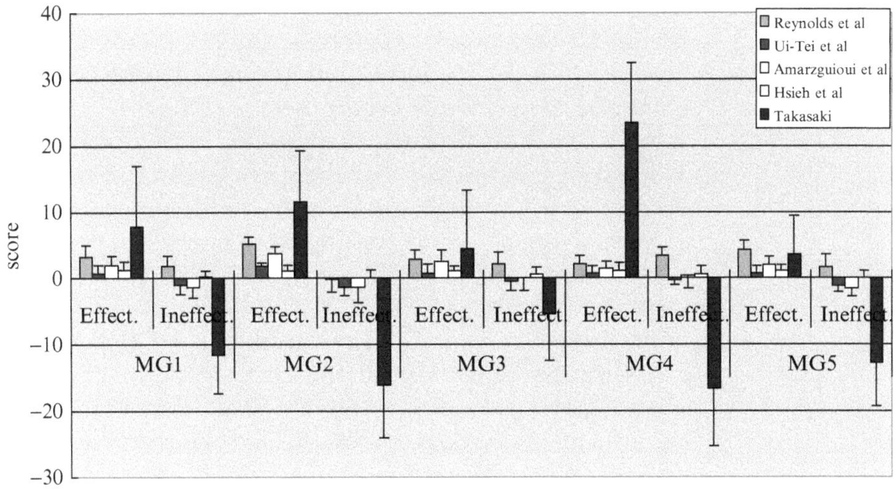

Fig. 1.2. Score comparisons between the proposed method and other methods. Scores of the effective and ineffective siRNA sequences were computed on the basis of positional scores of the individual reported guidelines shown in Saetrom and Snove, *(33)*.

other hand, as the range of the proposed score is 45 to −44.1 from **Table 1.3**, it is easy to distinguish the priority for the candidate sequences. The proposed scores, for example, indicate clear correspondences for the effective and ineffective siRNAs of MG1, MG2, MG4, and MG5. This therefore implies that the proposed score can easily be used for selecting high-potential siRNA candidates.

4.2. siRNA Scoring Method Using Mathematical Clustering Techniques

As it can be hypothesized that there are individual nucleotide features in previously reported effective siRNA sequences, it is possible to extract these features from those sequences by using mathematical clustering techniques – the SOM and the RBF network (*see* **Sects. 3.2.1** and **3.2.2**).

4.2.1. siRNA Classification by Using SOM

The SOM is unsupervised learning that uses only effective siRNA sequences in this case. This means that it is not necessary to use ineffective siRNAs for classifying effective siRNA sequences. If this technique could classify the large number of reported effective sequences into several sets of clustered sequences with similar nucleotide features, those sequences could be used as new guideline factors for selecting new siRNA candidates. Here, a basic SOM classification method using 200 effective siRNA sequences is described. It is also possible to apply for many other siRNA sequences using the same method *(30)*.

4.2.1.1. siRNA Sequence Selection Based on SOM

200 effective siRNA sequences from 129 genes were selected in the literature *(28)* and were classified by using the SOM technique (*see* **Sect. 3.2.1**). Although it was possible to classify 200 effective sequences into many detailed groups, four groups were

selected as a reasonable SOM classification. They were 54 into group#1 (SOM-C1), 51 into group#2 (SOM-C2), 48 into group#3 (SOM-C3), and 47 into group#4 (SOM-C4). The ratios of individual nucleotide occurrences at each of the positions from 1 to 19 are shown for the four groups in the four parts of **Fig. 1.3**, where it is evident that there are features of the nucleotide occurrences, i.e., high occurrences and low occurrences in the individual classes. These features were selected as preferred (high occurrences) and unpreferred (low occurrences) nucleotides shown in **Tables 1.4 a, b, c, d**. Although these positional nucleotide features are useful for selecting new siRNAs, candidate extraction procedures from the target gene based on the classified features are time-consuming tasks. Furthermore, it might be difficult to decide a few candidate sequences for synthesizing siRNAs. What we need is a more simple measure for selecting the siRNAs. The integration of the positional nucleotide features was therefore considered. Interestingly, there are some similarities among nucleotides at individual positions of each group. In SOM-C2, SOM-C3, and SOM-C4, for example, there are high ratios of the nucleotide G at position 1 and the nucleotide A at position 2. Therefore, if there were specific features common to individual groups, it might be possible to generate a more general feature of nucleotides for effective siRNA sequences by integrating individual nucleotide features. These features could be extracted from the classified groups by the following steps:

Step 1: Extraction of individual nucleotide features (significances) for each of the groups SOM-C1, -C2, -C3, and -C4.

The significance of each nucleotide at individual positions (i.e., 1 to 19) is obtained by using Eq. 1.1. Significant nucleotides and their corresponding z values for individual classes SOM-C1, SOM-C2, SOM-C3, and SOM-C4 are also shown in **Tables 1.4a–d**.

Step 2: Integration of the individual nucleotide significances.

As the four individual classified groups have their own nucleotide features, it can be hypothesized that the integration of the features might be another feature for the entire effective siRNA sequences. On the basis of this idea, the upper and lower levels of significance of each nucleotide at individual positions (i.e., 1–19) were first accumulated for the four groups. The upper and lower level accumulations for individual nucleotides (A, G, C, and T) are carried out using the following Eqs. 1.14

$$UN_p^{nv} = \sum_{i=1}^{4} UN_{ip}^{nv} \quad \text{for } n = A, G, C, \text{ and } T,$$
$$p = 1, 2, \ldots, 19, v = 95 \text{ or } 99 \qquad (1.14)$$

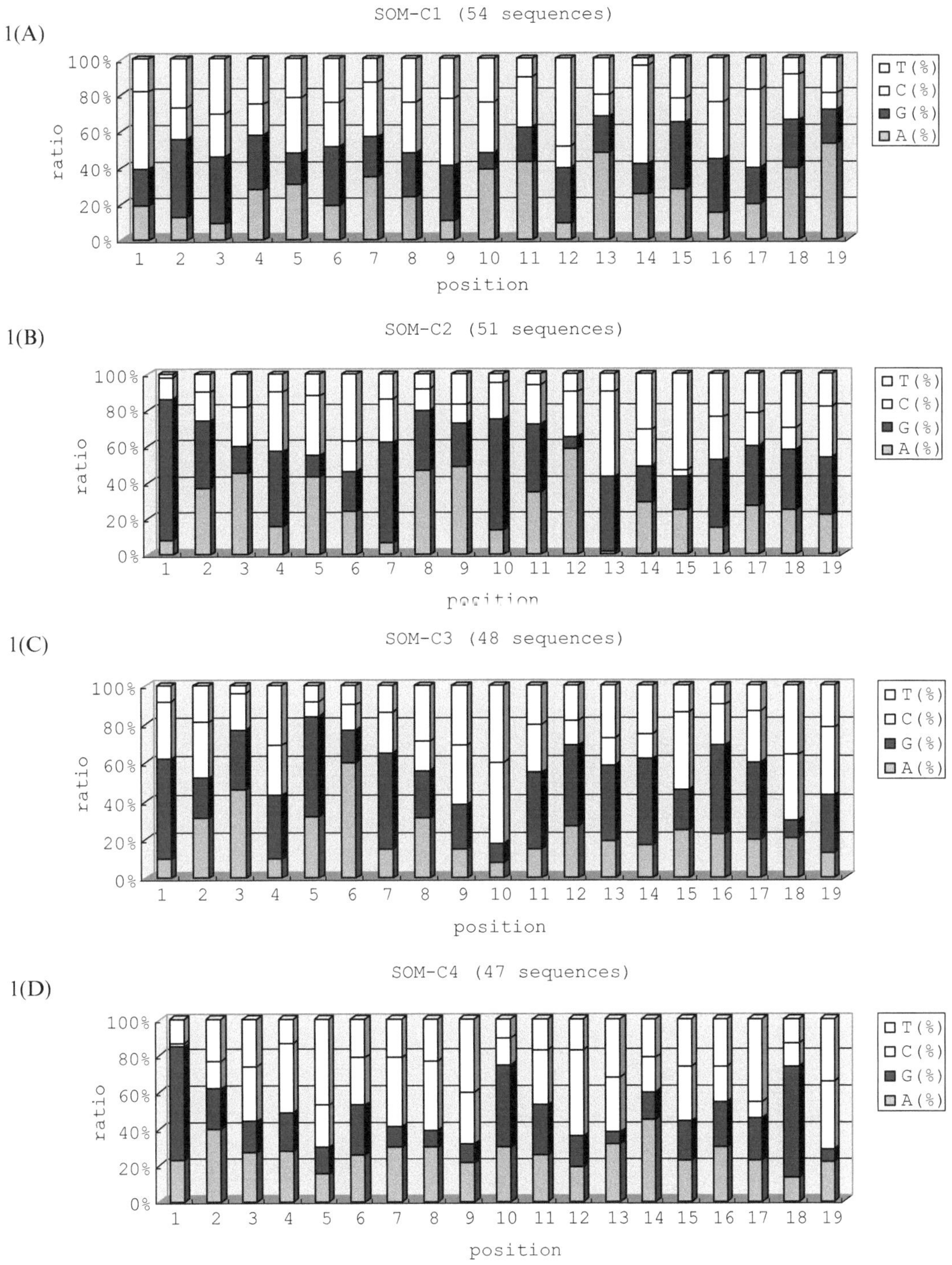

Fig. 1.3. Classifications of effective siRNA sequences derived by the SOM techniques. Two hundred effective siRNA sequences were classified into four groups: SOM-C1–SOM-C4. The ratios of the nucleotides A, G, C, and T at positions 1–19 for individual groups are shown in four parts.

Table 1.4
Significant nucleotides and their corresponding z values

a. SOM-C1 (54 sequences)

Position	1	2	3	4	5	6	7	8	9	10	11	12	13	14	15	16	17	18	19
$UN_1 \uparrow$	C	G	G						C	A	A	T	A	C	G		C	A	A
Z	2.6	2.8	1.9						1.7	2	2.6	4.6	3.4	4.4	1.9		2.6	2.1	4.2
$LN_1 \downarrow$		A	A						A	G	T	A,C	C	T	C	A		T	C
Z		2.3	2.9						2.6	2.7	2	2.9,2.4	2.4	3.1	2.2	1.9		2.2	2.9

b. SOM-C2 (51 sequences)

Position	1	2	3	4	5	6	7	8	9	10	11	12	13	14	15	16	17	18	19
$UN_2 \uparrow$	G		A	C	A,C	T	G	A	A	G		A	C	T	T			T	
Z	6.5		2.7	2	2.3,2	3.5	3.2	3	3.3	4		4.8	4.4	2.4	6.2			2.2	
$LN_2 \downarrow$;	A,T		G	A	G	G	A	T	C	A,T	T	G	A	G	G,C	A			
Z	3.1,2.9		2.7	1.9	3.2	1.7	3.3	1.8	1.9	2.2,2.4	2.2	4.1	4.1	2	2.3,3	2			

c. SOM-C3 (48 sequences)

Position	1	2	3	4	5	6	7	8	9	10	11	12	13	14	15	16	17	18	19
UN_3↑	G		A	T	G	A	G		T	C,T				G	C	G		T	C
Z	2.7		3.6	1.9	2.7	5.8	2.4		1.9	2.8,3.3				1.7	2.5	1.9		2.7	1.7
LN_3↓	A,T	G	T	A	C,T	G,C,T				A,G		C		C	G	T		G	
Z	2.1,2.0	1.7	2.7	2.1	2.6,2	2.3,1.8,1.7				2.5,3.3		1.8		1.8	1.7	1.7		3.5	

d. SOM-C4 (47 sequences)

Position	1	2	3	4	5	6	7	8	9	10	11	12	13	14	15	16	17	18	19
UN_4↑	G	A		C	T		C	C	T	G		C		A			T	G	C
Z	6	2		1.9	3.4		1.9	1.9	2.4	3.4		3.2		2.8			3.1	5.8	1.7
LN_4↓	C,T			T	A		G	G	G	T			G				C	A,C,T	G
Z	3.7,1.8			1.8	1.8		2	2.3	2	2.3			2.8				2.6	2.1,2,1.8	2.6

UNi and LN*i* are upper (preferred) and lower (unpreferred) nucleotides for class I (i.e., 1: SOM-C1, 2: SOM-C2, 3: SOM-C3 and 4: SOM-C4) and *Z* is the corresponding static value.

$$LN_p^{nv} = \sum_{i=1}^{4} LN_{ip}^{nv} \quad \text{for } n = A, G, C, \text{ and } T, \\ p = 1, 2, \ldots, 19, v = 95 \text{ or } 99 \tag{1.15}$$

The results obtained for individual nucleotides at positions 1–19 are listed in **Table 1.5a**.

A nucleotide quantitative measure at position p was then defined as a score S_p based on the accumulated upper and lower levels of significance. This score S_p for individual nucleotides was calculated by using the following equation:

$$S_p^n = UN_p^{nv} - LN_p^{nv} \quad \text{for } n = A, G, C, \text{ and } T, \\ p = 1, 2, \ldots, 19, v = 95 \text{ or } 99 \tag{1.16}$$

The individual scores at positions 1–19 that were obtained using this equation are listed in **Table 1.5b**.

The score S for a candidate siRNA sequence is defined as the summation of individual position scores and is calculated as follows:

$$S = \sum_{p=1}^{19} \{S_p^A, S_p^G, S_p^C, S_p^T\} \tag{1.17}$$

This equation means that the sequence score S is determined from the result of summing up individual nucleotide scores from positions 1 to 19. Therefore, if an siRNA candidate is expressed as a nucleotide sequence GCCAGGTTCTAGAGGATGA, the score S is computed in the following way:

$$S = 15.1 - 3.9 - 0.47 - 4 - 1.79 - 0.22 - 1.3 + 2.6 - 4.09 \\ -0.68 - 0.27 - 2.14 - 3.92 + 3.12 + 2.32 \\ +4.17 = 4.53$$

4.2.1.2. siRNA Sequence Selection Method Using SOM

The method using the score of the SOM classification to select siRNA sequences for the target gene is summarized as follows.

Step 1: Select as candidate sequences those 21-nt sequences with "AA" on the upstream end.

Step 2: Use Eq. 1.17 to calculate the scores for individual candidate sequences.

Step 3: Rank all candidate sequences according to their scores.

Step 4: Select several of the highest-ranked sequences as target sequences for dsRNA syntheses by considering how close to 50% their GC ratios are and also considering off-target genes.

Table 1.5
Upper and lower level significances and their integrated scores

a. Upper and lower significances for individual nucleotides

Position		1	2	3	4	5	6	7	8	9	10	11	12	13	14	15	16	17	18	19
A	*UN*↑		2.03	6.26		2.34	5.76		2.95	3.26	1.96	2.6	4.77	3.39	2.77				2.12	4.17
	LN↓	5.23	2.25	2.91	3.9	1.75		3.27		2.58	4.61		2.9	4.07			3.92		2.06	
G	*UN*↑	15.14	2.84	1.88		2.71	0	5.59		0	7.23		0		1.72	1.88	1.86		5.8	
	LN↓		1.73	2.73		3.18	4	1.96	2.29	1.96	6.07		4.09	2.77	1.99	4.02			3.48	2.61
C	*UN*↑	2.63			3.91	2.03		1.88	1.88	1.68	2.81		3.23	4.35	4.37	2.5		2.63		3.46
	LN↓	3.67				2.56	1.75			1.9			4.14	2.39	1.75	5.17		2.57	1.95	2.88
T	*UN*↑					3.42	3.45			4.22	3.34		4.62		2.39	6.2		3.12	4.89	
	LN↓	6.76		2.73	1.81	2.04	1.69		1.79		4.64	4.13			3.05		1.69		3.96	

b. The integrated individual nucleotide scores at positions 1–19

Position	1	2	3	4	5	6	7	8	9	10	11	12	13	14	15	16	17	18	19
A	−5.23	−0.22	3.35	−3.9	0.59	5.76	−3.27	2.95	0.68	−2.65	2 6	1.87	−0.68	2.77		−3.92		0.06	4.17
G	15.1	1.11	−0.85		−0.47	−4	3.63	−2.29	−1.96	1.16		−4.09	−2.77	−0.27	−2.14	1.86		2.32	−2.61
C	1.04			3.91	−0.53	−1.75	1.88	1.88	−0.22	2.81		−0.91	1.96	2.62	−2.67		0.06	−1.95	0.58
T	−6.8		−2.73	−1.81	1.38	1.76		−1.79	4.22	−1.3	−4.13	4.62		−0.66	6.2	−1.69	3.12	0.93	

4.3.2.1.3. Evaluation of the SOM Selection Method

The effectiveness of the proposed SOM selection method was evaluated for MG1–MG5 (*see* **Sect. 3.3.1**) by using Eq. 1.17 and Table 1.5b. The average scores of the effective sequences for MG1–MG5 were 10.4, 12.1, 13.9, 33.4, and 11.6, respectively, whereas those of ineffective sequences were –3.4, –5.5, 2, –11.8, and –5.2, respectively. The results indicate that the average scores of effective sequences for MG1–MG5 were distinctly greater than those of the ineffective sequences. The effective sequence scores for MG1–MG5 are more than 10, whereas the ineffective sequence scores are small or negative values. These score features correspond to the indication of the degree of gene-silencing efficacy. That is, it might be inferred that the higher the score, the higher the probability of effective gene silencing.

The average scores for MG1–MG5 were also examined by the previously reported score method of Reynolds et al., Ui-Tei et al., Amarzguioui and Prydz, and Hsieh et al. *(33)*. The relative average score relations between scores of the previous methods and those of the proposed SOM method are shown in **Fig. 1.4**.

Although the SOM selection method was shown on the basis of 200 effective siRNA sequences, if the method were a useful one, it would be evident more clearly for many other siRNAs. This issue was evaluated using 833 effective siRNAs (*see* **Sect. 3.3.2**). SOM classifications were first carried out for 833 siRNAs. As a result, they were classified into eight groups. Individual nucleotide significances were then integrated using Eqs. 1.14–1.17 and scores were computed for MG1–MG5. Relations between effective and ineffective sequence scores are also shown in SOM-833 of **Fig. 1.4**. The results indicate that the average scores using 833 siRNAs are clearer differences than those of 200 siRNAs. That is, it might be inferred that the larger effective siRNAs, the clearer the differences.

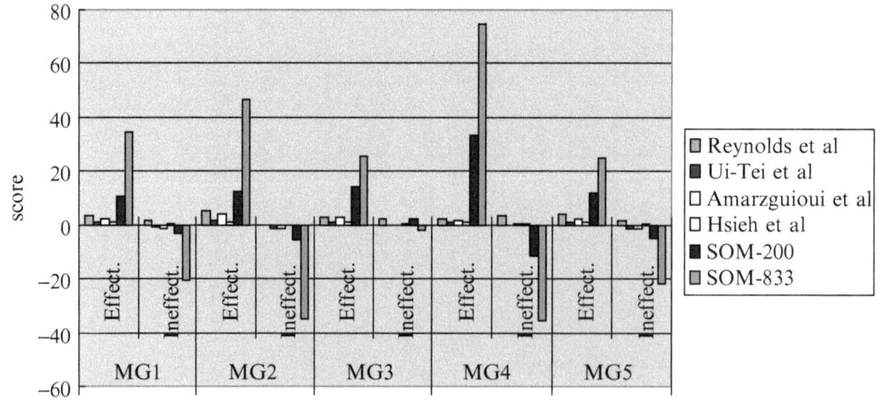

Fig. 1.4. Score comparisons between the proposed SOM method and other methods. Scores of the effective and ineffective siRNA sequences were computed on the basis of positional scores of the individual reported guidelines shown in Saetrom and Snove *(33)*. SOM-200: 200 effective siRNAs used; SOM-833: 833 effective siRNAs used.

4.2.2. siRNA Classification by Using RBF Network

The siRNA classification (the RBF network training and validation) was also carried out by using Eqs. 1.5–1.10 and a 2-to-1 ratio of training data to validation data (*see* **Sects. 3.2.2** and **3.3**) *(32)*. As a result, the eight clusters (C1–C8) were obtained as listed in **Table 1.6**. Since the percentage of effective siRNAs was highest for cluster C1, this cluster can be considered to most clearly show the nucleotide frequency features useful for designing effective siRNAs and therefore used it to examine the nucleotide occurrence distribution at individual positions (**Fig. 1.5**).

4.2.2.1. Nucleotide Feature Extraction from the Characterized Cluster

Although the trained RBF network can predict the probabilities of gene silencing for new siRNA candidate sequences (data not shown), it cannot extract individual nucleotide characteristics from the candidates. It was therefore considered that the scoring system based on RBF clusters can extract nucleotide features

Table 1.6
Clusters generated by the RBF network

Cluster ID	No. of sequences	Percentage of effective siRNAs
C1	134	94
C2	150	70.7
C3	125	70.4
C4	147	61.9
C5	141	43.3
C6	158	32.3
C7	143	27.1
C8	148	8.1

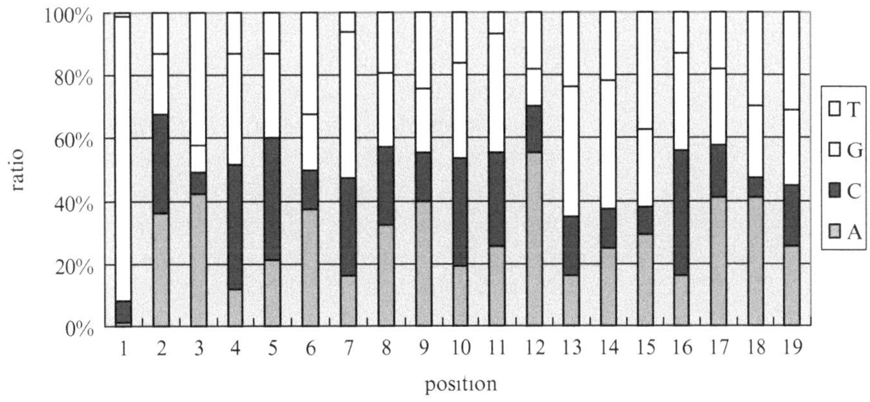

Fig. 1.5. Nucleotide distribution in the siRNA in the best cluster obtained by the RBF classification.

more clearly than the previously reported methods. To utilize the nucleotide features effectively for selecting effective siRNAs, the significance testing for the siRNAs in C1 was carried out using Eq. 1.1. The preferred and unpreferred nucleotides (*see* **Sect. 3.1**) for the siRNAs in C1 are listed in **Table 1.7**.

The gene degradation measure *S* was then defined based on the preferred and unpreferred nucleotides. If an siRNA candidate has many preferred nucleotides and few unpreferred nucleotides, it might have a high potential for gene silencing. The measure (score) *S* was therefore defined as follows:

$$S = \sum_{C1} \text{Preferred} - \sum_{C1} \text{Unpreferred}, \qquad (1.18)$$

where *Preferred* and *Unpreferred* are the preferred and unpreferred nucleotides, respectively, determined according to the C1 significances listed in **Table 1.7**. The score *S* for an siRNA candidate (cDNA) with the nucleotide sequence ACGCCAAAAACATAAAGAA is thus computed as follows:

$$\begin{aligned} S &= \sum_{C1} \text{Preferred} - \sum_{C1} \text{Unpreferred} \\ &= (2.8 + 5 + 4.7 + 2.3 + 3 + 2.3 + 6.7 + 3.3) \\ &\quad - (6.9 + 5.3 + 3 + 2.2 + 3) \\ &= 30.1 - 20.4 = 9.7 \end{aligned}$$

4.2.2.2. Evaluation of the RBF Selection Method

The effectiveness of the proposed RBF method for MG1–MG5 was evaluated by using Eq. 1.18 and **Table 1.7**. The average score of the effective sequences for MG1 was 9, whereas the average score of the ineffective sequences was −5.9. As there is a clear difference between the scores of the effective and ineffective sequences, the score features correspond to the effectiveness indication of the proposed method. The scores of effective and ineffective sequences for MG2, MG3, MG4, and MG5 were also computed by using Eq. 1.18. The average scores of effective sequences for MG2, MG3, MG4, and MG5 were respectively 7.9, 9.6, 30.1, and 3.8, whereas the average scores of the corresponding ineffective sequences were −10.9, −3.7, −21.2, and −7.7. As there are also clear differences between the averages of the effective and ineffective sequences for these genes, the individual scores indicate the effectiveness of the proposed method. Relations between the effective and ineffective sequence scores in the recently reported siRNAs are shown in RBF: C1 of **Fig. 1.6**.

The entire average of 103 effective sequences for these genes is 9.54, whereas that of 93 ineffective ones is −7.98. The results indicate that there are clear differences between the average scores of the effective and ineffective sequences of siRNAs for the reported genes (MG1–MG5). MG4 shows an especially clear

Table 1.7
Preferred and unpreferred nucleotides for the siRNAs in C1

Position	1	2	3	4	5	6	7	8	9	10	11	12	13	14	15	16	17	18	19
Preferred	G	A,C	A,T	C	C	A,T	C,G		A	C	C	A	G	G	T	C	A	A,T	T
Sig. value	14.2	2,2.8	3.5,6.1	5	4.7	2.3,3.4	2.6,4		3	3.4	2.3	6.7	2.8	2.5	4.7	5	3.3	3.3,2.8	3.1
Unpreffered	A,C,T	G,T	C,G	A,T	T	C,G	A,T		G	A	T	G	A	C	C	A,T		C	
Sig. value	6.9,4.1,5.5	2.8,2	4.1,5.3	4,2	2	2.4,3.1	3,4		2.4	2.2	3.7	4.6	3	2.7	3.5	3,2		4.3	

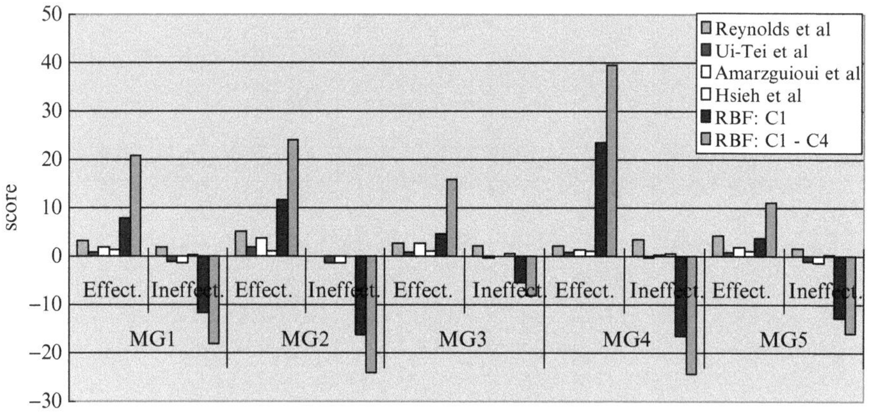

Fig. 1.6. Score comparisons between the proposed RBF method and other scoring methods. C1: C1 used, C1–C4: C1–C4 used.

score difference. These features correspond to the indication of the degree of gene-silencing efficacy. It might thus be inferred that the higher the score, the higher the probability of effective gene silencing.

The average scores of siRNAs for MG1–MG5 were also examined by using the previously reported scoring methods of Reynolds et al., Ui-Tei et al., Amarzguioui and Prygz, and Hsieh et al. *(33)*. The relative average score relations between scores obtained with those previous methods and the scores obtained with the proposed method were also shown in **Fig. 1.6**. The results indicate that the scores obtained with the previous methods do not always clearly differ between effective and ineffective siRNA sequences. The proposed RBF method can clearly indicate differences between effective and ineffective siRNAs as shown in **Fig. 1.6**. This therefore implies that the proposed method can easily be used for selecting high-potential siRNA sequences.

4.2.2.3. Integration of the Individual Class Nucleotide Significances

Although the proposed scoring method using the highest cluster (C1) clearly indicated differences between effective and ineffective siRNAs for various genes, it is also possible to use the other cluster groups that are more than 50% effective siRNAs: C2 (70.7%), C3 (70.4%), and C4 (61.9%). As these groups have their own nucleotide frequency features, it can be hypothesized that the integration of the features might yield other useful features for selecting effective siRNA candidates. The integrated measure (score) S_T was accordingly defined by accumulating the preferred and unpreferred nucleotides for the individual groups as follows *(32)*:

$$S_T = \sum_{\text{C1,C2,C3,C4}} \text{Preferred} - \sum_{\text{C1,C2,C3,C4}} \text{Unpreferred}. \tag{1.19}$$

For simplicity, the preferred and unpreferred significances were first computed for C1, C2, C3, and C4. They are listed in **Table 1.8**.

To evaluate the effectiveness of the integrated measure S_T, the average scores for MG1–MG5 were computed by using Eq. 1.19 and **Table 1.8**, and were compared with the previous results shown in **Fig. 1.6**. The score relations between the best cluster (C1) and accumulated clusters (C1 to C4) are also shown in **Fig. 1.6** for various genes. It is clear that the accumulated scores make clearer distinctions between the effective and ineffective siRNAs than do the scores obtained using the single cluster (C1). This implies that the integration of nucleotide features at individual classes might yield other useful features for selecting effective siRNA candidates. In addition, as the same integration of the clusters (C5–C8) less than 50% effective sequences might provide enrichment for unpreferred nucleotides, it could increase the score differences even more by using unpreferred nucleotide characteristics (data not shown).

4.2.2.4. Effectiveness of the Training and Validation Data

A method for selecting effective siRNA candidates was proposed by using a RBF network and significance testing. To avoid overtraining by the RBF network, the training and validation data were used in the proposed method. It is therefore expected that the ratios of the training and validation data might affect the efficiency of selecting effective siRNA candidates. Cases in which the ratios of the training and validation data are 2-to-1 (reported above), 3-to-1, and 10-to-1 were therefore evaluated. As it has already been found that the accumulated scores obtained using multiple clusters make clearer distinctions than do scores obtained using only the best cluster, the accumulated score relations among three cases were examined for the recently reported genes. As shown in **Fig. 1.7**, although there are not major score differences among three cases, the 10-to-1 ratio seems to yield clearer distinctions than do the 2-to-1 and 3-to-1 ratios.

In addition, as the main purpose of the score setting is to show clear priority for effective and ineffective siRNA sequences, the differences between the effective and ineffective scores were analyzed for the reported genes in 2-to-1, 3-to-1, and 10-to-1 ratios. The differences among these ratios are listed in **Table 1.9**. The results indicate that the 10-to-1 ratio also yields clearer distinction than do the 2-to-1 and 3-to-1 ratios. These results imply that a large amount of training data might yield better classifications of siRNA candidates.

4.3. Considerations for Individual Scoring Methods

Three scoring methods – statistical, SOM, and RBF network methods – were reviewed. The statistical scoring method for siRNA design was based on many effective siRNA sequences, and features of nucleotide frequencies for each position were extracted from the significance test. On the other hand, the

Table 1.8
The accumulated preferred and unpreferred significances for C1, C2, C3, and C4

	Position	1	2	3	4	5	6	7	8	9	10	11	12	13	14	15	16	17	18	19
Preferred	A↑	1.7	3.9	9.8	1.9		5.7			3		2.2	6.7	4.8	6.6			3.3	5.3	7.8
	C↑	2.8	2.8	4.4	6.8	6.4	2	2.6	3	8.2	9.4	2.3	4.4		3.1		7.2	2.5		3.7
	G↑	35		3.3	4.4		2.5	4	6.3			2.7		6	2.6	1.7	2.5		3.6	
	T↑			6.1		3.5	3.4		2.9	1.8	4.1	1.8	2.7			7.7	3.7	3.2	4.6	3.1
Unpreferred	A↓	17			9.1	1.7	2.4	3		2.9	5.9	2	6	3	2.1		4.8		2.4	
	C↓	7.6	2.1	6.6		4.7	5.2	3.4	7.7	1.8			3.8	4.3	6.7	7.6			4.3	3.4
	G↓	4.8	2.9	9.4	1.7	1.8	3.1	2.3	5.9	6.5	4.9	3.4	4.6	3.2	4			7.3	6.9	10
	T↓	14	4.2	6.7	4.2	3.9	2.1	4				3.7					5.3			

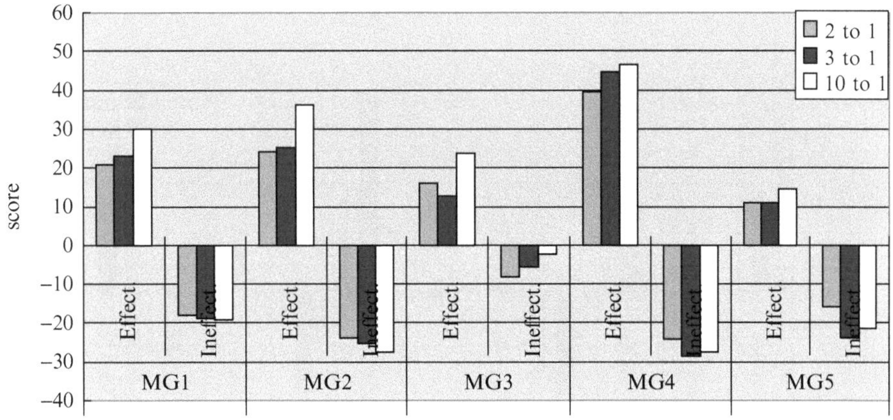

Fig. 1.7. The accumulated score relations for 2-to-1, 3-to-1, and 10-to-1 ratios of training data to validation data.

Table 1.9
The differences between the effective and ineffective scores for the reported genes

Gene	2-to-1	3-to-1	10-to-1
MG1	39.03	41.68	48.96
MG2	47.94	50.52	63.74
MG3	24.07	18.19	25.91
MG4	63.81	73.28	74.01
MG5	25.9	33.26	34.1

scoring methods using the SOM and the RBF network utilized mathematical clustering techniques. The SOM classification was carried out by using only effective siRNA sequences – unsupervised learning – whereas the classification of the RBF network was carried out by using effective and ineffective siRNA sequences – supervised learning. As the clustered siRNA sequence sets include similar nucleotide features based on the evaluation functions of individual clusters, the extracted nucleotide features for each position may be inferred to be a clearer distinction than those for nonclustered sequence sets. The relations among the above three methods were examined by the average score values for MG–MG5, shown in **Fig. 1.8**.

From **Fig. 1.8**, it is clear that the average scores of clustering methods are clearer differences than those of the statistical one. The result indicates that the clustered sequence sets with similar nucleotide features are useful for extracting individual nucleotide

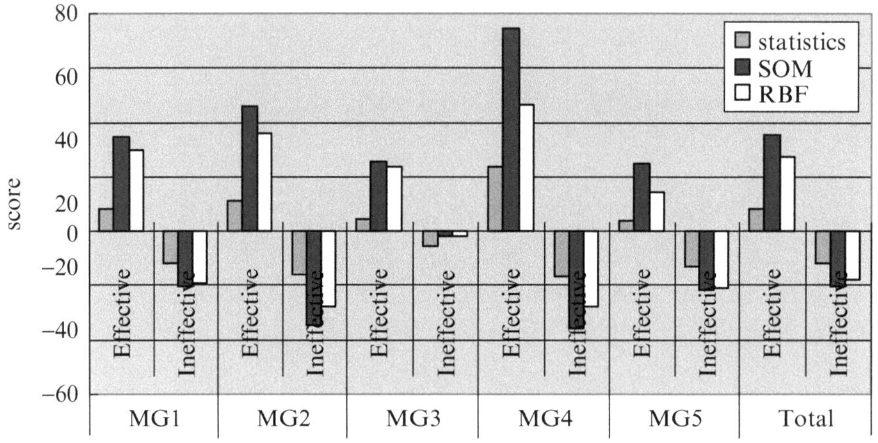

Fig. 1.8. The relations among statistics, SOM, and RBF network methods. Total: sum from MG1 to MG5.

frequencies for each position. Although the average scores of the SOM method seems to be clearer differences than those of the RBF network one, there is no big difference in the scores of total genes. In addition, as the scores of the RBF method were computed by the integration of the preferred clusters (C1–C4) shown in **Table 1.6**, it is possible to increase the score differences even more by using the unpreferred clusters (C5–C8) (data not shown).

4.4. Other Considerations for Effective siRNA Designs

Reynolds et al. reported at least three A or T residues at positions 15 to 19 *(18)*, Ui-Tei et al. reported at least five A or T residues from positions 13 to 19 *(19)*, and Jagla et al. reported more than three A or T between positions 13 and 19 *(22)*. In our analyses of 833 effective siRNA sequences, however, there were no distinct differences between the occurrences of A or T and other nucleotides at positions 13–19, or 15–19. The average ratios for these ranges of positions are 52.6 and 53.5%, respectively, and are not much different from the average occurrence of A and T in the entire population of 833 effective siRNA sequences: 49.3%. The relations between A or T contents of previous design rules and those of 833 effective sequences are listed in **Table 1.10a**.

The result may indicate that there is no preference for or against A or T occurrences of A and T at positions 13–19.

Reynolds et al. and Chalk et al. reported a GC content of 30–52% and 36–53%, respectively, as a criterion. The GC content of effective and ineffective sequences was also examined for reported genes. The distribution of GC content for the sets of reported genes is listed in **Table 1.10b**. The results indicate

Table 1.10a
The relations between the previous design rules and 833 effective sequences

	Positions	Nucleotides	No. of nucleotides	Content ratio (%)
Reynolds et al.	15–19	A or T	At least three	60
Ui-Tei et al.	13–19	A or T	At least five	71.4
Jagla et al.	13–19	A or T	More than three	53.1
833 siRNAs	15–19	A or T		53.5
833 siRNAs	13–19	A or T		52.6
833 siRNAs	1–19	A or T		49.3

Table 1.10b
GC content distribution

	MG1 (%)	MG2 (%)	MG3 (%)	MG4 (%)	MG5 (%)	833 seqs (%)
Effective seqs.	47.6	42.1	49.1	52.6	46.1	50.7
Ineffective seqs.	47.6	62.3	52.4	48.1	43.9	

that there is no big difference between effective and ineffective sequences. It could therefore be inferred that gene-silencing effectiveness does not depend on GC content. As the average GC content for 833 effective sequences is 50.7%, this value could be used as a guideline.

References

1. Fire, A., Xu, S., Montgomery, M.K., Kostas, S.A., Driver, S.E. et-al. (1998) Potent and specific genetic interference by double-stranded RNA in *Caenorhabditis elegans*. *Nature*, 391, 806–811.
2. Sharp, P.A. (2001) RNA interference-2001. *Genes Dev.*, 15, 485–490.
3. Elbashir, S.M., Harborth, J., Lendeckel, W., Yalcin, A., Weber, K. et al. (2001) Duplexes of 21-nucleotide RNAs mediate RNA interference in mammalian cell culture. *Nature*, 411, 494–498.
4. Elbashir, S.M., Lendeckel, W., and Tuschl, T. (2001) RNA interference is mediated by 21- and 22-nucleotide RNAs. *Genes Dev.*, 15, 188–200.
5. Dykxhoorn, D.M., Navia, C.D., and Sharp, P.A. (2003) Killing the messenger: Short RNAs that silence gene expression. *Nature Rev.*, 4, 457–467.

6. Hannon, G.J. (2002) RNA interference. *Nature*, 418, 244–251.
7. Holen, T., Amarzguioui, M., Wiiger, M.T., Babaie, E., and Prydz, H. (2002) Positional effects of short interfering RNAs targeting the human coagulation trigger Tissue Factor. *Nucleic Acids Res.*, 30, 1757–1766.
8. Elbashir, S.M., Martinez, J., Patkaniowska, A., Lendeckel, W., and Tuschl, T. (2001) Functional anatomy of siRNAs for mediating efficient RNAi in *Drosophila melanogaster* embryo lysate. *EMBO J.*, 20, 6877–6888.
9. Kumar, R., Conklin, D.S., and Mittal, V. (2003) High-throughput selection of effective RNAi probes for gene silencing. *Genome Res.*, 13, 2333–2340.
10. Mittal, V. (2004) Improving the efficiency of RNA interference in mammals. *Nat. Rev. Genet.*, 5, 355–365.
11. Schwarz, D.S., Hutvagner, G., Du, T., Xu, Z., Aronin, N. et al. (2003) Asymmetry in the Assembly of the RNAi Enzyme Complex. *Cell*, 115, 199–208.
12. Khvorova, A., Reynolds, A., and Jayasena, S.D. (2003) Functional siRNAs and miRNAs Exhibit Strand Bias. *Cell*, 115, 209–216.
13. Chalk, A.M., Wahlestedt, C., and Sonnhammer, E.L.L. (2004) Improved and automated prediction of effective siRNA. *Biochem. Biophys. Res. Commun.*, 319, 264–274.
14. Teramoto, R., Aoki, M., Kimura, T., and Kanaoka, M. (2005) Prediction of siRNA functionality using generalized string kernel and support vector machine. *FEBS Lett.*, 579, 2878–2882.
15. Naito, Y., Yamada, T., Ui-Tei, K., Morishita, S., and Saigo, K. (2004) siDirect: Highly effective, target-specific siRNA design software for mammalian RNA interference. *Nucleic Acids Res.*, 32, W124–W129.
16. Santoyo, J., Vaguerizas, J.M., and Dapozo, J. (2004) Highly specific and accurate selection of siRNAs for high-throughput functional assays. *Bioinformatics*, 21, 1376–1382.
17. Truss, M., Swat, M., Kielbasa, S.M., Schafer, R., Herzed, H. et al. (2005) HuSiDa – the human siRNA database: an open-access database for published functional siRNA sequences and technical details of efficient transfer into recipient cells. *Nucleic Acids Res.*, 33, D108–D111.
18. Reynolds, A., Leake, D., Boese, Q., Scaringe, S., Marshall, W.S. et al. (2004) Rational siRNA design for RNA interference. *Nat. Biotech.*, 22, 326–330.
19. Ui-Tei, K., Naito, Y., Takahashi, F., Haraguchi, T., Ohki-Hamazaki, H. et al. (2004) Guidelines for the selection of highly effective siRNA sequences for mammalian and chick RNA interference. *Nucleic Acids Res.*, 32, 936–948.
20. Amarzguioui, M. and Prydz, H. (2004) An algorithm for selection of functional siRNA sequences. *Biochem. Biophys. Res. Commun.*, 316, 1050–1058.
21. Hsieh, A.C., Bo, R., Monola, J., Vazquez, F., Bare, O. et al. (2004) A library of siRNA duplexes targeting the phosphoinositide 3-kinase pathway: determinants of gene silencing for use in cell-based screens. *Nucleic Acids Res.*, 32, 893–901.
22. Jagla, B., Aulner, N., Kelly, P.D., Song, D.A., Volchuk, A. et al. (2005) Sequence characteristics of functional siRNAs. *RNA*, 11, 864–872.
23. Huesken, D., Lange, J., Mikanin, C., Weiler, J., Asselbergs, F. et al. (2005) Design of a genome-wide siRNA library using an artificial neural network. *Nat. Biotech.*, 23, 995–1001.
24. Snove, O. Jr, Nedland, M., Fjeldstad, S.H., Humberset, H., Birkeland, O.R. et al. (2004) Designing effective siRNAs with off-target control. *Biochem. Biophy. Res. Commun.*, 325, 769–773.
25. Poggio, T. and Girosi, F. (1990) Networks for approximation and learning. *Proc. of IEEE*, 78: 1481–1497.
26. Wu, C.H. and McLarty, J.W. (2000) *Neural Networks and Genome Informatics.* Elsevier Science Ltd., NY.
27. Kohonen, T. (1990) The self-organizing map. *Proc. IEEE*, 78, 1464–1480.
28. Takasaki, S., Kotani, S., and Konagaya, A. (2004) An effective method for selecting siRNA target sequences in mammalian cells. *Cell Cycle*, 3, 790–795.
29. Takasaki, S., Kotani, S., and Konagaya, A. (2005) Selecting effective siRNA target sequences for mammalian genes. *RNA Biol.*, 2, 21–27.
30. Takasaki, S., Kawamura, Y., and Konagaya, A. (2006) Selecting effective siRNA sequences based on the self-organizing map and statistical techniques. *Comput. Biol. Chem.*, 30, 169–178.
31. Takasaki, S. and Konagaya, A. (2006) Comparative analyses for selecting effective siRNA sequences. *Chem-Bio Info. J.*, 6, 69–84.

32. Takasaki, S. and Kawamura, Y. (2007) Using radial basis function networks and significance testing to select effective siRNA sequences. *Comput. Stat. Data An.*, 51, 6476–6487.

33. Saetrom, P. and Snove, O. Jr. (2004) A comparison of siRNA efficacy predictors. *Biochem. Biophy. Res. Commun.*, 321, 247–253.

34. Ladunga, I. (2007) More complete gene silencing by fewer siRNAs: Transparent optimized design and biophysical signature. *Nucleic Acids Res.*, 35, 433–440.

35. Holen, T. (2006) Efficient prediction of siRNAs with siRNArules 1.0: An open-source JAVA approach to siRNA algorithms. *RNA*, 12, 1620–1625.

36. Saetrom, P. (2004) Predicting the efficacy of short oligonucleotides in antisense and RNAi experiments with boosted genetic programming. *Bioinformatics*, 20, 3055–3063.

37 Heale, B.S.E., Sifer, H.S., Bowers, C., and Rossi, J.J. (2005) siRNA target site secondary structure predictions using local stable substructures. *Nucleic Acids Res.*, 33, e-30.

38. Luo, K.Q. and Chang, D.C. (2004) The gene-silencing efficacy of siRNA is strongly dependent on the local structure of mRNA at the target region. *Biochem. Biophy. Res. Commun.*, 318, 303–310.

39. Bohula, E.A., Salisbury, A.J., Sohail, M., Playford, M.P., Riedemann, J. et al. (2003) The efficacy of small interfering RNAs targeted to the type I insulin-like growth factor receptor (IGFIR) is influenced by secondary structure in the IGFIR transcript. *J. Biol. Chem.*, 278, 15991–15997.

40. Elbashir, S.M., Harborth, J., Weber, K., and Tueschl, T. (2002) Analysis of gene function in somatic mammalian cells using small interfering RNAs. *Methods*, 26, 199–213.

Chapter 2

Deciphering the Code of Innate Immunity Recognition of siRNAs

Mouldy Sioud

Abstract

Small interfering RNAs (siRNAs) have been widely used for knocking down gene expression in a variety of organisms. Although experiments in cancer cell lines indicate that siRNAs are usually not detected by innate immunity, lipid-mediated delivery of siRNAs into blood cells is often accompanied by the activation of immunity. Recent studies indicated that certain siRNA sequences engage Toll-like receptor TLR7/8 signalling resulting in the activation of a large number of host defense genes including interferons (IFNs), proinflammatory cytokines, Mx proteins, chemokines, chemokine receptors, costimulatory molecules, RNA helicases, galectins, and ubiqitin ligases. In addition to immune activation, most siRNA sequences, if not all, can silence multiple genes in addition to the intended target gene, a phenomenon known as "off-target effects." Hence, one of the major challenges for therapeutic applications of siRNAs is to decipher the mechanisms involved in siRNA recognition by the immune system and to identify strategies that can evade immune activation. In this respect, the replacement of only uridines with their 2′-modified counterparts such as 2′-*O*-methyl uridines abrogates immune recognition of siRNAs. Interestingly, 2′-*O*-methyl-modified RNAs not only evade TLR7/8-sensing pathways, but also reduce siRNA off-target effects and antagonize with a variety of immunostimulatory RNAs to activate TLR7/8 signalling. RNA oligonucleotides and duplex siRNAs with 2′-deoxy uridines or thymidines exhibited no significant immunostimulatory effects and binding potency to TLRs. Therefore, I recommend the use of these modifications in order to evade immune sensing of siRNA and off-target effects. This chapter addresses the current state of knowledge regarding the molecular and cellular mechanisms of RNA recognition by the immune system and proposes a range of strategies allowing the design of siRNAs with minimal or maximal immunostimulatory potency for therapeutic applications.

Key words: RNA interference, small interfering RNA, innate immunity, Toll-like receptors, 2′-ribose modifications, off-target effects.

M. Sioud (ed.), *Methods in Molecular Biology, siRNA and miRNA Gene Silencing, vol. 487*

DOI: 10.1007/978-1-60327-547-7_2

1. Introduction

RNA interference (RNAi) was originally described in the nematode worm *Caenorhabditis elegans* as a response to double-stranded RNA (dsRNA), in which target mRNAs are degraded in a sequence-specific manner *(1)*. A similar phenomenon termed posttranscriptional gene silencing (PTGS) had been described years ago in plants, and now it is believed to function as a surveillance system for blocking the function of harmful RNAs such as viral RNAs *(2)*. Unlike invertebrates, vertebrates react to long dsRNA by activating the IFN pathway *(3, 4)*. Long double-stranded RNAs are synthesized during the replication of many viruses and it is a potent activator of innate immunity. Detailed biochemical studies of RNAi in vitro identified short RNA duplexes of 21 nucleotide (nt) in length, known as small interfering RNAs (siRNAs), as the main effector of RNAi in *Drosophila* cells *(5, 6)*. These findings led to the demonstration that siRNAs, normally generated from long dsRNA during RNAi in all cell types, could be used to inhibit gene expression in mammalian cells without triggering the IFN response *(7)*. Subsequent to this landmark observation by Tuschl and colleagues, the use of siRNAs to study gene functions in mammalian systems became a standard laboratory technique *(8, 9)*. Also, several studies have demonstrated the efficacy of siRNAs in animal models of human diseases upon local or systemic administration *(8, 9)*. Theoretically, the mRNA encoding any protein that is associated with a disease can be cleaved selectively by siRNAs.

In all organisms, the RNase Dicer initiates RNAi by cleaving long dsRNA substrates into siRNAs of about *21– 24* nt in length. Like all RNase III enzymes, Dicer leaves 2 nt 3′-overhangs and 5′-phosphate groups. These siRNA duplexes are then incorporated into a multiprotein complex, the RNA-induced silencing complex (RISC). In contrast to long dsRNAs, synthetic siRNAs enter directly into the RNAi pathway (**Fig. 2.1**). Subsequent to strand separation, the antisense strand guides the RISC to recognize and cleave target mRNA sequences *(7)*. The catalytic activity of RISC is mediated by the Argonaute 2 (AGO2) protein *(10, 11)*. The protein members of the argonaute family are highly basic proteins that contain two common domains, PAZ and PIWI domains. The PIWI domain is essential for interaction with Dicer and contains the nuclease activity that cleaves target mRNAs. AGO2 is also responsible for cleavage of the siRNA passenger strand, thus facilitating the formation of functional RISC complexes *(12, 13, see* **Fig. 2.1**). Analysis of the crystal structures of a siRNA guide strand associated with an AGO2 PIWI domain identified a seed sequence (nucleotides 2 to 8) that directs target mRNA recognition by RISC *(14)*.

Although the discovery of RNAi has provided a new biological strategy to investigate gene function and drug target validation, recent studies have shown that siRNA duplexes can activate innate

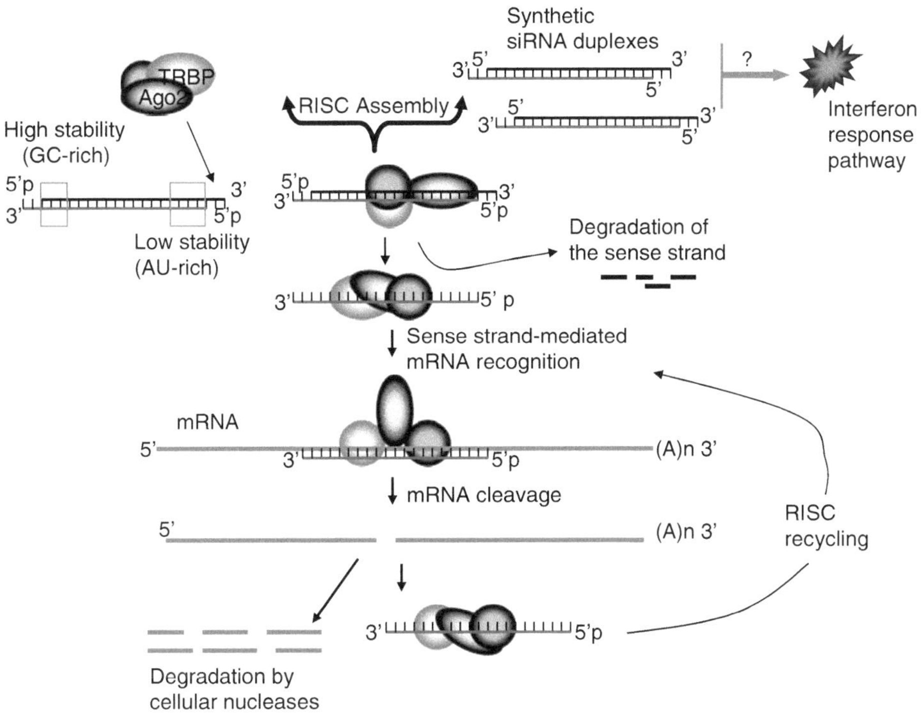

Fig. 2.1. Schematic representation of gene silencing by siRNAs. In contrast to long double-stranded RNAs, siRNAs are directly loaded into a multiprotein complex termed RNA-induced silencing complex (RISC) where the sense strand with high 5′-stability is cleaved by the nuclease AGO2 resulting in strand separation. Subsequently, the RISC containing the antisense strand (guide strand) seeks out and binds to complementary mRNA sequences. Bound mRNA molecules are then cleaved by AGO2 and cleaved mRNA fragments are rapidly degraded by cellular nucleases. Following dissociation, the active RISC is able to recycle and cleave additional mRNA molecules.

immunity by inducing a large number of immune response genes *(15–18)*. Additionally, several groups have shown that siRNAs can silence a variety of genes in addition to the intended target genes *(16)*. Therefore, the development of strategies that block siRNA unwanted effects is crucial to their therapeutic applications. However, if we view immune activation as beneficial for certain diseases, immunostimulatory siRNAs may contribute to the activation of innate and adaptive immunity against cancer and viral infected cells (*see* **Sect. 8**).

2. Immune Recognition of Pathogens

To distinguish pathogens from self-components, the innate immune system uses a wide variety of relatively invariable receptors to sense conserved structures expressed by a large group of microorganisms, known as pathogen-associated molecular patterns (PAMPs) *(19, 20)*. PAMPs are unique to microorganisms and can be found in lipopolysaccharide, peptidoglycan, capsular structures, bacterial flagellin, bacterial DNA, bacterial lipids, viral RNAs, and viral glycoproteins. Among the pattern-recognition receptors (PRRs), Toll-like receptors (TLRs) are crucial for pathogen-derived products and activation of innate and adaptive immunity *(21, 22)*. They owe their name to the *Drosophila melanogaster* protein Toll, which controls the dorsal–ventral patterning of the fly embryo *(23)*. In flies such as *Drosophila melanogaster*, the innate immune response to fungal and gram-positive bacteria is mostly under the control of the Toll signalling pathways *(24)*. TLRs are classical transmembrane proteins whose ligand-binding domains, composed of leucine-rich repeats (LRRs), are capable for recognizing either extracellular or membrane-enclosed foreign organisms *(25)*. In humans, ten functional TLRs (TLR-1-10) have been identified and shown to detect pathogen-derived compounds *(25)*. All TLRs contain extracellular LRR domains, which recognize pathogen-specific structures, a cytoplasmic signalling domain known as Toll-interleukin receptor (TIR) domain, which links the recognition signal with intracellular pathways. Signalling is dependent on the engagement of one or more number of adaptor proteins, particularly myeloid differentiation primary-response protein 88 (MyD88) and the TIR domain containing adaptor proteins. Upon activation, the TIR domain on the intracellular region of the TLR binds to the MyD88 TIR domain, resulting in the recruitment and activation of IRAKl and IRAK4, and subsequently TRAF6 *(25)*. This signalling pathway eventually leads to the activation of signal transduction cascades involving the recruitment of adaptor molecules, tyrosine phosphorylation, activation of transcription factors, and subsequent activation of immune response genes.

3. RNA Sensors

The immune system has evolved cellular and molecular strategies to discriminate between foreign and self-nucleic acids. Among the cytoplasmic sensors of long dsRNA is the dsRNA-dependent protein kinase (PKR) that phosphorylates serine and threonine residues of target proteins *(26)*. Most human cells constitutively express a low level of PKR, which remains inactive. However, upon binding to dsRNA, PKR forms a homodimer resulting in its autophosphorylation and activation. Activated PKR phosphorylates various substrates including the protein synthesis initiation factor elF-2α resulting in inhibition of viral and cellular protein translation, an essential step in antiviral resistance *(26)*. Also, PKR can phosphorylate IKK-β, leading to the activation of the NF-κB signalling pathway. It should be noted that recognition of dsRNAs by PKR is sequence-independent and the presence of IFN upregulates PKR expression. A second protein that is stimulated by long dsRNA is 2′-5′-oligoadenylate synthetase (OAS), which is expressed constitutively and upregulated through IFN-α and β signaling during antiviral responses *(27)*. This IFN-induced enzyme catalyses the formation from ATP of 2′-5′-linked oligoadenylates that activate a latent ribonuclease called RNase L, which degrades both cellular and viral RNAs. Although both OAS and PKR are implicated in antiviral immunity, PKR and RNase L are mainly IFN effectors and not absolutely required for IFN production. Therefore, other kinases may be involved. More recently, two additional intracellular helicases, retinoic-acid-inducible gene I (RIG-I) and melanoma differentiation-associated gene 5 (MDA-5), were identified *(28)*. These helicases are widely expressed in an inactive form, and like other antiviral proteins they are upregulated by IFN α/β. RIG-I encodes a caspase recruitment domain (CARD) at the N terminus, in addition to an RNA helicase domain. The RNA helicase domain requires ATPase activity and is responsible for viral dsRNA recognition and induction of conformational changes leading to the interaction of the RIG-I CARD domain with another CARD-containing adaptor protein, known as IPS-1, MAVS, Cardif, or VISA. *(29)*. IPS-1 is an outer mitochondrial membrane binding protein. IPS-1 activates IRF3 and IFR-7 through TBK1/IKKi, resulting in IFN-β production *(29*, **Fig. 2.2**). Mitochondrial retention of IPS-1 is essential for IRF3, IRF7, and NF-kB activation by RIG-1 *(29)*.

Although RIG-I seems to be an essential sensor of viral RNAs, microbial nucleic acids are also recognized by TLRs, especially in immune cells *(22)*. It should be noted that the subcellular localization of TLRs correlates with the nature of their ligands, rather than their sequence similarity. Whereas most TLRs are expressed in the plasma membrane for detecting bacterial components, TLR3, TLR7, TLR8, and TLR9 are expressed in intracellular compartments (endosomes, lysosomes) *(21)*. The immune function of this cellular localization is more likely to sense viral RNAs. TLR3 is also expressed on the cell surface and it is believed to recognize extracellular viral dsRNAs *(30)*. TLR7 and TLR8 recognize viral ssRNA and small synthetic antiviral compounds referred to as imidazoquinolines *(31)*. TLR9 recognizes unmethylated CpG-DNA motifs that exist in both viral and bacterial DNA, but are suppressed or methylated in vertebrate genomes *(32)*. However, the structural differences of eukaryotic versus prokaryotic DNA are presumably not the only mechanism for distinguishing self from nonself nucleic acids because under certain conditions TLRs can recognize self-RNA resulting in the induction of autoimmune diseases *(33)*. It should be noted that intracellular NOD-like receptors detect bacteria, whereas viruses are mainly detected by TLRs and RIG-like receptors. The virus-detecting TLRs operate mainly in plasmacytoid dendritic cells by respond-

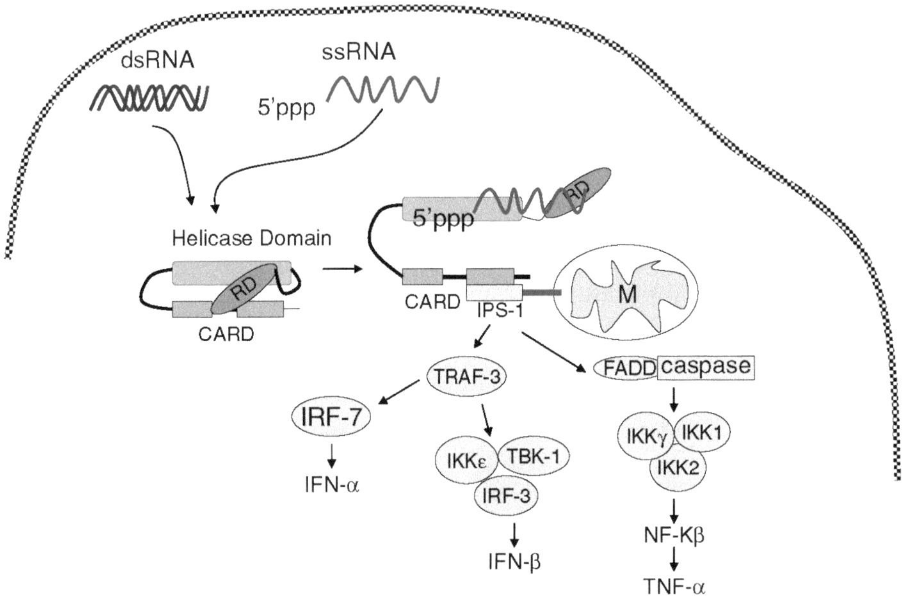

Fig. 2.2. Schematic representation of RIG-I-mediated immune activation. In resting state, RIG-I is maintained as a monomer. Following binding to viral ssRNA or dsRNA, it undergoes a conformational change allowing the CARD interaction with the downstream adaptor IPS protein that is expressed on the outer mitochondrial membrane. CARD–CARD interactions between activated RIG-I and IPS-I induces the recruitment of downstream signaling molecules resulting in the activation of IRF-7, IFR-3, and NF-kB (RD = repressor domain).

ing to viral nucleic acids that enter the cell via endocytosis. In these cells, the major immune response is the production of type 1 IFN *(34)*. During our studies with immunostimulatory RNAs, we have included some non-modified. CpG oligonucleotides (phosphodiester backbone) as controls for TLR9 activation. Upon transfection with DOTAP, however, none of the tested CpG induced a significant TNF-α production in human adherent PBMC cells (Furset and Sioud, unpublished data). Given that most of the work, if not all, with CpG has been done with phosphorothioate-modified CpG oligonucleotides, our observation suggests that TLR9 may recognize DNA structures that are present in phosphorothioate-modified CpG oligonucleotides.

4. Molecular Basis for Immune Sensing of Single- and Double-Stranded RNAs

Although siRNAs were initially thought to bypass the IFN response because they are too short to be sensed by dsRNA sensors *(7)*, we and others have shown that they could activate innate immunity in mammalian cells *(35–38)*. Early studies indicated sequence-independent activation of PKR and TLR3 signalling pathways by siRNAs *(18, 39)*. However, recently it was demonstrated that PKR and TLR3 do not represent the major pathways by which chemically synthesized siRNAs activate immunity in immune cells *(35–38)*. Indeed, a group of siRNA sequences stimulated monocytes or dendritic cells to produce proinflammatory cytokines and type I IFNs. This response is mainly mediated through TLR7 in mice and TLR7/8 in humans. Indeed, TLR7 knockout mice did not mount immune activation in response to siRNAs *(36)*. Under our experimental conditions, ss siRNAs were found to be more effective than ds siRNAs in triggering TLR7 and TLR8 responses in human monocytes or PBMCs *(35, 38)*. The extent of siRNA unwanted effects has recently been confirmed by expression profiling using microarrays, which identified over 400 siRNA-affected genes in PBMCs *(40)*. Genes encoding for proinflammatory cytokines, IFNs, and Mx proteins, are among the genes that are significantly induced. Mx proteins are IFN-induced GTPases that form complexes with dynamin, disrupting trafficking or activity of viral polymerases, thereby interfering with viral replication. The microarray data also reveal the activation of new genes with no known functions in RNA/TLR responses, which remain to be characterized. These includes new TLR-induced helicases and potential transport proteins. Collectively, these data highlight the importance of analyzing the immunostimulatory potential of siRNAs prior to clinical applications.

5. TLR7/TLR8 Recognition RNA Motifs

TLR7 and TLR8 recognize certain siRNA sequences, provided they are delivered to the endosomes as illustrated in **Fig. 2.3**. Thus, what is the nature of the IFN-inducing motif present in one sequence but absent in another? Initial experiments indicate that some types of secondary structures and/or specific nucleotides are responsible for the activation of NF-kB signalling pathway by siRNAs in human monocytes *(15)*. Monocytes are circulating peripheral blood cells that can be differentiated by cytokines into macrophages of different phenotypes as well as into dendritic cells. As mentioned above, siRNA effects are sequence-dependent and can occur with ds siRNAs and ss siRNAs *(35)*. Although a defined and universal sequence motif recognized by TLR7 and/ TLR8, as described for TLR9 (the CpG motif) has not been identified yet, Judge and colleagues found that the 5′-UGUGU-3′ motif was indispensable for the immune activation by a siRNA in human blood cells *(37)*. Hornung et al. identified a 9-nt RNA motif (5′-GUCCUUCAA-3′) that is recognized by TLR7 in the context of siRNA duplexes and the activity does not depend on GU content *(36)*. We have found that IFN induction by siRNAs cannot be easily suppressed by selecting siRNA sequences without the GU dinucleotides. Indeed, several siRNA sequences without GU bases induced TNF-α production in human blood

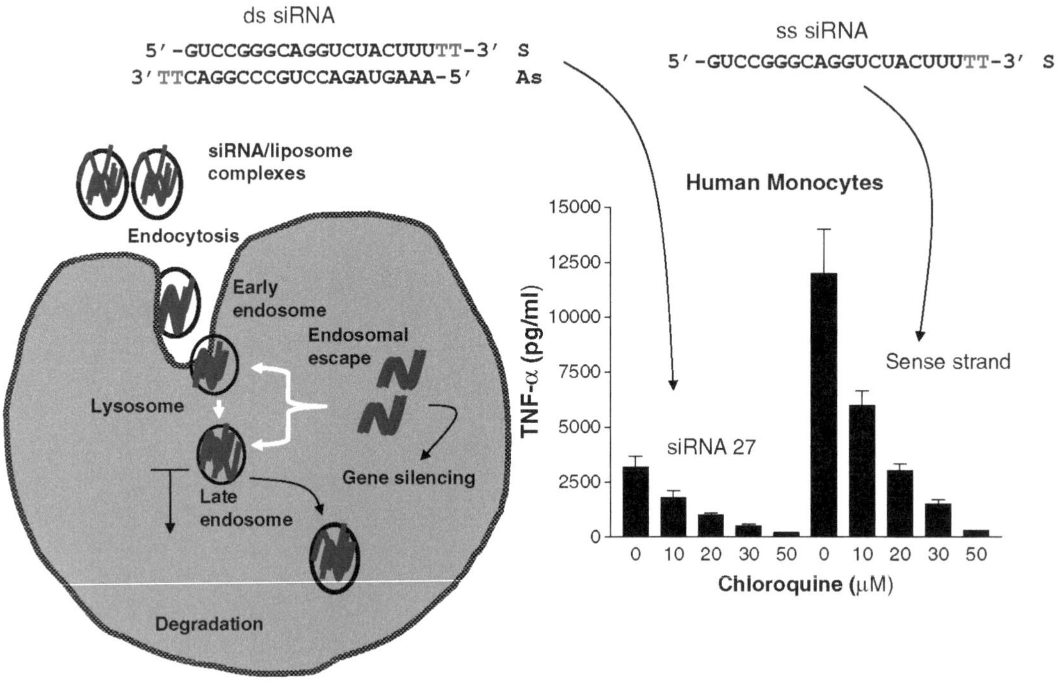

Fig. 2.3. Endosome maturation is required for siRNA activation of innate immunity. (**a**) Schematic representation of siRNA uptake by mammalian cells. (**b**) PBMCs were pretreated with chloroquine, an inhibitor for endosome maturation, for 2 h prior to transfection with either double- (ds) or single-stranded (ss) siRNA targeting mouse TNF-α. Subsequent to 18 h transfection time, TNF-α levels in culture supernatants were measured by ELISA.

cells. Although the precise nature of the RNA motifs responsible of innate immune activation is not known, we showed that the ability of either ss RNA or dsRNAs to trigger TNF-α production is largely dependent on the uridine content *(38)*. Replacement of uridines with adenosines abrogated immune activation *(38)*.

6. Overcoming siRNA Immune Activation

Much of the current interest in the mechanisms involved in RNA sensing or tolerance by the immune system was generated by the observation that siRNA can activate innate immunity *(15, 18)*. Considering the high frequency of uridines and/or GU dinucleotides in messenger RNAs it is more likely that a high proportion of self and nonself chemically made siRNA sequences will activate innate immunity. Therefore, it would be desirable to develop strategies that evade immune activation. At least two distinct ways to evade immune activation by siRNAs can be applied: first, the use of delivery agents that avoid the delivery of siRNA into intracellular endolysosomal compartments, and second, the use of chemically modified siRNAs.

However, the chemical modifications that block immune activation must be chosen carefully so as not to inhibit siRNA silencing potency. Thus, finding the appropriate chemical modifications for interfering with siRNA immune activation will be important for exploring their therapeutic applications. In this respect, Morrissey and colleagues showed that the incorporation of various 2′-modified nucleotides in siRNA sequences abrogated their immunostimulatory potency *(41)*. Fortunately, we have shown that replacement of only uridines with their 2′-fluoro, 2′-deoxy, or 2′-*O*-methyl-modified counterparts can abrogate immune recognition of siRNAs by TLRs without reducing their silencing potency *(38*, **Fig. 2.4a and b**). These findings have subsequently been confirmed by gene expression analysis using microarray technology (40, **Fig. 2.4c and d**). In accordance with our findings, Judge and colleagues demonstrated that the incorporation of 2′-*O*-methyluridine or 2′-*O*-methylguanosine residues into siRNAs can abrogate immune activation *(42)*. Collectively, current data offer the possibility of choosing the appropriate modifications that evade immune activation without reducing siRNA-silencing potency. When designing chemically modified siRNAs, it is important to consider some rules in order to avoid the inhibition of siRNA cleavage activity. First, the 5′-end of the sense strand must have a free hydroxyl or phosphate group. Second, the 5′-end of the sense strand can be modified in order to block its incorporation into the RISC. Third, the 3′-ends of the sense and antisense strands can be modified with any fluorochrome. These 3′-end modifications are expected to facilitate the examination of siRNA uptake and imaging in vivo (see chapter 5).

The finding that 2′-modified RNAs can evade immune activation suggest that naturally modified RNAs are not recognized by TLR7/8. Support for this view has been provided by Karikó and colleagues, who demonstrated that natural modifications that are frequently found in mammalian RNA (such as pseudouridine, 5′-methylcytidine, 2′-*O*-methyl) can interfere with the capacity of RNA to activate TLR7 in DC cells *(43)*. Thus, unmodified RNA corresponding to mammalian sequences would be expected to activate TLR7 or TLR8 more effectively than native RNAs provided they are delivered to the endosomes *(35)*.

In light of the finding that internalization of siRNA and endosomal maturation is a precondition for either double-stranded or single-stranded siRNA-based activation of the immune system *(35)*, one could use inhibitors of endosome acidification to block immune activation *(35)*. Because chloroquine and bafilomycin A1 blocked the immunological activity of siRNAs, we then investigated whether the RNAi pathway is active in chloroquine-treated human cells. The data indicate that the silencing activity of siRNAs was not inhibited, even enhanced, by chloroquine, which is known to induce the neutralization and the swelling of endosome vesicles. Similar results were obtained with bafilomycin A1.

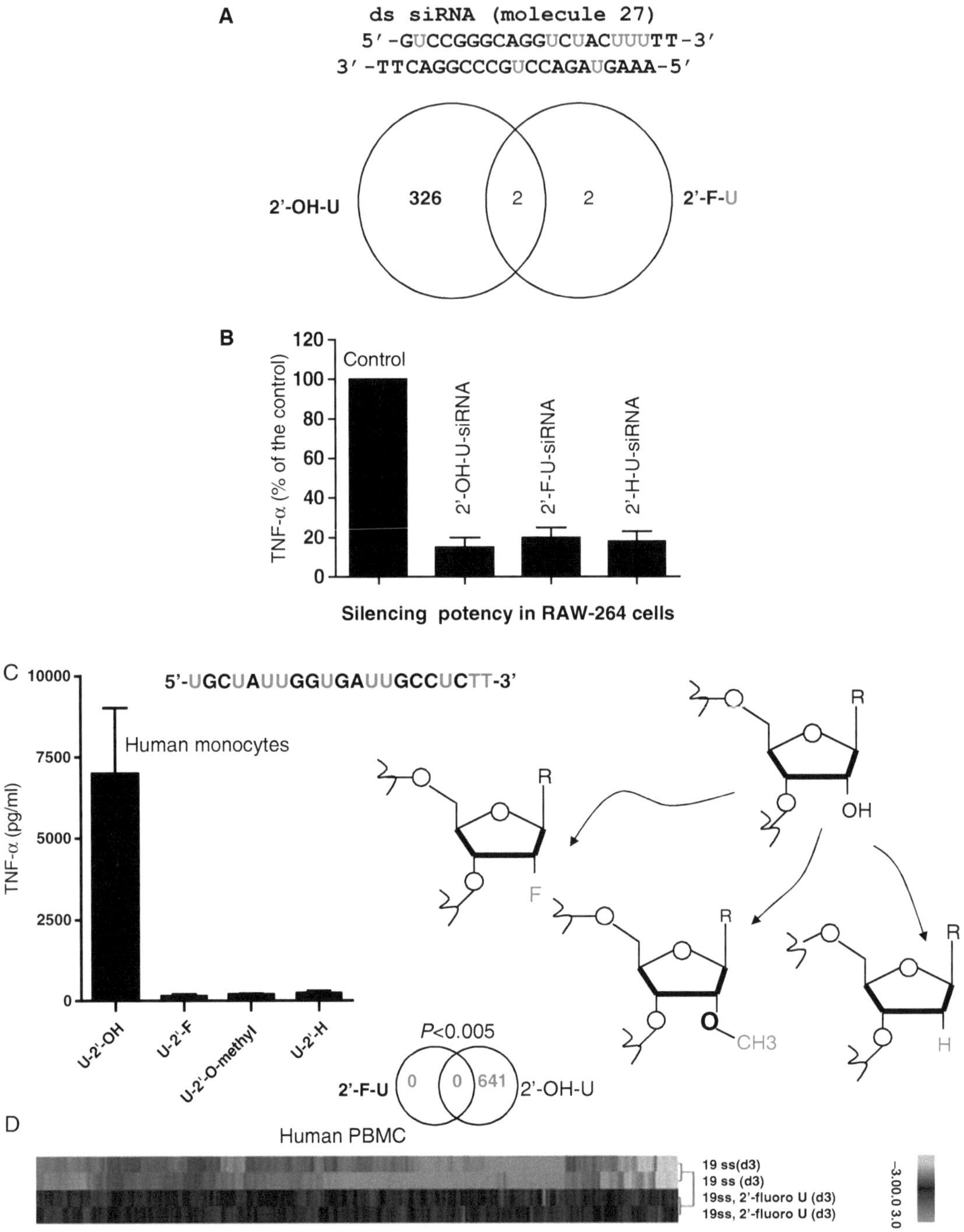

Fig. 2.4. Modified RNAs evade TLR recognition. (**a**) Double-stranded (ds) siRNA targeting mouse TNF-α activates innate immunity in human PBMC, whereas their 2′-uridine-modified counterparts did not. As shown unmodified siRNA altered the expression of 328 genes, whereas only 4 genes were affected by its 2′-fluoro-modified version ($P<0.001$). (**b**) Inhibition of TNF-α gene expression by the modified siRNA in RAW 264 mouse macrophages. (**c**) Analysis of TNF-α gene expression in human monocytes in response to unmodified and 2′-uridine-modified single-stranded (ss) siRNA targeting mouse basigin. (**d**) Global gene expression in human PBMC in response to unmodified and 2′-fluorouridine-modified basigin ss siRNA, molecule 19 *(38, 40)*.

7. Suppressive 2′-Modified RNAs

The finding that unmodified, but not 2′-modified, RNA is a potent trigger of innate immunity raised the question about the difference in their structures that might be relevant to TLR7/8 binding. So, which step is affected by 2′-modifications, and why cannot 2′-modified RNAs trigger immune activation? One way to address the first question is to assess whether 2′-modified RNA could antagonize with immunostimulatory RNAs to trigger TLR7/8 signaling. Studies of transfected human monocytes show that 2′-*O*-methyl-modified RNAs abrogates the activation of TLR7 by immunostimulatory RNAs *(44)*. Of considerable interest is that 2′-*O*-methyl-modified RNAs suppressed immune activation at very low concentrations *(45*, **Fig. 2.5**). In addition, we have shown that they can effectively inhibit

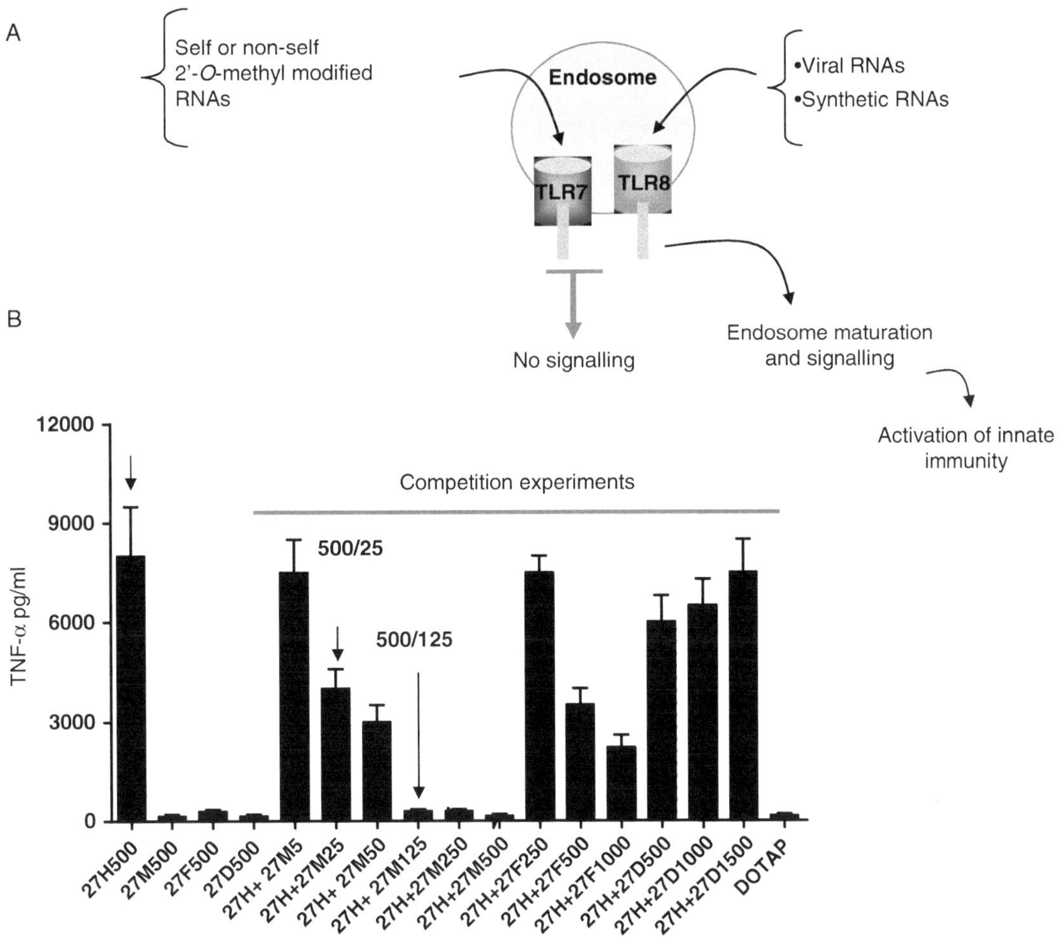

Fig. 2.5. 2′ *O*-Mehtyl-modified RNAs alter TLR signaling. (**a**) Schematic representation of TLR7/8 recognition of RNAs. (**b**) TNF-α production in human monocytes in response to unmodified ss siRNA 27 (27H) either alone or in combination with various concentrations of 2′-*O*-methyl (27M), 2′-fluoro (27F), or 2′-deoxy (27D). The numbers indicate the used RNA concentrations in ng/200 μl *(45)*.

immune activation by a variety of immunostimulatory sequences including bacterial and mitochondrial RNAs *(45)*. Also, chemically modified RNA can antagonize with immunostimulatory RNA to activate IDO, an immunosuppressive enzyme *(46)*. In accordance with our data, Robbins and colleagues have reported that 2′-modified immunostimulatory RNAs can function as TLR7/8 antagonists by inhibiting TLR7 signaling induced by immunostimulatory RNAs or loxoribine in both murine and human cells *(47)*. Suppressive 2′-modified RNAs should represent a new class of agents that may be useful in the treatment of autoimmunity triggered by TLR7 and TLR8 signaling. In contrast to 2′-O-methyl modified oligonucleotides, RNA oligonucleotides and duplex siRNAs with 2′-deoxy uridines or thymidines showed no significant immunostimulatory effects and they did not bind to TLRs (45, Sioud unpublished data). Furthermore, both modifications did not affect gene silencing. Given the important role of TLR7/8 in immunity, we recommend the use of 2′-deoxy modification in order to evade immune activation by siRNAs and not interfering with TLR biological function. The use of 2′-deoxy uridines or thymidines has also reduced siRNA off-target effects (40). This observation should support the incorporation of DNA bases into the siRNA antisense strand in order to reduce off-target effect.

8. The Molecular Basis of RNA Sensing by RIG-I

As discussed above, recent studies on the immune response to chemically made siRNAs have highlighted the involvement of endosomes. Indeed, cytoplasmic delivery of synthetic siRNAs by electroporation into adherent peripheral blood mononuclear cells did not induce significant levels of inflammatory cytokines and type I IFNs, whereas the same sequences when delivered by lipid did *(35)*. Thus, synthetic siRNAs are not detected by cytoplasmic sensors for viral RNAs. However, our data do not explain why ds siRNA are not sensed by RIG-I, a major cytoplasmic sensor of viral RNAs (35, **Fig. 2.2**). It has been demonstrated that innate immunity can be triggered by synthetic siRNA duplexes harboring 2-base 3′-overhangs, but not with siRNAs with blunt ends *(48)*. The authors showed that RIG-I can bind siRNAs with or without 2-base 3′-overhangs; however, only those with blunt ends could trigger its ATPase activity. This observation implies that endogenous shRNAs or miRNA harboring the Dicer signature, 2-base 3′-overhang, are not perfect stimulators of RIG-I. Thus, the structures of the 5′-ends between shRNAs (substrate for Dicer) and nonself dsRNAs such as viral RNAs are critical for self and nonself discrimination *(49, 50)*.

Notably, the predominant form of naturally occurring dsRNAs in mammalian cells is derived from endogenously expressed miRNAs that constitute a large class of noncoding small RNAs involved in gene regulation in a variety of organisms ranging from plants to mammalians (51, *see* Chaps. 18 and 19). Presently, more than 1000 potential human miRNAs have been identified and numerous others have been experimentally validated. Usually miRNAs are transcribed from endogenous genes by RNA polymerase II as long RNA precursors called primary miRNAs (pri-miRNA), containing one or more distinct miRNAs. In the nucleus, the RNA precursors are processed by Drosha to a 60–80 nt RNA hairpin intermediate, bearing 2-base 3′-overhang, called a pre-miRNA. The Drosha cleavage site was shown to be 11 base pairs from the stem single-stranded RNA junction *(52)*. Processed pre-miRNAs are then transported from the nucleus to the cytoplasm by exportin-5, where its 2-base 3′-overhang is recognized by Dicer, leading to the generation of mature miRNAs that can evade RIG-I activation.

During our studies, we have also found that innate immunity is not activated by synthetic single-stranded sequences (21 nt) when delivered to the cytoplasm via electroporation *(35)*. These RNAs do not contain 2-base 3′-overhang because they are single-stranded. To examine the contribution of RIG-I in sensing exogenous RNAs, adherent PBMCs were transfected with either T7-transcribed siRNAs or chemically synthesized siRNAs. The inhibition of endosome maturation by chloroquine abrogated the immunostimulatory activity of chemically made siRNAs, but not the T7-made siRNAs *(53)*. In addition, the immunostimulatory effect of the T7-made siRNAs was not inhibited with 2-aminopurine, a specific inhibitor of PKR *(53)*. So, which are the cytoplasmic factors that sense in vitro transcribed RNA but not in chemically made siRNAs? Additional studies from other investigators showed that RIG-I senses single-stranded RNA-bearing 5′-triphosphate, a specific signature of viral and in vitro transcribed RNAs *(54)*. Artificial capping or base modifications of the 5′-triphosphate abolished immune response. In general, self-RNA undergoes several modifications to eliminate or mask the 5′-triphosphate group. However, the data do not explain why certain endogenous RNA with 5′-triphosphates escape RIG-I recognition. Specific naturally occurring 2′-modifications might protect these RNA-bearing 5′-triphosphate from being detected by RIG-I. It should be noted that MDA5, the most closely related protein of RIG-I, is also an IFN-inducible protein. However, the exact mechanisms of RNA sensing by MDA5 have yet to be defined but seem to sense dsRNA structures from certain viruses. A summary of viral and synthetic RNA recognition by cellular sensors is shown in **Fig. 2.6**.

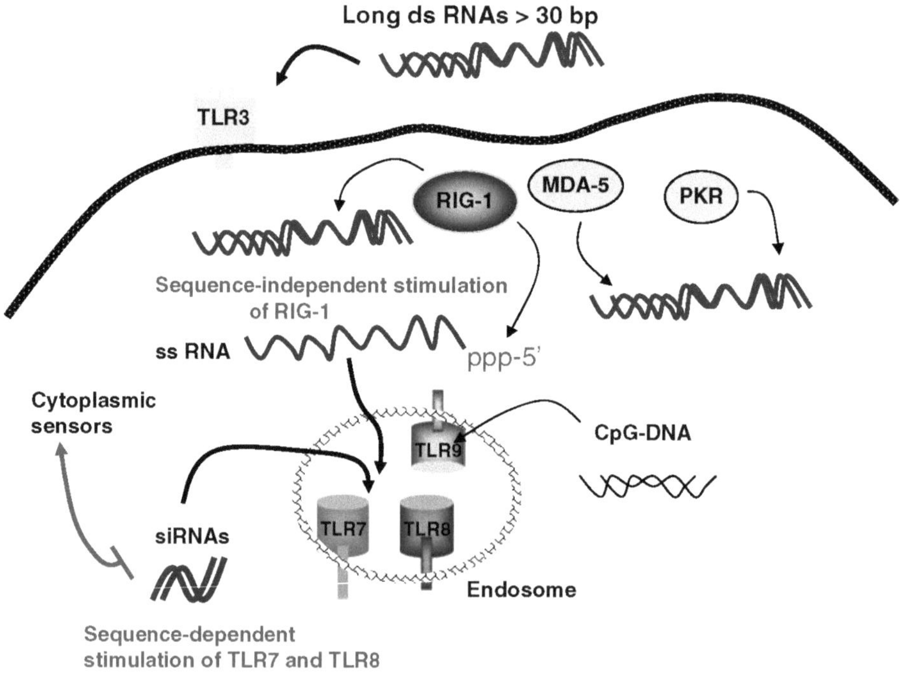

Fig. 2.6. Schematic representation of RNA sensing by cytoplasmic and Toll-like receptors. For details see text.

9. Effect of Chemical Modifications on siRNA Off-Target Effects

Another potential source of siRNA toxicity is the destruction of cellular mRNAs that share partial homology to the siRNA sequences, a consequence known as off-target effects *(17)*. Because the cellular pathways activated by miRNAs and siRNAs are overlapping, it is more likely that all siRNA sequences will exhibit a miRNA-like activity *(51)*. The most commonly used strategy to ensure siRNA target specificity is the basic local alignment search tool BLAST. However, short sequence stretches may not be detected by the BLAST program. In addition, the identification of such sequences does not necessarily indicate the occurrence of off-target effects. Similarly, the absence of short homologies will not rule out off-target effects. The best way to deal with this problem is to analyze global gene expression, specifically when siRNAs are going to be used in functional genomics or to develop therapeutics. During our studies with siRNAs, we have found that 2′-uridine modifications of siRNAs not only evade the activation of TLR7/8 but they can block most of TLR-independent effects including off-target effects *(40)*. Although the evading mechanisms remain to be investigated, it is possible that the interaction of siRNA sequences with unintended cellular mRNAs is affected by chemical modifications. Modifications of RNA might particularly be disruptive for siRNA binding to mismatched sequences. In accordance with our data, Jackson and colleagues found that the incorporation of 2′-*O*-methyl group at the second position of the guide strand can reduce most off-target gene-silencing effects without affecting siRNA silencing of the intended target gene *(55)*. Collectively, the data offer a simple strategy for reducing off-target effects.

10. Potential Beneficial Effects of Immunostimulatory siRNAs

Among the innate immune sensors that link innate and adaptive immunity, dendritic cells (DCs) play a crucial role in immune responses and are the only cell type capable of initiating adaptive immune responses by activating naïve T cells *(56, 57)*. These cells are generated from either myeloid or lymphoid bone marrow progenitors that home to sites of potential antigen entry, where they differentiate locally into immature DCs (**Fig. 2.7**).

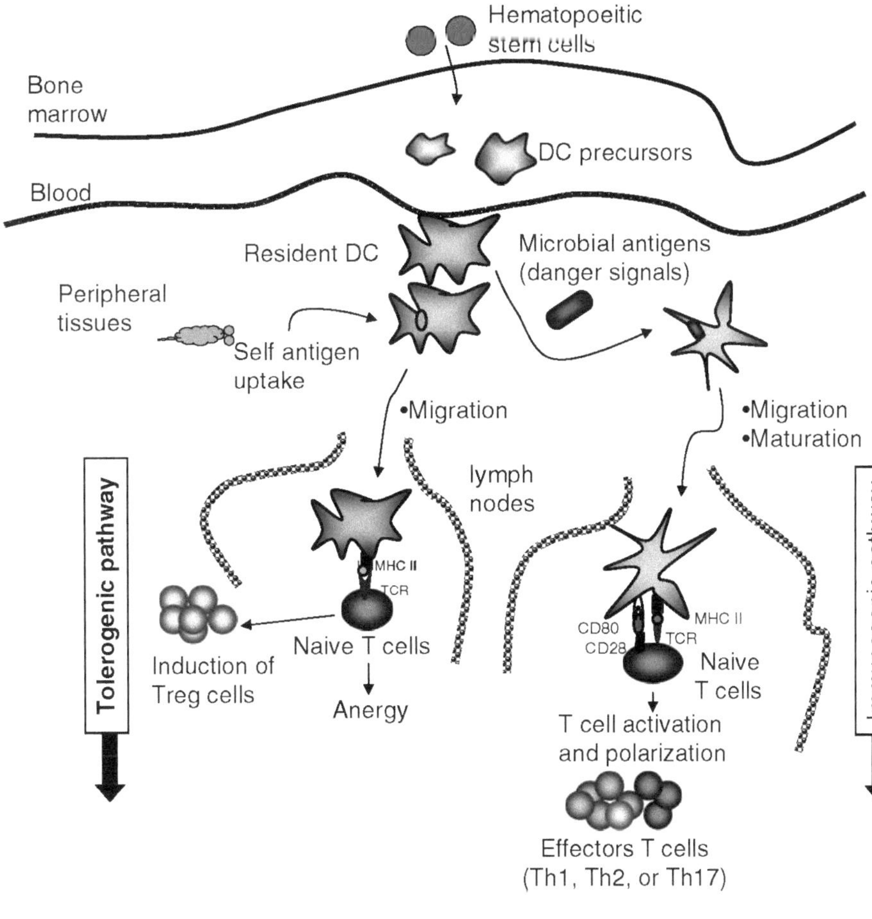

Fig. 2.7. Activation of antigen-presenting cells (APCs). APCs such as dendritic cells can capture antigens derived from self and nonself compounds with the same efficiency. However, only antigens captured in the presence of danger signals such as infections induce DC maturation and activation of naive T cells.

Recent studies demonstrated that both immunity and tolerance are controlled by DCs. In the absence of "danger signals" immature DCs mediate peripheral tolerance, leading to T cell anergy and/or deletion due to the absence of appropriate costimulation by CD80/CD86 molecules *(56)*. However, following antigen capture in the presence of maturation signals DCs undergo a complex maturation that entails upregulation of major histocompatibility complex class I and class II molecules, costimulatory molecules such as CD40, CD80, and CD86, and the production of IL12. Among the signaling pathways that activate DC maturation, TLR signaling pathway is the most effective. Unlike infectious pathogens, tumours do not induce an effective inflammatory response leading to DC activation. In addition, tumour microenvironment can protect tumour cells from immune destruction. In this respect, soluble immunosuppressive factors and membrane-bound molecules including transforming growth factor β, interleukin IL10, prostaglandin E2, and CTLA-4 represent a barrier for antitumour immunity *(58)*. Interfering with the expression of these factors might potentiate antitumour T cell effector function in vivo.

Notably, pathogen-mediated maturation of DCs is mediated mainly through the TLRs that are expressed on immature DCs. Optimal DC maturation might therefore require a combination of both cytokines and TLR ligands. Previously, we have shown that stimulation of DCs with immunostimulatory siRNAs can induce their maturation to secrete cytokines, including IL6 and IL12 *(35)*. While IL12 is required for a Th-1-type response, IL6 may render CD4+ effector T cells refractory to T reg cell-mediated suppression. The inappropriate expression of immunosuppressive cytokines and other negative regulators is expected to hamper immunity against tumours and virus-infected cells *(59)*. Therefore, the development of agents that stimulate DCs and subsequently suppress the expression of negative regulators, such as IL10, TGF-β, and SOCS proteins would facilitate the development of effective cancer vaccines. In a recent study, we have assessed the possibility of combining gene silencing and immunostimulation in one siRNA molecule. Immature monocyte-derived dendritic cells incubated with anti-IL10 siRNAs produced cytokines (e.g., IL6, TNF-α, IL12), and upregulated the expression of the costimulatory molecules (e.g., CD80, CD86), MHC class II molecules, and the chemokine receptor CCR7. Also, IL10 siRNAs enhanced the ability of DCs to activate T cells in MLR assays *(53)*. Thus, the possibility of triggering endogenous IL12 and IL6 production through deliberate activation of the TLR7 or TLR8 pathway via IL10 siRNAs is of considerable interest. In addition to IL10, bifunctional siRNAs against other key factors involved in immune suppression were developed. These include SOCS1, STAT-3, and TGF-β. The SOCS proteins have been identified as inhibitors of cytokine signaling and shown to function in a classical feedback loop *(60)*. They regulate signaling via the Jak/Stat pathway and modulate DC function by switching off IFN g and/or IL12 signaling during immune responses. By targeting SOCS1 with conventional siRNAs in DCs, recently Chen and colleagues demonstrated that antigen-specific antitumour immunity can be enhanced *(61)*. The use of siRNA to modulate the function of antigen-presenting cells (APCs) has just begun and potentially will guide us into a new era of molecular-mechanism-based development of cancer vaccines.

References

1. Jorgensen, R. (1990) Altered gene expression in plants due to trans interactions between homologous genes. *Trends Biotechnol*, 8, 340–344.
2. Fire, A., Xu, S., Montgomery, M. K., et al. (1998) Potent and specific genetic interference by double-stranded RNA in *Caenorhabditis elegans*. *Nature*, 391, 806–811.
3. Svoboda, P., Stein, P., Hayashi, H., et al. (2000) Selective reduction of dormant maternal mRNAs in mouse oocytes by RNA interference. *Development*, 127, 4147–4156.
4. Sen, G. C. (2001) Viruses and interferons. *Annu Rev Microbiol*, 55, 255–281.
5. Zamore, P. D., Tuschl, T., Sharp, P. A., et al. (2000) RNAi: double-stranded RNA directs the ATP-dependent cleavage of mRNA at 21 to 23 nucleotide intervals. *Cell*, 101, 25–33.
6. Elbashir, S. M., Lendeckel, W., and Tuschl, T. (2001) RNA interference is mediated by 21- and 22-nucleotide RNAs. *Genes Dev*, 15, 188–200.
7. Elbashir, S. M., Harborth, J., Lendeckel, W., et al. (2001) Duplexes of 21-nucleotide RNAs mediate RNA interference in cultured mammalian cells. *Nature*, 411, 494–498
8. Sioud, M. (2004) Therapeutic siRNAs. *Trends Pharmacol Sci*, 25, 22–28.
9. Hannon, G. J. and Rossi, J. J. (2004) Unlocking the potential of the human genome with RNA interference. *Nature*, 431, 371–378
10. Liu, J., Carmell, M. A., Rivas, F. V., et al. (2004) Argonaute2 is the catalytic engine of mammalian RNAi. *Science*, 305, 1437–1441.
11. Song, J. J., Smith, S. K., Hannon, G. J., et al. (2004) Crystal structure of Argonaute and its implications for RISC slicer activity. *Science*, 305, 1434–147.
12. Matranga, C., Tomari, Y., Shin, C., et al. (2005) Passenger-strand cleavage facilitates assembly of siRNA into Ago2-containing RNAi enzyme complexes. *Cell*, 123, 607–620.
13. Rand, T. A., Petersen, S., Du, F., et al. (2005) Wang: Argonaute2 cleaves the anti-guide strand of siRNA during RISC activation. *Cell*, 123, 621–629.
14. Ma, J. B., Yuan, Y. R., Meister, G., et al. (2005) Structural basis for 5′-end-specific recognition of guide RNA by the *A. fulgidus* Piwi protein. *Nature*, 434, 666–670.
15. Sioud, M. and Sorensen, D. R. (2003) Cationic liposome-mediated delivery of siRNAs in adult mice. *Biochem Biophys Res Commun*, 312, 1220–1225.
16. Jackson, A. L., Bartz, S. R., Schelter, J., et al. (2003) Expression profiling reveals off-target gene regulation by RNAi. *Nat Biotechnol*, 21, 635–637.
17. Semizarov, D., Frost, L., Sarthy, A., et al. (2003) Specificity of short interfering RNA determined through gene expression signatures. *Proc Natl Acad Sci U S A*, 100, 6347–6352.
18. Sledz, C. A., Holko, M., de Veer, M. J., et al. (2003) Activation of the interferon system by short-interfering RNAs. *Nat Cell Biol*, 5, 834–839.
19. Parkin, J. and Cohen, B. (2001) An overview of the immune system. *Lancet*, 357, 1777–1789.

20. Janeway, C. A. and Medzhitov, R. Jr. (2002) Innate immune recognition. *Annu Rev Immunol*, 20, 197–216.
21. Takeda, K. and Akira, S. (2005) Toll-like receptors in innate immunity. *Int Immunol*, 17, 1–14.
22. Sioud, M. (2006) Innate sensing of self and non-self RNAs by Toll-like receptors. *Trends Mol Med*, 12, 167–716.
23. Brennan, C. A. and Anderson, K. V. (2004) Drosophila: The genetics of innate immune recognition and response. *Annu Rev Immunol*, 22, 457–483.
24. Lemaitre, B., Nicolas, E., Michaut, L., et al. (1996) The dorsoventral regulatory gene cassette spatzle/Toll/cactus controls the potent antifungal response in Drosophila adults. *Cell*, 86, 973–983.
25. Akira, S. and Takeda, K. (2004) Toll-like receptor signaling. *Nat Rev Immunol*, 4, 499–511.
26. Li, S., Peters, G. A., Ding, K., et-al. (2006) Molecular basis for PKR activation by PACT or dsRNA. *Proc Natl Acad Sci U S A*, 103, 10005–10010.
27. Samuel, C. E. (2001) Antiviral actions of interferons. *Clin Microbiol Rev*, 14, 778–809.
28. Kato, H., Takeuchi, O., Sato, S., et al. (2006) Differential roles of MDA5 and RIG-I helicases in the recognition of RNA viruses. *Nature*, 441, 101–105.
29. Meylan, E., Tschopp, J., and Karin, M. (2006) Intracellular pattern recognition receptors in the host response. *Nature*, 442, 39–44.
30. Alexopoulou, L., Holt, A. C., Medzhitov, R., and Flavell, R. A. (2001) Recognition of double-stranded RNA and activation of NF-kappaB by Toll-like receptor 3. *Nature*, 413, 732–438.
31. Heil, F., Hemmi, H., Hochrein, H., et al. (2004) Species-specific recognition of single-stranded RNA via toll-like receptor 7 and 8. *Science*, 303, 1526–1529.
32. Krieg, A. M. (2002) CpG motifs in bacterial DNA and their immune effects. *Annu Rev Immunol*, 20, 709–760
33. Brentano, F., Kyburz, D., Schorr, O., et al. (2005) The role of Toll-like receptor signaling in the pathogenesis of arthritis. *Cell Immunol*, 233, 90–96.
34. Cao, W. and Liu, Y. J. (2007) Innate immune functions of plasmacytoid dendritic cells. *Curr Opin Immunol*, 19, 24–30.
35. Sioud, M. (2005) Induction of inflammatory cytokines and interferon responses by double-stranded and single-stranded siRNAs is sequence-dependent and requires endosomal localization. *J Mol Biol*, 348, 1079–1090
36. Hornung, V., Guenthner-Biller, M., Bourquin, C., et al. (2005) Sequence-specific potent induction of IFN-alpha by short interfering RNA in plasmacytoid dendritic cells through TLR7. *Nat Med*, 11, 263–270.
37. Judge, A. D., Sood, V., Shaw, J. R., et al. (2005) Sequence-dependent stimulation of the mammalian innate immune response by synthetic siRNA. *Nat Biotechnol*, 23, 457–462.
38. Sioud, M. (2006) Single-stranded small interfering RNA are more immunostimulatory than their double-stranded counterparts: a central role for 2′-hydroxyl uridines in immune responses. *Eur J Immunol*, 36, 1222–1230.
39. Kariko, K., Bhuyan, P., Capodici, J., et al. (2004) Small interfering RNAs mediate sequence-independent gene suppression and induce immune activation by signaling through toll-like receptor 3. *J Immunol*, 172, 6545–6549.
40. Cekaite, L., Furset, G., Hovig, E., et al. (2007) Gene expression analysis in blood cells in response to unmodified and 2′-modified siRNAs reveals TLR-dependent and independent effects. *J Mol Biol*, 365, 90–108.
41. Morrissey, D. V., Lockridge, J. A., Shaw, L., et al. (2005) Potent and persistent in vivo anti-HBV activity of chemically modified siRNAs. *Nat Biotechnol*, 23, 1002–1007.
42. Judge, A. D., Bola, G., Lee, A. C., et al. (2006) Design of noninflammatory synthetic siRNA mediating potent gene silencing in vivo. *Mol Ther*, 13, 494–505.
43. Kariko, K., Buckstein, M., Ni, H., et al. (2005) Suppression of RNA recognition by Toll-like receptors: The impact of nucleoside modification and the evolutionary origin of RNA. *Immunity*, 23, 165–175.
44. Sioud, M. (2007) RNA interference and innate immunity. *Adv Drug Deliv Rev*, 59, 153–163.
45. Sioud, M., Furset, G., and Cekaite, L. (2007) Suppression of immunostimulatory siRNA-driven innate immune activation by 2′-modified RNAs. *Biochem Biophys Res Commun*, 361, 122–126.
46. Furset, G., Floisand, Y., and Sioud, M. (2008) Impaired expression of indoleamine 2, 3-dioxygenase in monocyte-derived dendritic cells in response to Toll-like receptor-7/8 ligands. *Immunology*, 123, 263–271.

47. Robbins, M., Judge, A., Liang, L., et al. (2007) 2′-*O*-methyl-modified RNAs act as TLR7 antagonists. *Mol Ther*, 15, 1663–1669.
48. Marques, J. T., Devosse, T., Wang, D., et al. (2006) A structural basis for discriminating between self and nonself double-stranded RNAs in mammalian cells. *Nat Biotechnol*, 24, 559–565.
49. Robbins, M. A., Li, M., Leung, I., et al. (2006) Stable expression of shRNAs in human CD34 + progenitor cells can avoid induction of interferon responses to siRNAs in vitro. *Nat Biotechnol*, 24, 566–571.
50. Sioud, M. (2006) RNA interference below the immune radar. *Nat Biotechnol*, 24, 521–522
51. Bartel, D. P. (2004) MicroRNAs: genomics, biogenesis, mechanism, and function. *Cell*, 116, 281–297.
52. Han, J., Lee, Y., Yeom, K. H., et al. (2006) Molecular basis for the recognition of primary microRNAs by the Drosha-DGCR8 complex. *Cell*, 125, 887–901.
53. Furset, G. and Sioud, M. (2007) Design of bifunctional siRNAs combining immunostimulation and gene-silencing in one single siRNA molecule. *Biochem Biophys Res Commun*, 352, 642–649.
54. Hornung, V., Ellegast, J., Kim, S., et al. (2006) 5′-Triphosphate RNA is the ligand for RIG-I. *Science*, 314, 994–997.
55. Jackson, A. L., Burchard, J., Leake, D., et al. (2006) Position-specific chemical modification of siRNAs reduces "off-target" transcript silencing. *RNA*, 12, 1197–205.
56. Banchereau, J. and Steinman, R. M. (1998) Dendritic cells and the control of immunity. *Nature*, 392, 245–252.
57. Rossi, M. and Young, J. W. (2005) Human dendritic cells: Potent antigen-presenting cells at the crossroads of innate and adaptive immunity. *J Immunol*, 175, 1373–1381.
58. Pardoll, D. (2003) Does the immune system see tumors as foreign or self? *Annu Rev Immunol*, 21, 807–839.
59. Grutz, G. (2005) New insights into the molecular mechanism of interleukin-10-mediated immunosuppression. *J Leukoc Biol*, 77, 3–15.
60. Alexander, W. S. and Hilton, D. J. (2004) The role of suppressors of cytokine signaling (SOCS) proteins in regulation of the immune response. *Annu Rev Immunol*, 22, 503–529.
61. Shen, L., Evel-Kabler, K., Strube, R., et al. (2004) Silencing of SOCS1 enhances antigen presentation by dendritic cells and antigen-specific anti-tumor immunity. *Nat Biotechnol*, 22, 1546–1553.

Chapter 3

Targeted Delivery of Antisense Oligonucleotides and siRNAs into Mammalian Cells

Mouldy Sioud

Abstract

RNA interference (RNAi) is a natural mechanism for gene silencing that can be harnessed for the development of RNA-based drugs. Although synthetic small interfering RNA (siRNAs) can be delivered in vitro to virtually all cell types using lipid-based transfection agents or electroporation, efficient strategies for achieving either systemic or targeted delivery remains one of the major in vivo challenges. Among the targeting strategies, receptor-targeted delivery provides an innovative strategy to selectively direct therapeutics to cancer cells. Receptor-binding peptides can be incorporated into gene-delivery vesicles or directly conjugated to siRNAs in the hope of promoting their localization in target cells expressing the cognate receptors. This chapter discusses the current status of siRNA-targeting strategies using either peptides identified through iterative screening of random peptide phage libraries or naturally occurring peptides. Also, transcriptional targeting strategies and detailed protocols for the selection of cancer cell-binding peptide from random peptide libraries are described.

Key words: RNAi, siRNA, random peptide libraries, hormone peptides, peptide analogues, endocytose, cell surface receptors.

1. Introduction

The main goal of any therapy is to eradicate pathogenic cells such as malignant cells, while sparing normal cells. The advantages of being able to avoid delivering bioactive agents (cytotoxic drugs, radionuclides, cytokines) to healthy tissues/cells are numerous. For example, specific delivery of cytotoxic drugs to cancer cells could alleviate the problem of side effects because high concentrations of the drug within tumors could be attained without affecting normal tissues. Various experimental approaches have been used in order to achieve a preferential accumulation of

M. Sioud (ed.), *Methods in Molecular Biology, siRNA and miRNA Gene Silencing, vol. 487*

DOI: 10.1007/978-1-60327-547-7_3

bioactive agents in target tissues/cells *(1)*. To achieve specific delivery to target cells, different strategies have been pursued, including the use of random peptides and antibody libraries to identify ligands (peptides or antibodies) that bind to membrane receptors expressed on target cells *(2, 3)*. Furthermore, changes in the tumor environment such as increased expression of surface proteases and angiogenesis can be taken advantages of to improve the therapeutic index of existing and new therapeutic approaches such as siRNAs.

In vivo, siRNA delivery can be achieved by a number of strategies including lipid-based formulations (*see* Chaps. 9–17), atelocollagen (*see* Chap. 4) and nanoparticles (*see* Chap. 5), and magnetofection (*see* Chap. 6). As the used delivery agents can enter all cell types, specificity must be built into the delivery agents or the expressed short hairpin RNA (shRNAs). Therefore, identification of strategies that mediate the targeted delivery of nucleic acids (ASO, siRNAs, miRNAs) to specific cell types remain the most challenging technical step towards the transformation of RNA technology into modern medicine *(4)*. One of our research activities is aimed at improving antisense DNA and siRNA delivery by using peptides whose receptors are exclusively or preferentially expressed by cancer cells *(2, 5)*.

1.1. Selection of Cancer Cell Targeting Peptides

Genetic and epigenetic changes in cancer cells are expected to lead to changes in gene expression that could be used to classify various types of tumors and identify protein receptors for targeting and diagnostic imaging through receptor-binding antibodies or peptides. Traditionally, receptor- or cell-binding peptides are identified rationally via structure–activity studies that involve the synthesis of a large number of peptides for in vitro and in vivo testing *(6)*. However, the advances in combinatorial and biological peptide libraries have made it possible to select specific binding peptides for membrane receptors expressed by tumor cells *(7, 8)*. Random peptide libraries include virtually all possible sequences of small peptides that can mimic conformational structures of both continuous and discontinuous epitopes *(9, 10)*. Biopanning of these libraries on purified proteins or on whole cells has led to the selection of a large number of targeting peptides with high binding specificity *(2)*. Some of the selected peptides are listed in **Table 3.1**. In principle, these cell-binding peptides can be linked to any therapeutic agent to increase their accumulation in the cells that express the cognate peptide receptors.

In the case of cancer cells, targeting peptides are often selected for their binding to known cell surface receptors involved in biological functions such as cell proliferation, motility, evasion, and metastasis *(11–13)*. Among the attractive receptors, fibroblast growth factor (FGFR) and epidermal growth factor receptor (EGFR) families are often overexpressed in human tumors *(14)*.

Table 3.1
Target-specific peptides selected from random peptide phage libraries

Peptide	Cellular targets
Vasculature of various tumors	
CD**RGD**CFC, ACDC**RGD**CFCG	$\alpha_v\beta_3$, $\alpha_v\beta_5$ integrins
C**NGR**CVSGCAGRC, CVC**NGR**MEC, **NGR**AHA	Aminopeptidase N
CPGPEGAGC	Aminopeptidase P
TAASGVRSMH, LTLRWVGLMS	NG2 proteoglycan
C**GSL**VRC, C**GLS**DSC,	Tumor vasculature
LRIKRKRRKRKKTRK, NRSTHI	IC-12 rat trachea
SMSIARL, VSFLEYR	Mice prostate
ATWLPPR, RRKRRR, ASSSYPLIHWRPWAR	VEGF
CTTHWGFTLC	Gelatinase
Cell surface of various tumors	
KNGPWYAYTGRO, NWAVWXKR, YXXEDLRRR, XXPVDHGL	Surface idiotype of SUP-88 human B-cell lymphoma
LVRSTGQFV, LVSPSGSWT, ALRPSGEWL, AIMASGQWL, QILASGRWL, RRPSHAMAR, DNNRPANSM, LQDRLRFAT, PLSGDKSST	Surface idiotype of human chronic lymphocytic lymphoma (CLL)
IELLQAR	HL-60 human lymphoma and B-16 mouse melanoma
CVFXXXYXXC, CXFXXXYXYLMC, CVXYCXXXX-CYVC, CVXYCXXXXCWXC	Prostate specific antigen (PSA)
DPRATPGS	LNCaP prostate cancer
HLQLQPWYPQIS	WAC-2 human neuroblastoma
VPWMEPAYQRFL	MDA-MB435 breast cancer
TSPLNIHNGQKL	Head and neck cancer cell lines
SPLW/F,R/K,N/H,**S**, V/H,**L**	ECV304 endothelial cell line
RLTGGKGVG	HEp-2 human laryngeal carcinoma
LTVX**PW**X	Human breast tumor
KCCYSL	ErbB2 Tyrosine kinase type 1 receptor

The epidermal growth factor receptor type 2, known as ErbB2/HER-2, is a carcinoma-associated tyrosine kinase receptor whose activation is responsible for cancer cell survival, proliferation, and metastases *(14)*. Thus, ErbB2 receptor-binding peptides should be important for drug- or siRNA-targeted strategies. In this respect, several ErbB2-binding peptides have been selected from random peptide-phage libraries and one the selected peptides (KCCYSL) bound to the extracellular domain of the human purified ErbB2 receptor *(15, 16)*.

As mentioned above, an understanding of tumor angiogenesis, oxygenation, and related issues involving the tumor–host relationship is becoming essential to studies of cancer biology as well as to the design of more effective forms of cancer therapies *(17)*. The differentially expressed cell surface markers on endothelial cells in angiogenic vessels of tumors should be excellent targets for site-specific targeting. In addition, the accessibility of endothelial cell markers by the bloodstream makes them more attractive targets than their counterparts expressed on tumor cells. To identify endothelial cell-binding peptides, random peptide-phage libraries have been screened for binding to the vasculature of various organs by injecting the phage libraries intravenously into live animals *(18–20)*. This in vivo selection strategy has the advantage over in vitro selection strategies in that one can select whole animal peptides that bind to tumors. In this respect, Ruoslahti and colleagues have demonstrated that this innovative in vivo screening method is feasible leading to selection of $\alpha_v\beta_3$ and $\alpha_v\beta_5$ integrin receptor-binding peptides *(18, 19)*. A phage displaying a peptide containing the Arg-Gly-Asp (RGD) motif homed to tumors when injected intravenously into tumor-bearing mice. Also, phages displaying a cyclic CDCRGDCFC peptide (RGD-4C) peptide exhibited 10–20 times higher tumor-homing ability than the negative control phages *(20, 21)*. Interestingly, coupling of doxorubin to the RGD-4C targeting peptide led to the design of effective and less toxic agents than the use of doxorubin alone *(21, 22)*. Furthermore, the RGD-4C targeting peptide and other peptides were conjugated to several therapeutic agents such as apoptotic peptides *(23)* and radioactive agents for tumor imaging and therapy *(24)*. Because of these encouraging results, it would be attractive to conjugate siRNAs to RGD-4C peptide for tumor targeting.

1.2. Targeting Antisense ASOs and siRNAs into Cancer Cells Through Peptides

Towards the identification of cancer cell-targeting peptides for nucleic acid delivery, we have described the selection of breast cancer cell-specific peptides from random peptide-phage libraries *(5)*. In these experiments, the breast cancer cell line SKBR3 was used as the affinity matrix for selection. Subsequent to biopanning, internalized phage were selected, amplified, and the sequences of their displayed peptides were deduced from the DNA sequences (**Table 3.2**).

Notably, one of the requirements of receptor-mediated delivery is the internalization of the receptor after binding to its peptide ligand. Consistent with the biopanning protocol, the

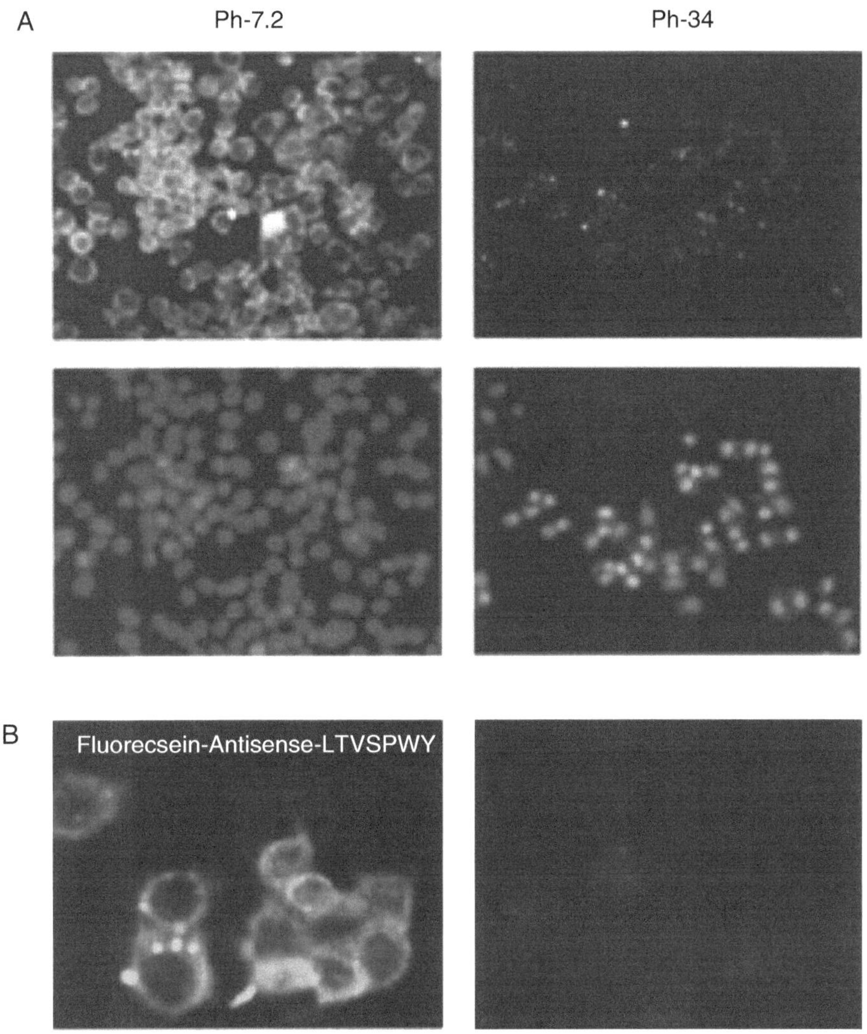

Fig. 3.1. Characterization of the phage displaying the LTVSPWY sequence.(**a**) Cellular uptake of the phage displaying the LTVSPWY sequence. To test for specificity, the binding and internalization of the phage displaying the LTVSPWY sequence (ph7-2) to SKBR3 were investigated by an epifluorescence microscope. A phage displaying a random sequence was included as a control (ph-34). (**b**) Peptide delivery of antisense oligonucleotides to SKBR3 cells. The cells were incubated with either fluorescein–antisense or fluorescein–antisense–peptide conjugates (0.5 μM) for 4 h at 37°C and then visualized with an epifluorescence microscope *(5)*.

Ph7-2 phage displaying the LTVSPWY peptide was internalized by SKBR3 cells when compared to a phage displaying the HTSPLSV peptide (**Fig. 3.2**). More importantly, the LTVSPWY peptide was able to deliver antisense oligonucleotides to breast cancer cell line SKBR3 (**Fig. 3.1b**). In these experiments, a phosphorothioate antisense RNA against the ErbB2 receptor was conjugated through the disulfide bridge to the peptide and then the purified conjugates were added to SKBR3 cells growing at 37°C.

Table 3.2
The peptide sequence displayed by positive phages

Phage	Peptide sequence	Frequency	Binding[a]
Ph-31	TLTVLPW	5	+++
Ph-3	LTVEPWL	1	++++
Ph-8	LTVSPWY	6	++++
Ph13	LTVSPLWD	1	+++
Ph52	LTVTPWL	1	+++
PH-54	LTVQPWP	3	++
Ph-40	LTVSPWT	1	++++
Ph-81	VLTVQPW	3	++
Ph-18	LTVSLWT	2	++++
Ph-22	PGVIPWN	2	+
Ph-9	LTYQTWP	1	+
Ph-4	ELYVSRL	2	+++
Ph-1	NLYVASW	1	++

After five rounds of biopanning as illustrated in Fig. 3.1,*E. coli* ER2537 cells were infected with the recovered phages and then plated onto LB agar plates. Single-phage plaques were picked and tested by flow cytometry for binding to SKBR3 and positive phages were identified. The amino acid sequences of the peptides displayed by the positive phages were deduced from the DNA sequences.

[a]The binding of the positive phages to SKBR3 was tested by flow cytometry *(5)*.

In contrast to the antisense RNA alone, the peptide–antisense conjugates (AP7-2) were effectively taken up and internalized by SKBR3 cells.

More recently, Wang and colleagues have conjugated vitamin E to the LTVSPWY peptide and demonstrated that the conjugates were preferentially taken up by the ErbB2-expressing cancer cells *(25)*, thus further validating the potential use of the LTVSPWY peptide as a targeting agent. It should be noted that the ErbB2 receptor is overexpressed in 30–50% of primary breast cancers and its expression is low in most normal adult tissues, thus making it an attractive target for peptide-mediated delivery into ErbB2-positive tumor cells.

Given the importance of directing RNA-drugs to target cells, we have established a targeting system based on human protamine, LTVSPWY peptide, and siRNAs. Previous studies have shown that protamine-conjugated antibodies specific for

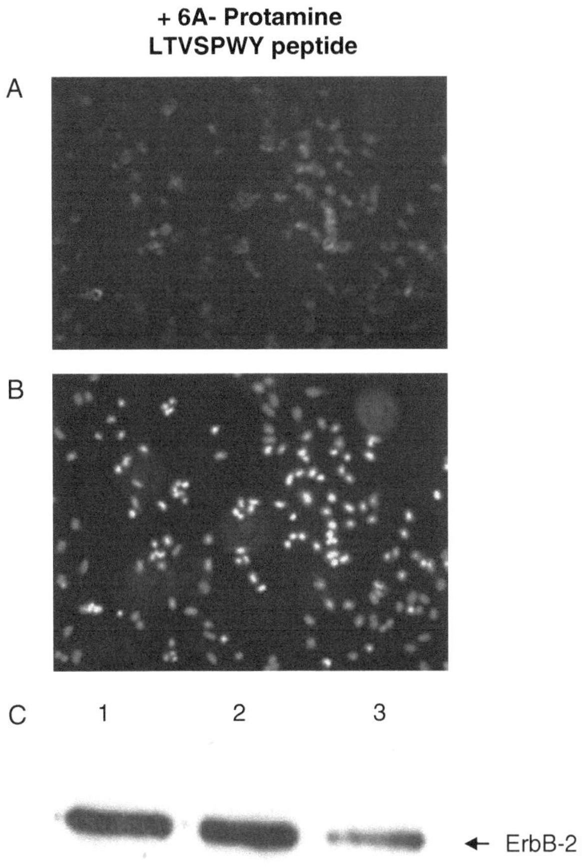

Fig. 3.2. Cancer cell targeting using the LTVSPWY peptide. (**a**) Binding of the protamine LTVSPWY peptide to the breast cancer cell line SKBR3. Cells were incubated with 6A-conjugated peptides for 1 h at 37°C, and subsequently analyzed with an epifluorescence microscope. The protamine LTVSPWY peptide sequence is 6A-**C**RSQSRSRYYRQRQRSRRRRRRS*A*LTVSPWY. (**b**) Cell nuclei were counterstained with Hoechst staining. (**c**) Downregulation of ErbB2 expression in SKBR3 cells. The siRNA molecules targeting the ErbB2 mRNA were incubated with the protamine LTVSPWY peptide for 1 h at room temperature and then added to SKBR3 cells growing in X-vivo 15 medium (lane 3). As a control, free siRNA molecules were added to the cells (lane 2). Following 48 h transfection time, cytoplasmic proteins were prepared and the expression of the ErbB2 gene was analyzed with Western blotting. Lane 1 = Untreated cells (Sioud and shadidi, unpublished data).

ErbB2 or LFA-1 receptors can deliver siRNA to cells expressing these receptors *(26, 27)*. Indeed, an ErbB2-specific single-chain antibody was fused to protamine and shown to facilitate siRNA binding and delivery to cells expressing the ErbB2 receptor. In addition to peptides and antibodies, McNamara and colleagues showed that RNA aptamers that bind specific cellular receptors can be covalently linked to siRNA to achieve cell-selective silencing in vivo *(28)*. In contrast to antibodies, small peptides would represent a central delivery agent because of their excellent tissue penetration and easy synthesis as well as conjugation to drugs and small oligonucleotides such as siRNAs and ASO. **Fig. 3.2a** and b show the cell uptake efficacy using 6-IAF-conjugated protamine LTVSPWY peptide. To test the feasibility of this delivery system, a siRNA targeting ErbB2 was incubated with the protamine LTVSPWY peptide for 1 h at RT and then added to SKBR3 cells. Subsequent to 48-h transfection time, protein extracts were prepared and the expression of ErbB2 was analyzed by Western blotting (**Fig. 3.2c**). A significant reduction in ErbB2 expression was evident in cells treated with the siRNA–peptide formulations.

To enhance tumor specificity, one might use bi-specific or tri-specific branched peptides targeted to different receptors expressed by the same cancer types. Therefore, we have evaluated the possibility of selecting additional targeting peptides. In these experiments, the random peptide-phage library was biopanned on cryosections from a breast ductal carcinoma tumor. Subsequent to three rounds of biopanning, sequence analysis identified a consensus sequence PQTP, suggesting the specificity of the selection protocol (**Fig. 3.3a**). The ability of the selected phages from the third round of biopanning to react with tumor cells was

A

Peptide sequence
I **Q** S **P** H F F
M V F **Q** P **P** N
I **S P Q T P** T
S T **P Q T P** R
L **P Q** Q K S P
Y P D **S** F **Q** V
M V F **Q** P **P** N
S P P Y L T R

B

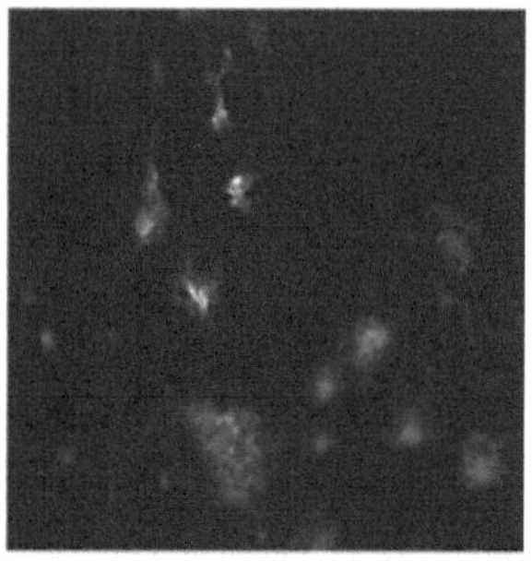

Fig. 3.3. *In situ* biopanning. (**a**) Cryosections from breast ductal carcinoma tumor were fixed with methanol and then rehydrated in PBS. Subsequently, the Ph.D.-7 random peptide library was incubated with tumor tissue sections for 1 h at 4°C. After six consecutive washings with PBS (pH 5.0), bound phages were eluted with 0.5 ml of PBS buffer pH 2.2, and subsequently neutralized with 1 μl of 2M Tris-base. Recovered phages were titered and amplified as described in Sect. 3.1.1, Steps 15–19. After three rounds of biopanning, phage clones were randomly picked out from the plate and sequenced. (**b**) To assess the binding of the selected phages to tumor cells, a cryosection was incubated with the amplified phages from the third round and then tumor cell-binding phages were detected using anti-M13 mouse monoclonal antibody in combination with PE-conjugated anti-mouse IgG antibody. Cell nuclei were counterstained with Hoechst staining (Sioud and shadidi, unpublished data).

examined by immunohistochemistry. **Fig. 3.3b** shows that the selected phages can bind to breast cancer tumors. Collectively, the examples described above indicated that efficient delivery of nucleic acids and cytotoxic drugs can be achieved by coupling them to cancer cell-specific peptides identified through a method that did not require any knowledge about their cognate receptors. This strategy does not require prior knowledge of the targeted cell membrane receptors and it is possible to identify new ligands to unknown receptors for tumor targeting.

1.3. Peptide–Hormone Conjugates

As indicated above, cancer cell targeting is usually achieved by adding to the drug delivery vesicles a ligand moiety specifically directed to the receptor expressed on cancer cells. Recent studies indicated that several hormone receptors are not only expressed by normal human tissues/cells but are also overexpressed in a large variety of human cancers, thus permitting an in vivo targeting of tumors for diagnostic and therapeutic purposes *(29–31)*. For example, the receptors for luteinizing hormone-releasing hormone (LHRH) are overexpressed in breast, ovarian, and prostate cancers *(32)*. The NPY receptors are mainly expressed in specific endocrine tumors and epithelial malignancies, particularly breast cancer cells *(29)*. The human gastrin-releasing peptide receptor (GRPr) is overexpressed on a variety of human cancer cells, including prostate, breast, lung, and pancreatic cancers *(31)*. Somatostatin receptors are overexpressed in many neuroendocrine tumors, including small-cell lung cancer, carcinoid tumor, insulinoma, gastrinoma, and medullary thyroid cancer *(30)*. The epidermal growth factor receptor is highly expressed by glioblastomas when compared to surrounding nontumor cells such as astrocytes *(33)*. Collectively, the examples described above clearly show that different hormone and growth factor receptors are overexpressed on the cell membrane of cancer cells. The coupling of their peptide ligands to siRNAs is anticipated to target systemically RNA-based drugs to specific cell populations.

Notably, the concept of using radiolabeled peptide hormones or derivatives for molecular imaging and therapy has been established by several groups *(29–31)*. For instance, the coupling of cytotoxic drugs to the LHRH peptide enhanced their accumulation in tumor cells expressing the LHRH receptor *(34)*. Peptide analogues of the human gastrin-releasing peptides exhibited an excellent targeting potency of cancer cells. Similarly, somatostatin analogues were developed and used for diagnostic imaging of somatostatin receptor-positive tumors. It should be noted that the somatostatin receptor is the first peptide hormone receptor identified for receptor-targeted diagnosis and therapy of tumors. Other peptides targeting epidermal growth factor (EGF), glutathione, laminin receptor, estrogen receptors, and the EGF receptor were also developed *(35)*. More recently it has been shown that cell-penetrating peptides (CPPs) can facilitate siRNA delivery into human cells *(36, 37)*. It should be noted that CPPs do not exhibit target specificity, and therefore enter all cell types.

1.4. Transcriptional Targeting of siRNAs

Short synthetic siRNA duplexes can be introduced into cells by transfection and have emerged as a valuable tool for analysis of gene function. However, the silencing effect is transient. To circumvent this problem, several groups have developed DNA expression vectors for RNAi in human cells that express shRNAs under the control of a RNA polymerase III or a pol II promoter *(38)*. To date, pol III promoters are used most frequently because it is possible to express small RNAs that carry the structural feature of siRNAs (**Fig. 3.4a** and **b**).

Although shRNA vectors are widely used as strategy for the analysis of gene function, currently siRNA vectors have some limitations: (1) siRNA expression cannot be controlled in a time- or tissue-specific manner because pol III promoters are constitutively active in all mammalian cells, and regulated expression

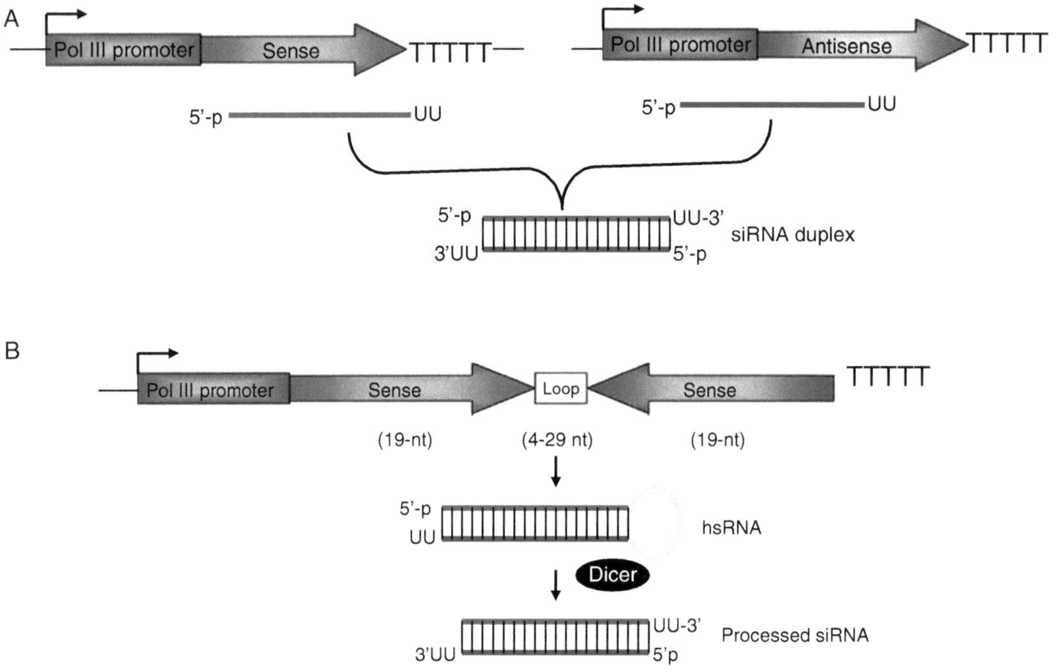

Fig. 3.4. Pol III expression strategy. (**a**) Cellular expression of small-interfering RNAs (siRNAs), tandem-type expression strategy. Both the sense and the antisense strands are expressed from different RNA polymerase (pol) III promoters (H1 or U6). The transcriptional termination of five thymidines is added to the 3′ of the sense and antisense sequences. After transcription, sense and antisense RNAs hybridize and form a duplex siRNA in the transfected cells. (**b**) Hairpin-type expression strategy. Sense sequence and its inverted sequence are expressed from H1 or U6 promoters. Both sequences can be separated by a loop sequence of 4–29 nucleotides (nt). The transcriptional termination of five thymidines is added to the 3′-end of the sense-inverted sequence. In the case of the H1 promoter, the transcript ends with two uridine 3′-overhangs. The transcribed RNA forms a hairpin with a loop structure that is processed by Dicer into a functional siRNA.

from pol III promoters is more difficult in comparison with pol II promoters. (2) Only one siRNA sequence is expressed from each promoter. (3) Constitutive expression of siRNAs might abrogate the cytoplasmic transport of miRNAs, a new class of small non-coding RNAs with essential biological functions (*see* Chap. 17). In this respect, a recent study has reported fatality in mice owing to competition between shRNAs and miRNAs for limiting cellular factors such as exportin 5 *(39)*.

In addition to sequence-specific inhibition of the target gene, siRNAs also have off-target effects that could be detrimental for healthy cells *(40, 41)*. Indeed, transcriptional profiling studies have showed that siRNA duplexes can potentially silence multiple genes in addition to the intended target. Therefore, there is an urgent need for controlled expression systems of shRNAs. To overcome this potential problem, regulated H1 and U6 expression vectors have been developed *(42)*. For example, a doxycycline-regulated form of the H1 promoter to drive the expression of siRNAs against β-catenin in colorectal cancer was designed *(43)*. Although these modified H1 and U6 promoters have provided significant insights into the regulation of siRNA expression, they do not restrict the expression of the shRNAs to pathogenic cells, and therefore their use in gene therapy is limited.

One way to target pathogenic cells and avoid healthy cells is to place the shRNA under the control of a promoter that is transcriptionally active only in pathogenic cells such as cancer cells, but not normal cells. Several studies have shown that a number of genes are expressed exclusively or predominantly in cancer cells when compared to normal counterparts. Moreover, various promoters have already been evaluated for transcriptional targeting in cancer gene therapy including the prostate-specific antigen (PSA) promoter for prostate cancer and the tyrosinase gene promoter for melanoma *(44)*. Among the genes that are preferentially expressed in tumors, survivin is a versatile candidate because it is expressed in many human cancer types but not in normal adult tissues *(45)*.

To test whether the survivin promoter could express siRNA in cancer cells, we first PCR-amplified the promoter sequence, and then modified the sequence to allow the cloning of shRNAs *(46)*. **Fig. 3.5a** illustrates the cloning strategy. When a siRNA targeting GFP was expressed under the survivin promoter, an inhibition of the target gene was seen only in cancer cells but not in normal cells *(46)*, thus confirming the specificity of the designed promoter (**Fig. 3.5b** and **c**). More, recently Song and colleagues described a vector for RNAi in which a synthetic siRNA is expressed under the PSA promoter *(47)*. Interestingly, reduced gene expression was achieved in a tissue-specific and hormone-dependent manner. Both approaches offer a number of important potential advantages when contrasted with modified U1 and U6 vectors. These include

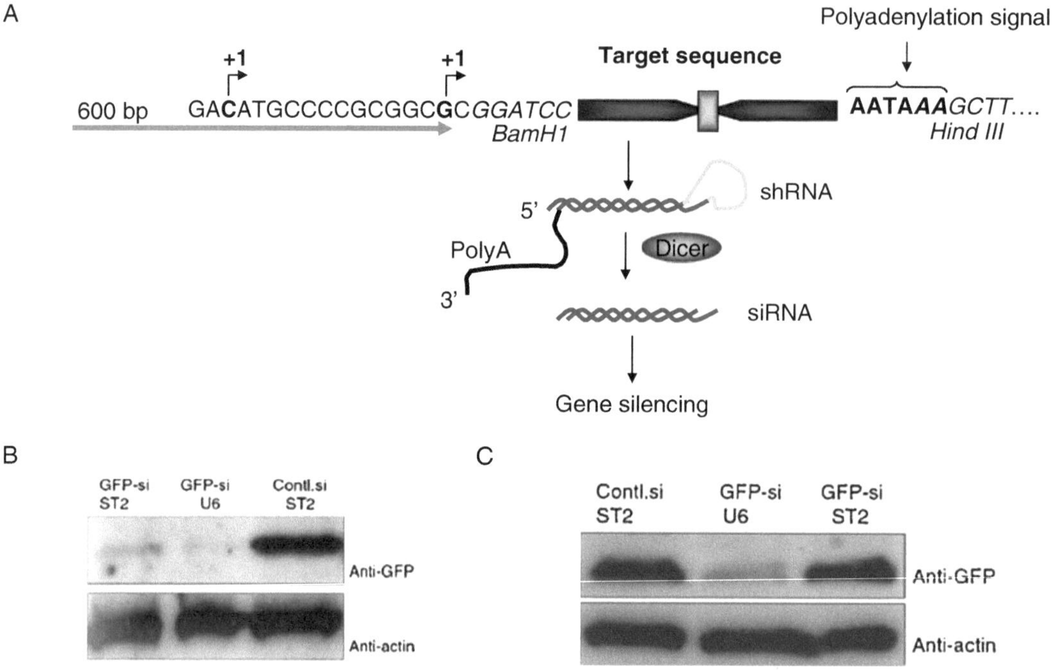

Fig. 3.5. Expression of siRNA under the survivin promoter. (**a**) A partial sequence of the survivin promoter showing the *BamHI* and *HindIII* cloning sites. The natural transcription starts are retained and a polyA stop signal is added at the end of the construct. The survivin promoter (ST2)-mediated siRNA expression inhibited the expression of GFP in (**b**) survivin-positive cancer cell lines, but not in (**c**) survivin-negative normal cells. In these experiments, cells were co-transfected with GFP-encoding plasmid in combination with anti-GFP shRNA under the control of the survivin promoter *(46)*.

(1) siRNA can be expressed only or preferentially in cancer cells and (2) it is possible to express a long, single transcript resembling the miRNA primary transcripts, allowing the generation of several siRNAs from each survivin promoter. Optimized tissue-specific promoters for shRNA expression are expected to facilitate the therapeutic applications of siRNAs.

As discussed in Chap. 2, long double-stranded RNAs (dsRNAs), frequently expressed in virus-infected cells, activate several antiviral cellular pathways leading to cell death. Therefore, this nonspecific effect of long dsRNAs as a death inducer can be used to selectively kill cancer cells by expressing long dsRNAs under the control of the survivin promoter. In this respect, our preliminary data indicate that the expression of 40-nt long dsRNA targeting the survivin mRNA can block the expression of the survivin gene and trigger the interferon pathway resulting in enhanced cell killing (Huynh et al, in preparation). This method which explores the specificity of the survivin promoter should be applicable to any gene.

In order to facilitate the selection of cancer cell-binding peptides from random peptide-phage libraries, a detailed experimental protocol is described below. A schematic representation

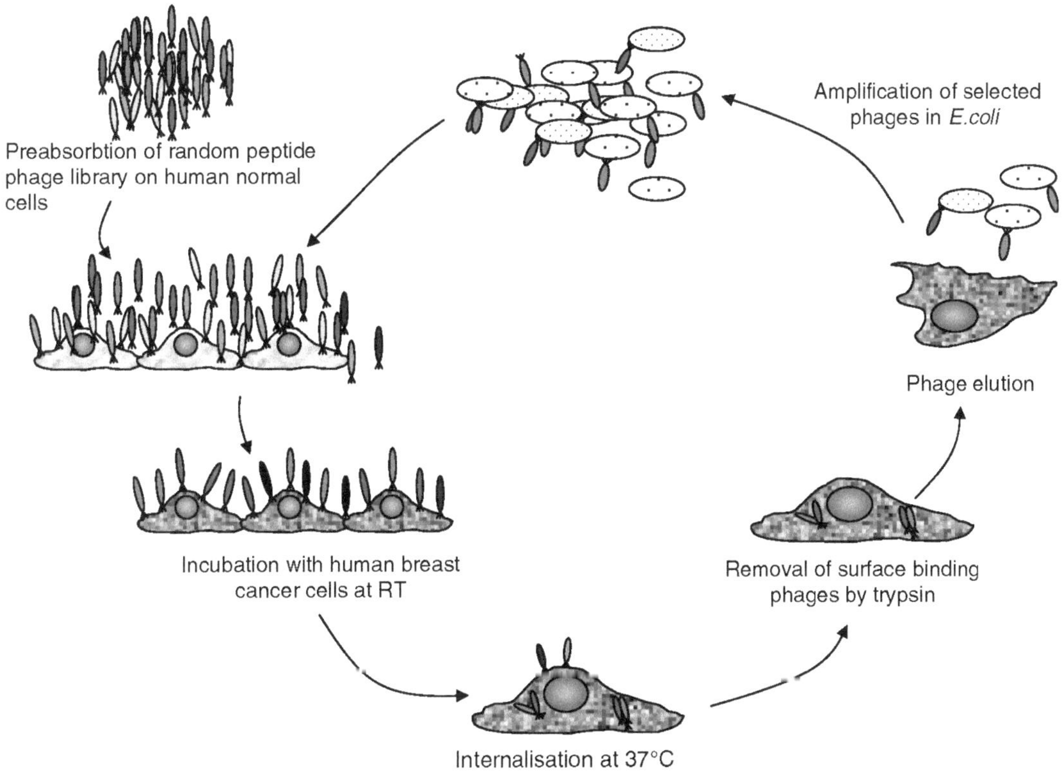

Fig. 3.6. Schematic representation of the biopanning protocol using as affinity matrix SKBR3 cancer cell line The Ph.D-7 phage display library, with a complexity of approximately 2×10^9 (New England BioLabs) was pre-absorbed on human mammary epithelial cells (HMECs) and peripheral blood mononuclear cells (PBMCs). Subsequently, pre-absorbed library was added to 70% confluent SKBR3 cells cultured in a T-25 tissue culture flask and then incubated for 1 h at 4°C. After incubation, cells were washed six times with PBS and then incubated with 10 ml pre-warmed RPMI medium for 15 min at 37°C. This step is expected to mediate the internalization of bound phages. After phage internalization, cells were washed five times with PBS and surface-bound phages were removed by 10 min trypsination at 37°C, followed by five washes with PBS. Internalized phages were recovered by lysing the cells as described in Sect. 3.1.1.

of the protocol is shown in **Fig. 3.6**. Using this protocol, cancer cell-binding peptides were selected from random peptide-phage libraries (*see* **Note 1**). When one of the selected 7-mer peptides was coupled to an antisense oligonucleotide targeting the ErbB2 receptor, specific delivery to breast cancer cells was demonstrated.

2. Materials

2.1. Buffers

1. Phosphate-buffered saline (PBS): 137 mM NaCl, 3 mM KCl, 8 mM Na_2HPO_4, 1.5 mM KH_2PO_4, pH 7.5.
2. PBST: PBS with 0.1% Tween 20.

2.2. Cells

1. Human mammary epithelial cells (HMECs) (*see* **Note 2**).
2. Human peripheral blood mononuclear cells (PBMCs).
3. Human breast cancer cell line SKBR3 (American Type Culture Collection, ATCC).

2.3. Media

1. Mammary epithelial-cell growth medium (MEGM) supplemented with 10% fetal bovine serum (FBS).
2. RPMI-1640 medium supplemented with 10% FBS.

2.4. Biopanning

1. Random peptide-phage libraries 7 (New England Biolabs, USA) (*see* **Note 3**).
2. Luria–Bertani medium (LB): 10 g bacto-tryptone, 5 g bacto-yeast extract, 10 g NaCl. Adjust pH to 7.0.
3. Tetracycline stock solution: 20 mg/ml in ethanol. Store at –20°C in the dark and vortex before use.
4. LB-tetracycline plates. Keep at 4°C in the dark.
5. *E. coli* strain ER2738.
6. PEG/NaCl: 20% (w/v) Polyethylene glycol 8000 and 2.5 M NaCl.
7. Phage elution buffer: 0.2 M Glycine-HCl, pH 2.2 with 1 mg/ml bovine serum albumin (BSA).
8. Neutralization buffer: 2 M Tris-base.
9. Top agarose: 10 g Bacto-tryptone, 5 g yeast extract, 5 g NaCl, 1 g $MgCl_2$. Dissolve in 900 ml of H_2O, add 7 g agarose, adjust the volume to 1 l, and autoclave.
10. IPTG/X-gal stock solution: 1.25 g isopropyl β-d-thiogalactoside (IPTG) and 1 g 5-bromo-4-chloro-3-indolyl-β-d-galactoside (X gal) in 25 ml dimethylformamide. Store at –20°C in the dark.
11. LB/IPTG/X-gal plates: LB medium with 15 g/l agar and 1 ml of IPTG/X-gal stock solution. Store at 4°C in the dark.

2.5. Flow Cytometry

1. 96-well microtiter plates with conical bottom.
2. 10 mM ethylenediaminetetraacetic acid (EDTA) in PBS, pH 8.0.
3. Flow cytometry buffer: 1% FBS or FCS in PBS containing 0.005% azide (FC buffer).
4. Anti-M13 mouse monoclonal antibody (mAb)
5. Goat phycoerythrin (PE)-conjugated anti-mouse IgG.
6. Fluorescence-activated cell sorter (FACS) caliber.

2.6. DNA Sequencing

1. BigDye sequencing kit (ABI PRISM).
2. pIII oligonucleotide hybridizing to 96 position of the DNA insert.

3. ABI PRISM 310 Genetic Analyzer sequencing machine.

2.7. Antisense– and Fluorescein–Peptide Conjugates

1. Synthetic peptides: LTVSPWY**K** and LTVSPWY**C.**
2. 6-iodoacetamidoflurescein (6-IAF)
3. A synthetic antisense oligonucleotide against ErbB2 5′-CTC-CATGGTGCTCACSSpy-3′. As control, a scrambled oligonucleotide: 5′-CGCCTTATCCCGTAGCSSpy-3′ (*see* **Note 4**). Both oligonucleotides contain dithiodipyridine (SSpy) groups at their 3′-ends for peptide conjugation.

3. Methods

3.1. Biopanning

3.1.1. Biopanning on Whole Cells

The breast cancer cell line SKBR3 was used as an affinity matrix to select phages from a 7-mer random peptide library as described in the following steps (*see* **Note 5**).

1. Plate SKBR3 and HMECs in T-25 tissue culture flasks, so they are 70–80% confluent the day of the biopanning.
2. Remove the medium, wash the cells once with 1X PBS, and add prewarmed complete RPMI medium containing 10% FCS (*see* **Note 6**).
3. Add 1–5 × 10^6 freshly isolated PBMCs to the T-25 flask containing HMECs. This step can be omitted (*see* **Note 2**).
4. Add 1 × 10^9 pfu of the phage library to the HMECs/PBMCs and incubate for 1 h or longer at 4°C. This step is expected to eliminate the phages that bind normal epithelial cells and blood cells.
5. Wash SKBR3 cells growing in T-25 tissue culture flasks with PBS, add prewarmed RPMI medium containing 10% FCS, and incubate at 37°C for 30 min.
6. Remove the medium from HMECs/PBMCs and centrifuge for 10 min at 4000g at 4°C in order to pellet the cells.
7. Transfer the supernatant that contains the unbound phages to a new falcon tube.
8. Remove the medium from the flask containing SKBR3 cells, and add the supernatant containing the unbound phages to SKBR3 cells.
9. Incubate the flask for 1 h at 4°C with gentle shaking. This step can be performed at room temperature or 37°C.
10. Remove the medium and wash SKBR3 cells around 10 times (5 min each) with 1X PBS 0.2% Tween 20 (PBST buffer) pH 5.0, twice with PBST pH 7.5, and twice with only PBS buffer pH 7.5 (*see* **Note 7**).

11. Add 0.5 ml of trypsin/EDTA to the cells and incubate at 37°C for 5–10 min in order to detach the cells and eliminate cell-surface binding phages.
12. Subsequent to trypsin treatment, add 5 ml of RPMI with 10% FCS to the detached cells and centrifuge at 1200g for 5 min at 4°C.
13. Wash the cells twice with 10 ml PBS and resuspend the cell pellet in 100–200 μl of water. For complete cell lysis, vortex vigorously.
14. To disrupt the binding of the phages to intracellular proteins, add 450 μl of elution buffer (pH 4) to the cell lysates, and incubate for 10 min at room temperature (*see* **Note 8**).
15. Neutralize the cell lysates with 1 μl of Tris-base (2M).
16. Infect 5 ml of exponentially growing ER2537 (OD_{600} ~ 0.5) with 100 μl of the phage eluate and incubate 30 min at 37°C without shaking.
17. Dilute the culture to 1/50 with LB medium and incubate overnight at 37°C with shaking at 200 g/min.
18. In parallel, determine the titer of the phage eluate by infecting 250 μl of ER2537 with 1 microL of phage eluate. Incubate for 5 min at 37°C, add 3 ml of melted top agar, and then plate the *E. coli* cells on an IPTG/X-gal plate.
19. Invert the plate and incubate at 37°C overnight.

3.1.2. Phage Preparation

Spin down the overnight culture (**Sect. 3.1.1**, Step 17) for 10 min at 8000g.

1. Transfer 80% of the cleared supernatant to a new tube. Add 1/5 volume of PEG/NaCl solution to the supernatant, mix well, and leave at 4°C for 2 h or more.
2. Centrifuge the precipitated phages 15 min at 12,000g at 4°C, aspirate, and discard the supernatant.
3. Resuspend the pellet in 1 ml of PBS.
4. Transfer the phage solution to a new microcentrifuge tube and spin for 5 min at 4000g at 4°C, to remove cellular debris.
5. Transfer the supernatant to a new microcentrifuge tube and, if necessary, repeat the PEG precipitation and the subsequent centrifugation steps.
6. Dry the phage pellet and resuspend in 250 μl of PBS.
7. Store phages at 4°C.

A single round of phage selection is described above. However, in an in vitro selection protocol, it is recommended that more than two rounds be performed in order to enrich for positive clones (*see* **Fig. 3.6**). Use around 10^{10} pfu for additional rounds of selection.

3.1.3. Phage Titration

After each round of selection determine the phage titers using the subsequent steps.

1. Grow a single colony of ER2738 in 10 ml LB medium until exponential phase ($OD_{600} \approx 0.5$).
2. Prewarm LB/IPTG/X-gal plates at 37°C.
3. Melt top agarose in the microwave oven, pour 3 ml per Falcon tube, and incubate the tubes at 60°C until use.
4. Prepare 10-fold dilutions of each phage preparation in LB medium.
5. Add 10 µl of each dilution to 3 ml melted top agarose containing 250 µl of exponentially growing ER2738. Mix well and pour onto a LB/IPTG/X-gal plate.
6. After 10 min incubation at room temperature, invert the plates and incubate at 37°C overnight.
7. The next day, count the blue plaques of plates with 100 plaques, and multiply each number by the dilution factor.

3.1.4. Single Plaque Phage Amplification

After the desired round of selection, random phage clones should be tested for binding to cells using the following steps.

1. From the last round of biopanning, prepare a dilution series, up to 10^{12}-fold, of the amplified phages.
2. Plate the 10^{8}–10^{12} phage dilutions onto LB/IPTG/X-gal plates as described in **Sect. 3.1.3**, Steps 2–7.
3. Invert the plates and incubate overnight at 37°C.
4. The same day inoculate 2 ml of LB with a single colony of ER2738 and grow overnight.
5. The next day, dilute the overnight culture 1/100 in LB medium.
6. Pick out single blue plaques from the plates (Step 2), inoculate 2 ml culture of the dilute ER2738 cells (Step 6), and incubate for 4–5 h at 37°C.
7. Spin down the bacteria and transfer 1 ml of the supernatant to a new microcentrifuge tube.
8. Add 200 µl of PEG/NaCl and incubate at 4°C overnight.
9. Centrifuge the precipitated phages at 12,000g for 15 min at 4°C. Remove the supernatant and dry the phage pellets.
10. Dissolve the phage pellets in 100 µl of PBS.
11. To remove the rest of bacteria, spin briefly at 4000g for 5 min.
12. Transfer 80 µl of the phage solution to a new microcentrifuge tube without touching the bacteria pellet.
13. Keep the amplified phages at 4°C.

3.2. Flow Cytometry

To evaluate the binding of the selected phages to human cancer cells, single phage clones from the enriched library should be purified (*see* **Sect. 3.1.4**) and tested for binding to cells by flow cytometry using the following steps.

1. Grow SKBR3 and HMECs in T-75 tissue culture flasks.
2. Wash the cells once with PBS, and then de-attach them from the culture flasks by treatment with 500 μl of 10 mM EDTA in PBS, pH 8.0. Incubate for 5 min at 37°C.
3. Add 5 ml of complete medium (RPMI + 10% FCS) to the flasks and transfer the de-attached cells to a Falcon tube.
4. Centrifuge at 1200g for 5 min.
5. Remove the medium and resuspend the cell pellet in 5 ml of fresh, complete RPMI medium.
6. Incubate the cells for 1 h at 37°C in order to recover from ETDA treatment.
7. Plate the cells into a conical 96-well microtiter plate at 10^5 cells/well. Wash the cells with FC buffer.
8. Centrifuge the plate at 1200g for 5 min and discard the supernatant.
9. Add to each well 10^8 pfu phage particles in 50–100 ml FC buffer, and then incubate 30 min on ice.
10. Wash the cells three times with FC buffer.
11. Add 50 ml of mouse IgG anti-M13 monoclonal diluted 1/250 in FC buffer, and incubate the plate for 30 min on ice.
12. Wash three times with the FC buffer.
13. Add 50 ml of PE-conjugated polyclonal anti-mouse IgG diluted 1/250 in FC buffer, and incubate the plate for 30 min on ice.
14. Wash three times with FC buffer.
15. Resuspend the cell pellet in 200 μl of PBS and analyze the phage-binding cells with flow cytometry.

3.3. Analysis of the Peptide Sequences Displayed by the Positive Phages

Single-strand DNAs were prepared from positive phages *(34)*, and sequencing of the DNA inserts was carried out by automatic sequencing (ABI PRISM 310 Genetic Analyzer). The peptide sequences displayed by the positive phage clones shared a major core motif (LTVXPWY) that was not found in negative phages *(32)*. The phage displaying LTVSPWY exhibited the strongest binding and internalization. Therefore, this peptide was used for the delivery of antisense oligonucleotides into breast cancer cell lines as described in **Sect. 3.4**.

3.4. Synthetic Peptides and Antisense-Peptide Conjugates

To determine whether the SKBR3-binding peptide LTVSPWY could be used to enhance the delivery of antisense oligonucleotides, a fluorescein-conjugated antisense oligonucleotide against the ErbB2 receptor was designed and coupled via a disulfide bridge to the peptide (*see* **Note 9**). SKBR3 cells incubated with the antisense–peptide conjugates for 4 h at 37°C exhibited intracellular binding, whereas no significant staining was obtained with the antisense alone (**Fig. 3.1c**). In contrast to the antisense alone, the antisense–peptide conjugates inhibited the ErbB2 gene expression *(32)*.

4. Notes

1. Phage display libraries are constructed both in lytic and non-lytic bacteriophages. The nonlytic M13 filamentous phage is the most used vector. This phage is from the fd phage family, which infects gram-negative bacteria by recognizing the F-pilus of their host strain. The coat proteins of M13, which are often used for fusion of library protein and peptides, are pIII, pVI, and pVIII (**Fig. 3.1A**). The pIII coat protein is often used for constructing short peptide libraries. The M13 phage contains five copies of the pIII coat protein, but only one copy of this protein is required for infection. A classic M13 short peptide library constructed on the pIII coat protein contains —three to five copies of the displayed foreign peptide, each fused to the N terminus of each pIII protein.
2. The HMECs were used as a normal counterpart of SKBR3 cancer cells. Both the HMECs and SKBR3 cells originate from breast tissue and have epithelial morphology. The PBMCs are intended to eliminate peptides binding to normal blood cells. However, once cancer-binding peptides are selected they can be tested for binding to blood leukocytes.
3. The Ph.D.7 phage library was purchased from New England Biolabs. The biopanning conditions were modified for selection of peptides that are taken up by the cells via receptor-mediated internalization. Notably, optimized panning conditions are important for successful ligand selection.
4. The phosphorothioate oligonucleotide against ErbB2 was described by Vaughn et al. *(36)*. The antisense and the control-scrambled oligonucleotide were fused to the synthetic peptides via the 3′ SSpy group. The S–S bridge between the oligonucleotide and the synthetic peptide is expected to break up in the reducing conditions of the cytoplasm.

5. Various phage display libraries were designed and found to be useful to study receptor–ligand interactions for a variety of target molecules. Successive selection of ligands from a phage display library is called "biopanning," which involves screening of a random peptide-phage library against a target molecule. High-affinity phage binders are then captured and amplified in —three to four rounds of panning. Subsequently, individual plaques of phage are analyzed for their specific binding and sequenced. The target molecules for a biopanning assay can be known immobilized proteins or unknown such as surface receptors expressed by certain cell types (*see* **Fig. 3.7**).
6. The RPMI medium containing 10% FCS was used in the biopanning experiments. This would eliminate the phages that bind to serum proteins. Additionally, the serum will also function as a blocker of unspecific binding. It is recommended that the number of input phages do not exceed 1×10^{10} pfu because high phage concentrations will result in much higher background and increased unspecific binding.
7. Acidic washing conditions were used to minimize the background and unspecific binding. It is noteworthy that many cell types do not endure these harsh conditions and require a much more gentle treatment. The cells were treated with trypsin to remove the surface binding phages and to select only the phages that are internalized.
8. M13 phage is extremely stable in an acidic environment, but elution in pH 2.2 must not exceed 10 min because it can destroy the phage. The eluted phages must be neutralized immediately with 2M Tris-base.
9. A lysine residue was added to the C terminus for antisense conjugation to the peptide. Peptides with a cysteine residue at the C terminus were directly conjugated to 6-1AF as described by the manufacturer (Molecular Probes).

Acknowledgments

We thank Dr Anne Dybwad for critical reading of the manuscript and the group members for their contribution to this work.

References

1. Garanger, E., Boturyn, D., and Dumy, P. (2007) Tumor targeting with RGD peptide ligands-design of new molecular conjugates for imaging and therapy. *Anticancer Agents Med. Chem.* 7, 552–558.
2. Shadidi, M. and Sioud, M. (2003) Selective targeting of cancer cells using synthetic peptides. *Drug Resist Updat.* 6, 363–371.

3. Aina, O.H., Sroka, T.C., Chen, M.L., et al. (2002) Therapeutic cancer targeting peptides. *Biopolymers* 66, 184–199.
4. Sioud, M. (2005) On the delivery of small interfering RNAs into mammalian cells. *Expert Opin. Drug Deliv.* 2, 639–651.
5. Shadidi, M. and Sioud, M. (2003) Identification of novel carrier peptides for the specific delivery of therapeutics into cancer cells. *FASEB J.* 17, 256–258.
6. Patel, D.S., Dessalew, N., Iqbal, P. et al., (2007) Structure-based approaches in the design of GSK-3 selective inhibitors. *Curr. Protein Pept. Sci.* 8, 352–364.
7. Sioud, M., Førre, Ø., and Dybwad, A. (1996) Selection of ligands for polyclonal antibodies from random peptide libraries: potential identification of (auto)antigens that may trigger B and T cell responses in autoimmune diseases. *Clin. Immunol. Immunopathol.* 79, 105–114.
8. Romanov, V.I. (2003) Phage display selection and evaluation of cancer drug targets. *Curr. Cancer Drug Targets.* 3, 119–129.
9. Falciani, C., Lozzi, L., Pini, A.et al., . (2005) Bioactive peptides from libraries. *Chem. Biol.* 12, 417–426.
10. Houghten, R.A., Pinilla, C., Blondelle, S.E. et-al., (1991) Generation and use of synthetic peptide combinatorial libraries for basic research and drug discovery. *Nature* 354, 84–86.
11. Fukuda, M.N., Ohyama, C., Lowitz, K., et al. (2000) A peptide mimic of E-selectin ligand inhibits sialyl Lewis X-dependent lung colonization of tumor cells. *Cancer Res.* 60, 450–456.
12. Lee, J.H., Engler, J.A., Collawn, J.F., et al. (2001) Receptor mediated uptake of peptides that bind the human transferrin receptor. *Eur. J. Biochem.* 268, 2004–2012.
13. Campa, M.J., Serlin, S.B., and Patz, E.F. (2002) Development of novel tumor imaging agents with phage-display combinatorial peptide libraries. *Acad. Radiol.* 9, 927–932.
14. Alaoui-Jamali, M.A. and Qiang, H. (2003) The interface between ErbB and non-ErbB receptors in tumor invasion: clinical implications and opportunities for target discovery. *Drug Resist. Updat.* 6, 95–107.
15. Urbanelli, L., Ronchini, C., Fontana, L., et al. (2001) Targeted gene transduction of mammalian cells expressing the HER2/neu receptor by filamentous phage. *J. Mol. Biol.* 313, 965–976.
16. Karasseva, N.G., Glinsky, V.V., Chen, N.X., et al. (2002) Identification and characterization of peptides that bind human ErbB-2 selected from a bacteriophage display library. *J. Protein Chem.* 21, 287–296.
17. Folkman, J. (2002) Role of angiogenesis in tumor growth and metastasis. *Semin. Oncol.* 29, 15–18.
18. Pasqualini, R. and Ruoslahti, E. (1996) Organ targeting in vivo using phage display peptide libraries. *Nature* 380, 364–366.
19. Pasqualini,R.,Koivunen,E.,andRuoslahti,E. (1997) Alpha v integrins as receptors for tumor targeting by circulating ligands. *Nat. Biotechnol.* 15, 542–546.
20. Pasqualini, R., Koivunen, E., Kain, R., et al. (2000) Aminopeptidase N is a receptor for tumor-homing peptides and a target for inhibiting angiogenesis. *Cancer Res.* 60, 272–722.
21. Arap, W., Pasqualini, R., and Ruoslahti, E. (1998) Cancer treatment by targeted drug delivery to tumor vasculature in a mouse model. *Science* 279, 377–380.
22. de Groot, F.M., Broxterman, H.J., Adams, H.P., et al. (2002) Design, synthesis, and biological evaluation of a dual tumor-specific motive containing integrin-targeted plasmin-cleavable doxorubicin prodrug. *Mol. Cancer Ther.* 1, 901–911.
23. Cheng, J.Q., Jiang, X., Fraser, M., et al. (2002) Role of X-linked inhibitor of apoptosis proteins in chemoresistance in ovarian cancer: possible involvement of the phosphoinositide-3 kinase/Akt pathway. *Drug Resist. Updat.* 5, 131–146.
24. Su, Z.F., Liu, G., Gupta, S., et al. (2002) In vitro and in vivo evaluation of a Technetium-99m-labeled cyclic RGD peptide as a specific marker of alpha(v)beta(3) integrin for tumor imaging. *Bioconjug. Chem.* 13, 561–570.
25. Wang, X.F., Birringer, M., Dong, L.F., et al. (2007) A peptide conjugate of vitamin E succinate targets breast cancer cells with high ErbB2 expression. *Cancer Res.* 67, 3337–3344.
26. Song, E., Zhu, P., Lee, S.K., et al. (2005) Antibody mediated in vivo delivery of small interfering RNAs via cell-surface receptors. *Nat. Biotechnol.* 23, 709–717.
27. Peer, D., Zhu, P., Carman, C.V., et al. (2007) Selective gene silencing in activated leukocytes by targeting siRNAs to the integrin lymphocyte function-associated antigen-1. *Proc. Natl. Acad. Sci. U S A* 104, 4095–4100.
28. McNamara, J.O., Andrechek, E.R., Wang, Y., et al. (2006) Cell type-specific delivery of siRNAs with aptamer-siRNA chimeras. *Nat. Biotechnol.* 24, 1005–1015.

29. Körner, M. and Reubi, J.C. (2007) NPY receptors in human cancer: A review of current knowledge. *Peptides* 28, 419–425.
30. Riccabona, G. and Decristoforo, C. (2003) Peptide targeted imaging of cancer. *Cancer Biother. Radiopharm.* 18, 675–687.
31. Hoffman, T.J., Quinn, T.P., and Volkert, W.A. (2001) Radiometallated receptor-avid peptide conjugates for specific *in vivo* targeting of cancer cells. *Nucl. Med. Biol.* 28, 527–539.
32. Dharap, S.S. and Minko, T. (2003) Targeted proapoptotic LHRH-BH3 peptide. *Pharm. Res.* 20, 889–896.
33. Shir, A. and Levitzki, A. (2001) Gene therapy for glioblastoma: Future perspective for delivery systems and molecular targets. *Cell Mol. Neurobiol.* 21, 645–656.
34. Dharap, S.S., Wang, Y., Chandna, P. et al., (2005) Tumour-specific targeting of an anticancer drug delivery system by LHRH peptide *PNAS* 102,12962–12967
35. Huang, P.S. and Oliff, A. (2001) Drug-targeting strategies in cancer therapy. *Curr. Opin. Genet. Dev.* 11, 104–110.
36. Muratovska, A. and Eccles, M.R. (2004) Conjugate for efficient delivery of short interfering RNA (siRNA) into mammalian cells. *FEBS lett.* 558, 63–68.
37. Crombez, L., Charnet, A., Morris, M.C., et al. (2007) A non-covalent peptide-based strategy for siRNA delivery. *Biochem. Soc. Trans.* 35, 44–46
38. Hannon, G.J. and Rossi, J.J. (2004) Unlocking the potential of the human genome with RNA interference. *Nature* 431, 371–378.
39. Grimm, D., Streetz, K.L., Jopling, C.L., et al. (2006) Fatality in mice due to oversaturation of cellular microRNA/short hairpin RNA pathways. *Nature* 441, 537–541.
40. Jackson, A.L., Bartz, S.R., Schelter, J., et al. (2003) Expression profiling reveals off-target gene regulation by RNAi. *Nat. Biotechnol.* 21, 635–637.
41. Lin, X., Ruan, X., Anderson, M.G., et al. (2005) siRNA-mediated off-target gene silencing triggered by a 7 nt complementation. *Nucleic Acids Res.* 33, 4527–4535.
42. McIntyre, G.J. and Fanning, G.C. (2006) Design and cloning strategies for constructing shRNA expression vectors. *BMC Biotechnol.* 6, 1.
43. Van de Wetering, M., Oving, I., Muncan, V., et al. (2003) Specific inhibition of gene expression using a stably integrated, inducible small-interfering-RNA vector. *EMBO Rep.* 4, 609–615.
44. Saukkonen, K. and Hemminki, A. (2004) Tissue-specific promoters for cancer gene therapy. *Expert Opin. Biol. Ther.* 4, 683–696.
45. Altieri, D.C. (2003) Validating survivin as a cancer therapeutic target. *Nat. Rev. Cancer* 3, 46–54.
46. Huynh, T., Wälchli, S., and Sioud, M. (2006) Transcriptional targeting of small interfering RNAs into cancer cells. *Biochem. Biophys. Res. Commun.* 350, 854–859.
47. Song, J., Pang, S., Lu, Y., et al. (2004) Gene silencing in androgen-responsive prostate cancer cells from the tissue-specific prostate-specific antigen promoter. *Cancer Res.* 64, 7661–7663.

Chapter 4

Local and Systemic Delivery of siRNAs for Oligonucleotide Therapy

Fumitaka Takeshita, Naomi Hokaiwado, Kimi Honma, Agnieszka Banas, and Takahiro Ochiya

Abstract

RNA interference (RNAi) is a relatively new found phenomenon of posttranscriptional gene silencing to regulate the expression of multiple genes involved in a wide range of biological processes. The gene-silencing technology via RNAi has also been developed into a commonly anti-gene method. Furthermore, in vivo data indicate that small interfering RNAs (siRNAs) may be used to treat human diseases. However, the most challenging issue to a successful in vivo application is the development of a delivery system that can transport siRNA molecules into the tissues and/or the cells of interest. Also, the evaluation of siRNA potency in vivo is central for the selection of therapeutic siRNAs. In this chapter, the effects of atelocollagen-delivered siRNAs in live animals were monitored using bioluminescence imaging.

Key words: RNA interference, small interfering RNA, bioluminescence imaging, delivery, atelocollagen.

1. Introduction

Recently, transfection of the synthetic small interfering RNA (siRNA) or short hairpin RNA (shRNA) expression vector into mammalian cells has become a widely used methodology for suppression of target genes in basic research. Furthermore, built on the rapid accumulated knowledge of RNAi, scientists are actively pursuing pharmacological applications of RNAi for human diseases. RNAi has the potential to be more selective and, as a result, more effective and less toxic than traditional drugs. Effective in vivo delivery of siRNA to the appropriate target cells or tissues is an essential component of these RNAi-based applications. Some

M. Sioud (ed.), *Methods in Molecular Biology, siRNA and miRNA Gene Silencing, vol. 487*

DOI: 10.1007/978-1-60327-547-7_4

reports showed that animal studies using siRNA have either used no additional formulation (naked siRNA) or have delivered siRNA formulated as complexes with liposome, peptides, polymers, or antibodies *(1, 2)*. Direct injection of siRNA has been evaluated for some tissue sites, and local treatment of siRNAs are being tested in ongoing clinical studies in the eye for age-related macular degeneration (AMD) by inhibition of angiogenesis behind the retina and in the lung for inhibition of respiratory syncytial Virus (RSV) infection. When siRNAs are administered locally, lower doses are sufficient since nonspecific delivery to other organs as well as renal or hepatic elimination is reduced. However, systemic delivery of siRNA is thought to be useful for treatment of chronic, inflammatory disease or metastatic cancer.

For the therapeutic use of siRNA in cancer, although the efficacy of siRNA has to be validated in animal models, evaluation in cultured cancer cells is required before any in vivo study. When target genes are selected, optimal design of the siRNA sequences are required (*see* Chap. 1). Public domains as well as commercial entities (e.g., Dharmacon, Ambion, Qiagen, and Invitrogen) provide designing tools using proprietary algorithms.

The most challenging issue to a successful in vivo application is a delivery system that transports the siRNA into the target tissues and the cell cytoplasm, or shRNA expression cassette to the nucleus much like the dependence of gene therapy on appropriate delivery methods (*see* Chaps. 3, 5, 6, 7, and 8). However, double-stranded siRNA molecules are unstable in the serum in vivo and they can be degraded by RNase activity within a short period of time. One interesting study reported that liver-targeted delivery and the stability of siRNA may be enhanced using chemical modifications of the oligonucleotide, for example, with cholesterol conjugates. These conjugates are more resistant to nuclease degradation, the cholesterol attachment stabilizing the siRNA molecules in the blood by increasing binding to human serum albumin and increasing the uptake of siRNA molecules by the liver in mice *(3)* and nonhuman primates *(4)*.

We recently developed an atelocollagen-mediated siRNA delivery system in vivo.

Atelocollagen is generated from type I collagen of calf dermis by pepsin treatment with removal of telopeptides at the N- and C-termini of collagen molecules *(5, 6)*. Atelocollagen is low in immunogenicity because it is free from telopeptides that contain most of collagen's antigenicity. The clinical safety of atelocollagen for a wide range of purposes including wound-healing, vessel prosthesis, and also as a bone cartilage substitute and as a hemostatic agent has been shown in vivo. Because the surface of atelocollagen molecules is positively charged, the complexes of the atelocollagen and negatively charged nucleic acid molecules can be prepared by simply mixing, whence they bond electrostatically.

Atelocollagen has the potential to protect the siRNA from digestion by endogenous nucleases, i.e., preventing the degradation of siRNA by RNase *(7)*.

The efficacy of atelocollagen for nucleotide delivery has been demonstrated, such as plasmid DNA and antisense oligonucleotides in vitro and in vivo *(5, 6, 8 – 12)*. Some reports also showed that atelocollagen has potential as carrier material for siRNA in vivo for both local *(7, 13–16)* and systemic treatment *(17–19)* in mouse models of cancer.

For the use of RNAi in living animals, how to evaluate the efficacy of downregulation of target mRNA by siRNA treatment is also important. In the case of local treatment, removing subcutaneous tumors or target organs for analysis of mRNA or protein level is relatively common. We use an assay system based on luminescence imaging for monitoring siRNA delivery in vivo and evaluating RNAi treatments of tumors, which is noninvasive, can be performed in real-time, is and quantitative *(7, 17, 20)*. In this method, the cancer cells to be injected to the animal are engineered to express luciferase, and siRNA that inhibits luciferase expression is used to quantitate the efficiency of siRNA delivery. By measuring the luciferase activity (photon number), we can determine the efficacy of siRNA administered that reaches the target cells or tissues (**Fig. 4.1**). Furthermore, the use of target-

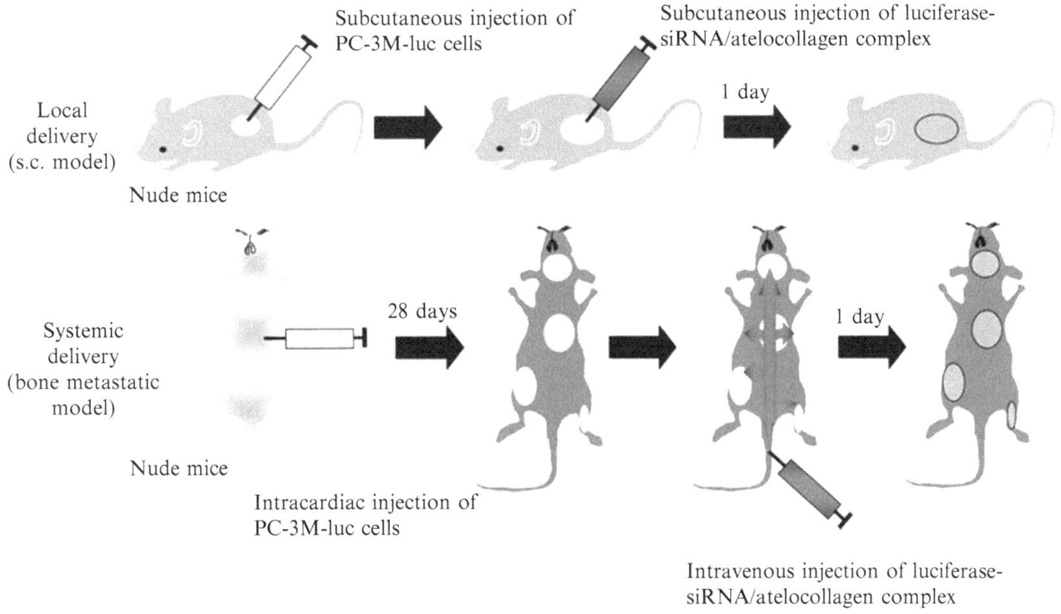

Fig. 4.1. Schematic representation of the assessment of RNAi molecules to regulate gene expression in live animals using bioluminescence imaging. Atelocollagen is useful for both local and systemic delivery of siRNA, since the siRNA/atelocollagen complex is stable in vivo. For the evaluation of systemic treatment of siRNA/atelocollagen, a mouse model of bone metastatic human prostate cancer was prepared. In this model, bone metastases developing in the jaws and/or legs of the mice were detected by noninvasive in vivo bioluminescence imaging analysis.

ing siRNA to a gene associated with the proliferation of cancer cells enables us to quantify the inhibition of cancer cell growth using luciferase as an indicator.

In this chapter, the assessment of siRNA molecules to regulate gene expression in live animals using bioluminescence imaging is introduced, taking atelocollagen-mediated siRNA delivery system as an example.

2. Materials

2.1. Formation of siRNA/Atelocollagen Complex

1. 3.5% atelocollagen solution, (Koken Co., Ltd., Tokyo, Japan) (see **Note 1**).
2. Diethylpyrocarbonate (DEPC)-treated Dulbecco's phosphate buffered saline without calcium and magnesium [DPBS(−)].
3. Synthetic luciferase-siRNA (Luciferase GL3 Duplex, Dharmacon, (Lafayette, CO) in deprotected, desalted, and annealed form. The sequence for luciferase-siRNA is 5′-CUU ACG CUG AGU ACU UCG A dTdT-3′, and 3′-dTdT GAA UGC GAC UCA UGA AGC U-5′ *(21)*.

2.2. Determination of Intensity of Luciferase Expression Using Bioluminescence Imaging

1. D-luciferin, Firefly potassium salt, 1.0 g/vial (Xenogen). Prepare a stock solution of luciferin at 15 mg/ml in DPBS(−). Filter sterilize through a 0.2 μm filter.
2. Isoflurane (Forane; Abbott Laboratories,. North Chicago, IL).
3. IVIS imaging system (Xenogen).
4. LivingImage software (version 2.50, Xenogen).

2.3. Evaluation for Local and Systemic Injection of siRNA/Atelocollagen Complex Using In Vivo Imaging

1. The bioluminescent human prostate carcinoma cell line PC-3M-luc-C6 (Xenogen Corp., Alameda, CA).
2. Culture medium for PC-3M-luc-C6 cells; minimum essential medium Eagle (Invitrogen-Life Technologies, Inc., Carlsbad, CA) supplemented with 10% heat-inactivated fetal bovine serum (Equitech-Bio, Kerrville, TX), nonessential amino acids (Sigma-Aldrich, St. Louis, MO), L-glutamine (ICN Biomedicals Inc., Costa Mesa, CA), 1 mM sodium pyruvate (Sigma-Aldrich), MEM vitamin solution (Sigma-Aldrich), and 200 μg/ml zeocin (Invitrogen-Life Technologies).
3. Eight- to ten-week old male athymic nude mice.
4. Syringe (1 ml) and needles (26 × 1/2" gauge and 27 × 1/2" gauge).

2.4. Detection of siRNA in Tumor and Normal Tissues by RNase Protection Assay

1. ISOGEN (NipponGene, Tokyo, Japan).
2. The sequence of the DNA template to synthesize antisense probe for luciferase-siRNA is 5′-TCG AAG TAC TCA GCG TAA GCC TGT CTC-3′.
3. α-^{32}P UTP (PerkinElmer, Waltham, MA).
4. mirVana microRNA Probe Construction Kit (Ambion, Austin, TX).
5. mirVana miRNA Detection Kit (Ambion).

3. Methods

Biophotonic imaging of living animals is also applied to assess whether a siRNA is being delivered to the target tissues (**Fig. 4.1**). For the success of any in vivo study of RNAi, careful evaluation in cultured cells is important. These include the selection of optimal siRNAs and examination of mRNA and protein levels of the targeted genes.

3.1. Formation of siRNA/Atelocollagen Complex

1. 3.5% atelocollagen is diluted to 1% solution for local treatment or 0.1% solution for systemic treatment with DEPC-treated PBS(−). Diluted atelocollagen is mixed by rotation at 4°C for two to three nights until the solution becomes homogenous.
2. On the day of treatment, an equal volume of diluted atelocollagen and siRNA solution is combined and mixed by rotation at 4°C for 20 min. The complex is then kept at 4°C until use. The final concentration of atelocollagen is 0.5% for local treatment, and 0.05% for systemic treatment. The final concentration of luciferase-siRNA is 15 μg/200 μl/mouse for local treatment, and 25 μg/200 μl/mouse for systemic treatment (*see* **Note 2**).

3.2. Determination of Intensity of Luciferase Expression Using Bioluminescence Imaging

1. The bioluminescent tumor-bearing mice are injected with firefly luciferin (150 mg/kg) intraperitoneally (*see* **Note 3**).
2. Mice are anesthetized by 3% isoflurane gas.
3. Ten minutes after luciferin injection, photons from tumor cells are counted using the IVIS imaging system (acquisition time, 10–30 s; f-stop, 1; binning, 1–8, field of view, 20–15). Data are analyzed using LivingImage software.
4. To control for mouse-to-mouse variability, the bioluminescence ratio for each mouse is normalized by dividing by the one-day-post-siRNA treatment/pre-siRNA treatment ratio of luciferase intensity for that mouse.

3.3. Evaluation for Local Injection of siRNA/Atelocollagen Complex Using In Vivo Imaging

1. PC-3M-luc-C6 cells are trypsinized from 15 cm dishes, and resuspended in DPBS to a concentration of 3×10^6 cells in 50 μl.
2. Mice are anesthetized by 3% isoflurane gas.
3. PC-3M-luc-C6 cells are injected subcutaneously (s.c.) into the dorsal region rear the thigh of the mouse.
4. After 5–8 days when the tumor has reached a volume of 50 mm^3, photons from tumor cells are counted using the IVIS imaging system (photon counts of the pretreatment).
5. The mice are injected s.c. with 200 μl of 0.5% atelocollagen containing 15 μg of luciferase-siRNA to cover the tumor.
6. The mice are re-imaged one day after treatment with siRNA/atelocollagen complexes (photon counts of one day post-treatment). An example of the data obtained is shown in **Fig. 4.2a**.

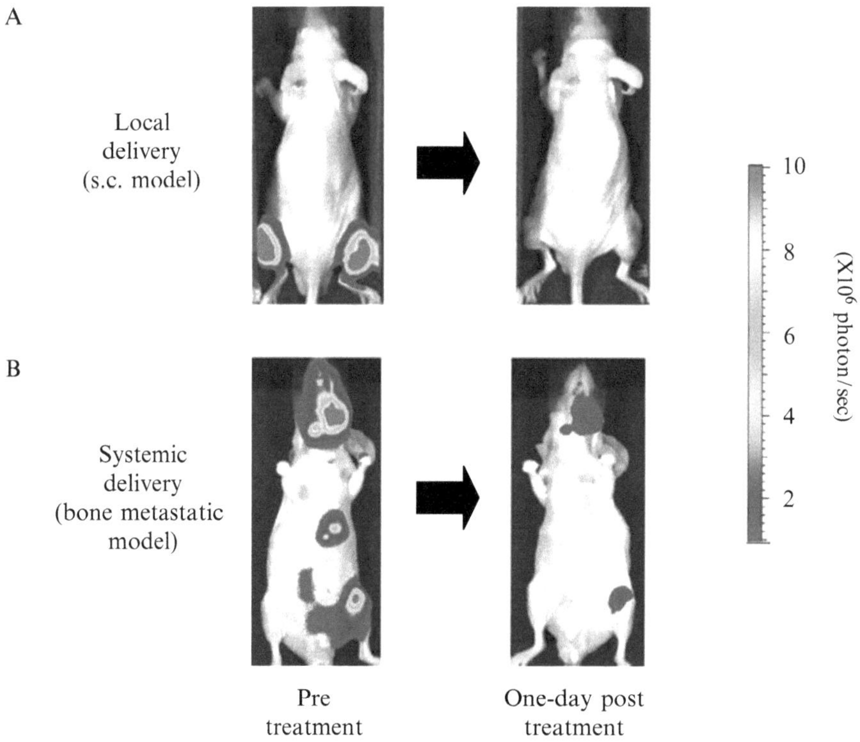

Fig. 4.2. Monitoring luciferase inhibition by siRNA treatment using bioluminescent imaging. (**a**) Nude mice were inoculated subcutaneously with 3×10^6 PC-3M-luc-C6 cells. Seven days after tumor injection, the animals were administered with luciferase-siRNA (15 μg) complexed with 0.05% atelocollagen by s.c. to cover the tumor. On the next day, the bioluminescent signals from s.c. tumor was inhibited by 90% when compared with pretreatment. (**b**) Images of nude mice injected with 3×10^6 PC-3M-luc-C6 cells suspended in 100 μl sterile DPBS into the left ventricle of the heart. Four weeks after tumor injection, the animals were administered with luciferase-siRNA (25 μg) complexed with 0.05% atelocollagen. On the next day, the bioluminescent signals of most of the metastatic sites were inhibited by 80% in the whole body when compared with pretreatment.

3.4. Evaluation for Systemic Injection of siRNA/Atelocollagen Complex Using In Vivo Imaging

1. To set up a mouse model of bone metastasis, the human prostate cancer PC-3M-luc-C6 cells (3×10^6 cells in 100 μl) are injected into the left heart ventricle of nude mice using a 27 × 1/2"gauge needle (*see* **Note 4**).
2. After injection of tumor suspension, photons from animal whole bodies are counted using the IVIS imaging system. A successful intracardiac injection is indicated by day zero images showing a systemic bioluminescence distributed throughout the animal, and only those mice showing a satisfactory injection are used in the experiments.
4. Four weeks after tumor injection, photons are counted using the IVIS imaging system (photon counts of the pretreatment).
5. The mice are injected with 200 μl of 0.05% atelocollagen containing 25 μg of luciferase-siRNA intravenously into the tail vein.
6. The mice are re-imaged one day after treatment with siRNA/atelocollagen complexes (photon counts of the one day posttreatment). An example of the obtained data is shown in **Fig. 4.2b**.

3.5. Detection of siRNA in Tumor or Normal Tissues by RNase Protection Assay

1. PC-3M-luc-C6 cells are injected s.c. into the dorsal region near the thigh of the mouse.
2. After 5–8 days, when the tumor has reached a volume of 50 mm^3, tumor-bearing mice received 200 μl of 0.05% atelocollagen containing 25 μg of luciferase-siRNA by intravenous tail vein injection.
3. The mice are sacrificed one day after treatment with siRNA/atelocollagen complexes, and s.c. tumors and normal tissues are removed and weighed.
4. Total RNA is extracted from the tumor and normal tissues by ISOGEN.
5. The RNase protection probe is made with a mirVana microRNA Probe Construction Kit. The complementary RNA probe specific for the antisense strand of luciferase-siRNA is generated using T7 RNA polymerase and ^{32}P-labeled UTP.
6. Total RNAs are utilized in an RNase protection assay using the mirVana miRNA detection kit as recommended by the manufacturer.
7. Protected fragments are separated by electrophoresis in 15% polyacrylamide 8 M urea gels.
8. The gels are exposed to x-ray films for 30 min, and then scanned and analyzed using NIH image software. An example of the obtained data is shown in **Fig. 4.3**.

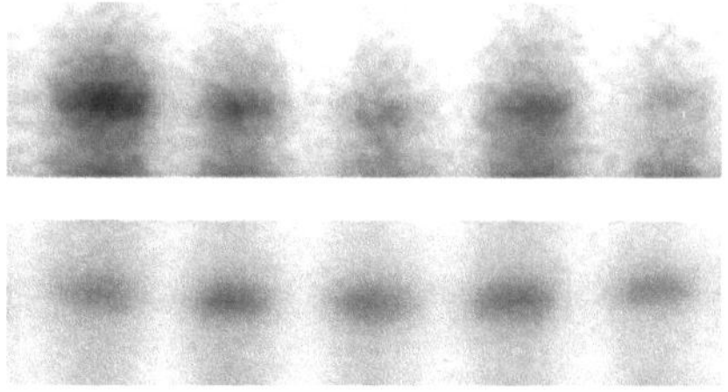

Fig. 4.3. Distribution of siRNA delivered with atelocollagen in tumor and normal tissues. Nude mice were inoculated subcutaneously with 3 × 10^6 PC-3M-luc-C6 cells. Once tumors had reached 50–100 mm^3, tumor-bearing mice were injected with 200 μl of 0.05% atelocollagen containing 25 μg of luciferase-siRNA by intravenous tail vein injection. The mice were sacrificed one day after treatment with siRNA/atelocollagen complexes, and total RNA was extracted from tumor and selected tissues. Detection of luciferase siRNA was performed by an RNase protection assay. Significant amounts of siRNAs were detected in tumor and several normal tissues such as liver, lung, spleen, kidney, with atelocollagen-mediated delivery.

9. Luciferase-siRNA levels are corrected for wet tissue weights.

4. Notes

1. Atelocollagen has to be kept at low temperature (4°C). If atelocollagen changes to a solid form caused by warming or freezing, it must not be used. Atelocollagen suitable for local or systemic siRNA delivery is supplied by Koken Bioscience Institute (Tokyo, Japan, http://www.kokenmpc.co.jp/english/index.html).
2. Optimizations of dose, injection volume, and number of injections are necessary because siRNA effective concentrations are expected to vary with target genes, cells, and animal models (*see* Chaps. 1 and 11–17)
3. Animal experiments must be performed in compliance with the guidelines of the institute for laboratory animal research of each institute.
4. When the needle is correctly positioned into the left ventricle, bright red oxygenated blood influxes into needle center should be seen *(22, 23)*. In this model, bone metastases developing in the jaws and/or legs of the mice are detected by noninvasive in vivo bioluminescence imaging analysis.

5. In addition to siRNA, microRNA (miRNA) is likely to play a key role in future cancer therapies (*see* Chaps. 18 and 19). miRNA is a 18–24-nucleotide single-stranded RNA and it is thought to control protein expression at the translational level, and more than 600 molecular species have been identified in humans. According to conservative estimates up to 30% of human genes are regulated by microRNAs. Depending on the progression of future cancer studies, miRNA may become an effective tool to elucidate disease mechanisms and novel treatments. Therefore, the application of an atelocollagen-mediated delivery of siRNA targeting miRNA might identify mechanisms of diseases that can be controlled by miRNA and establish effective therapies.

Acknowledgments

The authors would like to thank Dr. Shunji Nagahara of Formulation Research Laboratories, Technology Research and Development Center, Dainippon Sumitomo Pharma Co., Ltd. for technological support. This work was supported in part by a Grant-in-Aid for the Third-Term Comprehensive 10-Year Strategy for Cancer Control, a Grant-in-Aid for Scientific Research on Priority Areas Cancer from the Ministry of Education, Culture, Sports, Science and Technology, and the Program for Promotion of Fundamental Studies in Health Sciences of the National Institute of Biomedical Innovation (NiBio).

References

1. de Fougerolles, A., Vornlocher, H.P., Maraganore, J., and Lieberman, J. (2007) Interfering with disease: A progress report on siRNA-based therapeutics. *Nat. Rev. Drug Discov.* 6, 443–453.
2. Aigner, A. (2007) Applications of RNA interference: current state and prospects for siRNA-based strategies *in vivo*. *Appl. Microbiol. Biotechnol.* 76, 9–21.
3. Soutschek, J., Akinc, A., Bramlage, B., Charisse, K., Constien, R., Donoghue, M., et-al. (2004) Therapeutic silencing of an endogenous gene by systemic administration of modified siRNAs. *Nature* 432, 173–178.
4. Zimmermann, T.S., Lee, A.C., Akinc, A., Bramlage, B., Bumcrot, D., Fedoruk, M.N., et al. (2006) RNAi-mediated gene silencing in non-human primates. *Nature* 441, 111–114.
5. Ochiya, T., Nagahara, S., Sano, A., Itoh, H., and Terada, M. (2001) Biomaterials for gene delivery: atelocollagen-mediated controlled release of molecular medicines. *Curr. Gene Ther.* 1, 31–52.
6. Sano, A., Maeda, M., Nagahara, S., Ochiya, T., Honma, K., Itoh, H., et al. (2003) Atelocollagen for protein and gene delivery. *Adv. Drug Deliv. Rev.* 55, 1651–1677.
7. Minakuchi, Y., Takeshita, F., Kosaka, N., Sasaki, H., Yamamoto, Y., Kouno, M., et al. (2004) Atelocollagen-mediated synthetic small interfering RNA delivery for effective gene silencing in vitro and in vivo. *Nucleic Acids Res.* 32, e109.

8. Ochiya, T., Takahama, Y., Nagahara, S., Sumita, Y., Hisada, A., Itoh, H., et al. (1999) New delivery system for plasmid DNA *in vivo* using atelocollagen as a carrier material: The Minipellet. *Nat. Med.* 5, 707–710.
9. Honma, K., Ochiya, T., Nagahara, S., Sano, A., Yamamoto, H., Hirai, K., et al. (2001) Atelocollagen-based gene transfer in cells allows high-throughput screening of gene functions. *Biochem. Biophys. Res. Commun.* 289, 1075–1081.
10. Hirai, K., Sasaki, H., Sakamoto, H., Takeshita, F., Asano, K., Kubota, Y., et al. (2003) Antisense oligodeoxynucleotide against HST-1/FGF-4 suppresses tumorigenicity of an orthotopic model for human germ cell tumor in nude mice. *J. Gene Med.* 5, 951–957.
11. Takei, Y., Kadomatsu, K., Matsuo, S., Itoh, H., Nakazawa, K., Kubota, S., et al. (2001) Antisense oligodeoxynucleotide targeted to Midkine, a heparin-binding growth factor, suppresses tumorigenicity of mouse rectal carcinoma cells. *Cancer Res.* 61, 8486–8491.
12. Takei, Y., Kadomatsu, K., Itoh, H., Sato, W., Nakazawa, K., Kubota, S., et al. (2002) 5′-,3′-inverted thymidine-modified antisense oligodeoxynucleotide targeting midkine. Its design and application for cancer therapy. *J. Biol. Chem.* 277, 23800–23806.
13. Takei, Y., Kadomatsu, K., Yuzawa, Y., Matsuo, S., and Muramatsu, T. (2004) A small interfering RNA targeting vascular endothelial growth factor as cancer therapeutics. *Cancer Res.* 64, 3365–3370.
14. Fujii, T., Saito, M., Iwasaki, E., Ochiya, T., Takei, Y., Hayashi, S., et al. (2006) Intratumor injection of small interfering RNA-targeting human papillomavirus 18 E6 and E7 successfully inhibits the growth of cervical cancer. *Int. J. Oncol.* 29, 541–548.
15. Banno, H., Takei, Y., Muramatsu, T., Komori, K., and Kadomatsu, K.J. (2006) Controlled release of small interfering RNA targeting midkine attenuates intimal hyperplasia in vein grafts. *Vasc. Surg.* 44, 633–641.
16. Nozawa, H., Tadakuma, T., Ono, T., Sato, M., Hiroi, S., Masumoto, K., et al. (2006) Small interfering RNA targeting epidermal growth factor receptor enhances chemosensitivity to cisplatin, 5-fluorouracil and docetaxel in head and neck squamous cell carcinoma. *Cancer Sci.* 97, 1115–1124.
17. Takeshita, F., Minakuchi, Y., Nagahara, S., Honma, K., Sasaki, H., Hirai, K., et al. (2005) Efficient delivery of small interfering RNA to bone-metastatic tumors by using atelocollagen *in vivo*. *Proc. Natl. Acad. Sci. U S A* 102, 12177–12182.
18. Takeshita, F. and Ochiya, T. (2006) Therapeutic potential of RNAi against cancer. *Cancer Sci.* 97, 689–696.
19. Hanai, K., Takeshita, F., Honma, K., Nagahara, S., Maeda, M., Minakuchi, Y., et al. (2006) Atelocollagen-mediated systemic DDS for nucleic acid medicines. *Ann. N Y Acad. Sci.* 1082, 9–17.
20. McCaffrey, A.P., Meuse, L., Pham, T.T., Conklin, D.S., Hannon, G.J., and Kay, M.A. (2002) RNA interference in adult mice. *Nature* 418, 38–39.
21. Elbashir, S.M., Harborth, J., Lendeckel, W., Yalcin, A., Weber, K., and Tuschl, T. (2001) Duplexes of 21-nucleotide RNAs mediate RNA interference in cultured mammalian cells. *Nature* 411, 494–498.
22. Arguello, F., Furlanetto, R.W., Baggs, R.B., Graves, B.T., Harwell, S.E., Cohen, H.J., et-al. (1992) Incidence and distribution of experimental metastases in mutant mice with defective organ microenvironments (genotypes Sl/Sld and W/Wv). *Cancer Res.* 52, 2304–2309.
23. Jenkins, D.E., Yu, S.F., Hornig, Y.S., Purchio, T., and Contag, P.R. (2003) *In vivo* monitoring of tumor relapse and metastasis using bioluminescent PC-3M-luc-C6 cells in murine models of human prostate cancer. *Clin. Exp. Metastasis* 20, 745–756.

Chapter 5

Imaging of siRNA Delivery and Silencing

Anna Moore and Zdravka Medarova

Abstract

The fast developing field of RNA interference (RNAi) requires monitoring of small interfering RNA (siRNA) delivery to targeted organs and evaluating the efficiency of target gene silencing. The molecular imaging approach fits perfectly to fulfill these needs and provides information in a fast, reproducible, and noninvasive manner. This review serves as a first attempt to summarize existing information on various imaging modalities and their application for siRNA imaging. It is noteworthy that new publications in this field appear almost on a weekly basis and the authors have made a sincere attempt to reflect the development of this area in their review.

Key words: siRNA, molecular imaging, contrast agents, bioluminescence imaging, optical imaging, quantum dots, radionuclide imaging, magnetic resonance imaging.

1. Introduction

Since their discovery in 1998 *(1)*, small interfering RNAs (siRNAs) have emerged as a powerful tool for therapeutic gene silencing because of their unique specificity and efficiency. During the past years, RNA interference (RNAi) has become an indispensable research instrument in virtually all fields of medical and biological sciences. Its broad applicability (virtually any gene can be silenced), superior efficiency (100–1000-fold compared to antisense oligonucleotides), and exquisite specificity (single nucleotide) could potentially be harnessed to develop a powerful novel treatment paradigm with global relevance to any disease amenable to manipulation at the level of gene expression. In order to develop siRNAs into successful therapeutic agents, it is crucial to monitor their delivery and silencing efficiency in vivo.

M. Sioud (ed.), *Methods in Molecular Biology, siRNA and miRNA Gene Silencing, vol. 487*

DOI: 10.1007/978-1-60327-547-7_5

Molecular imaging techniques represent a powerful tool for real-time noninvasive monitoring of various events at a near microscopic level and have superior advantages over conventional in vitro and cell culture research techniques in biology. It is anticipated that these techniques can be applied for the in vivo tracking of siRNA delivery and for the assessment of silencing efficiency. This chapter gives an overview of the existing noninvasive imaging modalities and their possible applications for the imaging of siRNA delivery and silencing.

2. In Vivo Molecular Imaging Techniques

Molecular imaging is a relatively new yet fast developing field of biological science focused on the noninvasive visualization of molecular phenotypes in whole organisms. Molecular imaging can produce images that reflect cellular and molecular pathways and in vivo mechanisms of disease present within the context of physiologically authentic environments *(2)*. Present imaging technologies rely mostly on nonspecific macroscopic physical, physiological, or metabolic changes that differentiate pathological from normal tissue. By contrast, molecular imaging can identify specific events on the molecular level, such as gene expression or specific protein–protein interactions responsible for disease. That is why molecular imaging is perfectly suited for the early detection of molecular abnormalities, before any morphological changes become apparent.

The latest advances in noninvasive molecular imaging have created the possibility of achieving several important goals of biomedical research, such as *(1)* to develop noninvasive imaging techniques that reflect specific cellular and molecular processes, *(2)* to monitor multiple molecular events near-simultaneously, *(3)* to follow trafficking and targeting of cells, *(4)* to optimize drug and gene therapy, *(5)* to image drug effects at a molecular and cellular level, *(6)* to assess disease progression at a molecular pathological level, and *(7)* to create the possibility of achieving these goals of imaging in a rapid, reproducible, and quantitative manner *(2)*. Successful in vivo imaging of specific molecules and/or events responsible for certain pathological conditions involves *(1)* design of high-affinity imaging probes with favorable pharmacokinetics, *(2)* the ability of the probe to overcome biological delivery barriers (vascular, interstitial, etc.), *(3)* utilization of amplification strategies (chemical or biological), and *(4)* sensitive, fast high-resolution imaging techniques *(3)*. The advantages of molecular imaging used in biomedical research are numerous and include the possibility to assess whole-body biological pathways, which is not possible with conventional in vitro microscopic and cell/tissue culture techniques.

2.1. Radionuclide Imaging

Molecular imaging techniques utilize different imaging modalities, each of which has its own advantages and drawbacks. Radionuclide imaging is exemplified by positron emission tomography (PET), a procedure that records rays emitted from within the subject. Various molecules can be labeled with a positron-emitting isotope. Positron-emitting isotopes frequently used include ^{15}O, ^{13}N, ^{11}C, and ^{18}F, and some others. Most of them are produced in a cyclotron and have relatively short half-lives. Labeled molecular imaging probes or tracers can be introduced into the subject (animal or human). PET imaging can follow the distribution and concentration of these injected molecules *(4)*. gamma-Emitting isotopes (e.g., ^{99m}Tc, ^{111}In, ^{123}I, ^{131}I) can also be used for imaging living subjects but require different types of scanners known as gamma cameras, which when rotated around the subject (a technique known as single-photon emission computed tomography, SPECT) result in the production of tomographic images *(5)*. After acquiring a signal, the reconstruction software produces an image, which "portrays" the region of interest that was scanned. The sensitivity of PET is relatively high (10^{-11} to 10^{-12} mol/l), which is at least a log order more sensitive than SPECT, and is independent of the location depth of the reporter probe of interest. Since a positron-emitting isotope is capable of producing two gamma-rays of the same energy through emission of a positron from its nucleus, it is not possible for PET to distinguish between two probes labeled with different isotopes used simultaneously. To investigate multiple molecular events, it would be necessary to allow one probe to decay prior to administration of the other. However, multiple isotopes with various energy gamma-rays can be used in SPECT. The spatial resolution of most clinical PET scanners is 3–12 mm, resulting from a combination of factors. The number and geometry of detectors in the scanner as well as the number of counts acquired in the image and their statistical imprecision each reduce PET image resolution. These aspects vary between tomographs of different designs as well as from study to study, owing to varying image-acquisition times and tissue-radioactivity levels. The ultimate theoretical limit of PET resolution, however, is the distance traveled by the positron in tissue before the annihilation reaction.

2.2. Optical Imaging

In vivo optical imaging methods originated from previously developed in vitro and ex vivo microscopic techniques based on fluorescence, absorption, reflectance, or bioluminescence as a source of contrast. Imaging systems can be based on diffuse optical tomography, surface-weighted imaging (reflectance diffuse tomography), phase-array detection, confocal imaging, multiphoton imaging, or microscopic imaging with intravital microscopy. Progress in optical molecular imaging strategies has come from the recent development of targeted bioluminescence probes, near-infrared fluorochromes, activatable near-infrared fluorochromes, and red-shifted fluorescent proteins *(6)*. In addition, the development of the sensitive charged coupled device (CCD) has made it possible to detect light in the visible to near-infrared range. Used for all kinds of optical imaging, these cameras have proven to be highly reproducible, whenever the exposure conditions are kept identical *(2)*. The cost, space, and time involved in optical imaging are less demanding compared to other imaging modalities. Furthermore, the advantages of optical imaging methods include the use of nonionizing low energy radiation, high sensitivity with the possibility of detecting micron-sized objects, and continuous data acquisition in real time and in an intact environment. The fast and relatively easy imaging procedure makes this modality attractive for potential clinical use. Fluorescence optical imaging utilizes various fluorochromes (dyes, fluorescent proteins, etc.), whereas bioluminescence imaging (BLI) exploits the reaction between luciferase and its substrate, luciferin. The family of luciferase enzymes, present in certain bacteria, marine crustaceans, fish, and insects, consists of proteins that can generate visible light through the oxidation of an enzyme-specific substrate in the presence of oxygen and, usually, ATP as a source of energy. During these reactions, part of the chemical energy is released as visible light. A significant advantage of luciferases as optical indicators in live mammalian cells and tissues is the inherently low background, given the near absence of endogenous light from mammalian cells and tissues. Luciferase has therefore become a frequently used reporter in many optical imaging applications *(7)*.

In contrast to PET, optical imaging has a limited depth penetration but, in the case of bioluminescence optical imaging, high sensitivity. In fluorescence imaging, an excitation light of one wavelength (in the visible light range of 395–600 nm) illuminates the living subject, and a CCD camera collects emission light of a shifted wavelength. Cells tagged with fluorescently labeled ligands or expressing fluorescent proteins can be detected by this technique. The two main advantages of the latter are that fluorescent proteins (for example, green fluorescent protein, GFP) can be visualized as reporters in both live and fixed cells/tissues and that no substrate is required for their visualization. Certain drawbacks include difficulties in quantitation as well as the surface-weighted nature of the image (objects closer to the surface will appear brighter than deeper structures *(6)*).

Optical imaging in the near infrared (NIR) region between 700 and 900 nm has a low absorption by intrinsic photoactive biomolecules and allows light to penetrate several centimeters into the tissue, a depth that is sufficient to image practically all small animals *(8)*. Hence, imaging in the NIR region has minimal tissue autofluorescence, which dramatically improves the target/background ratio *(9)*. Also, fluorescence-mediated tomography (FMT) has recently been shown to three dimensionally localize and quantify fluorescent probes in deep tissues at high sensitivity *(10)*. Therefore, it becomes an attractive tool for screening mouse models and for new drug development.

Semiconductor nanocrystals or quantum dots are gaining interest as fluorescent tags for biological molecules due to their large quantum yield and photostability. Research on quantum dots (or qdots) has evolved over the past two decades from electronic materials science to biological applications *(11)*. Quantum dots are semiconductor crystallites 2–10 nm in diameter that contain approximately 500–1000 atoms of materials as cadmium and selenide. Quantum dots fluoresce with a broad absorption spectrum and a narrow emission spectrum. The larger the quantum dot, the longer the wavelength emitted. The broad absorption spectrum allows many different quantum dots to be excited with one excitation source. The emission spectra for each dot are typically very narrow, on the order of 30 nm, which permits spectral resolution of adjacent dots. Recent examples of their experimental use include the observation of diffusion of individual glycine receptors in living neurons and the identification of lymph nodes in live animals by near-infrared emission during surgery. The new generations of qdots have far-reaching potential for the study of intracellular processes at the single-molecule level, high-resolution cellular imaging, long-term in vivo observation of cell trafficking *(12)*, tumor targeting *(13)*, visualization of gene expression *(14)*, and diagnostics *(11)*.

2.3. Magnetic Resonance Imaging

Magnetic resonance imaging is the modality that can deliver the highest-resolution images. On a practical level, magnetic resonance imaging involves placing the subject in a strong magnetic field. Nuclei of certain elements have a magnetic moment that aligns preferentially along the direction of the magnetic field. Applied radio frequency results in excitation with subsequent relaxation of the nuclear spin system, which can be detected and translated into an MR image. An MR image in clinical and biological systems is typically an image of the hydrogen atoms in water and fat because of the natural abundance of water hydrogen. MRI systems are now available at near-microscopic resolutions for small animals. A main advantage of MR imaging is in its high spatial resolution and the ability to acquire physiological and anatomical parameters simultaneously. Due to the intrinsic low sensitivity of magnetic resonance imaging, various amplification strategies are used to detect the desired molecular or cellular event of interest. Although increasing the field strength of the MR scanner can improve the detection issue, utilization of metal-containing contrast agents can significantly improve the detection limit even within a relatively low magnetic field. Contrast agents can act by changing the relaxation rates of neighboring water molecules and giving positive or negative contrast on T1- or T2-weighted images, respectively. A different class of contrast agents relies on magnetization transfer to provide negative contrast. Magnetization transfer and T1- and T2-weighted agents alter some property of water in a catalytic way, but it is still the water that is imaged. Other contrast agents use alternative nuclei such as fluorine or hyperpolarized nuclei such as carbon, helium, or xenon, and these nuclei are imaged directly *(15)*. T1-weighted contrast agents are typically gadolinium(III) complexes, manganese(II) complexes, or Mn^{2+} cation. Gd(III) and Mn(II) are used as T1 relaxation agents because their electronic relaxation is slow, the magnetic moment is large, and water exchange is typically fast *(15)*. T2 contrast agents are typically iron oxide nanoparticles with various coatings (polysaccharide, synthetic polymer, or monomer coating). Superparamagnetic iron oxide nanoparticles (SPIO) alone or conjugated to target-specific ligands are widely used in biomedical applications to image cell trafficking *(16, 17)*, gene expression *(18–21)*, cancer lesions and their response to therapy *(22–24)*, as well as siRNA delivery (discussed below).

2.4. Other Molecular Imaging Modalities

Another modality available to modern molecular imaging is computed tomography. Images in computed tomography (CT) are obtained when component tissues differentially absorb x-rays as they pass through the body *(25)*. A low energy x-ray source of 30–50 kVp (i.e., of considerably lower energy than in clinical CT scanners) and a detector rotate around the animal, acquiring volumetric data. Most mouse CT images are collected with high-resolution phosphor screen/CCD detectors to optimize image quality. The system spatial resolution is primarily limited by the pixel sampling rate, the x-ray source size, and blurring in the phosphor screen. The radiation dose, however, is not negligible (0.6 Gy per scan; 5% of the LD50 for mice), and this can limit repeated imaging of the same animal. Unlike MRI, CT has relatively poor soft tissue contrast, often making it necessary to administer iodinated contrast media to delineate organs or tumors *(2)*. The most widely used clinical imaging modality is ultrasonography. The appeal of this modality is its low cost, availability, and safety. Ultrasound images are obtained when high-frequency (>20 kHz) sound waves are emitted from a transducer placed against the skin and the ultrasound is reflected back from the internal organs under examination.

Contrast in the images obtained depends on the imaging algorithm used, backscatter, attenuation of the sound, and sound speed. Ultrasound imaging using diagnostic ultrasound instrumentation operating in the 7.5–15 MHz frequency range has been successfully applied to a variety of mouse models *(26)*, yielding images with a spatial resolution of 300–500 μm. The strengths of ultrasound in cardiac, obstetric, vascular, and abdominal imaging appear most likely to extend to the mouse as a model when the technology is scaled down to achieve high resolution and a level of practicality/functionality similar to that available with present clinical ultrasound systems. Another emerging concept is that of using targeted ultrasonic contrast agents for molecular imaging of specific cell-surface receptors, especially within the vascular compartment *(27)*.

All of the above modalities, theoretically, can be applied to imaging of siRNA delivery and function as well as monitoring its time-dependent therapeutic influence on disease progression in the same animal or patient. The choice of modality depends on the spatial and temporal resolution needed, and the availability of amplification strategies (molecular imaging probes).

3. Imaging of RNA Interference-Current Status

3.1. Imaging of siRNA Using Bioluminescence

The field of in vivo RNA interference imaging is still in its early stages of development. Studies that describe the imaging of RNA interference are mostly restricted to fluorescence or bioluminescence reporter imaging. A lot of the early proof-of-principle studies on in vivo RNAi utilized suppression of luciferase transgene expression by either synthetic siRNAs or small-hairpin RNAs transcribed in vivo, achieving impressive 80–90% silencing efficiencies in the liver *(28)*. In the same study, the authors fused luciferase RNA with an endogenous gene encoding nonstructural protein 5B from viral-polymerase-encoding region of hepatitis C virus. By measuring the bioluminescence signal, they demonstrated that siRNA targeting reduced luciferase expression from HCV protein–luciferase fusion by 75%. This result suggests that it may be feasible to use in vivo imaging to monitor siRNA as a therapeutic tool and establish a method for the in vivo quantitative imaging of the silencing effect. In a similar study, Lewis et al. *(29)* co-injected a plasmid expressing luciferase with antiluciferase synthetic siRNA and observed significant 80–90% silencing, as measured by bioluminescence imaging in a variety of organs. This investigation, for the first time, utilized in vivo imaging to identify, on a whole-body scale, the scope of siRNA biological activity. More recently, bioluminescence imaging of luciferase expression was used to demonstrate the dose- and time-dependence of vector-based in vivo RNAi after hydrodynamic injection *(30)*. Bioluminescence imaging has also been employed for the development of optimized methods for lipid-based siRNA delivery, where siRNA was complexed with polyethyleneimine-based polyplexes and cationic lipid-based lipoplexes. RNAi directed against luciferase resulted in nearly 80% inhibition with as little as 0.2 pmol siRNA *(31)*. This inhibition was dose dependent and specific. A group from Japan utilized BLI for imaging siRNA complexed with atelocollagen. They showed that site-specific in vivo administration of an antiluciferase siRNA/atelocollagen complex reduced luciferase expression in a xenografted tumor. Furthermore, atelocollagen-mediated transfer of siRNA in vivo showed efficient inhibition of tumor growth in an orthotopic xenograft model of a human nonseminomatous germcell tumor *(32)*. Another study from the same group utilized BLI for monitoring the effect of systemic administration of siRNA complexed with atelocollagen and directed against potential targets to bone metastasis *(33)*. This treatment resulted in an efficient inhibition of metastatic tumor growth in bone tissues. In addition, this delivery method was found to be nontoxic and did not cause upregulation of serum cytokine levels. Though not using BLI directly, Takabatake et al. employed luciferase-directed siRNA for in vivo delivery to the kidney. Their method utilized injection via the renal artery followed by electroporation *(34)*. Luciferase targeting by siRNA encapsulated in the interior of pegylated immunoliposomes modified with monoclonal antibody to the transferrin receptor resulted in 64–68% knockdown of the transcript *(35)*. In a comprehensive imaging study *(36)*, bioluminescence was applied for the noninvasive assessment of P-glycoprotein silencing. Multidrug resistance (MDR) remains a major obstacle to successful chemotherapeutic treatment of cancer and can be caused by overexpression of P-glycoprotein, the *MDR1* gene product. shRNA constructs against human *MDR1* mRNA synthesized in this study inhibited expression of P-glycoprotein by >90%. Furthermore, after somatic gene transfer by hydrodynamic infusion of a *MDR1-Firefly luciferase* (*MDR1-FLuc*) fusion construct into mouse liver, the effect of shRNAi delivered in vivo on P-glycoprotein-FLuc protein levels was documented with bioluminescence imaging using d-luciferin. shRNAi against *MDR1* reduced the bioluminescence output of the P-glycoprotein-FLuc reporter fourfold in vivo compared with mice treated with control or scrambled shRNAi *(36)* (**Fig. 5.1**).

Hu-Lieskovan et al. reported on the delivery of siRNA by a nonviral delivery system, which uses a cyclodextrin-containing polycation to bind and protect siRNA and transferrin as a targeting ligand for delivery to transferrin receptor–expressing tumor cells. Bioluminescence imaging was employed here to show the effect of siRNA targeting metastatic Ewing's sarcoma chimeric fusion gene *EWS-FLI1*, found in 85% of patients with Ewing's family of

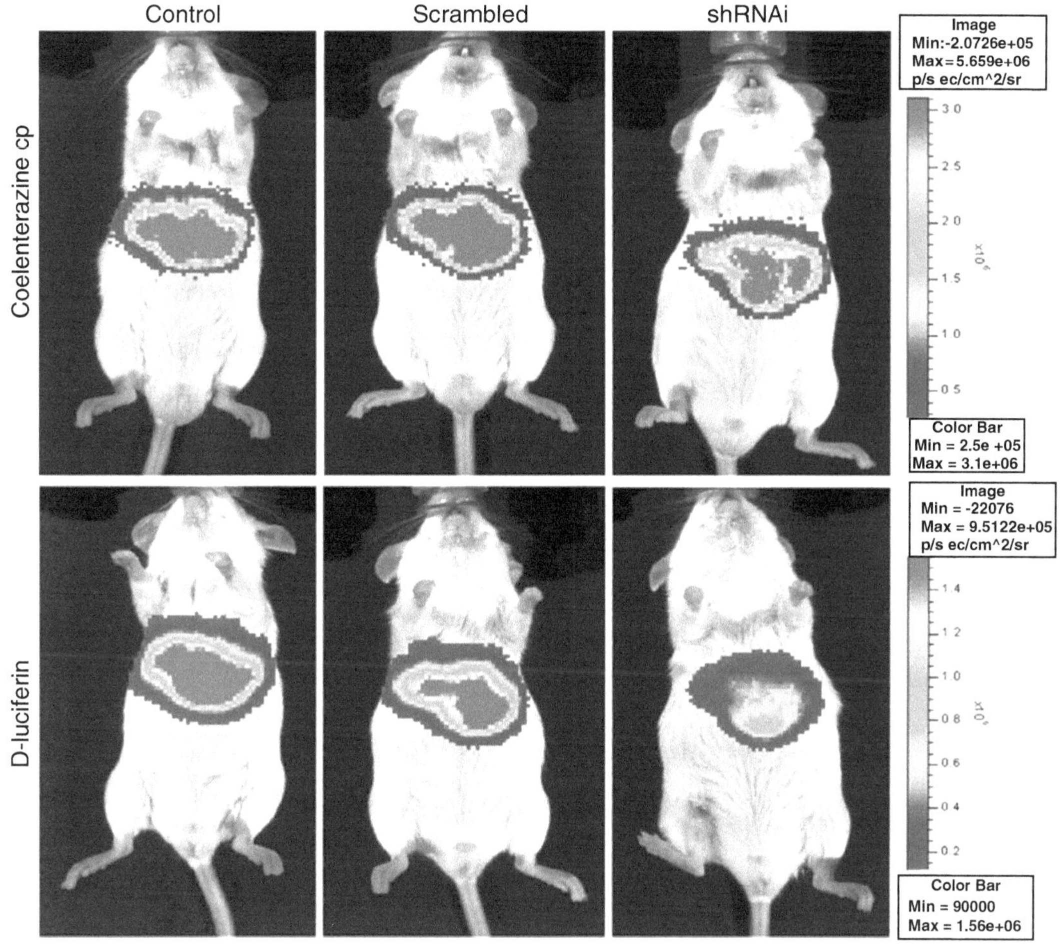

Fig. 5.1. Downregulation of P-glycoprotein in vivo: shRNAi-mediated reduction in P-glycoprotein protein levels. Representative bioluminescence images of RLuc expression with coelenterazine cp (*top*) and P-glycoprotein-FLuc expression with D-luciferin (*bottom*). Mice were hydrodynamically transfected with pMDR1-FLuc (1 μg) combined with pRLuc-N3 (1 μg; as transfection control) and treated as indicated with 10-fold excess of control (*left*), scrambled shRNAi (*middle*), or shRNAi against *MDR1* (*right*). (Reprinted by permission from AACR: *Clin Cancer Res.*, 11, 4487–4494 (2005)).

tumors (EFT). BLI showed that only the targeted, formulated siRNA against this gene achieves long-term tumor growth inhibition *(37)*. More recently, bioluminescence imaging was applied together with a mathematical model of siRNA delivery and function, in order to define, in a more general context, the effects of target-specific and treatment-specific parameters on siRNA-mediated gene silencing *(38)*. This particular study demonstrated that in rapidly dividing subcutaneous tumors, the silencing effect may persist ~10 days; in nondividing hepatocytes it may last as long as 3–4 weeks. More importantly, this study represents the origin of the current belief that siRNA dilution due to cell division, and not intracellular siRNA half-life, governs the duration of gene silencing. As a result of this study, now it is possible to apply this mathematical model in treatment design and to predict the dosing schedule required to maintain persistent silencing of target proteins with different half-lives in rapidly dividing or nondividing cells.

While many studies have explored bioluminescence imaging for the quantitative evaluation of RNAi efficiency *(32, 34, 35, 39)*, this imaging modality lacks clinical a equivalent, therefore precluding it from application in humans.

3.2. Imaging of siRNA Using GFP Reporter

Green fluorescent protein and its modifications have been used mostly for ex vivo imaging and microscopy to monitor the efficiency of siRNA action. As such, high-pressure hydrodynamic injection of siRNA–EGFP resulted in substantial reduction of EGFP expression in the liver in a large percentage of hepatocytes at 48 h after injection, as determined by fluorescent microscopy *(29)*.

Similar questions have been addressed using GFP reporters *(8, 36, 40, 41)*. For example, the utility of lentiviral vectors for the in vivo delivery of short-hairpin RNAs to the brain was demonstrated using an EGFP reporter *(42)*. However, this method has been restricted to ex vivo analysis with fluorescence microscopy or flow cytometry. An in vivo GFP fluorescence study reported by Rubinson et al. involved imaging of transgenic mice in which a lentiviral vector carrying shRNA also incorporated EGFP as a reporter gene permitting the evaluation of shRNA delivery *(43)*. Importantly the effect (94%) persisted into adulthood, due to integration of the vector into the host genome. Another in vivo imaging study involved intratumoral injection of siRNA against EGFP into EGFP-expressing B16F10 melanoma tumors followed by application of an external electric field. When the EGFP–siRNA was electrotransferred, a significant decrease in the GFP fluorescence of the tumor was observed by direct imaging of the animal (**Fig. 5.2**) and quantification of the fluorescence levels within 2 days following the treatment. The decrease was maximal at days 2–4 (reduction to about 30% of the physiological saline control) *(44)*. However, this is a very artificial system, requiring transgenesis and intratumoral injection, and only applicable in a research scenario. Besides this study, direct investigation of the delivery of siRNA to target tissues is exclusively based on ex vivo experiments *(45–50)*. These investigations provided answers to important new questions, such as the subcellular distribution of siRNAs delivered as part of various formulations, including tumor-targeted carriers *(46)* and their bioavailability.

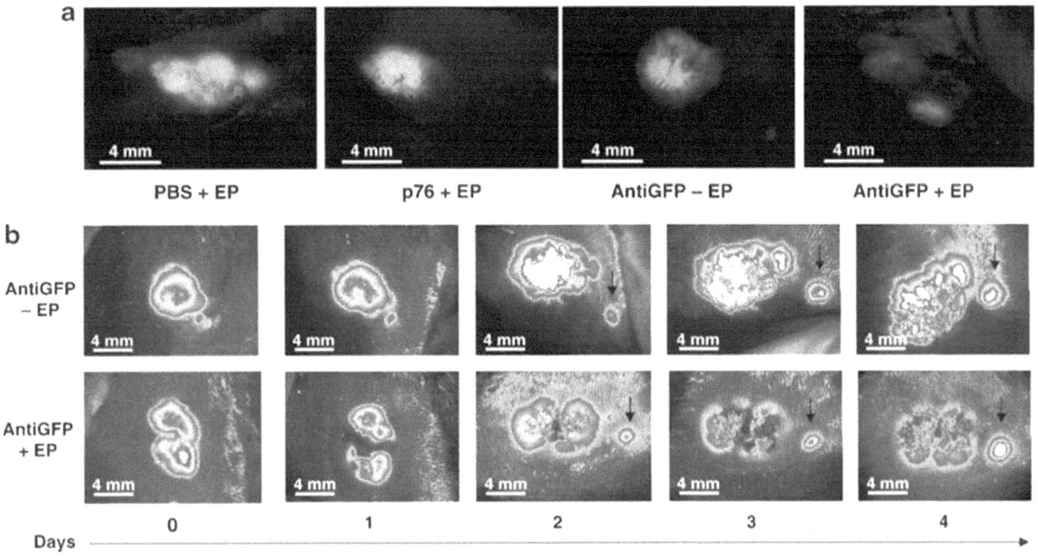

Fig. 5.2. Fluorescence analysis of tumors by in vivo stereomicroscopy in live animals. Ten days after the subcutaneous injection of 1×10^6 B16-F10 GFP cells, tumors having a mean diameter of 5–7 mm were treated. PBS + EP: intratumoral injection of 50 μl saline solution followed by application of EP. p76 + EP: intratumoral injection of 50 μl of saline solution containing 12 μg of p76 siRNA followed by application of EP. Anti-GFP – EP and anti-GFP + EP: intratumoral injection of 50 μl of saline solution containing 12 μg of egfp22 siRNA, followed (anti-GFP + EP) or not (anti-GFP – EP) by application of EP. B16-F10 GFP-derived tumors were clearly detected under the animal's skin upon fluorescence excitation and tumor margin could be easily defined. This enabled measurement of the tumor area and fluorescence intensity over a period of 5 days after treatment. However, owing to the growth characteristics of B16F10 tumors, necrosis occurred 3 days after the beginning of the treatment, as shown by the heterogeneity of the fluorescence pattern in the central region. (**a**) Representative images of EGFP fluorescence in B16-F10 tumors observed by noninvasive imaging in live animals 2 days after the different treatments. (**b**) Pseudocolor 8-bit images showing a representative analysis over time of tumor fluorescence (fluorescence levels: 256; white: most intense; black: least intense) upon treatment with the siRNA egfp22, which allow to see the different nodules that sometimes form the tumor. In the anti-GFP + EP sample a decrease in fluorescence can be observed in the nodules that were injected with the specific siRNA on day 0 but not in the uninjected small nodule emerging close to the main tumor (arrows). (Reprinted by permission from Macmillan Publishers Ltd.: *Gene Therapy*, 14, 752–759 (2007)).

3.3. Imaging of siRNA Using Other Optical Imaging Approaches

Application of quantum dots for imaging siRNA delivery and silencing is so far limited to in vitro studies with the possibility to translate them to in vivo applications. One study utilized cationic liposomes to co-deliver green quantum dots and siRNA targeting the lamin a/c gene (*Lmna*) into murine fibroblasts, followed by flow cytometry. The authors report that in co-transfected cells gene silencing correlated directly with intracellular fluorescence and resulted in about 90% knockdown in highly fluorescent cells *(51)*. Earlier this year the same group reported on a new system for siRNA delivery to tumor cells that consisted of a PEGylated quantum dot core as a scaffold with siRNA (to EGFP) and tumor-homing peptides attached to it. This system afforded intracellular localization of the quantum dot particle followed by endosomal escape mediated by the addition of cationic liposomes. A rather modest knockdown in EGFP fluorescence (about 29%) was achieved using this method *(52)*. In another study quantum dots were encapsulated into chitosan nanoparticles labeled with the HER2 antibody targeting the HER2 receptors to deliver HER2/neu siRNA. Targeted delivery of HER2 siRNA to HER2-overexpressing SKBR3 breast cancer cells was shown to be specific and confirmed by quantum dot microscopy *(53)*. While quantum dots remain an attractive tool for in vitro and animal testing, where fluorescence is the most accessible and common imaging modality, concerns over their cytotoxicity and the limited tissue penetration of light should be taken into account.

The studies on direct siRNA imaging so far have been limited to end-labeling of the molecule with fluorophores to monitor delivery of siRNA into cells. In the study by Chang et al., the authors propose to utilize molecular beacons (MBs) to leverage RNAi towards diagnostic and monitoring applications *(54)*. The MBs proposed here would rely on a fluorescence resonance energy transfer (FRET) fluorophore pair to generate signals upon binding to complementary sequences. The siRNA probe utilized in this study targeted the hTR sequence of telomerase mRNA. Cancer cells express high levels of telomerase mRNA, while normal cells have very limited expression of telomerase. The probe contained a Cy3-Cy5 fluorophore pair conjugated to the 3′ terminus of the antisense and 5′ terminus of the sense strand, respectively, and a highly flexible poly(ethyleneglycol) (PEG) molecule covalently linked to the RNA strands as the loop of an siRNA-based MB. Confocal imaging demonstrated the activation of the siRNA probe in cancer cells in contrast to normal cells. The authors reported that effective gene silencing (about 80%) of telomerase by the siRNA-based probe was achieved *(54)*. This approach could potentially provide promising new ways to evaluate mRNA expression in diseases and serve as useful reporter/gene silencing tools for target validation in basic science and for dual imaging and therapy in clinical research.

3.4. Imaging of siRNA Using SPECT

The only study that we are aware of utilizing SPECT imaging for visualizing cellular delivery and animal distribution of siRNA was published by a group from the University of Massachusetts Medical School *(55)*. siRNA in this study was chemically modified with an *S*-acetyl-*N*-hydroxysuccinimide (NHS) hydrazino nicotinamide (HYNIC) chelator to introduce ^{99m}Tc, a common isotope for SPECT imaging. Cellular accumulation of radiolabeled RNAs (both sense and antisense against the RIa mRNA of the Type I regulatory subunit a of cAMP-dependent protein kinase A) was tested in the ACHN kidney cancer cell line, which was selected for its high RIa mRNA expression. Cellular uptake of ^{99m}Tc-labeled small RNA was time dependent and reached 3×10^5 siRNA molecules after 24 h of incubation. There was no difference in cellular localization between radiolabeled sense and antisense RNA as confirmed by autoradiography.

The delivery of ^{99m}Tc-labeled RNAs in vivo was tested in nude mice bearing ACHN human kidney cancer xenografts. Both sense and antisense ^{99m}Tc-labeled RNAs accumulated in the tumors at the same rate following intravenous injection. The highest tumoral accumulation was observed at 4 h after injection, with radioactivity visualized in almost all tissues. While this study produced encouraging results regarding the absolute tumoral accumulation of radiolabeled RNAs, it is not clear whether the delivery of functional RNA was achieved and whether any therapeutic effect was observed. Nevertheless, this study paved the way for further investigations on the in vivo imaging of siRNA delivery.

3.5. Multimodal Imaging of siRNA Using MRI and Optical Imaging

From multiple recent studies, it has become apparent that the most promising strategies for the cancer therapeutic applicability of RNA interference would involve their complexing/conjugation to biocompatible carriers in order to improve their accumulation in target tissues and their bioavailability. A study from a group at Harvard was the first one to utilize the multimodal approach to image siRNA delivery and silencing in tumors *(24)*. To facilitate siRNA delivery to tumors and enable in vivo imaging of the delivery, the authors employed magnetic nanoparticles (MN), which had an optimal half-life of 8–12 h *(56)* and allowed visualization by magnetic resonance imaging. These nanoparticles were also labeled with Cy5.5 optical dye to enable correlative near-infrared optical imaging (NIRF). These imaging moieties were conjugated to a siRNA duplex targeting a gene of interest, as well as to myristoylated polyarginine peptides that facilitated translocation of the construct across the cell membrane into the cytoplasm. First, a model system involving animals bilaterally injected with 9L rat gliosarcoma tumors expressing GFP or RFP proteins was used. In vivo MR imaging of tumor-bearing animals followed by NIRF imaging of the same animals demonstrated that intravenous injection of MN-NIRF-siGFP resulted in successful tumoral delivery of the construct. In vivo optical imaging in the GFP channel confirmed that efficient silencing was achieved in 9L-GFP tumors (but not in 9L-RFP tumors) (**Fig. 5.3**). This was further confirmed by quantitative RT-PCR of *gfp* mRNA transcript levels.

In addition to the GFP model system, we have explored a therapeutic probe targeting the antiapoptotic gene *Birc5*, which encodes survivin. Survivin is a member of the inhibitor of apoptosis protein (IAP) family, which is highly expressed in most cancers and associated with chemotherapy resistance, increased tumor recurrence, and shorter patient survival, making antisurvivin therapy an attractive cancer treatment strategy *(57, see*

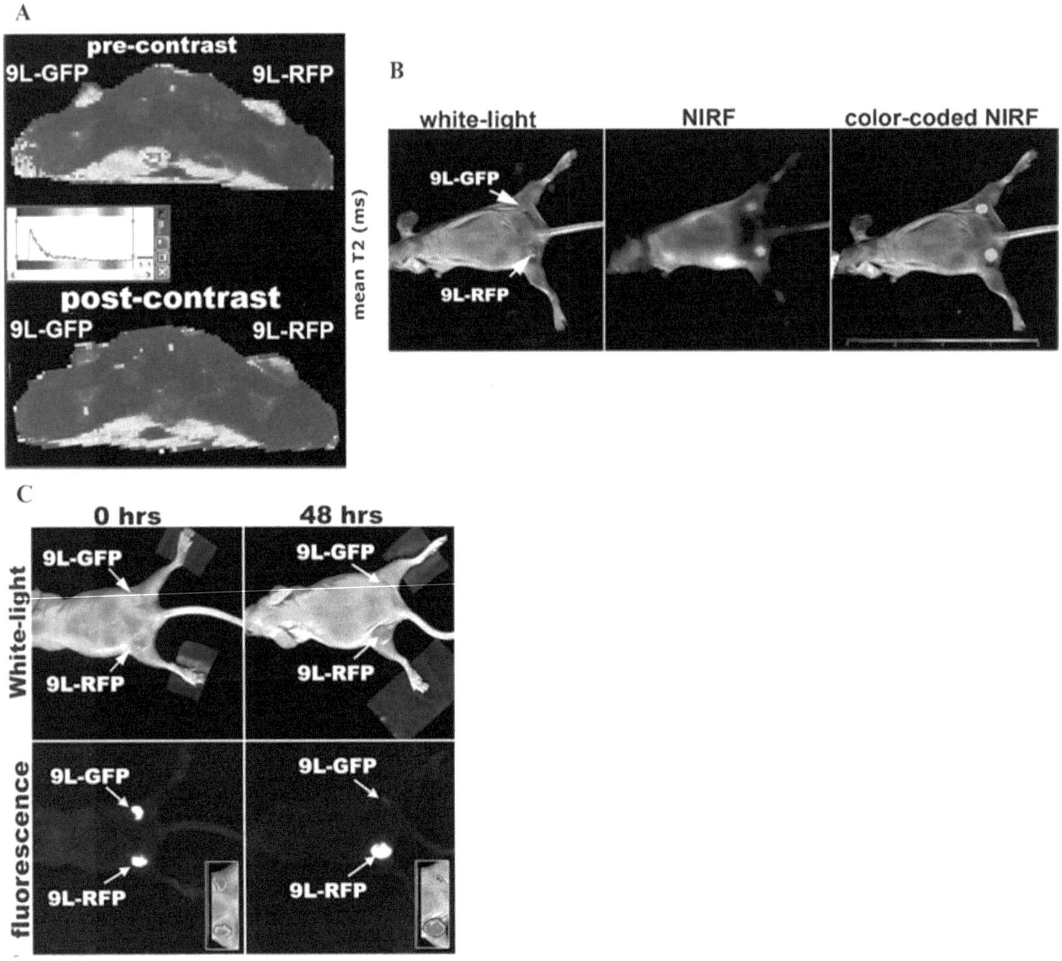

Fig. 5.3. In vivo imaging of siRNA (against GFP) delivery and silencing in tumors. (**a**) In vivo MRI was performed on mice bearing bilateral 9L-GFP and 9L-RFP tumors before and 24 h after MN-NIRF-siGFP administration. Following injection of the probe, there was a significant drop in T2 relaxivity associated with the tumors ($p = 0.0012$ for 9L-GFP, $p = 0.0049$ for 9L-RFP). Note that T2 relaxation times of muscle tissue remained unchanged. (**b**) In vivo NIRF optical imaging of the same animals as in (**a**) produced a high-intensity NIRF signal associated with the tumors confirmed the delivery of the MN-NIRF-siGFP probe to these tissues. (**c**) In vivo optical imaging of animals bearing bilateral 9L-GFP and 9L-RFP tumors 48 h after i.v. probe injection showing a dramatic decrease in 9L-GFP-associated fluorescence ($p = 0.0083$) and no change in 9L-RFP fluorescence. To generate GFP/RFP reconstructions, GFP and RFP images were acquired separately and then merged. (Reprinted by permission from Macmillan Publishers Ltd.: *Nature Medicine*, 13, 372–377 (2007)).

Chap. 14). MN-NIRF-siSurvivin nanoparticles injected intravenously in animals bearing subcutaneous human colorectal carcinoma xenografts accumulated in the tumors as evidenced by both MRI and NIRF imaging (**Fig. 5.4**). Time course injections in tumor-bearing animals resulted in significant silencing of the *Birc5* gene and increased apoptosis compared to animals injected with nanoparticles devoid of siRNA or nanoparticles bearing mismatched control siRNA. Further development of this approach

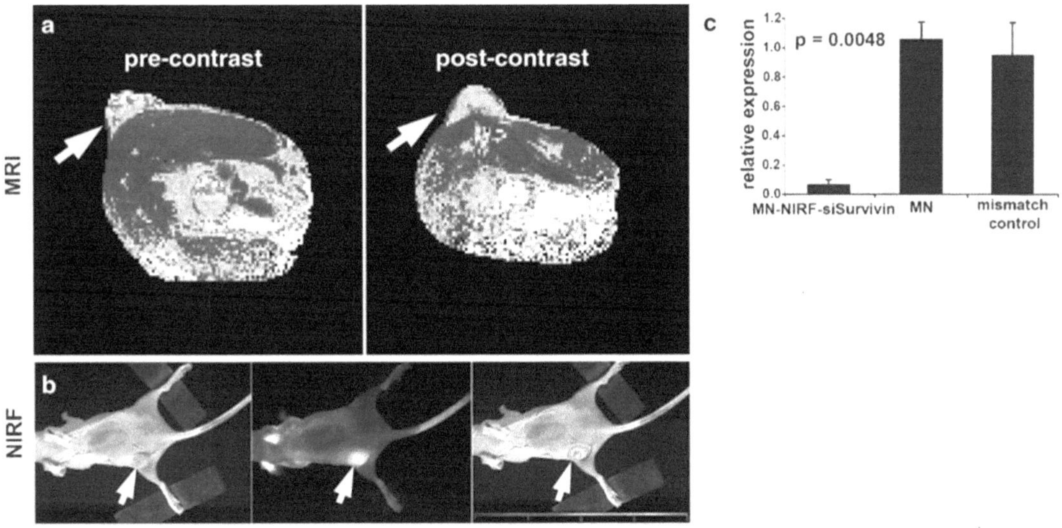

Fig. 5.4. Application of MN-NIRF-siSurvivin in a therapeutic tumor model. (**a**) In vivo MRI of mice bearing subcutaneous LS174T human colorectal adenocarcinoma (arrows) showed a significant drop in T2 relaxivity on postcontrast images of the tumors ($p = 0.003$) indicating probe delivery. (**b**) A high-intensity NIRF signal on in vivo optical images associated with the tumor (arrows) following injection of MN-NIRF-siSurvivin confirmed the delivery of the probe to this tissue (*left*, white light; *middle*, NIRF; *right*, color-coded overlay). (**c**) Quantitative RT-PCR analysis of survivin expression in LS174T tumors after injection with either MN-NIRF-siSurvivin, a mismatch control, or the parental MN nanoparticle. Survivin mRNA levels in tumors from MN-NIRF-siSurvivin treated animals were reduced by 97 ± 2%, compared to MN controls ($p < 0.01$) and 83 ± 2%, compared to mismatch controls ($p < 0.01$, data are representative of three separate experiments). (Reprinted by permission from Macmillan Publishers Ltd.: *Nature Medicine*, 13, 372–377 (2007)).

would include tumor-specific targeting and combination with conventional chemotherapeutics.

Overall, the in vivo imaging of siRNA delivery and silencing is a fast developing field (*see* Chap. 4). Clearly, to develop siRNAs as therapeutic agents, monitoring of the success of in vivo delivery is critical. Noninvasive molecular imaging techniques have great potential for this purpose and can potentially provide the most direct, noninvasive, and longitudinal measurements of the in vivo delivery of siRNAs as well as their direct effect on gene expression and indirect influence on biological parameters associated with silencing.

References

1. Fire, A., Xu, S., Montgomery, M., Kostas, S., Driver, S.E., and Mello, C. (1998) Potent and specific genetic interference by double-stranded RNA in *Caenorhabditis elegans*, *Nature* 391, 806–811.
2. Massoud, T. and Gambhir, S. (2003) Molecular imaging in living subjects: seeing fundamental biological processes in a new light, *Genes Dev.* 17, 545–580.

3. Allport, J. and Weissleder, R. (2001) In vivo imaging of gene and cell therapies, *Exp. Hematol.* 29, 1237–1246.
4. Gambhir, S. (2002) Molecular imaging of cancer with positron emission tomography, *Nat. Rev. Cancer* 2, 683–693.
5. Rosenthal, M., Cullom, J., Hawkins, W., Moore, S., Tsui, B., and Yester, M. (1995) Quantitative SPECT imaging: A review and recommendations by the Focus Committee of the Society of Nuclear Medicine Computer and Instrumentation Council, *J. Nucl. Med.* 36, 1489–1513.
6. Weissleder, R. and Mahmood, U. (2001) Molecular Imaging, *Radiology* 219, 316–333.
7. Contag, C. and Bachmann, M. (2002) Advances in in vivo bioluminescence imaging of gene expression, *Annu. Rev. Biomed. Eng.* 4, 235–260.
8. Lewis, J., Achilefu, S., Garbow, J., Laforest, R., and Welch, M. (2002) Small animal imaging: current technology and perspectives for oncological imaging, *Eur. J. Cancer* 38, 2173–2188.
9. Rudin, M. and Weissleder, R. (2003) Molecular imaging in drug discovery and development, *Nat. Rev.* 2, 123–131.
10. Ntziachristos, V., Tung, C., Bremer, C., and Weissleder, R. (2002) Fluorescence molecular tomography resolves protease activity in vivo, *Nat. Med.* 8, 757–760.
11. Michalet, X., Pinaud, F., Bentolila, L., Tsay, J., Doose, S., Li, J., Sundaresan, G., Wu, A., Gambhir, S., and Weiss, S. (2005) Quantum dots for live cells, in vivo imaging, and diagnostics, *Science* 307, 538–544.
12. Hardy, J., Edinger, M., Bachmann, M., Negrin, R., Fathman, C., and Contag, C. (2001) Bioluminescence imaging of lymphocyte trafficking in vivo, *Exp. Hematol.* 29, 1353–1360.
13. Bremer, C., Tung, C., Bogdanov, A. Jr., and Weissleder, R. (2002) Imaging of differential protease expression in breast cancers for detection of aggressive tumor phenotypes, *Radiology* 222, 814–818.
14. Contag, C., Spilman, S., Contag, P., Oshiro, M., Eames, B., Dennery, P., Stevenson, D., and Benaron, D. (1997) Visualizing gene expression in living mammals using a bioluminescent reporter, *Photochem. Photobiol.* 66, 523–531.
15. Caravan, P. (2007) in *Molecular and Cellular MR imaging*, Modo, M. and Bulte, J., Eds., pp. 37–59, CRC Press, Taylor & Francis Group, Boca Raton, FL.
16. Josephson, L., Tung, C., Moore, A., and Weissleder, R. (1999) High-efficiency intracellular magnetic labeling with novel superparamagnetic-Tat peptide conjugates, *Bioconjugate. Chem.* 10, 186–191.
17. Moore, A., Sun, P., Cory, D., Högemann, D., Weissleder, R., and Lipes, M. (2002) MR Imaging of insulitis in autoimmune diabetes, *Magn. Reson. Med.* 47, 751–758.
18. Weissleder, R., Simonova, M., Bogdanova, A., Bredow, S., Enochs, W., and Bogdanov, A. (1997) MR imaging and scintigraphy of gene expression through melanin induction, *Radiology* 204, 425–429.
19. Stegman, L., Rehemtulla, A., Beattie, B., Kievit, E., Lawrence, T., Blasberg, R., Tjuvajev, J., and Ross, B. (1999) Noninvasive quantitation of cytosine deaminase transgene expression in human tumor xenografts with in vivo magnetic resonance spectroscopy, *Proc. Natl. Acad. Sci. USA* 96, 9821–9826.
20. Weissleder, R., Moore, A., Mahmood, U., Bhorade, R., Benveniste, H., Chiocca, E., and Basilion, J. (2000) In vivo magnetic resonance imaging of transgene expression, *Nat. Med.* 6, 351–354.
21. Moore, A., Josephson, L., Bhorade, R., Basilion, J., and Weissleder, R. (2001) Human transferrin receptor gene as a marker gene for MR imaging, *Radiology* 221, 751–758.
22. Artemov, D., Mori, N., Okollie, B., and Bhujwalla, Z. (2003) MR molecular imaging of the HER-2/neu receptor in breast cancer cells using targeted iron oxide nanoparticles, *Magn. Reson. Med.* 49, 403–408.
23. Moore, A., Medarova, Z., Potthast, A., and Dai, G. (2004) In vivo targeting of underglycosylated MUC-1 tumor antigen using a multi-modal imaging probe, *Cancer Res.* 64, 1821–1827.
24. Medarova, Z., Pham, W., Kim, Y., Dai, G., and Moore, A. (2006) In vivo imaging of tumor response to therapy using a dual-modality imaging strategy, *Int. J. Cancer* 118, 2796–2802.
25. Dendy, P. and Heaton, B. (1999) in *Physics for Diagnostic Radiology*, pp. 249–278, Institute of Physics, Bristol, UK.
26. Turnbull, D. and Foster, F. (2002) In vivo ultrasound biomicroscopy in developmental biology, *Trends Biotechnol.* 20, S29–S33.
27. Lanza, G. and Wickline, S. (2001) Targeted ultrasonic contrast agents for molecular imaging and therapy, *Prog. Cardiovasc. Dis.* 44, 13–31.

28. McCaffrey, A.P., Meuse, L., Pham, T.T., Conklin, D.S., Hannon, G.J., and Kay, M.A. (2002) RNA interference in adult mice, *Nature* 418, 38–39.
29. Lewis, D.L., Hagstrom, J.E., Loomis, A.G., Wolff, J.A., and Herweijer, H. (2002) Efficient delivery of siRNA for inhibition of gene expression in postnatal mice, *Nat. Genet.* 32, 107–108.
30. Kobayashi, N., Matsui, Y., Kawase, A., Hirata, K., Miyagishi, M., Taira, K., Nishikawa, M., and Takakura, Y. (2004) Vector-based in vivo RNA interference: dose- and time-dependent suppression of transgene expression, *J. Pharmacol. Exp. Ther.* 308, 688–693.
31. Hassani, Z., Lemkine, G.F., Erbacher, P., Palmier, K., Alfama, G., Giovannangeli, C., Behr, J.P., and Demeneix, B.A. (2005) Lipid-mediated siRNA delivery down-regulates exogenous gene expression in the mouse brain at picomolar levels, *J. Gene Med.* 7, 198–207.
32. Minakuchi, Y., Takeshita, F., Kosaka, N., Sasaki, H., Yamamoto, Y., Kouno, M., Honma, K., Nagahara, S., Hanai, K., Sano, A., Kato, T., Terada, M., and Ochiya, T. (2004) Atelocollagen-mediated synthetic small interfering RNA delivery for effective gene silencing in vitro and in vivo, *Nucleic Acids Res.* 32, e109.
33. Takeshita, F., Minakuchi, Y., Nagahara, S., Honma, K., Sasaki, H., Hirai, K., Teratani, T., Namatame, N., Yamamoto, Y., Hanai, K., Kato, T., Sano, A., and Ochiya, T. (2005) Efficient delivery of small interfering RNA to bone-metastatic tumors by using atelocollagen in vivo, *Proc. Natl. Acad. Sci. USA* 102, 12177–12182.
34. Takabatake, Y., Isaka, Y., Mizui, M., Kawachi, H., Shimizu, F., Ito, T., Hori, M., and Imai, E. (2005) Exploring RNA interference as a therapeutic strategy for renal disease, *Gene Ther.* 12, 965–973.
35. Zhang, Y., Boado, R.J., and Pardridge, W.M. (2003) In vivo knockdown of gene expression in brain cancer with intravenous RNAi in adult rats, *J. Gene Med.* 5, 1039–1045.
36. Pichler, A., Zelcer, N., Prior, J.L., Kuil, A.J., and Piwnica-Worms, D. (2005) In vivo RNA interference-mediated ablation of MDR1 P-glycoprotein, *Clin. Cancer Res.* 11, 4487–4494.
37. Hu-Lieskovan, S., Heidel, J.D., Bartlett, D.W., Davis, M.E., and Triche, T.J. (2005) Sequence-specific knockdown of EWS-FLI1 by targeted, nonviral delivery of small interfering RNA inhibits tumor growth in a murine model of metastatic Ewing's sarcoma, *Cancer Res.* 65, 8984–8992.
38. Bartlett, D.W. and Davis, M.E. (2006) Insights into the kinetics of siRNA-mediated gene silencing from live-cell and live-animal bioluminescent imaging, *Nucleic Acids Res.* 34, 322–333.
39. Takahashi, Y., Nishikawa, M., Kobayashi, N., and Takakura, Y. (2005) Gene silencing in primary and metastatic tumors by small interfering RNA delivery in mice: quantitative analysis using melanoma cells expressing firefly and sea pansy luciferases, *J. Control Release* 105, 332–343.
40. Sorensen, D.R., Leirdal, M., and Sioud, M. (2003) Gene silencing by systemic delivery of synthetic siRNAs in adult mice, *J. Mol. Biol.* 327, 761–766.
41. Xia, H., Mao, Q., Paulson, H.L., and Davidson, B.L. (2002) siRNA-mediated gene silencing in vitro and in vivo, *Nat. Biotechnol.* 20, 1006–1010.
42. Van den Haute, C., Eggermont, K., Nuttin, B., Debyser, Z., and Baekelandt, V. (2003) Lentiviral vector-mediated delivery of short hairpin RNA results in persistent knockdown of gene expression in mouse brain, *Hum. Gene Ther.* 14, 1799–1807.
43. Rubinson, D.A., Dillon, C.P., Kwiatkowski, A.V., Sievers, C., Yang, L., Kopinja, J., Rooney, D.L., Ihrig, M.M., McManus, M.T., Gertler, F.B., Scott, M.L., and Van Parijs, L. (2003) A lentivirus-based system to functionally silence genes in primary mammalian cells, stem cells and transgenic mice by RNA interference, *Nat. Genet.* 33, 401–406.
44. Golzio, M., Mazzolini, L., Ledoux, A., Paganin, A., Izard, M., Hellaudais, L., Bieth, A., Pillaire, M., Cazaux, C., Hoffmann, J., Couderc, B., and Teissie, J. (2007) In vivo gene silencing in solid tumors by targeted electrically mediated siRNA delivery, *Gene. Ther.* 14, 752–759.
45. Sioud, M. and Sorensen, D.R. (2003) Cationic liposome-mediated delivery of siRNAs in adult mice, *Biochem. Biophys. Res. Commun.* 312, 1220–1225.
46. Schiffelers, R.M., Ansari, A., Xu, J., Zhou, Q., Tang, Q., Storm, G., Molema, G., Lu, P.Y., Scaria, P.V., and Woodle, M.C. (2004) Cancer siRNA therapy by tumor selective delivery with ligand-targeted sterically stabilized nanoparticle, *Nucleic Acids Res.* 32, e149.
47. Landen, C.N. Jr., Chavez-Reyes, A., Bucana, C., Schmandt, R., Deavers, M.T., Lopez-Berestein, G., and Sood, A.K.

(2005) Therapeutic EphA2 gene targeting in vivo using neutral liposomal small interfering RNA delivery, *Cancer Res.* 65, 6910–6918.

48. Santel, A., Aleku, M., Keil, O., Endruschat, J., Esche, V., Fisch, G., Dames, S., Loffler, K., Fechtner, M., Arnold, W., Giese, K., Klippel, A., and Kaufmann, J. (2006) A novel siRNA-lipoplex technology for RNA interference in the mouse vascular endothelium, *Gene Ther.* 13, 1222–1234.
49. Santel, A., Aleku, M., Keil, O., Endruschat, J., Esche, V., Durieux, B., Loffler, K., Fechtner, M., Rohl, T., Fisch, G., Dames, S., Arnold, W., Giese, K., Klippel, A., and Kaufmann, J. (2006) RNA interference in the mouse vascular endothelium by systemic administration of siRNA-lipoplexes for cancer therapy, *Gene Ther.* 13, 1360–1370.
50. Morrissey, D.V., Lockridge, J.A., Shaw, L., Blanchard, K., Jensen, K., Breen, W., Hartsough, K., Machemer, L., Radka, S., Jadhav, V., Vaish, N., Zinnen, S., Vargeese, C., Bowman, K., Shaffer, C.S., Jeffs, L.B., Judge, A., MacLachlan, I., and Polisky, B. (2005) Potent and persistent in vivo anti-HBV activity of chemically modified siRNAs, *Nat. Biotechnol.* 23, 1002–1007.
51. Chen, A., Derfus, A., Khetani, S., and Bhatia, S. (2005) Quantum dots to monitor RNAi delivery and improve gene silencing, *Nucleic Acids Res.* 22, E190–E97.
52. Derfus, A., Chen, A., Min, D., Ruoclahti, E., and Bhatia, S. (2007) Targeted quantum dot conjugates for siRNA delivery, *Bioconjug. Chem.* Web release 140July-2007, DOI: 10.1021/bc060367e.
53. Tan, W., Jiang, S., and Zhang, Y. (2007) Quantum-dot based nanoparticles for targeted silencing of HER2/neu gene via RNA interference, *Biomaterials* 28, 1565–1571.
54. Chang, E., Zhu, M., and Drezek, R. (2007) Novel siRNA-based molecular beacons for dual imaging and therapy, *Biotechnol. J.* 2, 422–425.
55. Liu, N., Ding, H., Vanderheyden, J., Zhu, Z., and Zhang, Y. (2007) Radiolabeling small RNA with technetium-99m for visualizing cellular delivery and mouse biodistribution, *Nucl. Med. Biol.* 34, 399–404.
56. Medarova, Z., Pham, W., Farrar, C., Petkova, V., and Moore, A. (2007) In vivo imaging of siRNA delivery and silencing in tumors, *Nat. Med.* 13, 372–377.
57. Wunderbaldinger, P., Josephson, L., and Weissleder, R. (2002) Crosslinked iron oxides (CLIO): a new platform for the development of targeted MR contrast agents, *Acad. Radiol.* 9(Suppl. 2), S304–306.
58. Fukuda, S. and Pelus, L.M. (2006) Survivin, a cancer target with an emerging role in normal adult tissues, *Mol. Cancer Ther.* 5, 1087–1098.

Chapter 6

Recent Advances in Magnetofection and Its Potential to Deliver siRNAs In Vitro

Olga Mykhaylyk, Olivier Zelphati, Edelburga Hammerschmid, Martina Anton, Joseph Rosenecker, and Christian Plank

Abstract

This chapter describes how to design and conduct experiments to deliver siRNA to adherent mammalian cells in vitro by magnetic force–assisted transfection using self-assembled complexes of small interfering RNA (siRNA) and cationic lipids or polymers that are associated with magnetic nanoparticles. These magnetic complexes are targeted to the cell surface by the application of a magnetic gradient field. In this chapter, first we describe the synthesis of magnetic nanoparticles for magnetofection and the association of siRNA with the magnetic components of the transfection complex. Second, a simple protocol is described in order to evaluate magnetic responsiveness of the magnetic siRNA transfection complexes and estimate the complex loading with magnetic nanoparticles. Third, protocols are provided for the preparation of magnetic lipoplexes and polyplexes of siRNA, magnetofection, downregulation of gene expression, and the determination of cell viability. The addition of INF-7 peptide, a fusogenic peptide, to the magnetic transfection triplexes improved gene silencing in HeLa cells. The described protocols are also valuable for screening vector compositions and novel magnetic nanoparticle preparations to optimize siRNA transfection by magnetofection in every cell type.

Key words: siRNA delivery in vitro, magnetic nanoparticles, magnetic transfection vectors.

1. Introduction

Magnetofection can be defined as a method for nucleic acid delivery under the influence of a magnetic field acting on nucleic acid vectors that are associated with magnetic (nano)particles. First reports on associating a vector (viral and nonviral) with magnetic particles date back to the year 2000 *(1,2)*. Several research

M. Sioud (ed.), *Methods in Molecular Biology, siRNA and miRNA Gene Silencing, vol. 487*

DOI: 10.1007/978-1-60327-547-7_6

groups have developed such methods independently (see, for example refs. *3–6)*. In the meantime, the term "magnetofection" is widely used in the scientific literature. We have mostly used electrostatic interactions to associate vectors with magnetic particles *(7)*; other groups have used biotin–streptavidin or antigen–antibody interactions *(5,6)*. We have demonstrated the suitability of magnetofection for magnetically localized gene delivery in vivo *(8)*. However, major improvements are still required to make the method efficient enough to be widely used in in vivo applications. In contrast, magnetofection is well established and widely used for in vitro applications and has been shown to potentiate viral as well as nonviral nucleic acid delivery *(8,9)*. Vectors associated with magnetic nanoparticles are added to cell culture supernatants. Cell culture plates are subsequently placed on magnetic plates which consist of an array of suitably positioned permanent magnets. Thus, the diffusion limitation to delivery is overcome, transfection/transduction is synchronized and greatly accelerated, and the vector dose requirement for efficient transfection/transduction is considerably reduced. These features together constitute a substantial improvement of transfection/transduction efficiency. The mechanism of magnetofection is probably the same as for standard transfection/transduction concerning vector uptake into cells *(10)*. Magnetic nanoparticles are co-internalized with vectors into cells and are biodegradable over long time periods. Importantly, magnetofection is applicable for both viral and nonviral vectors and among the latter for "large" nucleic acids (e.g., plasmids) or small constructs (e.g., synthetic siRNA and antisense oligonucleotides *(11–13)*) in a naked form or packaged as lipoplexes and polyplexes *(7)*.

The potential of magnetofection to efficiently deliver siRNA in vitro has been emphasized by several recent scientific publications. Gene silencing in primary cells are notoriously challenging due to the limited efficiency of most transfection reagents. Magnetofection has been shown to be a very effective way of transfecting siRNA in human primary endothelial cells derived from umbilical vein and from human cord blood *(14,15)*. It contributed to demonstrating the critical role of a transcription factor in angiogenesis *(15)* and of a selective estrogen receptor modulator in atherosclerosis *(14)*. Primary human gastric myofibroblasts are also difficult to transfect. Mc Caig and colleagues have successfully used magnetofection to deliver siRNA in these primary myofibroblast *(15)*. Another striking example of magnetofection potential for siRNA delivery in primary cells has been reported by Uchida and colleagues. Indeed, successful siRNA delivery was achieved in primary rat embryonic DRG neurons *(16)*. Efficiency of siRNA delivery mediated by magnetofection has also been reported for cell lines such as COS7 (monkey kidney) and 3Y1 (rat fibroblast), which allowed to demonstrate the novel role of phospho-b-catenin in microtubule growth *(17)*. Other successful applications of siRNA delivery with magnetofection have been reported in Vero E6 cells (monkey kidney) *(18)* and in the human microvascular endothelial cell line-1 (HMEC-1) *(19)*. Magnetofection has also been reported to be effective for siRNA transfection in suspension cells such as MOLT-4 and Jurkat (human T cell leukemia), which permitted to show the implication of RCAS1 (a receptor-binding cancer antigen) in T cell apoptosis induced by HIV infection *(20)*. The review by Bonetta et al. on "evaluating gene delivery methods" highlights the potential of magnetofection as a new tool to improve and target a broader range of cells and applications *(21)*. In the same way, the benefits of using magnetofection, and particularly SilenceMag (magnetofection-based reagent optimized for siRNA delivery), to concentrate and promote efficient siRNA transfection were assessed *(22)*.

All that is required to practice magnetofection are suitable magnetic nano- or microparticles and appropriate magnetic devices. In the meantime, these tools along with standardized application protocols for various vector types and cell culture formats are commercially available (OZ Biosciences, Marseille, France, http://www.ozbiosciences.com; Chemicell, Berlin, Germany, http://www.chemicell.com). The commercially available magnet array for magnetofection produces high-gradient magnetic fields (70–250 mT and a field gradient of 50–130 T m^{-1}) in the vicinity of the cells and sediments the full-applied vector dose on the cells to be transfected within minutes. The development of new magnetic nanoparticles is expected to lead to further improvements of the technique *(12)*.

Therefore, we focus here on a 96-well format screening procedure for magnetic nanoparticles to be used in nonviral magnetofection. We provide the protocols that comprise every step from magnetic nanoparticle synthesis and characterization to their use in siRNA magnetofection and instructions for data evaluation. We illustrate these protocols with magnetofection results obtained in adherent cancer cell lines stably transfected with GFP- and luciferase reporter genes.

We describe magnetite nanoparticles, which differ in their coating material and which are efficient in siRNA delivery to adherent cells in vitro by magnetofection and can be associated with a nucleic acid, or with a nucleic acid and an enhancer or vector (viral or nonviral lipoplex or polyplex *(23)*, which are stable enough to be stored over extended periods and are biocompatible enough for application in living cells). The compounds used here as components of the shell for the particles are known to be either useful in particle stabilization and/or in gene or drug delivery, and were found in our own experiments to be among the best for nucleic acid delivery *(24)*.

Subsequently, we describe how the binding of siRNA to the magnetic nanoparticles, either alone or in combination with a third agent that enhances transfection (known as an enhancer), can be characterized using radioactively labeled siRNA prepared according to the modified Terebesi procedure *(25)*. This can be used to determine suitable ratios and mixing orders of magnetic nanoparticles, nucleic acids, and third components to choose formulations that are potentially useful for siRNA magnetofection. This protocol uses as enhancers either the commercially available transfection reagent Metafectene (4 μl μg^{-1} DNA) or the branched polyethylene imine (25 kDa) ($PEI\text{-}25_{Br}$, nitrogen-to-DNA phosphate ratio N/P = 10). Other enhancers *(26)* known to be efficient in the transfection of a particular cell line can, however, also be combined with magnetic nanoparticles to construct magnetic vectors.

In principle, any nucleic acid can be delivered using magnetofection. For the screening purposes presented here, it is most useful to use reporter genes such as GFP and luciferase reporters. Both reporter gene systems allow rapid and sensitive result evaluation in cell lysates and even in living cells. In general, the luciferase reporter gene assay provides higher sensitivity and accuracy than the enhanced GFP (eGFP) assay. The advantage of using the GFP reporter gene is that the percentage of transfected cells can be easily determined.

Vectors are prepared in serum- and supplement-free medium and transferred to the cells in triplicates in a volume of 50 μl per well. We recommend performing serial dilutions of a given vector composition such that the highest siRNA dose transferred to the cell culture plate is 200 ng per well. Accordingly, the highest vector dose has to be prepared at a final siRNA concentration of 4 μg ml^{-1} in the complex. According to our results, optimum nanomaterial-to-siRNA ratio for magnetic vectors comprising enhancers described in this protocol is between 0.5 and 1 in terms of iron-to-siRNA wt/wt ratio. This has been tested in HeLa cells, H441 cells, and M1 cells. For magnetic duplexes comprising PEI-Mag2 nanomaterial and DNA, the optimal ratio is 1:1.

Furthermore, we describe a simple method to evaluate the magnetic responsiveness of the magnetic siRNA transfection complexes and to estimate the complex loading with core-shell magnetic nanoparticles. This can be used to determine whether the magnetic properties of the vectors are sufficient to sediment the vector at the cell surface in the applied magnetic fields. Data on magnetic responsiveness of the complexes can be also used to estimate the potential to target the complexes magnetically when applied in vivo. Cell culture and transfection procedures to adherent cell lines are described. Also, preparation of magnetic vectors and the magnetofection procedure are described.

siRNA transfection is known to provoke dose-dependent toxicity. Therefore, presenting results in terms of the absolute units of reporter gene expression normalized per weight of total protein in the examined cell lysate makes it possible to distinguish between gene downregulation and toxicity effects and is, therefore, especially important in siRNA transfection experiments. In fact, it has been shown that relatively low toxicity of the magnetic siRNA is one of the essential advantages of magnetofection *(10)*. In this respect, the quantification of complex internalization, evaluation of siRNA transfection efficiency in living cells and in cell lysates, and the MTT-based cell viability test are described.

INF-7 endosomolytic peptide derived from the N-terminal sequence of influenza virus hemagglutinin was shown to improve gene delivery with polyplexes *(27–30)*. Recently the potential of the INF-7 peptide to improve the silencing efficiency of siRNA targeting the epidermal growth factor receptor and the K-ras oncogenes in complexes with Lipofectamine2000 was demonstrated in human epidermis carcinoma cells *(31)*. We show an example INF7-mediated improvement of siRNA magnetofection in HeLa cells stably transduced with the GFP gene.

2. Materials

2.1. Synthesis of Magnetic Nanoparticles Suitable as Components of Magnetofection Complexes

1. Iron(II) chloride tetrahydrate (Sigma-Aldrich).
2. Iron(III) chloride hexahydrate (Sigma-Aldrich).
3. Argon 5.0 (Sauerstoffwerk Friedrichshafen GmbH).
4. 10% Hydroxylamine hydrochloride solution in water (*see* **Note 1**).
5. PEI-Mag2 precipitation/coating solution: 5 g polyethylenimine 25 kDa, branched (PEI-25_{Br}; Sigma-Aldrich, cat. no. 40,872-7) plus 25 ml 28–30% ammonium hydroxide solution (Sigma-Aldrich, cat. no. 320145) plus 2.5 ml lithium 3-[2-(perfluoroalkyl) ethylthio]propionate (Zonyl FSA; Sigma-Aldrich) filled up with water to a total volume 100 ml, degassed with argon/helium.
6. PL-Mag1 precipitation/coating solution 1: 4 g Pluronic F-127 (Sigma-Aldrich) filled up with water to a total volume 50 ml, degassed with argon/helium. PL-Mag1 precipitation/coating solution 2: 30 ml 28–30% ammonium hydroxide solution plus 15 ml ammonium bis[2-(perfluoroalkyl) ethyl] phosphate solution (Zonyl FSE; Sigma-Aldrich, cat. no. 421391) filled up with water to a total volume 50 ml, degassed with argon/helium.

7. PalD1-Mag1 precipitation/coating solution: 4 g palmitoyl dextran PalD1 (*see* **Note 2**) plus 30 ml 28–30% ammonium hydroxide solution filled up with water to a total volume 100 ml, degassed with argon/helium.

2.2. Determination of Magnetic Nanoparticle Concentration in Terms of Dry Weight and Iron Content

1. Ammonium acetate buffer for iron determination: Dissolve 25 g ammonium acetate (Sigma) in 10 ml water, add 70 ml glacial acetic acid, and adjust volume to 100 ml with water.
2. 10% Hydroxylamine hydrochloride (Sigma-Aldrich) in water.
3. 0.1% Phenanthroline solution: Dissolve 100 mg 1,10-phenanthroline monohydrate (Sigma, cat. no. 77500) in 100 ml water; add 2 drops concentrated hydrochloric acid (Fluka, cat. no. 84415). If necessary, warm up to obtain a clear solution.
4. Iron stock solution: Dissolve 392.8 mg ammonium iron(II) sulfate hexahydrate (Sigma, cat. no. F3754) in a mixture of 2 ml concentrated sulfuric acid and 10 ml water; add 0.05 N $KMnO_4$ dropwise until a pink color persists and adjust the volume to 100 ml with water.
5. Standard iron solution (make fresh as required): Dilute iron stock solution 25 to 1 with water just before calibration measurements.
6. 0.05 N $KMnO_4$ solution: Dissolve 0.790 g $KMnO_4$ in 100 ml water.

2.3. Radiolabeling (Iodination) of siRNA

1. siRNA solution: 5 nmol (78.3 µg) GFP-22 siRNA (Qiagen) is reconstituted with 39.1 µl of the siRNA suspension buffer at 2 µg siRNA/µl and stored in aliquots at –20°C.
2. Sodium ^{125}iodide in NaOH: Activity 1 mCi in 10 µl (Amersham Biosciences, cat. no. 1MS30). Caution: Radioactive material! Store at ambient temperature 15–20°C. Retains its iodination efficiency over 2 months storage.
3. 250 µM potassium iodide in water. Prepare on the day of DNA labeling from 25 mM potassium iodide.
4. 1 M sodium hydroxide in water.
5. 30 mM Thallium trichloride tetrahydrate (Sigma-Aldrich) solution in water. To obtain a clear solution, heat the tube to 70°C using a water bath. The solution is stable and can be stored at least for a year.
6. 1 M sodium sulfite in water. Prepare on the day of siRNA labeling. 1 M ammonium acetate buffer, pH 7.
7. Disposable Sephadex G25 PD-10 desalting columns (GE Healthcare).

2.4. Testing the siRNA Binding Capacity of the Magnetic Nanoparticles

1. Suspension of the magnetic nanoparticles: Dilute stock suspension of magnetic nanoparticles in water at a concentration of 288 μg iron ml^{-1}. Prepare just before the experiment.
2. Metafectene (Biontex Laboratories).
3. Polyethylenimine 25 kDa, branched ($PEI\text{-}25_{Br}$; Sigma-Aldrich).
4. ^{125}I-labeled siRNA solution: 4.8 μg ml^{-1} total siRNA (GFP-22 siRNA, Qiagen) comprising 2×10^5 CPM ml^{-1} ^{125}I-labeled siRNA (from Sect. 3.3) in RPMI medium without supplements, or whatever solvent of interest.

2.5. Cell Culture and Plating of Adherent Cells for Transfection

1. NCI-H441 human pulmonary epithelial (H441) cells derived from papillary carcinoma of the lungs (ATCC).
2. NCI-H441 cells stably expressing eGFP (H441-GFP cells).
3. NCI-H441 cells stably expressing luciferase (H441-Luci cells).
4. HeLa human cervical epithelial adenocarcinoma cells (ATCC).
5. HeLa cells stably expressing eGFP (HeLa-GFP cells).
6. HeLa cells stably expressing luciferase (HeLa-Luci cells).
7. NCI-3T3 mouse fibroblasts stably expressing GFP (3T3-GFP cells).
8. H441 culture medium: Modified RPMI 1640 medium with 2 mM L-glutamine, 10 mM HEPES, 1 mM sodium pyruvate, 4.5 g l^{-1} glucose, 1.5 g l^{-1} sodium bicarbonate supplemented with 10% heat-inactivated FCS, 100 U ml^{-1} penicillin, 100 μg ml^{-1} streptomycin, and 2 mM L-glutamine. Split the cells 1 to 4–5 when they are about 80–90% confluent.
9. HeLa culture medium: DMEM supplemented with 2 mM L-glutamine, 1 mM sodium pyruvate supplemented with 10% heat-inactivated FCS, 100 U ml^{-1} penicillin, and 100 μg ml^{-1}. Split the cells 1 to 5–7 when they are about 80–90% confluent.
10. 3T3 culture medium: DMEM medium supplemented with 10% heat-inactivated FCS, 100 U ml^{-1} penicillin, and 100 μg ml^{-1}.
11. Trypsin/EDTA solution, 0.25%/0.02% (wt/vol) (Biochrom).
12. PBS-Dulbecco's w\o Ca^{2+}, Mg^{2+} solution (Biochrom).

2.6. Preparation of Magnetic Nanoparticle–siRNA Transfection Complexes

1. Magnetic nanoparticles are synthesized according to the procedure described in **Sect. 3.1.** The commercially available magnetic nanoparticles are suspended in water at 36 μg ml^{-1} just before use (the concentration refers to the iron content). This will result in a magnetic nanoparticle

iron-to-siRNA ratio of 0.5:1 (wt/wt) according to the protocol described in **Sect. 3.6** (*see* **Note 3** and **4**).

2. Metafectene as an enhancer: mix 5.8 μl Metafectene with 34.2 μl water in a tube for each transfection complex to be tested (resulting finally in a Metafectene-to-siRNA vol/vol ratio of 4:1 when used according to the protocol described in **Sect. 3.6**.
3. PEI as an enhancer: Prepare a solution of PEI-25Br in water at 45.36 μg ml^{-1} (resulting finally in an N/P ratio of 10 when the protocol described in **Sect. 3.6** is used.
4. siRNA stock solution (100x): reconstitute 5 nmol siRNA, e.g., GFP-22 siRNA (Qiagen) or luciferase GL3 siRNA (Qiagen), at 480 μg siRNA ml^{-1} with 162.9 μl or 154.4 ml of the siRNA suspension buffer, respectively, and store in aliquots at –20°C.
5. siRNA solution: prepare by 1 to 100 dilution of the 100x siRNA stock solution with serum- and supplement-free medium (e.g., RPMI 1640).

2.7. Evaluation of Magnetic Responsiveness of the siRNA-Magnetic Nanoparticle Complexes

1. Reagents and solutions are as in **Sect. 2.6.**
2. 8 Ne-Fe-B permanent magnets 18.0 × 16.0 × 4.0 mm (IBS Magnets).

2.8. Magnetofection

1. Cells are plated 24 h prior to transfection according to **Sect. 2.5**.
2. Magnetic transfection complexes and appropriate controls including complexes of nonsilencing siRNA(s) if necessary are prepared just before magnetofection (**Sect. 3.6**).
3. 96-Magnets magnetic plate (magnetic plate; OZ Biosciences).

2.9. Evaluation of Transfection Complex Association with Cells and Internalization into Cells

1. GFP-22 siRNA labeled with AlexaFluor 555 (GFP-22 siRNA, Qiagen).
2. FACS buffer: PBS supplemented with 1% FCS.
3. Hoechst 33342 stock solution: Hoechst 33342, trihydrochloride trihydrate (Invitrogen) 1 mg ml^{-1} water. Store in the dark at 4°C.
4. YOYO-1 iodide (491/509) stock: 1 mM solution in DMSO (Invitrogen). Store in aliquots in the dark at –20°C.

2.10. Quantification of Transfection Complexes Internalization into Cells Using Radioactively Labeled siRNA

1. ^{125}I-labeled siRNA solution: 4.8 μg ml^{-1} total siRNA (GFP-22 siRNA, Qiagen) comprising 1×10^6 CPM ml^{-1} ^{125}I-labeled siRNA (from **Sect. 3.3**) in RPMI medium (without supplements, or whatever solvent of interest). This should be prepared fresh for each experiment. Other reagents are as described in **Sect. 2.6**.

2.11. Quantification of GFP Reporter Gene Downregulation in Cell Lysate and in Living Cells

1. Lysis buffer: 0.1% Triton X-100 in 250 mM Tris pH 7.8.
2. GFP stock solution: 500 ng GFP per μl PBS. Store in small portions at −70°C.
3. BioRad protein assay reagent (BioRad).
4. BSA stock solution: 1.5 mg ml^{-1} BSA (Sigma) in PBS. Store at 4°C.

2.12. Quantification of Luciferase Reporter Gene Downregulation in Cell Lysates

1. Luciferin buffer: 35 μM d-luciferin (Roche Diagnostics), 60 mM DTT (Sigma-Aldrich, cat. no. D9779), 10 mM magnesium sulfate, 1 mM ATP, in 25 mM glycyl–glycine–NaOH buffer, pH 7.8.
2. Luciferase standard stock: 0.1 mg luciferase (Roche Diagnostics) per ml and 1 mg BSA (Sigma) per ml in 0.5 M Tris–acetate buffer, pH 7.5. Store in aliquots at −70°C.

2.13. MTT-Based Test for Toxicity of the Transfection Complexes

1. MTT solubilization solution: 10% Triton X-100 in 0.1 N hydrochloric acid in anhydrous isopropanol (solution can be stored at room temperature, 15–25°C).
2. MTT solution: 1 mg thiazolyl blue tetrazolium bromide (MTT; Sigma) per ml and 5 mg ml^{-1} glucose in PBS-Dulbecco's solution (solution must be stored at −20°C).

2.14. Improvement of Reporter Gene Downregulation by Magnetic siRNA Vectors Modified with INF-7 Fusogenic Peptide

1. INF-7 stock: 10 mg ml^{-1} in 20 mM HEPES, pH 7.4. Store in aliquots at −70°C.
2. INF-7 solution: 2.33 μg/10 μl in 20 mM HEPES, pH 7.4. Prepare before starting the experiment and keep at 4°C.
3. Other reagents are described in Sect. 2.6.

3. Methods

3.1. Synthesis of Magnetic Nanoparticles Suitable as Components of Magnetofection Complexes

1. Dissolve 0.05 mol (13.52 g) of iron(III) chloride hexahydrate and 0.025 mol (4.97 g) iron(II) chloride tetrahydrate in 300 ml water and filter using a 0.2 μm filtering flask or bottle-top filter (whenever possible, use fresh reagents); transfer the solution to a 500 ml round-bottom flask (make fresh as required). Remove dissolved oxygen by continuous argon or helium bubbling through the solution for ~10 min, (for PL-Mag1 nanomaterial add PL-Mag1 precipitation/coating solution 1); cool the solution to 2–4°C and continue bubbling argon/helium through the solution.
2. To obtain a primary precipitate, rapidly add precipitation/coating solution (for PL-Mag1 nanomaterial add PL-Mag1 precipitation/coating solution 2), heat the material to 90°C over a 15 min interval, and stir at this temperature for the

next 120 min. Cool the mixture down to 25°C (no more inert gas bubbling is needed), and then incubate for 2 h with continuous stirring.

3. Sonicate the product for 10 min using a resonance frequency of ~20 kHz, 75 mW, 60 s sonication/30 s break interval. Dialyze against water over 2 d using Spectra/Por 6 50 kDa cut-off dialysis membrane to neutralize the suspension and to remove excess unbound stabilizer. Sterilize the suspension using ^{60}Co gamma irradiation, dosage 25 kGy *(32)*. The particle types described here can be stored for at least 1 year without losing their properties that are required for magnetofection (**Table 6.1**, *see* **Note 5**).

3.2. Determination of Magnetic Nanoparticle Concentration in Suspension in Terms of Dry Weight, Iron Content, and Iron Concentration per Dry Weight of Magnetic Nanoparticles

1. To determine the magnetic nanoparticles concentration in suspension in terms of iron content, take 20 µl aliquots of the magnetic nanoparticle suspension, add 200 µl concentrated hydrochloric acid and 50 µl water. Wait until the magnetic nanoparticles are completely dissolved, then adjust the volume to 5 ml with water.
2. Transfer 20 µl of the resultant solution to a microcentrifuge tube, add 20 µl concentrated hydrochloric acid, 20 µl hydroxylamine hydrochloride solution, 200 µl ammonium

Table 6.1
Characteristics of the magnetic nanoparticles synthesized according to the Sect. 3.1 (see Note 6)

Parameter	Nanoparticles PEI-Mag2	PalD1-Mag1	PL-Mag1
Mean magnetite crystallite size <*d*> (nm)	9	8.5	10.6
Mean hydrated particle diameter *D* (nm)	63 ± 36	55 ± 10	101 ± 20
Iron content (g Fe/g dry weight)	0.56	0.526	0.47
Average iron weight per particle P^{Fe}_{part} (g Fe/particle)	1.4×10^{-18}	1.21×10^{-18}	2.35×10^{-18}
Saturation magnetization of the "core" M_s (A·m²/kgFe)	62	63	109
Effective magnetic moment of the insulated particle m_{eff} (A·m²)	5.5×10^{-20}	4.3×10^{-20}	9.1×10^{-20}
ξ-Potential in water (mV)	+55.4 ± 1.6	−15.6 ± 1.6	−13.3 ± 1.6

acetate buffer, 80 μl 1,10-phenanthroline solution, and 860 μl water. Mix well and allow to stand for 20 min.

3. Also prepare a blank sample by mixing 20 μl concentrated hydrochloric acid, 20 μl hydroxylamine hydrochloride solution, 200 μl ammonium acetate buffer, 80 μl 1,10-phenanthroline solution, and 880 μl water (*see* **Note 6**).
4. Measure the absorbance of the samples from Step 2 at 510 nm against the blank (Step 3) using a spectrophotometer, for example, a Beckman DU 640 spectrophotometer.
5. To construct a calibration curve for the determination of the iron concentration, add increasing amounts of iron standard solution to microcentrifuge tubes (e.g., 50, 70, and 90 up to 150 μl) and adjust the volume to 150 μl with water. Use 150 μl water instead of the iron solution to prepare a blank sample. To each tube, add 20 μl concentrated hydrochloric acid, 20 μl 10% hydroxylamine hydrochloride solution, 200 μl ammonium acetate buffer, 80 μl 0.1% 1,10-phenanthroline solution, and 730 μl water. Mix well and allow to stand for 20 min. Measure the absorbance at 510 nm against the blank. Plot the absorbance at 510 nm as a function of the iron concentration in the standard samples. Use linear regression as an approximation function for calculating the iron concentration in the magnetic nanoparticle samples.
6. To determine iron concentration per dry weight of magnetic nanoparticles, freeze-dry under high vacuum as follows: transfer 1 ml aliquots of magnetic nanoparticle suspensions into pre-weighed glass vials, freeze samples (at –80°C or in liquid nitrogen), and dry overnight under high vacuum (suitably using a lyophilizer). Weigh the vials again to calculate the dry weight. Add 1 ml concentrated hydrochloric acid. Wait until the magnetic nanoparticles are completely dissolved. Transfer 20 μl of the resultant solution to a microcentrifuge tube and determine iron content following the protocol described in Steps 3 and 5. Calculate iron concentration per dry weight of magnetic nanoparticles (*see* **Note 7**). Examples of the results are given in **Table 6.1** (*see* **Note 8**).

3.3. Radiolabeling (Iodination) of siRNA (see Note 9)

1. In vial 1 (ideally a conical screw cap microcentrifuge tube) prepare a mixture of 15 μl siRNA solution (2 μg siRNA/μl) and 15 μl 0.1 M ammonium acetate buffer, pH 5. In vial 2, prepare a mixture of 15 μl 250 μM potassium iodide, 2 μl sodium ^{125}I (0.2 mCi), and 10 μl 0.1 M sodium hydroxide.
2. Add 15 μl 30 mM thallium trichloride solution to vial 2 (*see* **Note 10**), quickly mix, and immediately transfer the content of vial 2 to vial 1, incubate the vial at 60°C for 45 min, then cool on ice.

3. Add 7.5 μl 0.1 M sodium sulfite, then 30 μl 1 M ammonium acetate buffer, pH 7, incubate for 60 min at 60°C, then cool on ice.
4. During the incubation time of the previous step, equilibrate a Sephadex G25 PD10 desalting column with water according to the instructions of the manufacturer. Apply the reaction mixture to the column and let it penetrate the column bed. Position a rack with 20 aligned microcentrifuge tubes under the column for fraction collection. Add 2 × 5 ml water for elution and collect 11 drops each (= 400–500 μl) in the microcentrifuge tubes aligned in the rack.
5. Using a handheld radiation monitor, determine the early eluting product fractions with the highest radioactivity (it is expected that the fractions will be among fractions 6 to 9) (*see* **Fig. 6.1a**).
6. Transfer a 10 μl aliquot of the product fraction to a scintillation vial and determine the radioactivity (CPM) using a gamma counter (e.g., Wallac 1480 Wizard 3″ automatic gamma counter). In another aliquot of the product fraction from Step 4, determine the DNA concentration by measuring the absorbance *D* at 260 nm (*see* **Fig. 6.1a**) and using the following formula: *siRNA concentration (μg/ml)* = (D_{260}) × *(dilution factor)* × *(50 μg siRNA/ml)*.

3.4. Testing the siRNA Binding Capacity of the Magnetic Nanoparticles

This procedure can be accomplished within 2 h.

1. For use as a transfection enhancer, mix 20.2 μl Metafectene and 119.8 μl water, or prepare a solution of 45.4 μg PEI per ml water (N/P = 10 (*see* **Note 11**)). This should be prepared fresh before the experiment. In general, any other transfection reagent can be tested as an enhancer instead.
2. In a 96-well round bottom plate (Techno Plastic Products) add 20 μl of magnetic nanoparticle suspension (from Step 1) into well A1 (corresponding to 5.76 μg iron of magnetic nanoparticles). Add 10 μl water into each well from A2 to A6.
3. Transfer 10 μl from A1 into A2, mix, from A2 into A3, etc., down to A5. Discard excess 10 μl from A5. The A6 well is a reference.
4. Add 20 μl of enhancer dilution to each well from A1 to A6; mix well with a pipette. To measure DNA association with magnetic nanoparticles in the absence of an enhancer, add 20 μl water to each well.
5. Add 150 μl ^{125}I-labeled siRNA solution (4.8 μg ml^{-1} in RPMI medium without supplements, or whatever solvent

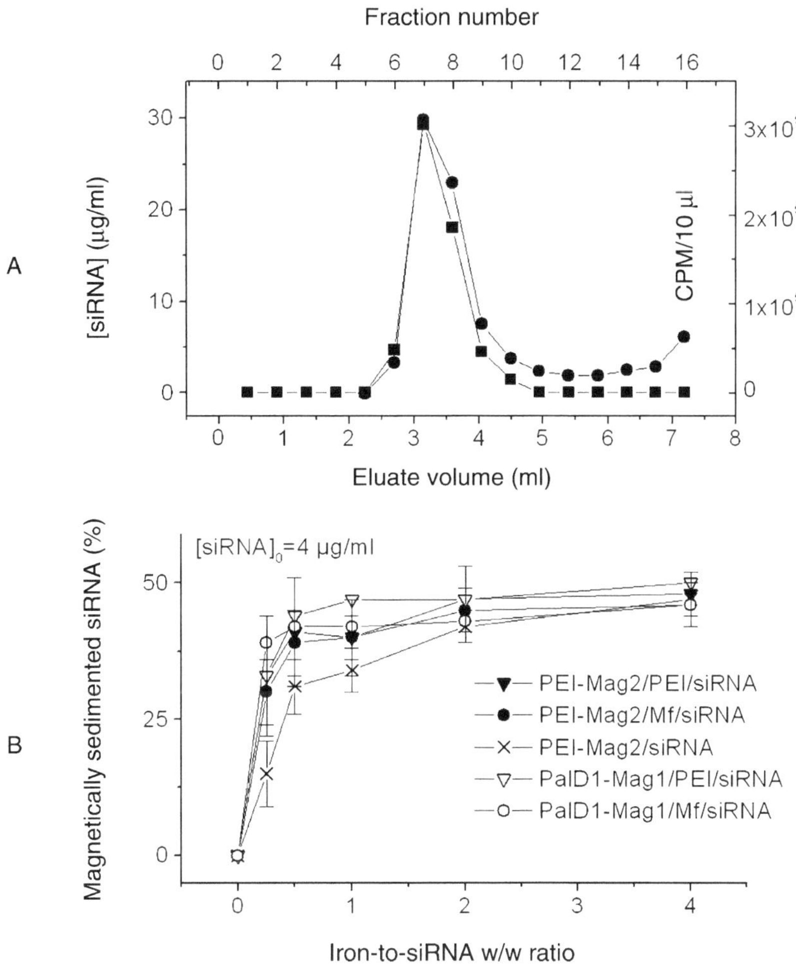

Fig. 6.1. Radiolabeling (iodination) of siRNA and testing siRNA association with magnetic nanoparticles in transfection complexes. (**a**) siRNA concentration (■) and ^{125}I-radioactivity in 10 µl aliquots (●) measured in the fractions after radiolabeled siRNA purification on Sephadex G25 PD-10 disposable columns. (**b**) siRNA associated and magnetically sedimented with PEI-Mag2 and PalD1-Mag1 magnetic nanoparticles in duplexes (duplex means magnetic nanoparticle plus siRNA only, without enhancer) and triplexes in the presence of PEI-25_{Br} (nitrogen-to-siRNA phosphate ratio N/P = 10) or Metafectene (4 µl Mf/1 µg siRNA) plotted against magnetic nanoparticle concentration (in terms of iron-to-siRNA weight ratio, starting siRNA concentration of 4 µg ml^{-1}). Plots assignment as shown in the figure. A range of iron-to-siRNA ratios (w/w) from 0.25 to 4, according to Sect. 3.4, has been examined with magnetic nanoparticles having a highly positive ξ-potential (PEI-Mag2) and with other particles having a negative ξ-potential (PalD1-Mag1). In the presence of both transfection reagents (4 µl Metafectene-Pro per µg DNA or PEI-25_{Br} at N/P = 10), both magnetic particle types efficiently form triplexes, showing a potential suitability for magnetofection. Positively charged PEI-Mag2 particles also form duplexes with siRNA.

of interest) comprising 2 × 10^5 CPM ml^{-1} ^{125}I-labeled siRNA (from **Sect. 3.3**) to each well from A1 to A6; mix well with a pipette.

6. Incubate for 15 min to allow siRNA binding to magnetic nanoparticles.

7. To sediment magnetic nanoparticles/siRNA transfection complexes, place the plate on the 96-Magnets magnetic plate (magnetic plate; OZ Biosciences) for 30 min.
8. Carefully sample 100 µl supernatant from each well using a pipette. Transfer each sample together with the pipette tip into scintillation vial. Take care to avoid disturbing magnetically sedimented complexes.
9. Measure the radioactivity (CPM) in every vial using the gamma counter.
10. Calculate magnetic sedimentation of the siRNA associated with magnetic nanoparticles (%; *see* **Fig. 6.1b**) as follows: *Magnetically sedimented siRNA (%)* $= [1 - CPM_{sample}/CPM_{ref}] \times 100$, where CPM_{ref} is the measured radioactivity from well A6 if the assay is carried out to the above protocol.

3.5. Cell Culture and Plating of Adherent Cells for Transfection

1. Culture H441 cells (human adenocarcinoma bronchial epithelial cells) at 37°C in a 5% CO_2 atmosphere. Split cells at a ratio of 1:4 to 1:5 every 4–5 days before reaching 100% confluence. Seed plates 24 h before transfection (*see* **Note 12**). H441 cells are used as an example, but other cell lines could equally be used.
2. For plating, wash the cells with PBS, aspirate supernatant, and add 2 ml trypsin–EDTA (0.25%) solution per 75 cm^2 cultivation flask. Shake gently so that the solution can cover the area of the cells, and then take out all of trypsin with a Pasteur pipette and incubate the flask at 37°C for 2–3 min. Observe the cells under microscope and when the cells are detached immediately add 10 ml H441 culture medium to arrest trypsin action.
3. Count the cells using a microscope counting chamber (hemocytometer) and resuspend in H441 culture medium at a density of 1.67×10^5 cells per ml, before transferring to a reagent reservoir.
4. Transfer 150 µl of the cell suspension per well to the 96-well flat bottom plate (Techno Plastic Products, cat. no. 92096) or to a clear bottom black-walled plate, 96-well (Greiner Bio-One) using a multichannel pipette (*see* **Note 13**). 25,000 cells per well provide a confluence of 50% before magnetofection 24 h later.
5. Store the plate in a cell culture incubator at 37°C in a 5% CO_2 atmosphere until transfection, usually 24 h later. The cells should be approximately 50% confluent at the time of transfection.
6. Adherent cells that divide more rapidly than H441 cells (NIH-3T3 or HeLa cells) would be plated at a density of 5,000–10,000 cells per well.

3.6. Preparation of Magnetic Nanoparticle–siRNA Transfection Complexes

1. To test the transfection complexes, add 20 μl of a suspension of magnetic nanoparticles to be tested (from Step 1 in **Sect. 2.6**) into wells A4, A7, and A10 and E1, E4, E7, and E10, respectively, of flat-bottom 96-well plates.
2. Add 40 μl each of the enhancer solution (from Step 2 or 3 in **Sect. 2.6**) to the same wells and mix using a pipette.
3. Add 300 μl each of siRNA solution from Step 4 in **Sect. 2.6** (4.8 μg ml^{-1} in serum- and supplement-free medium (e.g., RPMI 1640), which delivers 1.44 μg siRNA per well) to the same wells and mix well using a pipette. This would result in a final volume of 360 μl in wells A4, A7, and A10 (*see* **Note 14**).
4. For the untransfected control setup, add 300 μl serum- and supplement-free medium and 60 μl water to well A1. For other controls and references (e.g., magnetic nanoparticle–siRNA duplexes without enhancer, enhancer–siRNA complexes without magnetic nanoparticles), substitute the omitted component(s) with medium and/or water, respectively.
5. Incubate for 15 min at room temperature.
6. During the incubation time, fill the remaining wells of columns 1, 4, 7, and 10 with 180 μl each of serum- and supplement-free medium (RPMI 1640).
7. Prepare a 1:1 dilution series when the 15 min incubation time has expired as follows: Transfer 180 μl each from A1, A4, A7, and A10 to B1, B4, B7 and B10, respectively, using a multichannel pipette, mix, transfer 180 μl from the respective wells in row B to row C and so on down to row D. And 180 μl each from E1, E4, E7, and E10 to F1, F4, F7, and F10, respectively, using a multichannel pipette, mix, transfer 180 μl from the respective wells in row F to row G and so on down to row H (*see* **Note 15**). The characteristics for selected siRNA transfection complexes prepared according to the protocol described in **Sect. 3.6** are given in **Table 6.2**.

3.7. Evaluation of Magnetic Responsiveness of the siRNA–Magnetic Nanoparticle Complexes: Complex "Loading" with Magnetic Nanoparticles

To evaluate magnetically induced velocity (magnetic responsiveness) of a magnetic siRNA complex in a gradient magnetic field and estimate complex loading with magnetic nanomaterial:

1. Prepare magnetic nanoparticle–siRNA transfection complexes: add 60 μl of a suspension of magnetic nanoparticles into a tube; add 120 μl of the enhancer solution (from Step 2 or 3 in **Sect. 2.6**) to the same tube and mix using a pipette; add 900 μl of the siRNA solution from Step 4 in **Sect. 2.6** and mix well using a pipette. This results in a final volume of 1080 μl. For magnetic nanoparticle–siRNA duplexes with-

Table 6.2
Characteristics of selected siRNA transfection complexes

Complex	Iron-to-siRNA w/w ratio	ζ-Potential (mV)	Mean hydrated diameter *D* (nm)	Efficient velocity in magnetic fields[a] υ_z (μm/s)	Average magnetic moment of the complex M^a (10^{-16} A·m²)	Number of magnetic particles in a complex $N=M/m_{eff}$
Duplexes						
PEI/siRNA	–	+15.2 ± 1.8	413 ± 190	–	–	
Mf/siRNA	–	+36.1 ± 9.7	283 ± 133	–	–	
PEI-Mag2/siRNA	1:1	−14.0 ± 0.8	685 ± 242	1.2	17.2	20483
PEI-Mag2/siRNA	2:1	−10.1 ± 1.2	736	1.96	30.2	35946
Triplexes						
PEI-Mag2/PEI/siRNA	0.5:1	+2.0 ± 4	394 ± 70	1.19	9.5	11290
PalD1-Magl/PEI/siRNA	0.5:1	+7.2 ± 1.5	370 ± 115	1.49	11.6	15895
PEI-Mag2/Mf/siRNA	0.5:1	+36.4 ± 3.8	210 ± 86	0.86	3.8	4500
PalD1-Magl/Mf /siRNA	0.5:1	+12 ± 6.3	326 ± 175	0.72	30.2	12181

[a]For a magnetic field configuration as shown in Fig. 6.2.

out enhancer or enhancer–siRNA complexes without magnetic nanoparticles, substitute the omitted component(s) with medium and/or water, respectively. Incubate for ~15 min at RT. Use half of the volume to measure the mean hydrated diameter *D* of the complex by photon correlation spectroscopy using, for example, a Malvern Zetasizer 3000 (UK).

2. Put 500 μl of the complex suspension into an optical cuvette. Position two sets of 4 quadrangular Ne-Fe-B permanent magnets (18.0 × 16.0 × 4.0 mm) cantered beside a measuring window of the optical cuvette in a spectrophotometer (for example, Beckman DU 640 Spectrophotometer); put the cuvette with the suspension into the sample holder, and immediately start to measure the time course of the turbidity

(optical density) at 360 nm in a "kinetic mode." The set-up for the measurements is shown in **Fig. 6.2a**. The configuration of the magnetic field in the measuring window is shown in **Fig. 6.2b**. Examples of the experimental results for the kinetics of the magnetic sedimentation (responsiveness) for the selected complexes are given in **Fig. 6.2c**.

3. For evaluation of the efficient velocity υ_z of the complexes under a gradient magnetic field and further calculation of the average magnetic moment M of the magnetic complex(s) and estimation of the number of magnetic nanoparticles N associated with the complex *see* **Note 16**. Examples of results are shown in **Table 6.2**.

3.8. Magnetofection

Timing: ~30–40 min plus 48–72 h to allow reporter gene down-regulation. Magnetofection should be carried out under sterile conditions.

1. Check the plates prepared for transfection according to **Sect. 3.5** under the microscope for cell state and confluence. Cell confluence of ~40–50% before transfection is preferable for H441 cells.
2. Cells that divide more rapidly than H441 cells (NIH-3T3 or HeLa cells) can be transfected at lower confluence of ~30–40%. Aspirate the medium from the wells, and add 150 µl fresh cultivation medium per well.
3. Transfer 50 µl each of the transfection complexes prepared in Sect. 3.6 into the culture plates with the seeded cells as follows: Using a multichannel pipette, mix the dilutions of transfection complex prepared in column 1 of the complex preparation plate (from **Sect. 3.6**) by pipetting up and down, then transfer 50 µl to the wells of columns 1, 2, and 3 (to test each composition and dilution of transfection complex in triplicate) of the cell culture plate (from Step 1). Transfer 50 µl from each well of column 4 of the complex preparation plate to columns 4, 5, and 6 of the cell culture plate. Transfer 50 µl from each well of column 7 of the complex preparation plate to columns 7, 8, and 9 of the cell culture plate. Transfer 50 µl from each well of column 10 of the complex preparation plate to columns 10, 11, and 12 of the cell culture plate. This results in delivery of 200, 100, 50, and 25 ng siRNA per well in rows A (E), B (F), C (G) and D (H), respectively.
4. Place the cell culture plate on a 96-magnet magnetic plate for 15–30 min to create at the cell layer location a permanent magnetic field with a field strength and gradient of 70–250 mT and 50–130 T m^{-1}, respectively.

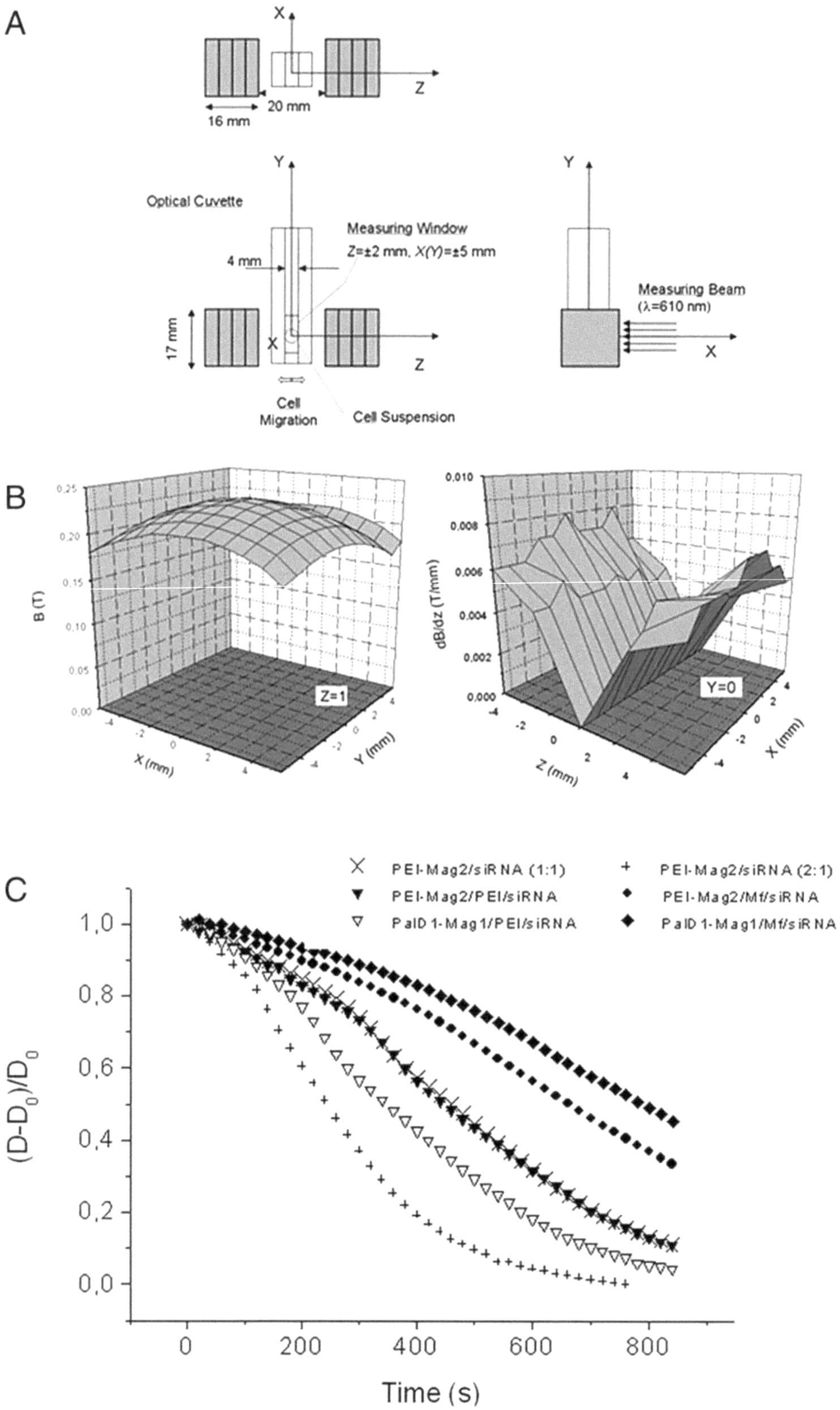

Fig. 6.2. Evaluation of magnetic responsiveness of the magnetic siRNA transfection complexes in the applied magnetic fields. (**a**) Setup for measurements of the magnetic responsiveness of the complexes: Gradient permanent magnetic field is created by two sets of four quadrangular Ne-Fe-B permanent magnets 17 × 4 mm positioned at a distance of 20 mm

5. Remove the magnetic plates after 20–30 min exposure of the cells to the magnetic field (*see* **Note 17**) and incubate the plate containing the transfected cells in a cell culture incubator at 37°C in a 5% CO_2 atmosphere until results are evaluated (usually 48–72 h post-transfection).

3.9. Evaluation of the Transfection Complex Association with Cells and Internalization into Cells

1. To evaluate siRNA transfection efficiency and transfection complex association with cells and internalization into cells, prepare the transfection complexes with GFP-siRNA labeled with AlexaFluor 555 according to **Sect. 3.6** and perform transfection of the cells according to **Sect. 3.8**.
2. After incubation at 37°C in a 5% CO_2 atmosphere, examine the plate using a fluorescence microscope at 490/509 nm (green fluorescence) to visualize the cells expressing the eGFP reporter gene and at 510/650 nm (red fluorescence) to visualize localization of the siRNA-Alexa555 complexes (*see* **Fig. 6.3a** and **b**).
3. To allow visualization of internalized complexes with respect to the cell nuclei, add 1 μl per well of the cell-permeable nuclear counterstain Hoechst 33342 (1 mg ml^{-1} stock solution in water). Incubate for 15–20 min. Examples of the results are shown in **Fig. 6.3**.
4. To analyze cell/association and internalization using a FACS Vantage microflow cytometer, wash the adherent cells prepared in Step 1 with 150 μl PBS per well; aspirate the supernatant with a Pasteur pipette; add 10 μl Trypsin–EDTA (0.25%) solution per well, and incubate the flask at 37°C for 2–3 min.
5. Observe the cells under a microscope. When the cells are detached, immediately add 200 μl cell culture medium to arrest trypsin activity.
6. Combine cells from triplicate wells of cell culture plates in a fluorescence-activated cell sorting (FACS) tube.
7. Centrifuge at 300g (1,200 r.p.m. on a Heraeus Megafuge 2.0) for 5 min; remove supernatants carefully, and add 1 ml PBS supplemented with 1% FCS (FACS buffer).

Fig. 6.2. (continued) and centered around the measuring window of an optical cuvette in a Beckman DU 640 Spectrophotometer. (**b**) The magnetic field and the gradient of the field applied: The magnetic field is rather uniform in a measuring window $Z = \pm 2$ mm, $X(Y) = \pm 5$ mm in the X–Y plane parallel to the surface of the magnets and measuring light beam with an average magnetic field $\langle B_z \rangle$value in the direction of the particles moving perpendicular to the measuring beam of 0.213±0.017 T with an uniform magnetic field gradient $\partial B/\partial z$ in direction of the complexes movement of 4 ± 2 T/m. (**c**) Time course of the normalized turbidity of the magnetic complex suspensions (optical density at 360 nm normalized to the initial one D/D^0) upon application of the gradient permanent magnetic field versus time for siRNA duplexes with PEI-Mag2 nanoparticles PEI-Mag2/siRNA at iron-to-siRNA wt/wt ratio of 1:1 and 2:1, and siRNA triplexes with PEI-Mag2 and PalD1-Mag1 nanoparticles and PEI-25_{Br} (PEI) or Metafectene (Mf) as enhancers at iron-to-siRNA wt/wt ratio of 0.5:1 prepared as described in Sect. 3.6. Plots assignment as shown in the figure.

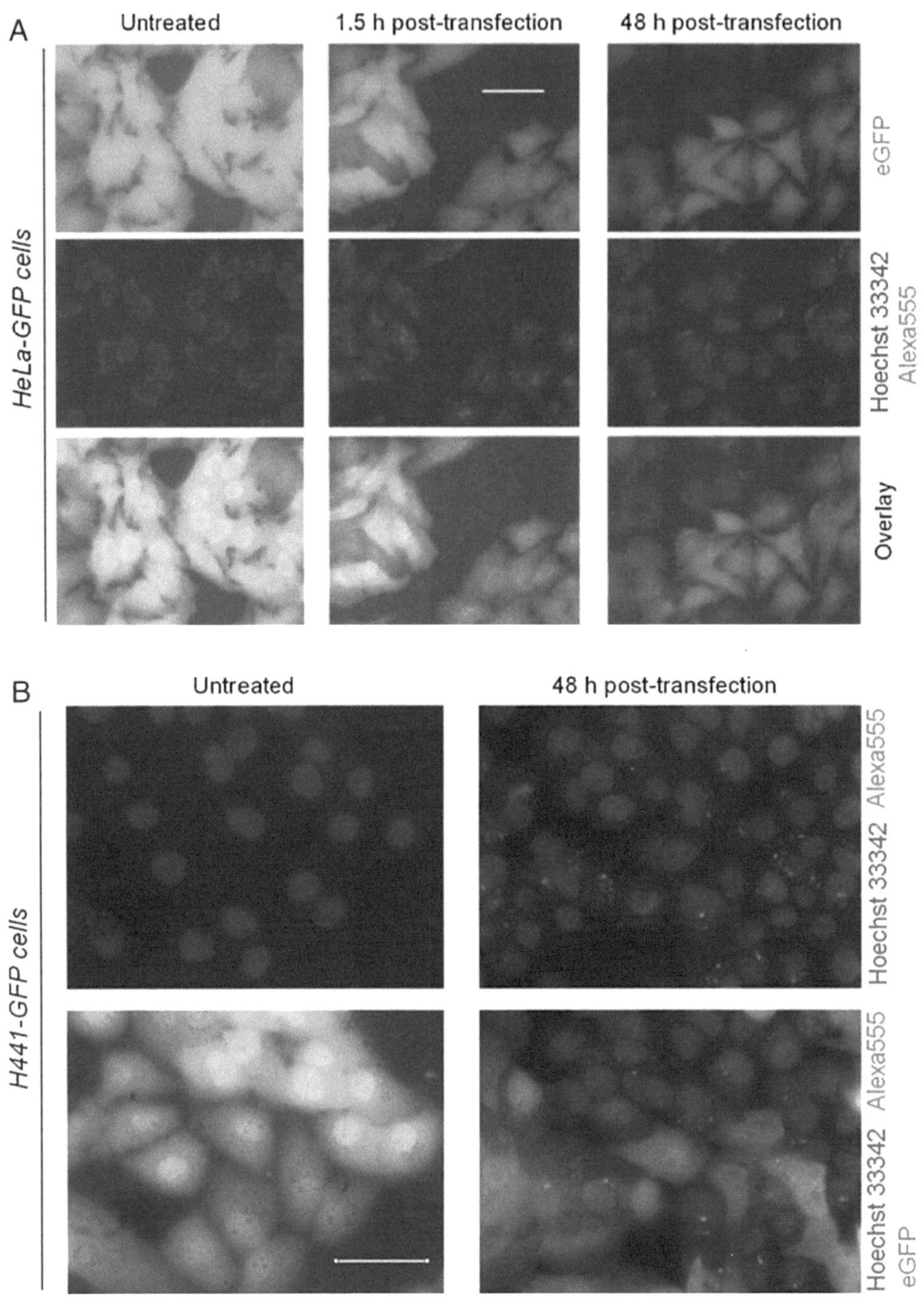

Fig. 6.3. Enhanced GFP (eGFP) expression and cell association/internalization of transfection complexes comprising magnetic nanoparticles with human cervical carcinoma cells and human pulmonary epithelial cells stably transfected with eGFP protein (HeLa-GFP and H441-GFP cells) detected by microscopy. HeLa-GFP and H441-GFP cells were incubated for 30 min at the magnetic plate with PEI-Mag2/PEI/GFP-siRNA-Alexa555 triplexes at siRNA concentration of 100 ng/10,000 cells/0.33 cm^2; iron-to-siRNA wt/wt ratio of 0.5, PEI/siRNA ratio of N/P = 10, and observed after 1.5 h and/or 48 h with a fluorescence microscope. Images were obtained at original magnification of 40. Bar = 50 μm. Hoechst 33342 was used as a nuclear counterstain. The pictures show fluorescence images taken at 490/509 nm (green fluorescence) for eGFP fluorescence, 510/650 nm (red fluorescence) for GFP-siRNA-Alexa555, and at 350/461 nm (blue fluorescence) for Hoechst 33342 nuclear staining, or overlays thereof. Fluorescence microscopy data prove the association of the magnetic transfection complexes with a majority of the cells already 1.5 h post-transfection and are indicative of internalization into cells. 48 h post-transfection fluorescently labeled siRNA triplexes comprising magnetic nanoparticles are localized predominantly around the nuclei.

8. Centrifuge again at 300g for 5 min; discard supernatants carefully; resuspend the cells in 0.5 ml FACS buffer.
9. To stain the siRNA complexes that are associated with cells but are not internalized into cells, add 1 μl of the cell-impermeable intercalating nucleic acid stain YOYO-1 iodide (1 mM in DMSO); incubate for 10 min.
10. Analyze the cells on a flow cytometer: excite fluorescence with an argon laser >488 nm, and detect eGFP or YOYO-1 fluorescence using a 530/30-nm bandpass filter and Alexa555 fluorescence using a 575/26-nm bandpass filter. Analyze a minimum total of 10,000 events per sample.
11. The percentages of cells with internalized or associated Alexa555-labeled transfection siRNA complexes are determined as a percentage of gated fluorescent events detected with a 575/26-nm bandpass filter (FL2) using untreated cells as a reference (*see* **Fig. 6.4a**). The percentage of cells either expressing eGFP or associated with YOYO-1-labeled transfection complexes is determined as a percentage of gated fluorescent events detected with a 530/30-nm bandpass filter (FL1) using untreated cells as a reference. Examples of the results are shown in **Fig. 6.4b**.

3.10. Quantification of Transfection Complex Internalization into Cells Using Radioactively Labeled siRNA

1. To quantify transfection complex internalization into cells, prepare the transfection complexes with ^{125}I-labeled siRNA solution from **Sect. 2.10** according to the protocols described in **Sect. 3.6** and perform transfection of the cells according to **Sect. 3.8**. Reserve 50 μl each of the transfection complexes as a reference.
2. After incubation at 37°C in a 5% CO_2 atmosphere, wash the cells with 150 μl PBS per well at different time points post-transfection; aspirate supernatant with a Pasteur pipette. To remove extracellularly bound complexes, add 100 μl per well heparin solution containing 75 mM sodium azide to inhibit endocytosis *(33)*.
3. After incubation at 37°C in a 5% CO_2 atmosphere for 30 min, wash the cells with 150 μl PBS per well; aspirate supernatant; add 10 μl Trypsin–EDTA (0.25%) solution per well, and incubate the flask at 37°C for 2–3 min.
4. Observe the cells under a microscope. When the cells are completely detached, add 200 μl cell culture medium.
5. Carefully collect the cell suspension from each well using a pipette. Transfer each sample together with the pipette tip into a scintillation vial. Measure the radioactivity (CPM) in every vial using the gamma counter.
6. Calculate the siRNA associated with magnetic nanoparticles as follows: $\textit{Internalized siRNA (\%)} = [CPM_{sample}/CPM_{ref}] \times 100$, where CPM_{ref} is the measured radioactivity from the reference

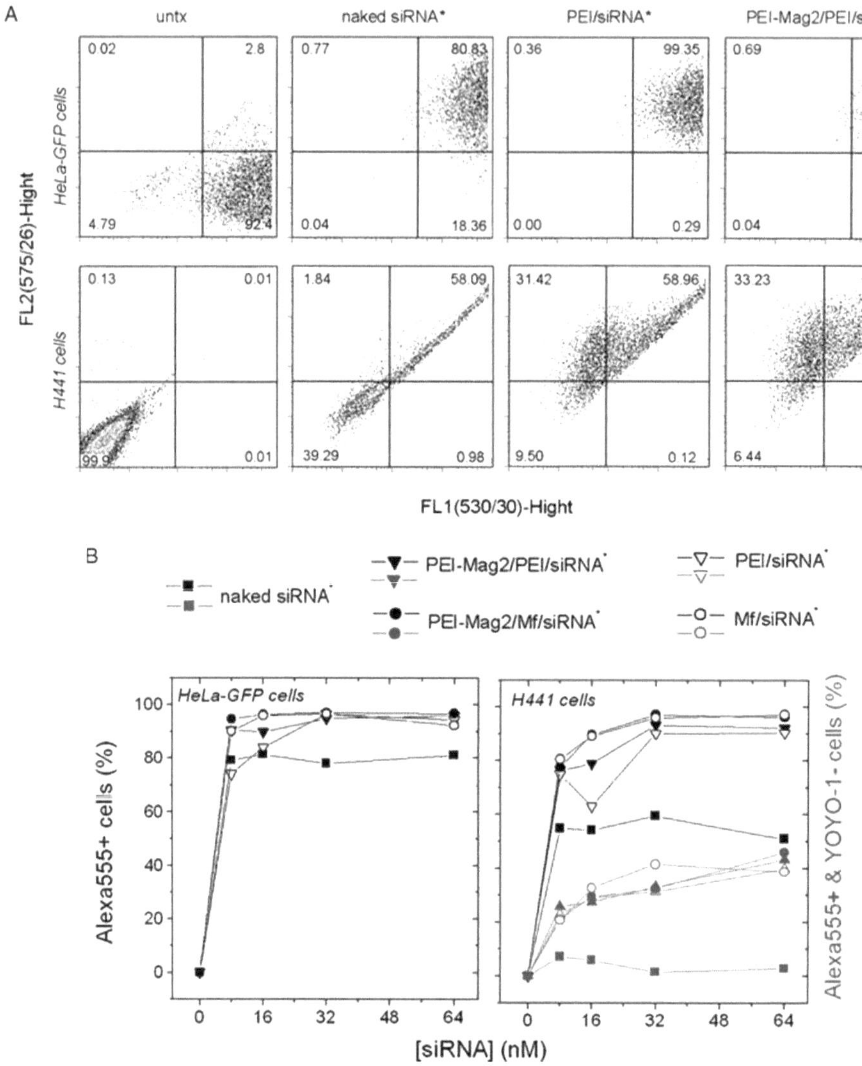

Fig. 6.4. Vector association/uptake in HeLa human cervical epithelial adenocarcinoma cells stably expressing eGFP (HeLa-GFP cells) and H441 human lung epithelial cells characterized by flow cytometry. HeLa-GFP cells and H441 cells were transfected in a 96-well plate as described in Sect. 3.9. 48 h post-transfection the cells were trypsinized, washed, and resuspended in 1% FCS in PBS. Vector cell/association/internalization was analyzed using a FACS Vantage microflow cytometer. (**a**) Density plots of untransfected cells (untx); cells transfected with siRNA-Alexa555 (siRNA*) alone without magnetic nanoparticles or enhancer (naked siRNA*); polyplexes of PEI-Br_{25} with siRNA-Alexa555 (PEI/siRNA*); magnetic triplexes comprising PEI-Mag2 magnetic nanoparticles, PEI-Br_{25}, and siRNA-Alexa555 (PEI-Mag2/PEI/siRNA*) for vector uptake analysis. Transfected H441 cells were additionally incubated with YOYO-1 to stain the siRNA transfection complexes associated with the cells but not internalized into cells. The siRNA dose was 100 ng per well in the examples in Fig. 6.4a. The numbers in squares indicate the percentages of gated cells with untreated cells as a reference. (**b**) Percentage of Alexa555 positive HeLa and H441 cells associated with siRNA* in dependence of the siRNA concentration in a well (black lines) and percentage of H441 cells that have internalized complexes (red lines) for triplexes of PEI-Mag2 nanoparticles with Metafectene (Mf) or PEI vs. nonmagnetic duplexes of siRNA* and naked siRNA*. Plots assignment as shown in the figure. The results given here clearly show that more than 80% of the HeLa-GFP and H441 cells are associated with transfection complexes, at the same time association with cells and internalization of naked siRNA is significantly lower compared to both magnetic and nonmagnetic transfection complexes. There is no considerable difference between the percentages of cells associated with/or internalized magnetic and nonmagnetic transfection complexes.

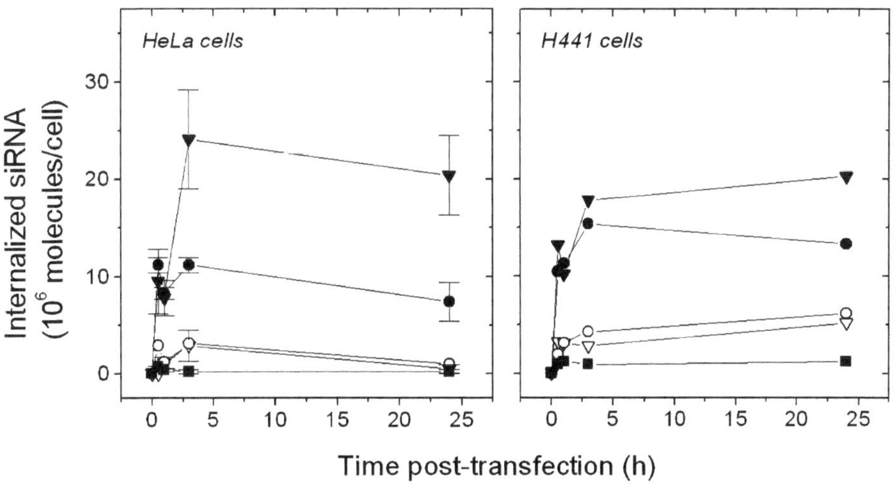

Fig. 6.5. Vector internalization in HeLa human cervical epithelial adenocarcinoma cells and H441 human lung epithelial cells. HeLa-GFP cells and H441 cells were transfected in a 96-well plate using ^{125}I-labeled siRNA complexes. The siRNA dose was 100 ng per well. At time points 0.5 h, 1 h, 3 h, and 24 h post-transfection the cells were incubated with heparin solution in the presence of sodium azide to remove extracellularly bound complexes, washed, trypsinized, and collected. Cell-associated radioactivity was measured with a gamma counter. The applied dose of the radioactively labeled siRNA complexes was used as a reference. The results were recalculated in terms of the siRNA molecules internalized per seeded cell. siRNA alone without magnetic nanoparticles or enhancer (naked siRNA, squares); Metafectene lipoplexes (open circles); polyplexes with PEI (open triangles); triplexes comprising Metafectene, siRNA, and PEI-Mag2 (black circles); triplexes comprising PEI, siRNA, and PEI-Mag2 (black triangles). In contrast to the FACS data on the percentage of the cells that have internalized the siRNA complexes (shown in Fig. 6.4b for H441 cells), the results given here clearly show that cell internalization of naked siRNA is negligible; polyplexes and lipoplexes of the siRNA are internalized better compared to naked siRNA. Magnetofection results in significantly higher internalization levels of the siRNA compared to lipo- or polyfection with the same vector type.

sample. In the example shown in **Fig. 6.5** the results are recalculated in terms of the siRNA molecules internalized per seeded cell.

3.11. Quantification of GFP Reporter Gene Downregulation in Living Cells and in Cell Lysate

1. To prepare cell lysates from adherent cells, wash transfected adherent cells (from **Sect. 3.8**) with 150 µl per well PBS using a multichannel pipette. Add 100 µl lysis buffer per well. Incubate for 10 min at RT, then place the culture plate on ice.
2. To quantify GFP expression in cell lysates, transfer 50 µl cell lysate from each well into a black 96-well plate with a transparent bottom (e.g., clear bottom black-walled plate, Greiner). Add 100 µl PBS per well and mix with the pipette. Measure the fluorescence intensity (485/535 nm, 1.0 s) using a microplate fluorescence reader, for example, a Wallac 1420 Multilabel counter. As blanks, measure wells with lysates of nontransfected cells.
3. To construct a calibration curve for the determination of the amount of GFP in transfected cell samples, in well A1

of a 96-well clear bottom black-walled plate, add 3 μl GFP stock solution to 147 μl lysis buffer, and mix well. Add 50 μl lysis buffer each to wells A2–A12. Transfer 100 μl from A1 to A2, mix well, transfer 100 μl from A2 to A3, mix and so on down to A11. Discard the surplus 100 μl from well A11. A12 is left as a blank. Add 100 μl PBS to each well of row A and mix well. Using a microplate fluorescence reader (e.g., Wallac 1420 Multilabel counter), measure the fluorescence intensity of GFP (excitation 485 nm, emission 535 nm, measuring time 1 s per well). Plot the measured fluorescence intensity as a function of GFP content per well. Use linear regression to derive a calibration function from which the GFP content in the samples can be calculated. Examples of the results are shown in **Fig. 6.6a**.

4. Use a calibration curve, constructed as described, to calculate the amount of GFP in the transfected cell samples (*see* **Note 18**).
5. To allow the results of the reporter gene expression assays to be presented as weight GFP (or luciferase) per weight unit total protein, the total protein content of the samples can be determined as follows: first, add 150 μl water to each well in a flat-bottom 96-well plate. Using a multichannel pipette, transfer 10 μl each of the cell lysates (from Step 1 or from Step 1 in **Sect.3.12**) into the corresponding wells of the protein assay plate. Add 40 μl BioRad protein assay reagent to each well; mix carefully using a plate shaker or a multichannel pipette. Measure the absorbance at 590 nm using a microplate reader (e.g., a Wallac 1420 Multi-label counter; measuring time set to 0.1 s).
6. To construct a calibration curve for the determination of the amount of total protein in the transfected cell sample, add 25 μl lysis buffer per well in one row (e.g., row A) of a flat-bottom 96-well plate. Add 50 μl BSA stock solution to well 1 (e.g., A1). Mix well using a pipette. Transfer 50 μl from well 1 to well 2, mix, transfer 50 ml from well 2 to well 3, and so on down to well 11. Well 12 is left as a blank. Add 150 μl water per well in another row (e.g., row B). Transfer 10 μl from row A to row B. Add 40 μl BioRad reagent to each well and mix carefully using a plate shaker or a multichannel pipette. Measure the absorbance at 590 nm (or 570 nm) using a microplate reader (e.g., a Wallac 1420 Multilabel counter; measuring time set to 0.1 s). Plot the measured absorbance versus the protein content per well. Use linear regression to derive a calibration function from which the protein content in the samples can be calculated.
7. Calculate the total protein content per 10 μl cell lysates for every sample using the calibration curve (from Step 6).

8. Calculate weight GFP per weight unit total protein (*see* **Note 19**); normalize results to the reference data determined for untransfected cells. The results can be plotted against time post-transfection to evaluate the time course of the silencing effect and to define the optimum exposition for screening experiments (*see* **Fig. 6.6a** and **b**) or against the siRNA concentration or dose per well in order to get a dose–response curve (examples of the results are given in **Figs. 6.6c** and **6.7a**).

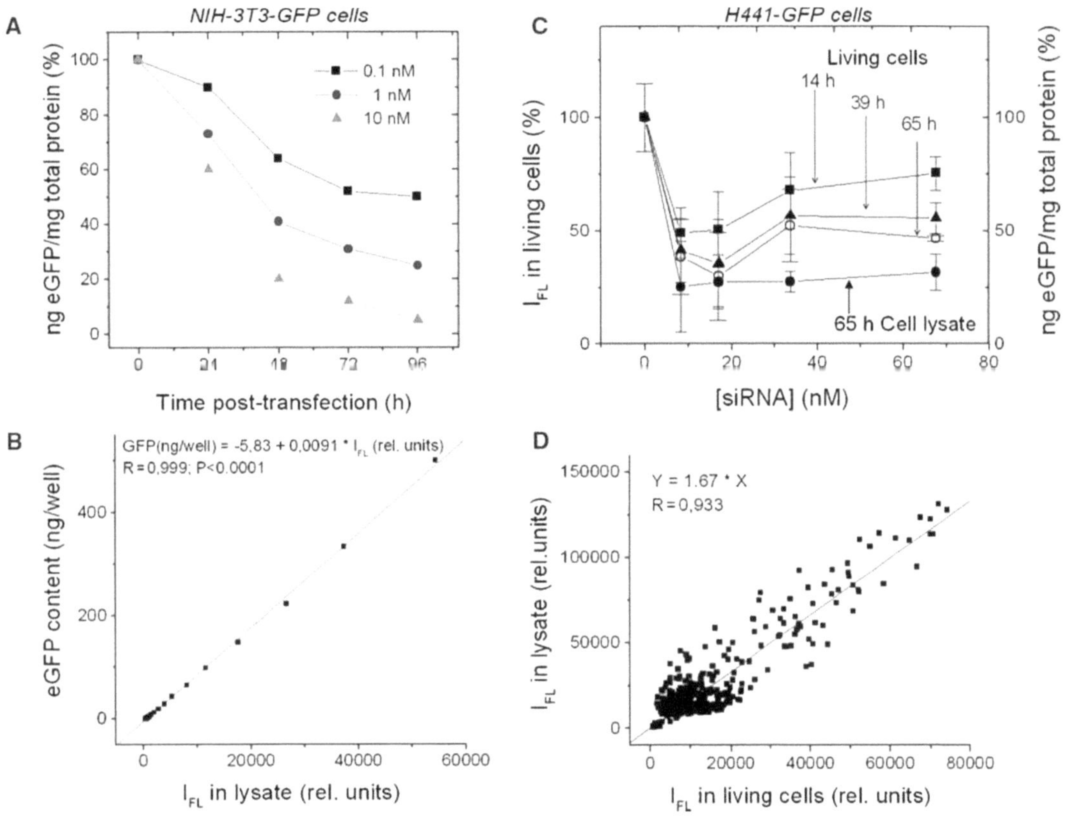

Fig. 6.6. Enhanced GFP (eGFP) reporter gene expression analysis in cell lysates and in living cells. (**a**) GFP stably transfected NIH-3T3 cells (NIH-3T3-GFP cells) were seeded in a 96-well plate and 24 h later transfected with 200 μl transfection volume of the magnetic complexes prepared with 0.5 μl of SilenceMag (OZ Biosciences) and 1 nM, 5 nM, or 10 nM siRNA (targeting GFP). GFP expression was monitored in cell lysates in function of post-transfection incubation time. Results show percentage of reporter gene expression. Untreated cells were used as a reference. (**b**) Shows a calibration curve for eGFP. (**c**) GFP stably transfected H441 human lung epithelial cells (H441-GFP cells) were seeded in a 96-well plate and 24 h later transfected with a 200 μl transfection volume of the magnetic triplexes PalD1-Mag1/PEI/siRNA (iron-to-siRNA ratio of 0.5 to 1, PEI/siRNA ratio of N/P = 10). GFP expression was monitored in living cells (fluorescence intensity, I_{FL}) at different time points post-transfection, and finally in cell lysate in function of the siRNA dose (concentration). Results show percentage of reporter gene expression. Untreated cells were used as a reference. (**d**) Shows the correlation between fluorescence intensity I_{FL} registered in cell lysates against I_{FL} registered for the same sample in wells with living cells for a set of samples. The results show minimum GFP expression (maximum down-regulation effect) 60–96 h post-transfection. Thus, in screening experiments for siRNA delivery, magnetic complex analysis of GFP expression can be performed after 60–90 (usually after 72) h post-transfection. The example given in (c) shows that GFP-expression monitoring in living cells provide realistic semiquantitative information on transfection efficiency.

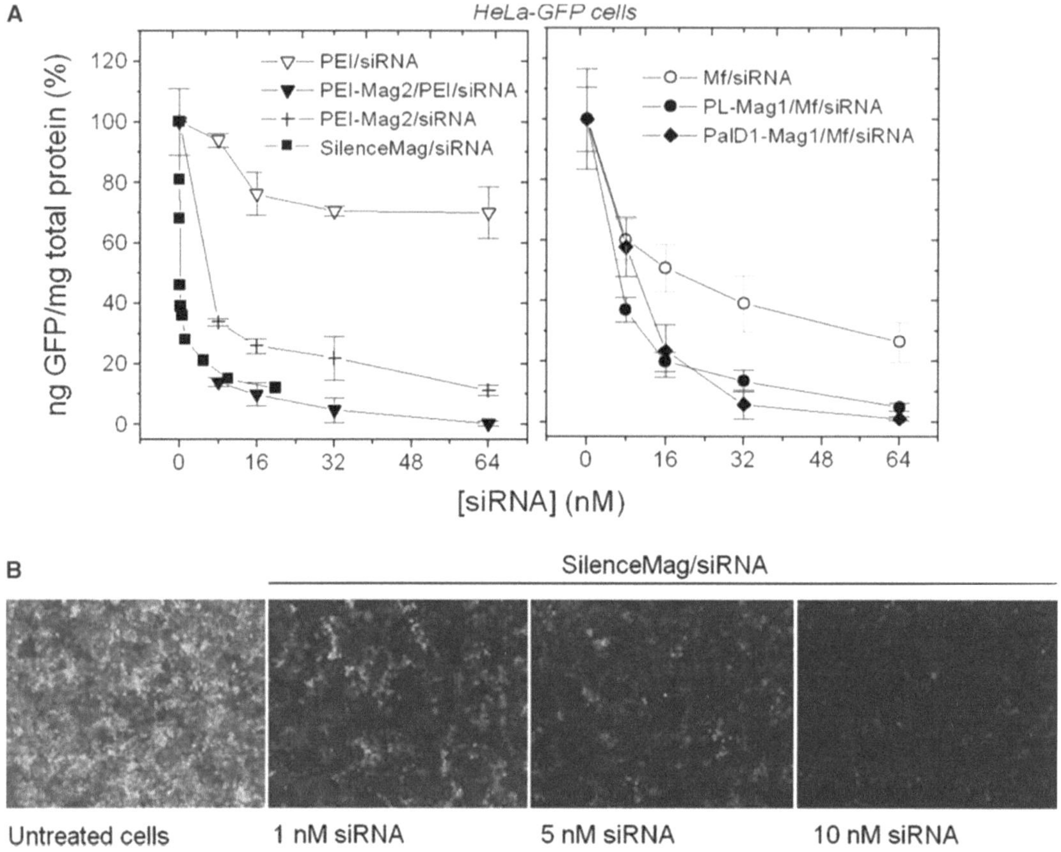

Fig. 6.7. Magnetofection versus lipofection and polyfection efficiency in HeLa-GFP cells. (**a**) GFP stably transfected HeLa cells (HeLa-GFP cells) were seeded in a 96-well plate and 24 h later transfected with a 200 μl transfection volume of the magnetic anti-GFP–siRNA complexes prepared with 0.5 μl of SilenceMag (OZ Biosciences) at different concentrations of siRNA or PEI/siRNA and Mf/siRNA poly- and lipoplexes, or magnetic duplexes PEI-Mag2/siRNA (Iron-to-siRNA ratio of 1) or magnetic triplexes PEI-Mag2/PEI/siRNA, PL-Mag1/Mf/siRNA, and PalD1/Mf/siRNA (iron-to-siRNA ratio of 0.5 to 1) (Mf-to-siRNA vol/wt ratio of 4, PEI-to-siRNA ratio N/P = 10). GFP expression was monitored 72 h post-transfection. Results show percentages of reporter gene inhibition. Plots assignments as shown in the figure. (**b**) GFP expression was monitored 72 h post-transfection by fluorescence microscopy in HeLa-GFP cells transfected with SilenceMag as shown in (a) at 1, 5, or 10 nM siRNA. The results show that magnetofection results in significantly lower expression levels of the GFP (i.e., more efficient target gene downregulation) compared to lipo- or polyfection with the same vector type. Efficiency of the PEI.Mag2/PEI/siRNA complexes is comparable with that of a magnetofection-based formulation of OZ Biosciences called SilenceMag. Magnetic duplexes PEI-Mag2/siRNA (at iron-to-siRNA ratio of 1) deliver siRNA rather efficiently, but less efficient compared to the PEI-Mag2/PEI/siRNA magnetic triplexes formulated at an iron-to-siRNA ratio of 0.5:1.

9. To evaluate the time course of the silencing effect, measurements of the GFP expression in living cells can be performed as follows (*see* **Note 20**): aspirate cell culture medium, wash the cells twice with 150 μl per well PBS, and measure the fluorescence intensity (485/535 nm, 1.0 s) using a microplate fluorescence reader. As blanks, measure wells with nontransfected cells. Change PBS for the complete cell growth medium and continue incubation. At the end of

observation after final measurements in living cells, perform cell lysis according to **Step 1**, and determine GFP concentration in cell lysates as described in **Steps 2, 3**, and **8**.

10. To calculate absolute values of GFP concentration from measurements in living cells, plot fluorescence intensity I_{FL} registered in cell lysates against I_{FL} registered for the same samples in wells with living cells for a set of samples. Examples of the results are shown in **Fig. 6.6d**. Use linear regression to derive a function for calculating I_{FL} in lysate. The function in combination with a calibration curve for lysates (shown in Fig. 6.6b) allows one to estimate GFP content in living cells. Examples of the results are shown in **Fig. 6.6c**.

3.12. Quantification of Luciferase Reporter Gene Downregulation in Cell Lysates

1. Prepare cell lysates from adherent cells as described in **Sect. 3.11**, Step 1.
2. To quantify luciferase reporter gene expression in cell lysates, transfer 50 μl cell lysate from each well into a 96-well black flat-bottom microplate. Add 100 μl luciferase buffer per well, optionally mix with a pipette. Measure the chemiluminescence intensity (count time 0.20 min with background correction) using a luminometer such as, for example, a microplate scintillation and luminescence counter (Canberra Packard) or a Wallac Victor 2 Multilabel Counter (Perkin Elmer).
3. To construct a calibration curve to determine the amount of luciferase in transfected cell samples, add 50 μl lysis buffer per well to columns 1 and 3 of a black 96-well plate and 40 μl lysis buffer per well in columns 2 and 4. To well A1, add 30 μl lysis buffer and 20 μl luciferase standard stock (0.1 mg luciferase per ml and 1 mg BSA per ml in 0.5 M Tris–acetate buffer, pH 7.5). Pipette 50 μl from A1 to B1, mix well, then from B1 to C1, etc., down to H1. From H1, continue the dilution series by transferring 50 μl to A3; continue in column 3 down to G3. H3 is a blank. Pipette 10 μl each from column 3 to 4, and from column 1 to 2. Add 100 μl luciferase buffer each to the wells of columns 2 and 4. Measure the chemiluminescence intensity as described above. Plot the logarithm of luciferase content in the dilution series as a function of the logarithm of measured luminescence intensity (light units). Use an approximation function (usually linear regression in this concentration range) for calculating the amount of luciferase in the transfected cell samples. An example of the results is given in **Fig. 6.8**.
4. Use a calibration curve, constructed as described in Step 3, to calculate the amount of luciferase in the transfected cell samples.

5. To allow the results of the luciferase expression assays to be presented as weight luciferase per weight unit total protein, determine total protein content in the lysate as described in **Sect. 3.11**, Steps 3–5.
6. Calculate weight luciferase per weight unit total protein; normalize results to the reference data determined for untransfected cells. The results can be plotted against time post-transfection or against the siRNA concentration or dose per well in order to get a dose–response curve. Examples of the results are given in **Fig. 6.8a** and **c**.

3.13. MTT-Based Test for Toxicity of the Transfection Complexes

1. Wash transfected adherent cells with 150 μl PBS per well using a multichannel pipette and discard wash solutions.
2. Add 100 μl per well of MTT solution, and incubate in a cell culture incubator for 1.5–2 h.
3. Observe the accumulation of the insoluble violet formazan crystals. When necessary, continue the incubation to obtain an optical density of ~0.3–1.0 at 550–590 nm for untreated cells (as a reference) after product solubilization.
4. Add 100 μl MTT solubilization solution to dissolve formazan.
5. Seal the plate with parafilm or an adhesive film to avoid liquid evaporation, and incubate overnight at RT until complete dissolution of the formazan crystals.
6. Measure the optical density *D* of the MTT–formazan solution after solubilization in the range of the wide absorption spectrum maximum (550–590 nm), for example, at 590 nm, using a microplate reader (e.g., Wallac Multi-label Counter; measuring time 0.1 s). Use untransfected cells as a reference. Register the absorbance for one or several wells with a mixture of 100 μl MTT solution and 100 ml solubilization solution as a blank.
7. Cell viability in terms of cell respiration activity *(34,35)* normalized to the reference data (%) is expressed as: *Cell viability (%)* $= (D_{sample} - D_{blank})/(D_{ref} - D_{blank})\cdot 100\%$, here

Fig. 6.8. (continued) for luciferase. (**c**) Luciferase stably transfected H441 cells (H441-Luci-cells) were seeded in a 96-well plate and 24 h later transfected with naked siRNA, PEI/siRNA, and Mf/siRNA poly- and lipoplexes, or magnetic duplexes PEI-Mag2/siRNA (iron-to-siRNA ratio of 1) or magnetic triplexes PEI-Mag2/PEI/siRNA, PL-Mag1/Mf/siRNA, and PalD1/Mf/siRNA. Luciferase expression in lysate and cell viability in terms of respiration activity was measured 48 h post-transfection as a percentage of a reference (untransfected cells). In (a) and (c) PEI/siRNA ratio of N/P = 10. Mf-to-siRNA vol/wt ratio of 4. Iron-to-siRNA ratio for magnetic triplexes 0.5:1. The results in (a) show minimum luciferase expression (maximum downregulation effect) in HeLa-Luci cells between 24 and 48 h post-transfection, i.e., in screening experiments for siRNA delivery by magnetic complex analysis of luciferase expression can be performed usually after 48 h post-transfection. The results in (c) imply that magnetic triplexes comprising PEI-Mag2 nanomaterial are considerably more efficient in downregulation of the target gene in H441-Luci cells compared to similar nonmagnetic vectors. The results of the MTT assay suggest relatively low toxicity of tested complexes within the tested siRNA concentration range.

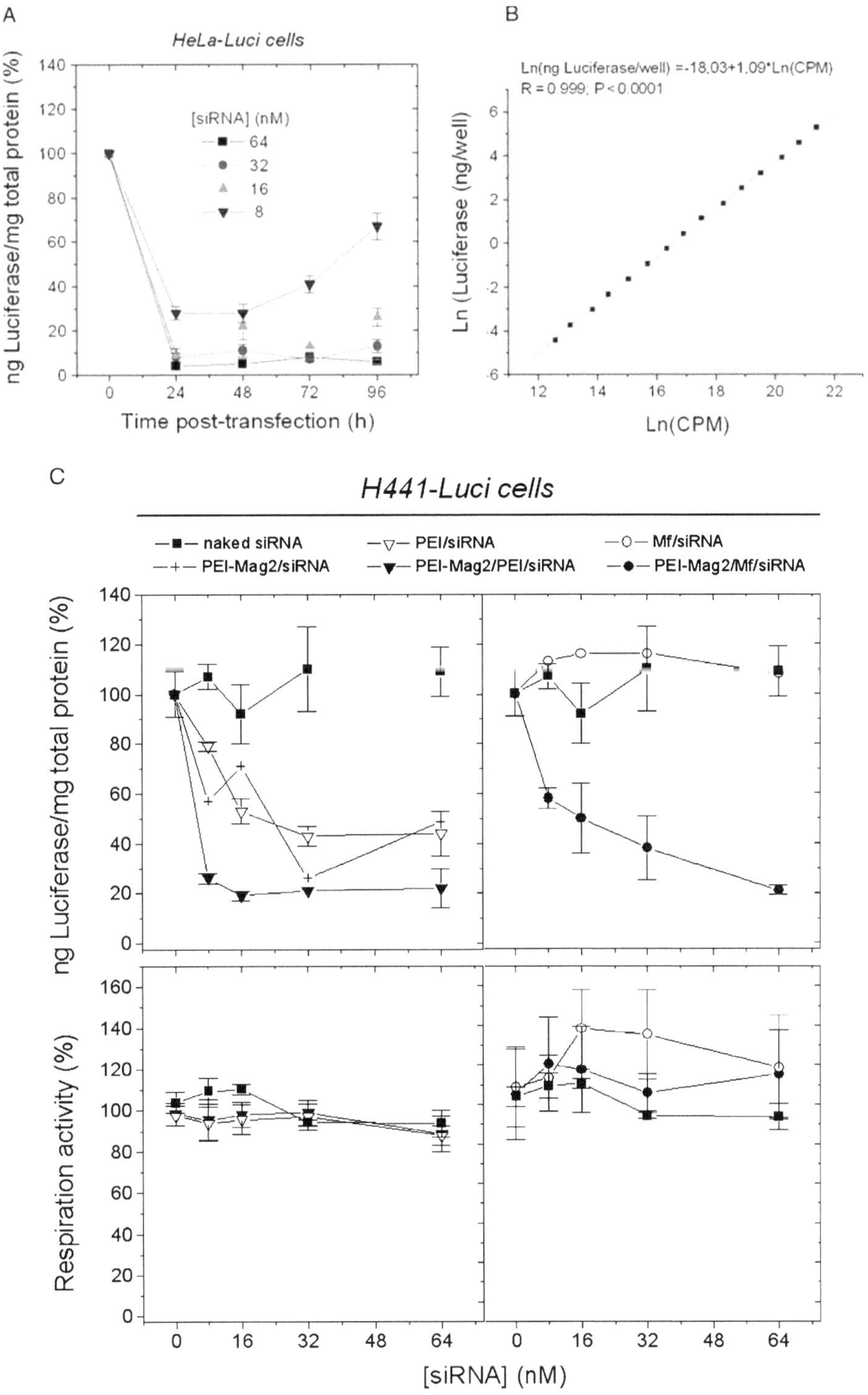

Fig. 6.8. Luciferase reporter gene expression analysis in cell lysates. Magnetofection versus lipofection and polyfection efficiency in H441-Luci cells and MTT-based toxicity test. (**a**) Luciferase stably transfected HeLa cells (HeLa-Luci cells) were seeded in a 96 well plate and 24 h later transfected with magnetic triplexes PL-Mag1/PEI/siRNA. Luciferase expression was monitored in cell lysates in function of post-transfection incubation time. (**b**) Shows a calibration curve.

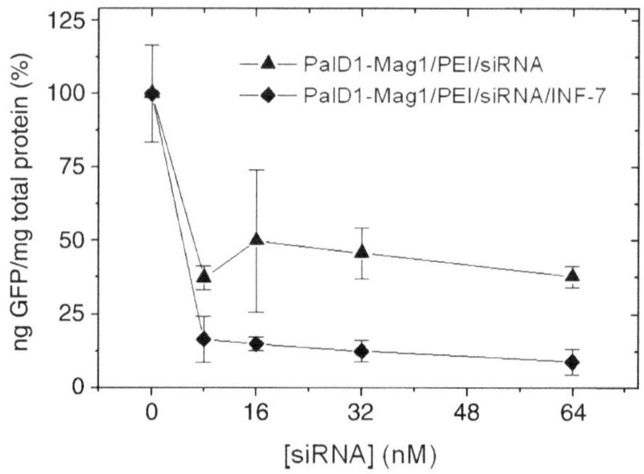

Fig. 6.9. Enhancement of the GFP downregulation by magnetic siRNA vectors modified with INF-7 fusogenic peptide in GFP stably transfected HeLa human cervical epithelial adenocarcinoma cells. The cells were seeded in a 96-well plate and 24 h later transfected with magnetic triplexes PalD1-Mag1/PEI/siRNA or modified with INF-7 peptide as described in Sect. 4.14 (INF-7-to-siRNA mol/mol ratio of 9.4) PalD1-Mag1/PEI/siRNA/INF-7. Iron-to-siRNA ratio of 0.5 to 1. PEI-to-siRNA ratio N/P = 10. GFP expression was monitored 72 h post-transfection. Results show percentages of reporter gene inhibition. Plots assigned as shown in the figure. The results clearly demonstrate considerable GFP-gene silencing improvement resulting from modification of the magnetic transfection triplexes with INF-7 peptide.

D_{sample}, D_{blank} and D_{ref} are optical densities at the maximum of the MTT–formazan absorption spectrum registered for a sample, blank and reference sample, respectively. Examples of the results are given in **Fig. 6.8c**.

3.14. Enhancement of the Reporter Gene Downregulation by Magnetic siRNA Vectors Modified with INF-7 Fusogenic Peptide

To test the efficiency of the magnetic transfection triplexes modified with INF-7 fusogenic peptide (*see* **Note 21**):

1. Add 10 μl INF-7 solution to the transfection complexes from **Sect. 3.6**, Step 4 (*see* **Note 22**).
2. Perform Steps 5 and 9, described in **Sect. 3.6** in order to generate dilutions of the INF-7 modified complexes.
3. Perform Magnetofection according to **Sect. 3.8** and evaluate the GFP expression in cell lysates as described in **Sect. 3.11**. An example of the data is shown in **Fig. 6.9**.

4. Notes

1. Unless stated otherwise, all solutions should be prepared in water that has a resistivity of 18.2 MΩ;·cm and total organic

content of less than five parts per billion. This standard is referred to as "water" in this text.

2. Palmitoyl dextran PalD1 for PalD1-Mag1 nanomaterial stabilization can be synthesized using a modification of Suzuki's procedure *(36)*: Dry Dextran 10 (20 g; $\bar{M}_w = 10500$; Amersham Biosciences) is suspended in anhydrous formamide (300 ml); the mixture is stirred in a water bath at 70°C for 1.5 h until the polysaccharide dissolves completely. Anhydrous tributylamine (30 ml; Sigma) and palmitoyl chloride (6 ml; Sigma) are added, and warming and stirring is continued for a further 2 h. The reaction mixture is cooled to 20°C. The resultant solution is diluted with 1500 ml methanol; a white precipitate is formed, collected by centrifugation, washed with methanol, and dried in vacuum. The crude product is dissolved in 400 ml formamide by stirring and heating at 50°C. The solution is poured into 2000 ml methanol. The white precipitate is collected by centrifugation, washed with methanol, and dried in vacuum. The PalD1 product with an esterification degree of 10 palmitoyl groups per 100 dextran units (determined as described in ref. 37) is water soluble.
3. This iron-to-siRNA weight ratio has turned out useful for triplexes of a variety of magnetic nanoparticle types. To determine the optimal weight ratio for an unknown particle type, it is useful to carry out this protocol also with magnetic nanoparticle stock suspensions of 18, 72, or more µg iron per ml (resulting in wt/wt ratios of 0.25 and 1 or higher).
4. The particles may aggregate due to magnetization; therefore, do not magnetize magnetic nanoparticles before transfection. Do not freeze magnetic nanoparticle suspensions. Before use, always vortex magnetic nanoparticle suspensions very thoroughly. Optionally, sonicate magnetic nanoparticle suspension after longer periods of storage using a water bath sonicator.
5. Gamma sterilization is preferred as heat sterilization in an autoclave in the presence of air can result in at least partial desorption of the coating components as well as surface oxidation of the magnetite nanocrystals. It is important to avoid freezing and magnetization of the suspensions prior to magnetofection.
6. Appropriate dilutions for measurement have a concentration between 0.5 and 6 µg iron per ml. The suggested final dilution for measurement of the original 20 µl magnetic nanoparticle with iron concentration of 10–90 mg iron per ml is 1:15,000.

7. The iron content of the magnetic nanoparticles varies from 0.41 to 0.56 g iron per g dry weight; aqueous suspensions after dialysis of the material contain usually ~10 mg iron per ml.
8. The method described in **Sect. 3.1** yields materials with magnetite cores with a mean crystallite size of 8–11 nm. The mean magnetite crystallite size was calculated from the broadening of the x-ray diffraction peak using the Scherer formula. The mean hydrodynamic diameters of these particles vary from 63 to 101 nm, and the x-potentials of the materials vary from highly positive (+55 mV) to negative (−16 mV), depending on the coating material used. The hydrodynamic diameter and *z*-potential of the magnetic nanoparticles and transfection complexes (given in **Table 6.2**) were determined by photon correlation spectroscopy using, for example, a Malvern Zetasizer 3000 (UK). The average iron weight per particle is evaluated taking into account the magnetite crystallite size (core diameter). The effective magnetic moment of the insulated particle is evaluated taking into account the experimentally determined saturation magnetization of the core and average iron weight per particle. The parameters given here are usually not supplied by manufacturers of magnetic nanomaterials. The physical and chemical characteristics of the particles are listed in **Table 6.1**. Remarkably, the ξ-potentials of these particles range from highly negative to highly positive. Particles with a negative ξ-potential are not suitable to bind nucleic acids on their own. For this purpose, either a combination with enhancers or divalent cations are required. These magnetic particles are used in combination with either siRNA alone (in the latter case, only positively charged particles such as PEI-Mag2 are used) or in formulations with nucleic acids and enhancers (either a lipid transfection reagent or PEI). We have found that particles with a magnetite crystallite size of 9–11 nm are superior to smaller particles with 3–4 nm crystallite size as components of the magnetic transfection vectors for magnetofection. For more details concerning the synthesis and characterization of the magnetic nanoparticles as components of gene vectors, see also ref. *(24)*.
9. This protocol has to be performed by authorized personnel and according to the rules and regulations for work with radioactive substances. Use pipette tips provided with an aerosol filter to avoid radioactive contamination of the pipette. This procedure can be accomplished during 2 h.
10. For complete dissolution of thallium chloride just before DNA labeling, heat the solution to 70°C using a water bath. Caution: Thallium chloride is highly toxic.

11. The N/P ratio is a measure of the ionic balance of the complexes and refers to the number of nitrogen residues of PEI per DNA phosphate. 1 μg of DNA contains 3 nmol of anionic phosphate. For N/P = 10, 1 μg DNA (3 nmol phosphate) corresponds to 30 nmol PEI units, i.e., to 1.3 μg PEI.
12. Cell culture and plating should be performed under sterile conditions. Timing: 30 min cell plating plus 24 h cell culture in a plate before transfection.
13. Cell seeding in clear bottom black-walled plates enables GFP expression measurements in living cells.
14. The order of reagent mixing and the medium for reagent dilution can be critical for the sizes, charges, and compositions of the complexes, and thus for final transfection efficiencies. To optimize the conditions for a given cell line, magnetic nanoparticle type and enhancer reagent, also other mixing orders as described above should/could be tested.
15. Timing: 60 min. Prepare transfection complexes just before transfection—all stages should be undertaken under sterile condition.
16. To evaluate the magnetic moment M of the magnetic complex and estimate the number of magnetic nanoparticles N associated with the complex, an efficient velocity υ_z of the complexes under a gradient magnetic field evaluated from the magnetic responsiveness curves as $\upsilon_z = \langle L \rangle / t_{0.1}$, where $\langle L \rangle = 1$ mm is an average path of the particle movement in an optical cuvette and $t_{0.1}$ is the time required for a 10-fold decrease in optical density. The magnetic moment M of the complexes is calculated from the efficient velocity υ_z of the complexes as $M = 3\pi\eta D\upsilon_z / \frac{\partial B}{\partial z}$, where D is the average hydrodynamic diameter of the complexes determined using the dynamic light scattering method, and $\eta = 8.9 \cdot 10^{-4}$ Pa·s (kg·m^{-1}·s^{-1}) is the viscosity of water. The total magnetic moment of the complex $\vec{M}$ is the product of the effective magnetic moment m_{eff} of the magnetic nanoparticle under the magnetic field $\vec{B}$ and the total number N of magnetic particles associated with the complexes as $M = N \cdot m_{\text{eff}}$. At a magnetic field of 213 mT, the magnetization of the magnetite nanoparticles according to the experimentally measured magnetization curve corresponds to 97% of its saturation value, M_s. Thus the effective magnetic moment m_{eff} of each particle is $m_{\text{eff}} = (0.97 M_s) P_{\text{part}}^{\text{Fe}}$, where M_s is the specific saturation magnetization per unit of iron weight and $P_{\text{part}}^{\text{Fe}}$ is the content of iron in one particle with the diameter of the magnetite core equal to the average crystallite size. Thus

determination of the complex velocity υ_z leads to evaluation of the number of magnetic particles associated with the complex. For more details on the physical background for the estimations see also ref. *(38)*.

17. The optimal incubation conditions may differ from one cell type to another and from one complex to another and must be determined experimentally.
18. Use the same type of 96-well clear bottom black-walled plate for both eGFP calibration curve measurements and experimental sample measurements. Make sure to measure equal volumes for the calibration curve and the experimental samples.
19. Bear in mind that the luciferase and GFP assays are carried out with 50 μl cell lysate, while the protein assay is carried out with 10 μl only. Correspondingly, the measured values for luciferase (or GFP) must be divided by 5 to obtain correct results, when normalizing per total protein determined in 10 μl cell lysate.
20. To allow repeated GFP expression measurements post-transfection in living cells, the cells have to be seeded in a 96-well clear bottom black-wall plate: Greiner Bio-One; catalogue no. 655090 (or similar clear bottom black-wall plate).
21. INF-7 endosomolytic peptide derived from the influenza virus INF7 containing 24 amino acids [GLFEAIEGFIENGWEGMIDGWYGG) *(39)*] can be synthesized using the published procedure *(27)* or purchased from NeoMPS (Strasbourg, France).
22. Addition of the 2.33 μg INF-7 (865 pmol) peptide to (1.44 μg) 92.1 pmol siRNA in a 360 μl transfection complex results in INF-7-to-siRNA mol/mol ratio of 9.4.

Acknowledgments

The authors would like to thank Dr. Bob Scholte for transduction of the H441 cells with eGFP and luciferase using lentiviral vectors. This work was supported by the European Union through the FP6-LIFESCIHEALTH Project "Improved precision of nucleic acid based therapy of cystic fibrosis" under contract no. 005213 as well as by the German Ministry of Education and Research, Nanobiotechnology grants 13N8186 and 13N8538. Financial support of the German Excellence Initiative via the "Nanosystems Initiative Munich (NIM)" is gratefully acknowledged.

References

1. Plank, C., Scherer, F., Schillinger, U., and Anton, M. (2000) Magnetofection: enhancement and localization of gene delivery with magnetic particles under the influence of a magnetic field. *J. Gene. Med.*, 2, 24.
2. Mah, C. Zolotukhin, I., Fraites, T.J., Dobson, J., Batich, C., Byrne, B.J. (2000) Microsphere-mediated delivery of recombinant AAV vectors in vitro and in vivo. *Mol. Ther.*, 1, S239.
3. Hughes, C., Galea-Lauri, J., Farzaneh, F., and Darling, D. (2001) Streptavidin paramagnetic particles provide a choice of three affinity-based capture and magnetic concentration strategies for retroviral vectors. *Mol. Ther.*, 3, 623–630.
4. Scherer, F., Anton, M., Schillinger, U., Henke, J., Bergemann, C., Kruger, A., Gansbacher, B., and Plank, C. (2002) Magnetofection: enhancing and targeting gene delivery by magnetic force in vitro and in vivo. *Gene Ther.*, 9, 102–109.
5. Mah, C., Fraites, T.J., Jr., Zolotukhin, I., Song, S., Flotte, T.R., Dobson, J., Batich, C., and Byrne, B.J. (2002) Improved method of recombinant AAV2 delivery for systemic targeted gene therapy. *Mol. Ther.*, 6, 106–112.
6. Pandori, M.W., Hobson, D.A., and Sano, T. (2002) Adenovirus-microbead conjugates possess enhanced infectivity: A new strategy to localized gene delivery. *Virology*, 299, 204–212.
7. Plank, C., Anton, M., Rudolph, C., Rosenecker, J., and Krotz, F. (2003) Enhancing and targeting nucleic acid delivery by magnetic force. *Expert Opin. Biol. Ther.*, 3, 745–758.
8. Schillinger, U., Brill, T., Rudolph, C., Huth, S., Gersting, S., Krotz, F., Hirschberger, J., Bergemann, C., and Plank, C. (2005) Advances in magnetofection - magnetically guided nucleic acid delivery. *J. Magn. Magn. Mater.*, 293, 501–508.
9. Huth, S., Lausier, J., Gersting, S.W., Rudolph, C., Plank, C., Welsch, U., and Rosenecker, J. (2004) Insights into the mechanism of magnetofection using PEI-based magnetofectins for gene transfer. *J. Gene Med.*, 6, 923–936.
10. Krotz, F., de Wit, C., Sohn, H.Y., Zahler, S., Gloe, T., Pohl, U., and Plank, C. (2003) Magnetofection--a highly efficient tool for antisense oligonucleotide delivery in vitro and in vivo. *Mol. Ther.*, 7, 700–710.
11. Plank, C., Schillinger, U., Scherer, F., Bergemann, C., Remy, J.S., Krotz, F., Anton, M., Lausier, J., and Rosenecker, J. (2003) The magnetofection method: using magnetic force to enhance gene delivery. *Biol. Chem.*, 384, 737–747.
12. Dobson, J. (2006) Gene therapy progress and prospects: magnetic nanoparticle-based gene delivery. *Gene Ther.*, 13, 283–287.
13. Doshida, M., Ohmichi, M., Tsutsumi, S., Kawagoe, J., Takahashi, T., Du, B., Mori-Abe, A., Ohta, T., Saitoh-Sekiguchi, M., Takahashi, K. et al. (2006) Raloxifene increases proliferation and up-regulates telomerase activity in human umbilical vein endothelial cells. *J. Biol. Chem.*, 281, 24270–24278.
14. Deleuze, V., Chalhoub, E., El-Hajj, R., Dohet, C., Le Clech, M., Couraud, P.O., Huber, P., and Mathieu, D. (2007) TAL-1/SCL and its partners E47 and LMO2 up-regulate VE-cadherin expression in endothelial cells. *Mol. Cell Biol.*, 27, 2687–2697.
15. McCaig, C., Duval, C., Hemers, E., Steele, I., Pritchard, D.M., Przemeck, S., Dimaline, R., Ahmed, S., Bodger, K., Kerrigan, D.D.et-al. (2006) The role of matrix Metalloproteinase-7 in redefining the gastric microenvironment in response to Helicobacter pylori. *Gastroenterology*, 130, 1754–1763.
16. Uchida, Y., Ohshima, T., Sasaki, Y., Suzuki, H., Yanai, S., Yamashita, N., Nakamura, F., Takei, K., Ihara, Y., Mikoshiba, K. et al. (2005) Semaphorin3A signalling is mediated via sequential Cdk5 and GSK3b phosphorylation of CRMP2: implication of common phosphorylating mechanism underlying axon guidance and Alzheimer's disease. *Genes Cells*, 10, 165–179.
17. Huang, P., Senga, T., and Hamaguchi, M. (2007) A novel role of phospho-[beta]-catenin in microtubule regrowth at centrosome. *Oncogene*, 26, 4357–4371.
18. Mizutani, T., Fukushi, S., Iizuka, D., Inanami, O., Kuwabara, M., Takashima, H., Yanagawa, H., Saijo, M., Kurane, I., and Morikawa, S. (2006) Inhibition of cell proliferation by SARS-CoV infection in Vero E6 cells. *FEMS Immunol. Med. Microbiol.*, 46, 236–243.
19. Sapet, C., Simoncini, S., Loriod, B., Puthier, D., Sampol, J., Nguyen, C., Dignat-George, F., and Anfosso, F. (2006) Thrombin-induced endothelial microparticle generation: identification of a novel pathway involving ROCK-II activation by caspase-2. *Blood*, 108, 1868–1876.

20. Minami, R., Yamamoto, M., Takahama, S., Miyamura, T., Watanabe, H., and Suematsu, E. (2006) RCAS1 induced by HIV-Tat is involved in the apoptosis of HIV-1 infected and uninfected CD4+ T cells. *Cell. Immunol.*, 243, 41–47.
21. Bonetta, L. (2005) The inside scoop—evaluating gene delivery methods. *Nat. Methods*, 2, 875–883.
22. Smith, C. (2006) Sharpening the tools of RNA interference. *Nat. Methods*, 3, 475–486.
23. Felgner, P.L., Barenholz, Y., Behr, J.P., Cheng, S.H., Cullis, P., Huang, L., Jessee, J.A., Seymour, L., Szoka, F., Thierry, A.R. et al. (1997) Nomenclature for synthetic gene delivery systems. *Hum. Gene Ther.*, 8, 511–512.
24. Mykhaylyk, O., Vlaskou, D., Tresilwised, N., Pithayanukul, P., Moller, W., and Plank, C. (2007) Magnetic nanoparticle formulations for DNA and siRNA delivery. *J. Magn. Magn. Mater.*, 311, 275–281.
25. Terebesi, J., Kwok, K.Y., and Rice, K.G. (1998) Iodinated plasmid DNA as a tool for studying gene delivery. *Anal. Biochem.*, 263, 120–123.
26. Azzam, T. and Domb, A.J. (2004) Current developments in gene transfection agents. *Curr. Drug Deliv.*, 1, 165–193.
27. Plank, C., Oberhauser, B., Mechtler, K., Koch, C., and Wagner, E. (1994) The influence of endosome-disruptive peptides on gene transfer using synthetic virus-like gene transfer systems. *J. Biol. Chem.*, 269, 12918–12924.
28. Plank, C., Zauner, W., and Wagner, E. (1998) Application of membrane-active peptides for drug and gene delivery across cellular membranes. *Adv. Drug Deliv. Rev.*, 34, 21–35.
29. Funhoff, A.M., van Nostrum, C.F., Lok, M.C., Fretz, M.M., Crommelin, D.J., and Hennink, W.E. (2004) Poly(3-guanidinopropyl methacrylate): a novel cationic polymer for gene delivery. *Bioconjug. Chem.*, 15, 1212–1220.
30. Jiang, X., Lok, M.C., and Hennink, W.E. (2007) Degradable-brushed pHEMA-pDMAEMA synthesized via ATRP and click chemistry for gene delivery. *Bioconjug. Chem.*, 18, 2077–2084.
31. Oliveira, S., van Rooy, I., Kranenburg, O., Storm, G., and Schiffelers, R.M. (2007) Fusogenic peptides enhance endosomal escape improving siRNA-induced silencing of oncogenes. *Int. J. Pharm.*, 331, 211–214.
32. Kowalski, J.B. and Tallentire, A. (1999) Substantiation of 25 kGy as a sterilization dose: a rational approach to establishing verification dose. *Radiat. Phys. Chem.*, 54, 55–64.
33. Gersdorff, K. von. (2006) PEG-shielded and EGF receptor-targeted DNA polyplexes: Cellular mechanisms. Ph D Thesis, Ludwig Maximilian University, Munich, Germany, Supervisor Prof. Dr. E. Wagner. 125 pp. http://edoc.ub.uni-muenchen.de/5485/.
34. Berridge, M.V., Herst, P.M., and Tan, A.S. (2005) Tetrazolium dyes as tools in cell biology: new insights into their cellular reduction. *Biotechnol. Annu. Rev.*, 11, 127–152.
35. Berridge, M.V., Tan, A.S., and Hilton, C.J. (1993) Cyclic adenosine monophosphate promotes cell survival and retards apoptosis in a factor-dependent bone marrow-derived cell line. *Exp. Hematol.*, 21, 269–276.
36. Suzuki, M., Mikami, T., Matsumoto, T., and Suzuki, S. (1977) Preparation and antitumor activity of O-palmitoyldextran phosphates, O-palmitoyldextrans, and dextran phosphate. *Carbohydr. Res.*, 53, 223–229.
37. Snyder, F. and Stephens, N. (1959) A simplified spectrophotometric determination of ester groups in lipids. *Biochim. Biophys. Acta*, 34, 244–245.
38. Wilhelm, C., Gazeau, F., and Bacri, J.C. (2002) Magnetophoresis and ferromagnetic resonance of magnetically labeled cells. *Eur. Biophys. J.*, 31, 118–125.
39. Esbjorner, E.K., Oglecka, K., Lincoln, P., Graslund, A., and Norden, B. (2007) Membrane binding of pH-sensitive Influenza fusion peptides. Positioning, configuration, and induced leakage in a lipid vesicle model. *Biochemistry*, 46, 13490–13504.

Chapter 7

In Vitro and In Vivo Gene Silencing by TransKingdom RNAi (tkRNAi)

Shuanglin Xiang, Andrew C. Keates, Johannes Fruehauf, Youxin Yang, Hongnian Guo, Thu Nguyen, and Chiang J. Li

Abstract

RNA interference (RNAi) is a potent and specific mechanism for eliminating the mRNA of specific genes. This gene silencing mechanism occurs naturally and is highly conserved from plants to human cells, holding promise for functional genomics and for revolutionizing medicine due to its unlimited potential to treat genetic, epigenetic, and infectious disease. However, efforts to unleash the enormous potential of RNAi have met with significant challenges. Delivery is problematic because short interfering RNAs (siRNA) are negatively charged polymers that inefficiently enter cells and undergo rapid enzymatic degradation in vivo. In addition, the synthesis of siRNAs is expensive for long-term research and therapeutic applications. Recently, we have shown that nonpathogenic bacteria can be engineered to activate RNAi in mammalian cells (TransKingdom RNA interference; *tk*RNAi). This new approach offers several advantages and has significant implications. First, this method allows the establishment of a long-term stable gene silencing system in the laboratory against genes of interests in vitro and in vivo, and enables high-throughput functional genomics screening in mammalian systems. RNAi libraries can be constructed, stored, reproduced, amplified, and used with the help of *E. coli* as currently done with gene cloning. Second, this technology provides a clinically compatible way to achieve RNAi for therapeutic applications due to the proven clinical safety of nonpathogenic bacteria as a gene carrier. *tk*RNAi also eliminates the siRNA manufacture issue, and may circumvent or mitigate host interferon-like responses since siRNA is produced intracellularly.

Key words: RNAi, TransKingdomRNA interference, *tk*RNAi, gene silencing, functional genomics, RNAi therapy, shRNA, bacteria-mediated RNAi.

1. Introduction

The recent discovery of RNA interference (RNAi), a potent gene silencing mechanism found in eukaryotic cells, promises to revolutionize medicine due to its unlimited potential to treat genetic,

M. Sioud (ed.), *Methods in Molecular Biology, siRNA and miRNA Gene Silencing, vol. 487*

DOI: 10.1007/978-1-60327-547-7_7

epigenetic, and infectious disease *(1–3)*. RNAi is triggered when endogenous micro RNA, or exogenous double-stranded RNA (dsRNA) or short hairpin RNA (shRNA) are processed by the cytoplasmic enzyme Dicer into 21- to 23-nucleotide short interfering RNA (siRNA) duplexes *(4–6)*. The processed siRNA duplexes are then loaded into a large multiprotein complex called RISC (RNA-induced silencing complex) where the siRNA duplex is unwound and the passenger (sense) siRNA strand is discarded *(7, 8)*. The RISC complex then locates target mRNA using the incorporated guide (antisense) siRNA strand and cleaves them using the slicer activity of the Argonaute protein, thereby preventing protein production.

RNAi has proven a powerful technology for laboratory research of gene functions. However, in vivo gene silencing and large-scale research use of siRNA are limited due to delivery challenge and cost of synthesis *(9, 10)*. To overcome these obstacles, we have developed a bacteria-based RNAi technology called TransKingdom RNAi (*tk*RNAi) (**Fig. 7.1**) for in vitro and in vivo gene silencing in mammalian cells *(11)*.

This approach utilizes genetically engineered, nonpathogenic *E. coli* to simultaneously manufacture silencing shRNA and deliver them to target cells. Our method makes use of a bacterial vector, pT7RNAi-*Hly-Inv*, termed TRIP (TransKingdom RNAi plasmid) that, when introduced into *E. coli* containing endogenous T7 RNA polymerase activity (e.g., BL21 DE3), can produce high levels of silencing shRNA. To enable delivery to the gastrointestinal tract, the TRIP plasmid was engineered to express the invasin gene (*Inv*) from *Yersinia pseudotuberculosis*, which allows noninvasive *E. coli* to enter β1-integrin positive epithelial cells *(12, 13)*. To facilitate efficient gene silencing following cell entry, the TRIP vector was also engineered to express the listeriolysin O gene (*HlyA*) from *Listeria monocytogenes*, which allows the bacterially produced shRNA to escape from entry vesicles *(14, 15)*. Using this approach, we have shown that *tk*RNAi directed against the colon cancer oncogene β-catenin can induce significant gene silencing in vitro and in vivo *(11)*.

This technique offers a number of advantages over chemically modified siRNA and viral vector-mediated shRNA methods for biomedical research and development of RNAi-based medical therapies.

First, *tk*RNAi has significant implications for high throughput functional genomics in mammalian systems. Bacteria, *E. coli* in particular, have served as a well-validated and versatile vector system for the revolution in molecular biology and biotechnology that has occurred over the last few decades. Using *tk*RNAi, a laboratory can easily establish *E. coli*-based RNAi against various genes of interest. Besides saving on the cost of siRNA manufacture, the bacterial gene silencing system can be reproduced and

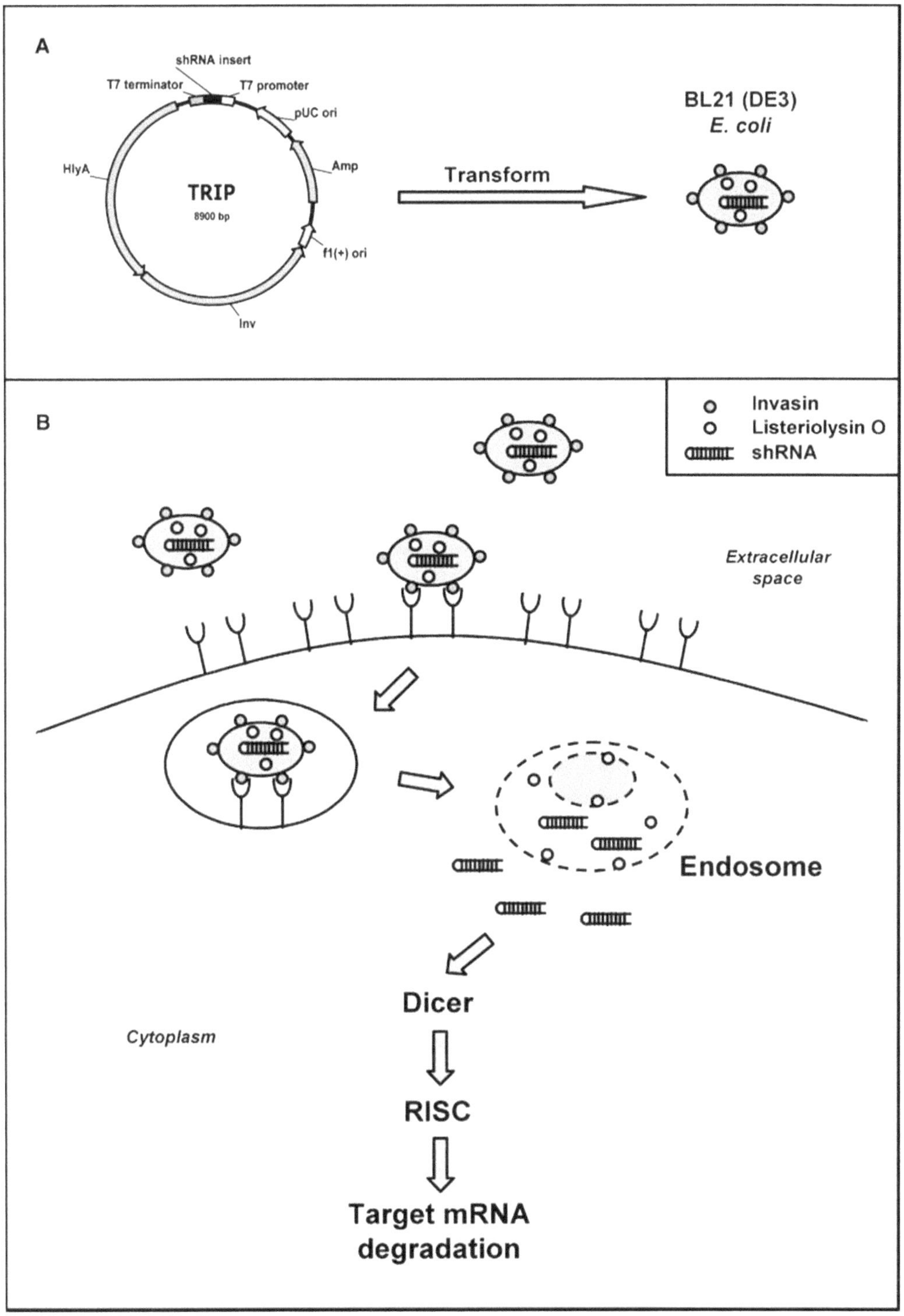

Fig. 7.1. Schematic representation of TransKingdom RNAi.

stored for long term use, offering a stable and consistent gene silencing tool. Further, the bacterial vector can be used in a routine biological laboratory, rather than a BL2 laboratory, which is required for viral vector-based systems. The bacterial system can also conveniently be translated to in vivo systems in order to verify in vitro observations in live animals. Other advantages of using bacteria as a delivery vector for siRNA include the ability to control the vector using antibiotics and/or auxotrophy, and the ease of engineering specific vectors for particular applications.

Second, the *tk*RNAi system provides a practical and clinically compatible way to achieve RNAi for medical therapies. Although RNAi can theoretically be employed to target any disease gene with specificity and potency, the development of RNAi-based therapeutics has been impeded by challenges in delivery, manufacture, and the activation of host interferon-like responses *(9, 10)*. Of these, delivery has proven to be the major stumbling block. In order to overcome these limitations, researchers have focused mainly on developing chemically modified siRNA to increase stability, and on complexing siRNA with liposomes or nanoparticles to promote cell uptake *(9, 16, 17)*. The main disadvantage of these pharmaceutical approaches, however, is that they tend to have a limited ability to target specific cell types or tissues. Moreover, they typically require large quantities of siRNA that is expensive to manufacture. Viral vectors have also been explored as a means of delivering RNAi in vivo *(9)*. While this approach has important research applications, problems associated with insertional mutagenesis, safety, lack of tropism, and the generation of host immune responses have significantly limited the utility of viral vectors for gene therapy.

Unlike viral vectors, nonpathogenic bacteria (including *E. coli*) have been used safely for many years as probiotics or as vaccine vectors, and do not integrate into the human genome *(18–21)*. Thus, the *tk*RNAi technique may help to achieve potent and therapeutic RNAi with versatility and less cost. This RNAi approach can be exploited clinically to silence genes of interest in the colonic mucosa, and possibly in other organs which can be colonized by commensal or nonpathogenic bacteria, including the oral cavity, urinary bladder, female genital tract, etc. Because shRNAs are released intracellularly by the engineered bacteria directly into the cytoplasm, this RNAi approach eliminates the siRNA manufacture issue and may have the advantage of mitigating the Toll-like receptor-mediated immunostimulatory effect of siRNA (*see* Chap. 8). Therefore, the TransKingdom system may provide a practical and clinically compatible way to achieve RNAi for medical indications.

2. Materials

2.1. Bacterial Culture

1. Brain Heart Infusion (BHI) Broth (Remel, M & M Industries, Inc., Chattanooga TN): Dissolve 37 g BHI in 1 l of tissue culture water, dispense into appropriate containers, and sterilize by autoclaving at 121°C for 15 min. Store at room temperature.
2. Sterile disposable round-bottom plastic tubes with dual-position cap (14 ml: VWR International, West Chester, PA); KIMAX brand baffled culture flasks (500 ml; Fisher Scientific, Pittsburgh, PA).
3. Ampicillin 100 mg/ml (Sigma, St Louis, MO). Store at –20°C.
4. Isopropyl β-D-1-thiogalactopyranoside (IPTG; Sigma, St Louis, MO). Prepare 1 M working solution, filter through a 0.22 μm filter, and store at –20°C.

2.2. Cell Culture

1. SW480 colonic epithelial cells (American Type Culture Collection, Manassas, VA) are maintained in complete growth media in an atmosphere of 5% CO_2 and 95% air. For long-term storage, cells are resuspended in complete growth medium supplemented with 5% (v/v) dimethyl sulfoxide and placed in liquid nitrogen.
2. Roswell Park Memorial Institute (RPMI) medium (Invitrogen, Carlsbad, CA) supplemented with 10% fetal bovine serum (Invitrogen, Carlsbad, CA). Store at 4°C.
3. Sterile plastic tissue culture flasks (75 cm^2; BD Falcon, Franklin Lakes, NJ) and dishes (6 cm; Corning Life Sciences, Lowell, MA).
4. Penicillin–streptomycin solution: 10,000 U/ml penicillin, 10 mg/ml streptomycin in 0.9% sodium chloride (Sigma, St Louis, MO). Store at –20°C.
5. Amphotericin B solution: 250 mg/ml in tissue culture water (Sigma, St Louis, MO). Store at –20°C.
6. 0.25% Trypsin–EDTA solution (Sigma, St Louis, MO) is stored in aliquots at –20°C.
7. Gentamycin solution: 10 mg/ml in tissue culture water. Store at 4°C.
8. Ofloxacin 10 mg/ml in tissue culture water. Store at –20°C. All working solutions are prepared by diluting the corresponding stock solution 1000-fold.

2.3. Oral Administration of Bacteria to Mice

1. Female C57BL/6 mice (Charles River Laboratories, Wilmington, MA) are housed under conventional conditions in isolator cages (4 mice per cage). Mice are fed standard chow

(Harlan Teklad, Indianapolis, IN) and provided with tap water ad libitum.

2. 10X phosphate buffered saline is stored at room temperature. A 1X working solution is prepared by diluting the stock solution 10-fold in sterile water. Store the solution at 4°C.
3. Feeding needle (PS 20; Poppers and Sons, Inc., New Hyde Park, NY).
4. 1 ml Norm-Ject tuberculin syringes (Henke Sass Wolf, Tuttlingen, Germany).

2.4. Intravenous Administration of Bacteria to Mice

1. Female nude Balb/c mice (Nu/Nu; Charles River Laboratories, Wilmington, MA) are housed under specific pathogen-free conditions in sterile isolator cages (4 mice per cage). Mice are fed with irradiated chow (Harlan Teklad, Indianapolis, IN) and provided with sterile water ad libitum
2. 1 ml Norm-Ject tuberculin syringes (Henke Sass Wolf, Tuttlingen, Germany).
3. 26G1/2 PrecisionGlide needles (Becton Dickinson, Franklin Lakes, NJ).

3. Methods

In vitro and in vivo gene silencing via *tk*RNAi takes advantage of a "Trojan Horse" strategy. In this system, interfering shRNA are transcribed (from the T7 promoter) inside nonpathogenic bacteria (or commensal bacteria) that are engineered to actively invade target cells. This process is comprised of four steps:

1. Cell entry
2. Bacterial lysis and rupture of the entry vesicle
3. Release of interfering shRNA into the host cell cytoplasm
4. Gene-silencing through RISC

As gene silencing relies on effective bacterial invasion and the intracellular release of interfering shRNA, the ability of the engineered bacteria to stimulate cellular uptake by endocytosis and subsequent intracellular are important.

For systemic treatment, intravenous injection and oral administration of shRNA-expressing bacteria can be used. Intravenous injection of nude mice carrying human colon cancer xenografts with 1×10^8 c.f.u. of engineered *E. coli* resulted in bacterial delivery to the liver, spleen, and tumor tissue during the first 24 h. It should be noted that during the first week after treatment, bacterial numbers continue to increase in the tumor tissue but decrease in the liver and spleen. Interestingly, injection of live engineered bacteria did not generate a significant endotoxin response in nude mice; thus, repetitive treatment is feasible. The enrichment of the engineered *E. coli* in the xenograft tissue likely reflects the ability of this facultative anaerobe strain to colonize and replicate in the hypoxic environment present in tumor tissues.

For oral administration of *tk*RNAi, attenuated versions of *E. coli* are developed, thus allowing the administration of large amounts of therapeutic bacteria (typically in the order of 10^{10} c.f.u.). Such attenuated *tk*RNAi bacteria do not colonize the gastrointestinal tract when administered orally. They are rapidly eliminated.

3.1. *In Vitro* tk*RNAi*

A typical in vitro *tk*RNAi experiment (*see* **Fig. 7.2**) consists of three experimental phases:

1. Preparation of shRNA-expressing bacteria
2. Target cell preparation and bacterial infection
3. Postinfection treatment and assessment of target gene silencing

3.1.1. Bacteria Preparation

1. Chemically competent *E. coli* BL21-Gold (DE3) (50~100 μl) cells are transformed with control or silencing TRIP plasmids (100 ng) according to the manufacturer's instructions (Stratagene; *see* **Note 1**). Bacteria are then grown on BHI plates containing 100 μg/ml ampicillin overnight at 37°C until colonies appear. A single colony is then inoculated into BHI medium containing 100 μg/ml ampicillin, and grown overnight at 37°C (*see* **Note 2**).
2. The next day, 5 ml of each overnight culture is diluted 1:40 into fresh BHI medium containing 100 μg/ml ampicillin

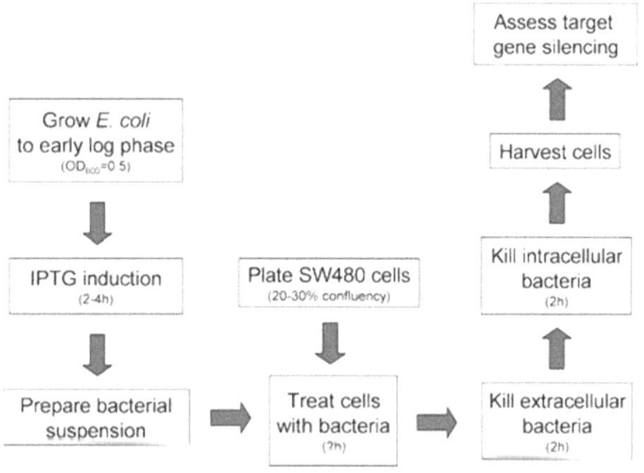

Fig. 7.2. Flow chart outlining in vitro TransKingdom RNAi.

and grown for a further 2–4 h (until the OD_{600} = 0.5). Each culture is then treated with IPTG (1 mM final concentration) for 2–4 h to induce transcription of interfering shRNA (*see* **Note 3**). To verify the identity of each bacterium, plasmid DNA is isolated from each overnight culture and sequenced using standard techniques.

3. After IPTG induction, the total number of bacteria in each culture is calculated by measuring the OD_{600} value (8×10^8 bacteria/ml culture has an OD_{600} = 1). The number of bacteria for cell treatment is then calculated according to cell confluency and the needed multiplicity of infection. We advise to use a range of 20:1 to 2000:1, bacteria to cells in appropriate reaction volumes.
4. The required volume of bacteria culture is then centrifuged at 2500g for 10 min at 4°C and the pellet is washed once with serum-free, RPMI 1640 medium containing 100 μg/ml ampicillin and 1 mM of IPTG, and resuspended in the same medium at the required density for bacterial infection.

3.1.2. Cell Preparation and Bacterial Infection

1. SW480 human colon cancer cells are cultured in an atmosphere of 95% air, 5%CO_2 at 37°C in RPMI 1640 medium containing 10% FBS, 10 U/ml penicillin G, 10 μg/ml streptomycin, and 250 μg/ml amphotericin. 24 h prior to bacterial infection, stock cell cultures are trypsinized, resuspended in complete RPMI 1640 medium, and plated on 6 cm tissue culture dishes at 20–30% confluency.
2. 30 min prior to bacterial infection, the cell culture medium is replaced with 2 ml of fresh serum-free RPMI 1640 medium containing 100 μg/ml of ampicillin and 1 mM IPTG.
3. Bacteria as prepared in **Sect. 3.1.1** are then added to the cells at the desired MOI for 2 h at 37°C.

3.1.3. Postinfection Treatment and Assessment of Target Gene Silencing

1. After the infection period, the cells are washed 3 times using serum-free RPMI 1640 medium (*see* **Note 4**). The cells are then incubated with 2 ml of fresh complete RPMI 1640 medium containing 100 μg/ml of ampicillin and 150 μg/ml of gentamycin for 2 h to kill any remaining extracellular bacteria.
2. After treatment with ampicillin and gentamycin, the cells are incubated with 3 ml RPMI 1640 medium containing 10 μg/ml of ofloxacin in order to kill any intracellular bacteria.
3. The cells are then harvested at various time points (from 24 to 96 h) in order to assess the extent of target gene silencing by real-time PCR (for mRNAs) and Western blotting (for proteins). To illustrate the protocol, some experimental data are presented in **Fig. 7.3**.

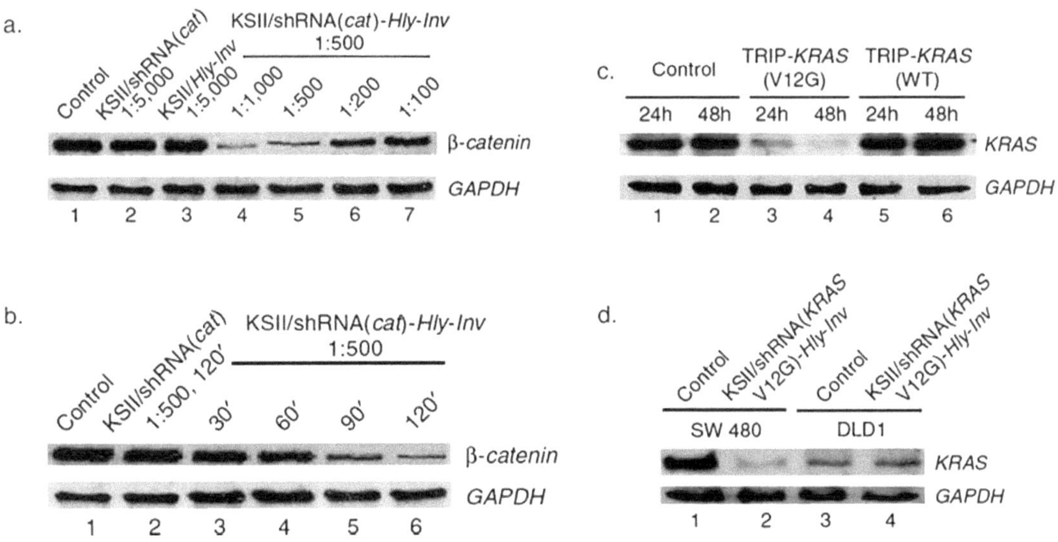

Fig. 7.3. TransKingdom RNAi in vitro. (**a**) The expression of beta-catenin protein was silenced in SW480 cells after treatment with *E. coli* expressing shRNAs against beta-catenin at various MOI (lanes 4–7). Lanes 2 and 3 show lack of gene silencing when either the hly-inv part of the plasmid (lane 2) or the shRNA encoding part (lane 3) of the plasmid are missing, even at very high MOI. (**b**) The gene-silencing effect depends on the co-culture time of human cells with the bacteria (lanes 3–6). *E. coli* containing a TRIP against wild-type k-Ras exerted no gene silencing effects against mutant k-Ras (V12G). (**c**) *E. coli* containing a TRIP against mutant k-Ras (V12G) silenced codon-matched mutant k-Ras expression in SW480 cells, but not in DLD1 cells containing a k-Ras mutation in a different codon.

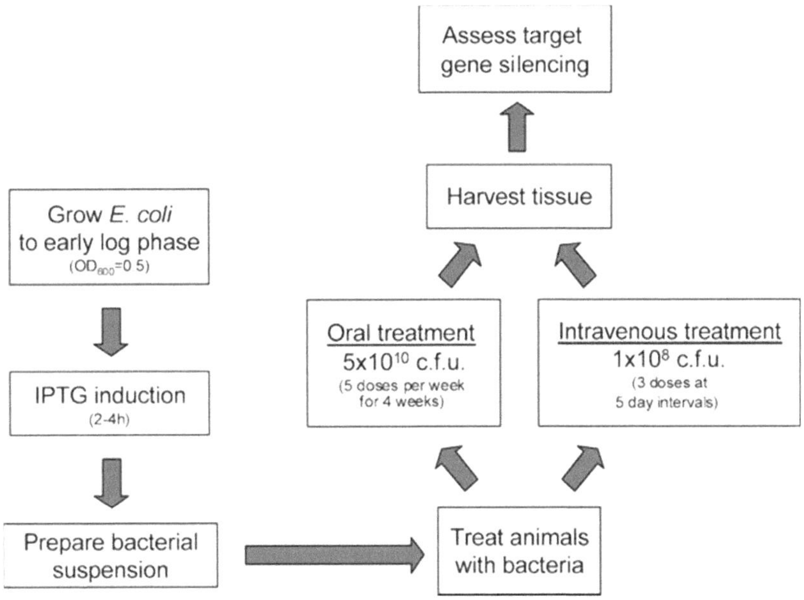

Fig. 7.4. Flow chart outlining in vivo TransKingdom RNAi.

3.2. In Vivo tkRNAi

Similar to the in vitro *tk*RNAi protocol (**Sect. 3.1**), a typical in vivo *tk*RNAi experiment (*see* **Fig. 7.4**) also consists of three experimental phases (**Fig. 7.4**):

1. Preparation of shRNA-expressing bacteria
2. Oral or intravenous delivery
3. Assessment of target gene knockdown

3.2.1. Bacteria Preparation

1. Transformed *E. coli* BL21 (DE3) bacteria containing control or silencing TRIP plasmids (verified by plasmid purification and DNA sequencing) are grown in BHI medium containing 100 μg/ml ampicillin at 37°C until early log phase (OD_{600} = 0.5). The bacteria are then harvested by centrifugation at 2500g for 10 min at 4°C, resuspended in 25 ml of BHI medium, aliquoted and stored in –80°C freezer as 15% glycerol stock (*see* **Note 5**).
2. One day prior to animal treatment, the bacteria stocks are thawed, inoculated into 50 ml of fresh BHI medium containing 100 μg/ml ampicillin, and incubated overnight with shaking at 37°C.
3. The next day, each overnight culture is re-inoculated into fresh BHI medium (at a 1:40 ratio) containing 100 μg/ml ampicillin and grown for a further 2–4 h (until OD_{600} = 0.5). IPTG is then added to a final concentration of 1 mM, and the bacteria are incubated at 37°C with shaking for another 2–4 h.
4. Subsequent to IPTG induction, the total number of bacteria in each culture is determined by measuring the OD_{600} value (8×10^8 bacteria/ml culture has an OD_{600} = 1). The volume of bacterial culture required for animal treatment is then centrifuged at 2500*g* for 10 min at 4°C and the pellet is washed once with 1X PBS, and then resuspended at the required density in 1X PBS. This preparation is ready for oral administration or intravenous injection. Keep the preparations at 4°C, but warm just prior to treatment.

3.2.2. Oral Treatment of Normal Mice

1. Age-matched female C57BL/6 mice are randomly divided into treatment and control experimental groups (typically consisting of 6–8 animals per group).
2. Animals in each treatment group are given 5×10^8 to 5×10^{10} c.f.u. of shRNA-expressing *E. coli* (in 200 μl PBS) via a PS-20 oral feeding needle fitted to a 1 ml syringe. Control animals are treated similarly except that the administered *E. coli* contains a TRIP vector without the shRNA insert, or an insert coding for an inactive (e.g., scrambled) shRNA sequence.
3. Oral administration of control or shRNA-expressing bacteria is then performed 5 days per week for a total of 4 weeks (*see* **Note 6**). Mice are sacrificed 2 days after the last treatment,

and colonic tissues are collected for analysis of target gene silencing.

3.2.3. Intravenous Treatment of Nude Mice Bearing Colon Cancer Xenografts

1. Age-matched female nude Balb/c mice are randomly divided into treatment and control experimental groups (typically consisting of 6–8 animals per group).
2. Three weeks before bacterial treatment, animals in each experimental group are subcutaneously implanted in the right flank with 1×10^7 SW480 colon cancer cells (in 100 μl PBS) using a 1 ml syringe fitted with a 26G needle.
3. Bacterial treatments are initiated when the xenograft tumors reach approximately 10 mm in diameter. Animals are treated intravenously with 1×10^8 c.f.u. of shRNA-expressing or control *E. coli* (in 100 μl PBS) by tail vein injection using a 1 ml syringe fitted with a 26G needle.
4. Control or shRNA-expressing bacteria are administered to the animals every 5 days for a total of three treatments (*see* **Note 7**). Mice are sacrificed 5 days after the final treatment, and the tumor tissues are collected for analysis of target gene silencing.

3.2.4. Assessment of Target Gene Knockdown

1. For analysis of target gene mRNA levels by real-time PCR, colon and xenograft tissues are frozen and stored at –80°C. Total RNA isolation and real-time PCR analysis are then performed according to standard protocols.
2. For analysis of target gene protein levels by immunohistochemistry, colon and xenograft tissues are fixed in paraformaldehyde, paraffin-embedded, sectioned, and then stained according to standard procedures. For illustration of the protocol, some experimental data are presented in **Fig. 7.5**.

4. Notes

1. As TRIP plasmids are relatively large (~8.9 kb), the transformation efficiency using *E. coli* BL21-Gold (DE3) is quite low. To guarantee successful transformation, newly constructed TRIP plasmids (in the ligation mixture) should be immediately transformed into competent BL21-Gold bacteria. Plasmids purified from BL21-Gold bacteria can then be used to synthesize other engineered BL21 (DE3) strains.
2. Bacteria containing TRIP plasmids tend to grow very slowly in LB medium. Brain heart Infusion (BHI) medium, which is nutrient-rich, allows for faster growth of TRIP-containing bacteria. The expression of invasin (*Inv*) and listeriolysin

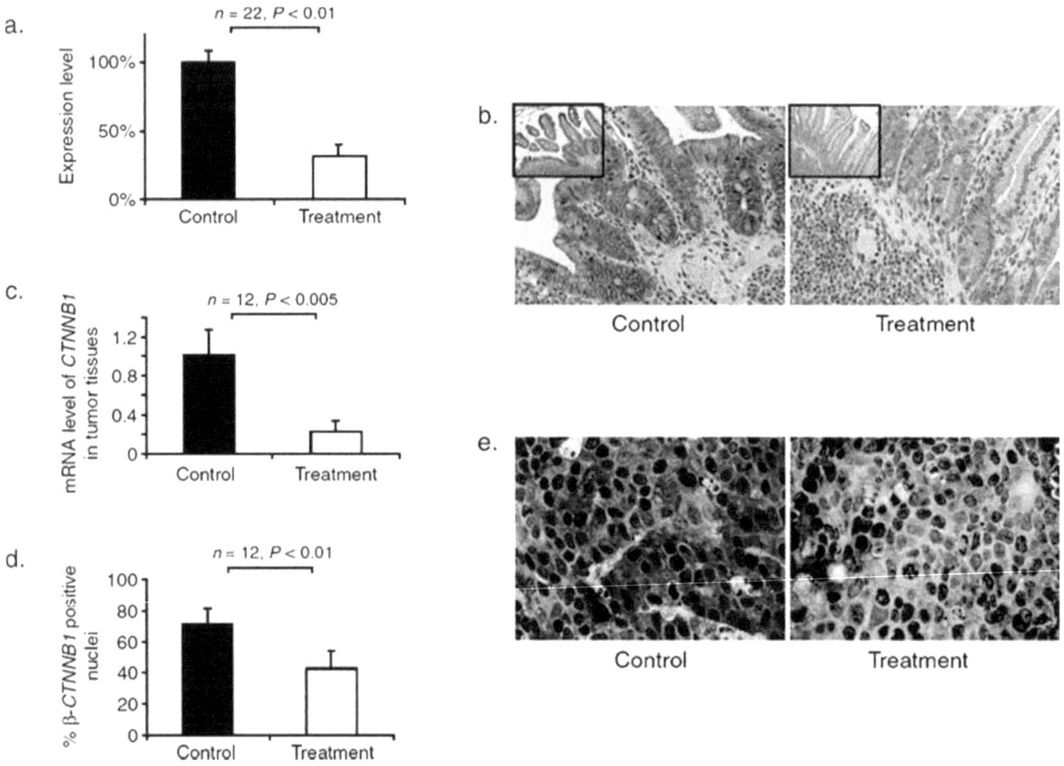

Fig. 7.5. TransKingdom RNAi in vivo. (**a**) Oral administration of *E. coli* expressing shRNAs against beta-catenin in mice leads to significant reduction of beta-catenin expression in normal intestinal epithelium, especially in the regions of, or adjacent to, Peyer's patches. (**b**) Representative view of intestinal epithelium from treated (*right*) and control (*left*) animals. (**c–e**) Intravenous administration of *E. coli* containing a TRIP against human beta-catenin in mice containing human colon cancer xenograft tumors resulted in a decrease in beta-catenin mRNA levels (c) and protein levels (d, e) in tumor tissues.

O (*HlyA*) is also better in BHI medium than LB medium. Maximal expression of both of these proteins is critical for stimulating bacterial endocytosis by target cells and endosomal escape of shRNA from entry vacuoles.

3. With most bacteria, IPTG induction is usually performed in the mid-log phase (OD_{600} > 0.7). However, with *E. coli* BL21-Gold (DE3), IPTG induction is best performed during early log phase (OD_{600} = 0.4 ~ 0.6).
4. Cells must be washed using serum-free RPMI 1640 medium. Washing the cells with PBS will cause the bacteria to firmly attach to the cell surface. This can negatively affect cell viability and enhance the probability of false positive data.
5. The use of aliquoted bacterial stocks gives more reproducible results for in vivo *tk*RNAi than growing single colonies from BHI-ampicillin plates.

6. Oral administration is well tolerated with no gross or microscopic signs of epithelial damage or ulcerations.
7. Intravenous injection is generally well tolerated without adverse effects. However, mice should be closely monitored during treatment. If the nude mice get sick during treatment, 40 mg/kg of ofloxacin can be applied. Alternatively, one treatment with bacteria can be omitted.

Acknowledgements

We thank Dr. J. T. LaMont for advice and discussions, and C. Griillot-Courvalin of the Pasteur Institute, Paris, France for providing the sequences for *Inv* and *Hly* (pGB2Ω) and also for helpful discussions.

References

1. Barik, S. (2005) Silence of the transcripts: RNA interference in medicine. *J Mol Med* 83, 764–773.
2. Cheng, J. C., Moore, T. B., and Sakamoto, K. M. (2003) RNA interference and human disease. *Mol Genet Metab* 80, 121–128.
3. Grunweller, A. and Hartmann, R. K. (2005) RNA interference as a gene-specific approach for molecular medicine. *Curr Med Chem* 12, 3143–3161.
4. Hannon, G. J. and Conklin, D. S. (2004) RNA interference by short hairpin RNAs expressed in vertebrate cells. *Methods Mol Biol* 257, 255–266.
5. Provost, P., Dishart, D., Doucet, J., Frendewey, D., Samuelsson, B., and Radmark, O. (2002) Ribonuclease activity and RNA binding of recombinant human Dicer. *EMBO J* 21, 5864–5874.
6. Tijsterman, M. and Plasterk, R. H. (2004) Dicers at RISC; the mechanism of RNAi. *Cell* 117, 1–3.
7. Gregory, R. I., Chendrimada, T. P., Cooch, N., and Shiekhattar, R. (2005) Human RISC couples microRNA biogenesis and posttranscriptional gene silencing. *Cell* 123, 631–640.
8. Hammond, S. M., Bernstein, E., Beach, D., and Hannon, G. J. (2000) An RNA-directed nuclease mediates post-transcriptional gene silencing in Drosophila cells. *Nature* 404, 293–296.
9. Li, C. X., Parker, A., Menocal, E., Xiang, S., Borodyansky, L., and Fruehauf, J. H. (2006) Delivery of RNA interference. *Cell Cycle* 5, 2103–2109.
10. Shankar, P., Manjunath, N., and Lieberman, J. (2005) The prospect of silencing disease using RNA interference. *JAMA* 293, 1367–1373.
11. Xiang, S., Fruehauf, J., and Li, C. J. (2006) Short hairpin RNA-expressing bacteria elicit RNA interference in mammals. *Nat Biotechnol* 24, 697–702.
12. Isberg, R. R., Voorhis, D. L., and Falkow, S. (1987) Identification of invasin: a protein that allows enteric bacteria to penetrate cultured mammalian cells. *Cell* 50, 769–778.
13. Young, V. B., Falkow, S., and Schoolnik, G. K. (1992) The invasin protein of *Yersinia enterocolitica*: internalization of invasin-bearing bacteria by eukaryotic cells is associated with reorganization of the cytoskeleton. *J Cell Biol* 116, 197–207.
14. Grillot-Courvalin, C., Goussard, S., Huetz, F., Ojcius, D. M., and Courvalin, P. (1998) Functional gene transfer from intracellular bacteria to mammalian cells. *Nat Biotechnol* 16, 862–866.
15. Mathew, E., Hardee, G. E., Bennett, C. F., and Lee, K. D. (2003) Cytosolic delivery of antisense oligonucleotides by listeriolysin O-containing liposomes. *Gene Ther* 10, 1105–1115.
16. de Fougerolles, A., Vornlocher, H. P., Maraganore, J., and Lieberman, J. (2007) Interfering with disease: a progress report on

siRNA-based therapeutics. *Nat Rev Drug Discov* 6, 443–453.

17. Gilmore, I. R., Fox, S. P., Hollins, A. J., and Akhtar, S. (2006) Delivery strategies for siRNA-mediated gene silencing. *Curr Drug Deliv* 3, 147–145.
18. Kolida, S., Saulnier, D. M., and Gibson, G. R. (2006) Gastrointestinal microflora: probiotics. *Adv Appl Microbiol* 59, 187–219.
19. Snelling, A. M. (2005) Effects of probiotics on the gastrointestinal tract. *Curr Opin Infect Dis* 18, 420–426.
20. Daudel, D., Weidinger, G., and Spreng, S. (2007) Use of attenuated bacteria as delivery vectors for DNA vaccines. *Expert Rev Vaccines* 6, 97–110.
21. Kruis, W., Fric, P., Pokrotnieks, J., Lukas, M., Fixa, B., Kascak, M., Kamm, M. A., Weismueller, J., Beglinger, C., Stolte, M., Wolff, C., and Schulze, J. (2004) Maintaining remission of ulcerative colitis with the probiotic *Escherichia coli* Nissle 1917 is as effective as with standard mesalazine. *Gut* 53, 1617–1623.

Chapter 8

Bacterial Delivery of siRNAs: A New Approach to Solid Tumor Therapy

De-Qi Xu, Ling Zhang, Dennis J Kopecko, Lifang Gao, Yueting Shao, Baofeng Guo, and Lijing Zhao

Abstract

RNAi is a powerful research tool for specific gene silencing and may also lead to promising novel therapeutic strategies. However, the development of RNAi-based therapies has been slow due to the lack of targeted delivery methods. The biggest challenge in the use of siRNA-based therapies is the delivery to target cells. There are many additional obstacles to in vivo delivery of siRNAs, such as degradation by endogenous enzymes and interaction with blood components leading to nonspecific uptake into cells, which govern biodistribution and availability of siRNA in the body. Naked unmodified synthetic siRNA including plasmid-carried-shRNA-expression constructs cannot penetrate cellular membranes, and therefore, systemic application is unlikely to be successful. The success of gene therapy by siRNAs relies on the development of safe, economical, and efficacious in vivo delivery systems into the target cells. Attenuated *Salmonella* have been employed recently as vectors to deliver silencing hairpin RNA (shRNA) expression plasmids into mammalian cells. This approach has achieved gene silencing in vitro and in vivo. The facultative anaerobic, invasive *Salmonella* have a natural tropism for solid tumors including metastatic tumors. Genetically modified, attenuated *Salmonella* have been used recently both as potential antitumor agents by themselves, and to deliver specific tumoricidal therapies. This chapter describes the use of attenuated bacteria as tumor-targeting delivery systems for cancer therapy.

Key words: Bacterial delivery vector, *Salmonella*, siRNA, shRNA, RNAi, solid tumor.

1. Introduction

RNAi is a eukaryotic intracellular process wherein a small interfering RNA (siRNA) directs a sequence-specific degradation of its target mRNA *(1)*. Within the cytoplasm of a given cell, this RNAi duplex is recognized by a multiprotein complex called RISC

M. Sioud (ed.), *Methods in Molecular Biology, siRNA and miRNA Gene Silencing, vol. 487*

DOI: 10.1007/978-1-60327-547-7_8

(RNA-induced silencing complex). One of the RNA strands (the antisense strand) remains bound to the RISC complex and binds to complementary mRNA sequences within the cytoplasm. This mRNA–siRNA–RISC complex induces cleavage of the mRNA, thereby preventing its translation into protein and lowering gene-specific expression.

Many tumors are caused by mutations that abnormally upregulate specific gene expression. RNAi therapy has been used successfully to knockdown this abnormal gene expression and the associated tumor growth in animal studies, but the application of RNAi therapy in humans has faced practical difficulties. The biggest challenge to the systemic use of siRNA-based therapies is the difficulty of delivery, specifically across the plasma membrane to reach the cytoplasm of target cells. There are additional obstacles to in vivo delivery of siRNAs, such as enzymatic degradation in blood, interaction with blood components, and nonspecific uptake by cells, which affect siRNA availability and distribution in the body. Naked unmodified synthetic siRNA including plasmid-encoded-shRNA-expression constructs cannot, by themselves, penetrate cellular membranes and, therefore, direct systemic administration of purified siRNA is unlikely to be successful. Alternatively, shRNA can be expressed inside host cells using DNA templates that direct the synthesis of RNA duplexes *(2, 3)*. In this latter case, the RNAi effect can be sustained for a longer term, depending upon the vector employed and optimal selective pressure for retention of the plasmid-borne expression cassettes. DNA-directed RNAi (ddRNAi) expression constructs can be directly administered intratumorally for surface nodes, but are not appropriate for systemic administration for the reasons indicated above. Systemic delivery systems that use viral vectors, such as retrovirus or adenovirus, have targeting advantages, but are limited to research use in animal models due to safety issues. Thus far, bacteria as a delivery system have great potential advantages over other delivery vectors and have already been employed in animals and humans (*see* Chap. 7). An ideal delivery system should be (a) nontoxic to normal cells, and (b) able to deliver the therapeutic efficiently and specifically to the tumor. Attenuated *Salmonella* have been employed recently as vectors to deliver silencing shRNA expression plasmids to mammalian cells and shown to achieve specific gene silencing in vitro in cultured cancer cell lines and in vivo in mice containing implanted tumors *(4)*. The facultative anaerobic, invasive *Salmonella* have a natural tropism for solid tumors including metastatic tumors. Genetically modified, attenuated *Salmonella* have been used as potential antitumor agents, either to elicit direct tumoricidal effects and/or to deliver tumoricidal molecules (e.g., siRNA) *(5–9)*. This chapter is focused specifically on the use of live bacterial vector systems for targeted delivery of cancer therapies.

1.1. Design of siRNA-Expression System

Bacterial delivery of siRNA is dependent upon the construction of shRNA-expression cassettes designed to induce sustained RNAi production. Plasmid-driven expression of siRNA in vivo has typically been based on the general design. A Pol III promoter is used to direct the synthesis of inverted RNA repeat sequences separated by a short spacer region. The expressed hairpin RNA contains two complementary nucleotide stems that encode a specific target sequence, a terminal loop sequence (4 to 9 nt) at one end, and a 3′-U overhang at the opposite end. These hairpin RNAs are processed in vivo by Dicer to leave only the duplex region (**Fig. 8.1**) *(10)*. Since poly(T) is used as a transcriptional terminator, stretches of >4 Ts or As in the target sequence should be avoided. siRNA target regions with 30–50% GC content appear to be more active than those with a higher GC content. The selection of the target site can be aided by using the BLAST search tool (www.ncbi.nlm.nih.gov/BLAST). RNAi-mediated gene silencing can be assessed in cultured mammalian cells by plasmid delivery and endogenous expression of shRNA harboring a fold-back stem–loop structure. The advantage of this shRNA expression system is that the RNAi effects are prolonged when plasmid-based siRNAs are used.

The selection of specific target sequence, the length of the inverted repeats that encode the stem of a putative hairpin, the order of the inverted repeats (i.e., sense vs. antisense), the length and the composition of the spacer sequence that forms the loop region, all can affect the functioning of the siRNA. It is necessary to check the specificity of the target sequence to confirm that there are no highly homologous sequences in other genes. The selected target site should be devoid of strong secondary structures, which could impair the binding of the siRNAs; computer-based folding programs can be helpful in this selection. However, the actual site accessibility can only be evaluated empirically in biological function studies. The shRNA cassette can be designed with either the sense or the antisense strand, placed immediately after the promoter. However, since U6 initiates transcription with a G, it is preferable that this be the first base of the sense strand, since a G in the first position of the sense strand of siRNA is positively correlated with siRNA functionality *(11)*. The length of the loop sequence of the hairpin-type siRNA is also important for silencing activity. Although a short 4-nt, 5′-UUCG-3′ could still support significant RNAi function *(12)* when a 19-bp stem sequence was used, the hairpin RNA with a standard 9-nt loop sequence (UUCAAGAGA) had greater silencing activity than the corresponding RNA with a shorter loop sequence *(12–14)*. However, it has been shown that the natural loop sequence of micro-RNAs, which are endogenous shRNAs, are preferable sequences for shRNA production.

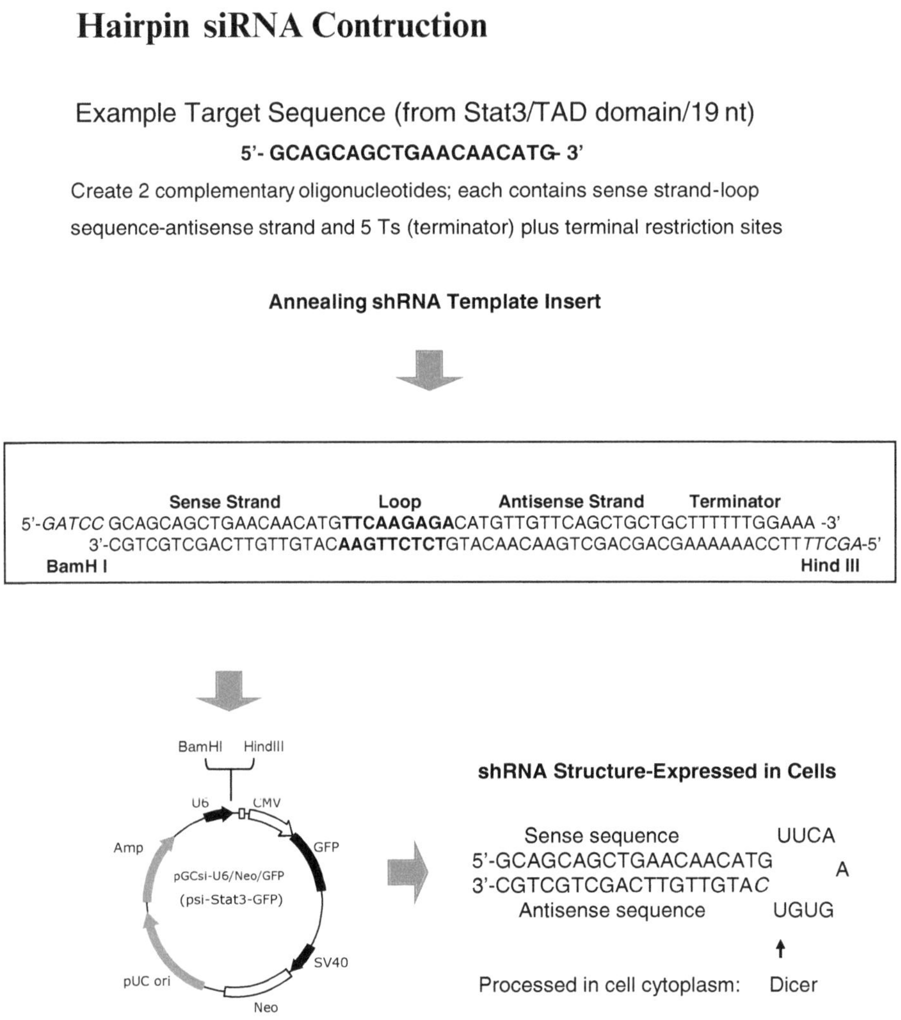

Fig. 8.1. Strategies for hairpin siRNA plasmid construction for bacterial delivery of design target siRNA sequence using the web-based converter at www.ambion.com/techlib/misc/psilencer_converter.html. Here we chose a target sequence GCAGCAGCTGAACAACATG from the TAD domain of human oncogene Stat3 (corresponding to nucleotides 2,144 to 2,162; GenBank accession number NM_003150). The expression vector psi-Stat3-GFP was constructed with pGCsiU6Neo-GFP vector containing a U6 promoter, *Bam*HI and *Hin*dIII cloning sites, a CMV promoter and GFP-encoding sequence, and a neomycin-resistant gene for selection. The details of construction are described in the text. The structure of psi-Stat3-GFP plasmid containing the sequence of Stat3-specific hairpin RNA (shRNA-Stat3) or Scramble sequence insert from Ambion's (negative control plasmid) are shown in the lower panel.

Vector-based RNAi synthesis also permits coexpression of reporter genes such as GFP or luciferase, which facilitate the tracking of transfected cells or evaluating the effect of a GFP-tagged shRNA expression vector. In any delivery system, coexpression of GFP can be very important to track bacterial distribution in vitro (in cell lines) or in vivo. For any RNAi experiment, it is necessary to include a negative control plasmid to rule out nonspecific gene knockdown. An optimal negative control plasmid should contain a scrambled sequence of your gene-specific siRNA.

1.2. Choice of Attenuated Bacterial Species or Strains

Any bacteria suitable for use as a tumor-specific delivery vector must possess several features: (i) It must be genetically or biologically traceable (e.g., through genetic markers or GFP production), (ii) it must be nonpathogenic with minimal toxicity, (iii) it must have reduced immunogenicity, (iv) it should have invasive and motile abilities (e.g., ability to invade epithelial cells and to spread from cell to cell), (v) it must have preference for tumor tissue over normal tissue, and (vi) it must be susceptible to antibiotics for eradication purposes. The discovery that genes vectored by bacteria can be functionally transferred to mammalian cells has suggested the possible use of bacterial vectors as vehicles for gene therapy. Several reports have demonstrated the utility of anaerobic bacteria (such as *Clostridium spp.*) as anticancer agents; their utility may be limited by the absolute requirement for anoxic conditions, restricting their use to large tumors *(15)*. These restrictions do not apply to facultative anaerobes (e.g., *Salmonella spp.*) such as the gram-negative, bacterium *Salmonella enterica* serovar Typhimurium (*S. typhimurium*). *S. typhimurium* is known to colonize various types of human and murine tumors, allowing treated mice to survive for periods long after untreated mice have died. *Salmonella* can grow in the presence or absence of oxygen and can colonize both large and small tumors *(16)*. Salmonellae have been shown to inhibit the number and size of melanoma micrometastases and to distribute throughout the tumor mass *(17)*.

Construction of attenuated bacterial strains is feasible with current DNA manipulation techniques and knowledge of molecular microbiology. A strain could be deleted for a single gene or mutated in multiple genes to avoid reversion to virulence. The following principles have played a role in the selection of bacterial genes to create attenuating mutations, which would be essential for use in cancer therapy. (A) Advances in microbiology and biotechnology have led to genetic modification of bacteria so that they can replicate and persist within tumors, but which limit their distribution to and survival in normal tissues. For example, bacterial mutations that result in nutritional auxotrophy generally restrict bacterial growth in normal tissues. Deletions in *purI* create a bacterial requirement for exogenous adenine and deletions in the *aro* pathway result in a requirement for aromatic amino acids. Auxotrophic *Salmonella* mutants disseminate to the nutrient-rich tumor environment where they replicate to levels exceeding 10^3–10^4 times the low concentration found in normal tissues *(9)*. Combining auxotrophic mutants with mutations in the antigenic bacterial LPS genes can minimize immunogenic/inflammatory potential. (B) Deletion of the *msbB* (renamed *waaN*) gene, which controls the terminal myristylation of lipid A, reduces strain toxicity. The *msbB* mutant is reduced in endotoxicity about 10,000-fold, which allows for the systemic administration of such attenuated strains. The mutation of this gene reduces tumor necrosis factor-α (TNF-α)-mediated septic shock and tremendously increases the LD50 of this pathogen in the mouse model *(18)*. The *msbB* mutant is of clinical interest because it allows *Salmonella* to be administered safely to mammals, which is essential for the development of safe, live, attenuated bacterial delivery systems. However, *S. typhimurium msbB*-deletion strains have severe growth defects, in many laboratory media, that can be suppressed by extragenic compensatory mutations which arise at high frequency in related genes. Suppressor mutations for *msbB* allow *Salmonella* to grow well in culture media but still have reduced lipid A myristylation to avoid the septic shock response. By choosing a suppressor mutation that confers the desired characteristics on a given product strain, (e.g., the tumor-targeting *Salmonella* strain VNP20009 is nontoxic and retains tumor targeting and tumor inhibition abilities) *(19)*. (C) However, another well-characterized transcriptional regulon required for *Salmonella* pathogenesis is the *phoP/phoQ* two-component regulatory system *(20–21)*. This regulatory system controls the expression of more than 40 different genes required for virulence in mice and which promote resistance to innate immune defenses (e.g., resistance to defensins and low pH), induces spacious phagosomes and survival in macrophages, and is involved in nutrient scavenging and lipid A modifications *(22)*. Due to the simultaneous altered expression of multiple virulence factors, *phoP/phoQ* mutants retain a high degree of attenuation in IFN-$\gamma^{/-}$ mice after oral challenge with 5 × 10^9 cfu *(23)*. Virulence attenuation of *phoP/phoQ* mutants can also be attributed to structural modification of the lipid A moiety. The structural modifications to lipid A, the host signaling portion of LPS, occur by the addition of aminoarabinose and 2-hydroxymyrisate to the mutant hexaacyl lipid A. This modification not only confers polymyxin resistance, but also affects LPS-induced host responses (i.e., decreased E-selectin and TNF-α expression *(24)*. The bacterial load of *S. typhimurium phoP/phoQ* mutants in mice in infected organs is elevated during the first week postinfection, but total bacterial clearance is achieved by day 14. Thus, even lacking a specific mutation in *msbB*, the *phoP/phoQ* deletion mutant has great reduced endotoxicity.

In our recent report, a *phoP/phoQ* deletion mutant of *S. typhimurium (4)* was employed as a carrier for shRNA or shRNA combined with an antitumor protein-expressing plasmid for use in prostate cancer therapy. In vivo mouse and in vitro tissue culture studies showed significant delivery efficacy for *Salmonella*-delivered therapy into mammalian tumor cells. Regardless of the route of administration (i.e., oral, intravenous, intratumoral, or intraperitoneal inoculation) *Salmonella*-delivered therapy resulted in significant antitumor efficacy with no toxic effects observed in the mouse model. By day 15, when no bacteria were detected in normal spleen or liver tissues, *Salmonella* remained within the tumor tissue. The *S. typhimurium phoP/phoQ* deletion mutant has been shown to accumulate preferentially >1,000-fold greater in tumors than in normal tissues and to disperse homogeneously in tumor tissues. *Salmonella*-delivered siRNA was found to downregulate expression of Stat3 and to exert significant antitumor effects including reduction in metastases and reversal of tumorigenesis *(4)*.

1.3. Administration of Vector-Based shRNA Expression Systems Delivered with Attenuated Salmonella

Tumor-targeting bacteria have been investigated intensively in recent years as anticancer agents by themselves or, more recently, as a delivery system for anticancer therapies. Systemic administration of live bacterial vectors by intravenous (i.v.) or intraperitoneal injection (i.p.) has been widely and effectively used in mouse tumor studies *(24–26)*. Direct local administration, such as intratumoral injection, has been demonstrated to be practical but still is limited to use on surface tumors. Oral administration would be ideal if a reproducible amount of *Salmonella* could reach the target tumor cells. In mice, oral administration of *S. typhimurium* results in reproducible systemic inoculation. The intravenous administration of attenuated Salmonellae has already been employed in human use and allows for bloodstream delivery of a specific therapeutic dose to both primary and secondary tumors *(18, 26)*. Bacterial motility should favorably disperse the vector throughout the tumor tissue, overcoming diffusion and distal pressure gradients to deliver the therapeutic agents to all regions of the tumor. Motility is a major advantage that bacteria have over viral vectors, as viruses exhibit a lack of tumor specificity and are poorly distributed throughout the tumor mass. In addition, metabolically active bacteria will replicate and continue to produce active antitumor agents, thus amplifying the dose to an effective level only within the target tumor.

2. Materials

2.1. Bacterial Strain, Growth Media, and Electrotransformation

1. Attenuated *S. typhimurium phoP/phoQ* null mutant LH 430 was derived from ATCC 14028 by deletion of the *phoP/phoQ* locus *(27, 28)*.
2. *E. coli* DH5α (Invitrogen).
3. Luria-Bertani (LB) broth, brain heart infusion (BHI) broth, (BD Diagnostic Systems).

4. SOC medium (Quality Biological, Inc.).
5. 10% glycerol, sterilized and stored at 4°C.
6. Gene Pulser apparatus and cuvettes (0.2 cm electrode gap) (Bio-Rad, Hercules, CA).

2.2. Culture of Cell Lines, Cell Transfection, and Stable Cell Line Establishment

1. The human prostate cancer cell line PC-3M *(29)* (Xenogen Corporation) (*see* **Note 1**).
2. Clonal HeLa cell line that stably expresses the cycle 3 variant of GFP introduced via Invitrogen's pTracer SV40 vector, which will be used to demonstrate the reduction of GFP expression after transfection with a plasmid containing a GFP-siRNA insert.
3. Iscove's modified Dulbecco's medium (IMDM) (Invitrogen).
4. Fetal bovine serum (FBS) (Invitrogen).
5. LipofectAMINE 2000 (Invitrogen).
6. Green fluorescent protein vector (pGCsiU6/Neo/GFP, Jikai Chemical, Inc).
7. Gentamicin, neomycin, tetracycline, penicillin, streptomycin (Sigma) and G418 sulfate (InvivoGen, San Diego, CA).
8. Trypsin/EDTA solution (Invitrogen).
9. Opti-MEM I medium (Invitrogen).
10. Tissue culture plate with 6, 24, or 96-well cluster flat bottom with lid (Corning Inc., NY).
11. 75 cm^2 cell culture flask (Corning Inc., NY).
12. 5% CO_2 humidified incubator.

2.3. Construction and Expression of shRNA and shRNA-GFP Expression Plasmids

1. Construct two complementary oligonucleotides containing target-siRNA-sequence referring to the principles described in **Sect. 1.1**. The target sequence, GCAGCAGCTGAACAACATG, corresponds to the nucleotides 2,144–2,162 of the TAD domain of oncogene Stat3 (GenBank accession no NM_003150). The two complementary oligo DNAs are the sense strand 5′-*GATCC*GCAGCAGCTG AACAACATG**TTCAAGAGA**CATGTTGTTCAGCTG-CTGCTTTTTGGAAA-3′, and the antisense strand 3′-C-GTCGTCGACTTGT-TGTACAAGTTCTCTGTACAACA AGTCGACGACGAAAAACCTT*TTCGA*-5′. After annealing, the DNA sequence should contain the sense strand, a short spacer (loop shown in bold), the antisense strand, five Ts (Terminator), and *Bam*HI and *Hind*III cutting sites (*see* **Fig. 8.1**).
2. The pGCsiU6/Neo/GFP vector contains a U6 promoter, polycloning sites, and the CMV promoter controlling the GFP gene (Jikai Chemical, Inc) (*see* **Note 3**)

3. The pSilencer 2.1-U6 neo siRNA Expression Vector Kit contains the scramble negative control, the linearized pSilencer 2.1-U6 neo vector, and the GFP-siRNA control insert for GFP-expression knockdown.
4. Molecular Grade Water (DNase, RNase, and Protease-free) (Mediatech, Inc.).
5. 1 X DNA annealing solution: 0.1 M potassium acetate, 30 mM HEPES-KOH, pH 7.4.
6. T4 DNA ligase, 10 X T4 DNA ligase buffer (Roche).
7. Restriction enzymes: *Bam*HI, *Hin*dIII (Roche).
8. 1 X TE: 10 mM Tris, 1 mM EDTA, pH 7.4.
9. Wizard Minipreps DNA Purification Systems (Promega).
10. ProbeQuant G-50 Micro Columns (Amersham Biosciences).

2.4. Titration and Distribution of Bacteria in Mouse Tumor Model

1. Luria-Bertani Broth (LB broth) and brain heart infusion (BHI broth) (BD Diagnostic Systems).
2. Waring Blender (Fisher Scientific).
3. Stirrer Motor Homogenizer with electronic speed controller (Cole Parmer).
4. Ultraviolet–visible spectrophotometer (Shimadzu, Japan).
5. Fluorescence microscope (Olympus, Japan).
6. Flow cytometer (Coulter, Hialeah, FL).
7. Frosted glass slides (Colorfrost/Plus—Fisher Scientific Co).
8. Tissue-Tek V.I.P (Electron Microscopy Science).
9. Sakura Tissue-Tek OCT Compound (International Equipment Inc).
10. 1 X PBS, pH 7.4, cold.

2.5. Northern Blot Assays for RNAi Effect Analysis

2.5.1. Isolation of Total RNA for Northern Blot

1. TRIzol Reagent (Invitrogen).
2. RNase-free deionized water.
3. Chloroform (Invitrogen).
4. 100% isopropanol.
5. 70% and 100% ethanol.
6. 3 M sodium acetate.
7. 1 X TE buffer: 10 mM Tris, 1 mM EDTA, pH 7.4.
8. MOPS (10 X) running buffer: 0.4 M morpholinopropanesulfonic acid, 0.1 mM.
9. Sodium acetate, 10 mM EDTA, adjust to pH 7.0 with 2 N NaOH. And store at room temperature protected from light.

10. Transfer buffer (Ambion, NorthernMax Kit) (cat #1940).
11. ULTRAhyb buffer (Ambion).
12. Formaldehyde gel-loading buffer: 50% glycerol, 1 mM EDTA, 0.25% bromophenol blue, 0.25% xylene cyanol, 60 μg/ml ethidium bromide (EB).
13. 1 and 1.2% agarose gels (Sigma).
14. Hybond-N + membrane (Amersham Pharmacia Biotech, Piscataway, NJ).
15. α-^{32}P dATP (6000 Ci/mmol) (Amersham).
16. X-ray film (Kodak).

2.6. Isolation for Small RNA and Detection of shRNAsby Northern Blot

2.6.1. Isolation of Small RNA from Total RNA Samples

1. *mir*Vana miRNA isolation kit (Ambion) (contains miRNA wash solution I, wash solution 2/3, collection tubes, filter cartridges, lysis/binding buffer, miRNA homogenate additive, acid-phenol: chloroform, gel loading buffer II, elution solution).
2. RNase-free 1.5 or 0.5 ml microfuge tubes.
3. 100% Ethanol (ACS grade) (Invitrogen).
4. Microcentrifuge capable of at least 10,000× g.
5. RNase-free water (Mediatec Inc. Herndon, VA).

2.6.2. Detection of shRNA by Northern Blot

1. 10 X TBE for 1 L: 109 g of Tris-base, 55 g of boric acid, 40 ml for 0.5 M EDTA (final concentration: 0.9 M, 0.9 M, and 20 mM, respectively).
2. BrightStar-plus nylon membrane (Ambion cat # 10100).
3. 50 X Denhardt's solution: 10 g Ficoll 400, 10 g bovine serum albumin, 10 g polyvinylpyrrolidone, add nuclease-free water to 1 L.
4. 20 X SSC: 175.3 g NaCl, 88.2 g sodium citrate, 800 ml nuclease-free water, adjust to pH 7.0 with Tris, and then add nuclease-free water to 1 L.
5. Prehybridization solution: 6 X SSC, 10 X Denhardt's solution, 0.2% SDS.
6. Hybridization solution: 6 X SSC, 5 × Denhardt's solution, 1–5 × 10^6 cpm 5′end-labeled antisense probes, 0.2% SDS.
7. Wash solution: 6 x SSC, 0.2% SDS.
8. 15% denaturing polycrylamide gel (see Ambion's reagents (www.ambion.com).
9. Novex mini-cell system (XCell II) (Invitrogen).
10. Oligonucleotide (refer to **Sect. 2.3**) containing 19-nt complementary to the antisense of target siRNA.

11. T4 polynucleotide kinase (Takara Shuzo Co., Kyoto, Japan).
12. γ^{32}P-ATP (6000 Ci/mmol) (Amersham).
13. PhosphorImager (Molecular Dynamics) or x-ray film (Kodark).

2.7. RT-PCR Assay for RNAi Effect Analysis at mRNA Level(see Note 2)

We utilized the Superscript first-strand synthesis system for RT-PCR (Invitrogen).

2.7.1. Total RNA Isolation (see Sect. 2.5.1)

2.7.2. Reverse Transcription and PCR Amplification of cDNA

1. The Superscript first-strand synthesis system kit (Cat #11904-018, Invitrogen)—this kit provides all components for first strand cDNA synthesis.
2. Ex-Taq DNA polymerase (Takara).
3. Two specific amplification primers selected from vector sites flanking the target mRNA sequence.
4. Human β-actin primers and probes were obtained from Applied Biosystems.

2.8. Western Blot Assay for tho RNAi Effects at the Protein Level

1. Blocking solution (10% nonfat dry milk powder in TBST, pH 7.4).
2. TBST (pH 7.4): 0.2% Tween 20, 20 mM Tris-HCl, 150 mM NaCl.
3. Running buffer for Tris-glycine gels (10 X): 250 mM Tris base, 1.92 M glycine, 1%.
4. Sample buffer (2 X): 125 mM Tris-HCl, pH 6.8, 4% SDS, 20% glycerol, 0.1% bromophenol blue, 5% ß-ME* (2-mercaptoethanol; omit for native proteins).
5. Transfer buffer (for PVDF membrane): Add 18.2 g Tris base, 86.5 g glycine to 4 l of H_2O. Add 1200 ml methanol and bring to 6 l with H_2O (pH should be ~8.3–8.4).
6. Protein samples prepared from siRNA-treated cultured cell lines or tissue specimens from mouse model.
7. Primary antibody specific for protein of interest diluted in TBST buffer.
8. Secondary antibody: Horseradish peroxidase (HRP)- or alkaline phosphatase (AP)-anti-Ig conjugate (Santa Cruz, CA).
9. PBS (pH 7.1): 137 mM NaCl, 7 mM Na_2HPO_4, 1.5 mM KH_2PO_4, 2.7 mM KCl.
10. Trypsin-EDTA solution (Invitrogen).
11. 2 X Laemmli SDS sample buffer (Bio-Rad).
12. Lysis buffer: 1 M Tris-HCl, 0.5 M NaCl, 0.5 M EDTA, 1 mM phenylmethylsulfonyl fluoride, and 10 μg/ml each of aprotinin, pepstatin, and leupeptin (Sigma).
13. PVDF (Millipore Immunbilon P).

14. ECL (enhanced chemiluminescent detection kit (Amersham).
15. Kodak X-OMAT (XAR-5, 18 × 24 cm).
16. Novex mini-cell system (XCell II) (Invitrogen)
17. 12–16% Tris-glycine gel (Pre-Cast, Noevx, Invitrogen)

2.9. Animal Model, Tumor Orthotopic Implantation, and Administration of Bacteria-Carrying siRNA-Expression Plasmid

1. 6-week-old male BALB/c nude mice weighing 18–24 g were raised in specific-pathogen free (SPF), and under controlled conditions of temperature (23 ± 3°C) and relative humidity (50 ± 20)%.
2. Dissecting microscope (Model MZ6, Leica, Nussloch, Germany).
3. Latex-free syringe (1–3 ml) and hypodermic needle (22–30G) (Becton Dickinson).
4. Isopropyl alcohol.
5. Ketamine hydrochloride (100 mg/ml) and xylazine (20 mg/ml) at a mixture ratio of 7:3, respectively.
6. Surgical tools such as forceps and scissors.
7. Stainless steel gavage tube.

3. Methods

3.1. Construction of shRNA and shRNA-GFP Expression Vectors

3.1.1. Cloning of the siRNA Target Insert into Vector

1. Two complementary oligonucleotides containing the sense and antisense siRNA sequences and other required elements to express shRNA (*see* **Sects. 1.1** and **2.3**) were synthesized and column-purified using ProbeQuant G-50 Micro Columns (Amersham). The oligonucleotides were designed with *Bam*HI and *Hind*III sites at opposite ends (*see* **Fig. 8.1**).
2. Prepare a 1 μg/μl solution of each single-strand oligonucleotide in 1 X TE.
3. Anneal the two complementary template oligonucleotides. Assemble the annealing mixture as follows: 2 μl of sense siRNA template, 2 μl of antisense siRNA template oligonucleotides, and 46 μl 1 X DNA annealing buffer. Heat the mixture at 90°C for 3 min, then anneal at 37°C for 1 h.
4. Prepare the pGCsiU6/Neo/GFP backbone vector by digestion with *Bam*HI and *Hind*III. (*see* **Note 3**).
5. Set up a ligation reaction containing insert double-stranded oligo-DNA-template and pGCsiU6/GFP vector-digested with *Bam*HI and *Hind*III.

6. Set up the positive control ligation reaction containing GFP-siRNA control insert, and linearized pSilencer 2.1-U6 neo vector using standard protocols, following manufacturer's instructions (Ambion, pSilencer neo cat #5764).

3.1.2. Prepare DNA-Uptake Competent Cells of E. coli *DH5α and* S. typhimurium *phoP/phoQ*

1. The bacteria were routinely grown in 100 ml LB at 37°C to an OD_{600} of ~0.2–0.6. Bacterial cells should be collected in early to mid-log phase, and therefore the growth time will vary by strain and growth conditions.
2. Chill cells on ice, transfer the cells to prechilled centrifuge tubes, and centrifuge in cold conditions at 4,000× g for 4 min.
3. Resuspend the pellets in 50 ml of ice-cold sterile water and centrifuge as above.
4. Resuspend the cell pellet in a total of 2 ml of ice-cold sterile 10% glycerol in water, transfer to two 1.5 ml microcentrifuge tubes, and then centrifuge at 2,500× g for 5 min at 4°C.
5. Resuspend the cells in ~1 ml of ice-cold sterile 10% glycerol in water. The cell concentration should be ~2 × 10^{10} cfu bacteria/ml.
6. Aliquot the cell suspension in 0.1 ml quantities into prechilled sterile microcentrifuge tubes, freeze in a dry ice/ethanol bath, and store at –70°C until use.

3.1.3. Transform Competent Cells with Ligation Products by Electroporation

1. Prechill all tubes and electroporation cuvettes on ice, and mix 40 μl of competent cells with 1 μl of ligation products (do not exceed 1 μl of the ligation products if the ligation products were not desalted). Mix well, but gently, and incubate on ice for ~1 min.
2. Transfer the cell–DNA mixture to a cold 0.2 cm cuvette, shake to the bottom, and place in the Gene Pulser cuvette holder.
3. Apply one pulse with the Gene Pulser set at 2.5 kV and 25 uF and the pulse controller at 200 Ω. If using a 0.1 cm cuvette, set the apparatus at 1.5–1.8 kV.
4. Immediately suspend the electroporated cells in 1–2 ml of SOC, and incubate at 37°C with agitation for 1 h.
5. Plate the cells on selective agar growth medium (containing ampicillin at 50 μg/ml) in a volume of 100–200 μl to select for cells containing the vector plasmid.

3.1.4. Identify Positive Colonies and Prepare Plasmid DNA

1. Select at least 10 positive colonies (i.e., containing target insert sequence with vector DNA) grown on the selective agar plates. Grow culture in LB broth overnight. Prepare plasmid DNA

using the standard method (Promega Mini-Prep), and check plasmid DNA for size on 1.0% agarose gel.

2. Digest the plasmid DNA with *Bam*HI and *Hin*dIII, and check the digestion products on a 2% agarose gel. The target DNA fragments should be ~60–70 bp.
3. Confirm the insert DNA sequences: sequencing should be conducted using standard methods (ready reaction dideoxy terminator cycle sequencing kits (Applied Biosystems)). PCR primers for amplifying DNA fragments containing inserts depend upon the primer designed binding sites denoted by arrows on the plasmid map (**Fig. 8.1**).

3.1.5. Positive and Negative Controls

1. We chose the pSilencer neo with GFP-siRNA control insert as a positive targeting sequence, and a clonal HeLa cell line stably expressing the cycle 3 variant of GFP as controls to monitor GFP expression knockdown.
2. We utilize the pSilencer neo negative control plasmid containing scrambled sequences which are not found in the mouse, human, or rat genome databases.
3. To assess siRNA efficacy, cells transfected with the pSilencer neo plasmid expressing target-specific siRNA should be compared to cells transfected with the scrambled vector control. Also, the GFP-specific shRNA expressed vector should be used in experiments to demonstrate the RNAi efficacy in the HeLa cell line expressing GFP protein.

3.2. Cell Culture, Cell Transfection, and Stable Cell Line Establishment

3.2.1. Cell Culture Maintenance

1. Grow the PC-3M cell line in a 5% CO_2 incubator at 37°C in IMDM (supplemented with 10% FBS, 100 μg/ml penicillin, and 100 μg/ml streptomycin).
2. Passage cells every 3–4 days by detaching with trypsin-EDTA solution treatment. Do not exceed 30 passages after unfreezing the stock cells. Over-density of cell layer or over-passage may affect siRNA effects.
3. At least 24 h after seeding the cells, ensure that a confluency of 60–80% is reached. Begin transfection with your siRNA components or expression plasmids.

3.2.2. Cell Transfection with shRNA-Expression Plasmid and Stable Cell Line Establishment

1. Dilute 0.8 μg of shRNA-expression vector DNA into 50 μl of Opti-MEM I medium for transfection.
2. Trypsinize 90% confluent cells grown in a 175 ml cell culture flask with 10 ml of trypsin-EDTA for 10 min at 37°C.
3. Dilute the cell suspension 1:5 with fresh medium and seed into new culture flask or culture plate without serum in the well of tissue culture plate. Mix gently.

4. After 5 min incubation, add the diluted Lipofectamine 2000 (2 µl of Lipofectamine in 50 µl of Opti-MEM I Medium) to the wells with the diluted shRNA expression vector DNA. Mix gently and incubate for 15 min at room temperature to allow complex formation to occur.
5. Add 100 µl complete growth medium without antibiotics with 20,000 cells to each well containing shRNA expression vector-Lipofectamine 2000 complexes. Mix gently by rocking the plate back and forth.
6. Incubate the cells at 37°C in a CO_2 incubator until you are ready to harvest cells.
7. 24–48 h after transfection, start to assay for target gene knockdown.
8. You may select a stable cell line which is constantly expressing the selected shRNA.
9. Cells were passaged twice weekly with fresh selection medium containing 400 µg/ml G418 (*see* **Note 4**). Within two weeks, when individual transfectant foci developed, and each of these were individually dissociated, transfer the cells to a single well (6-well dish) and culture in selection medium containing 400 µg/ml G418.
10. Expand the desired clones using minimal concentration (200–400 µg/ml) of G418.
11. After 6–8 rounds of G418 selection, culture the cells in nonselective medium, collected two times per week and preserved in liquid nitrogen.

3.2.3. Infection of PC-3M Cells with S. typhimurium *Carrying shRNA Expression Vector*

1. In a 24-well plate, seed PC-3M cells at a density of 0.5–2 × 10^5 cells per well in 500 µl IMDM with10% FBS and grow at 37°C, 5% CO_2 for 24 h.
2. Prior to coculture with bacteria, the cells should be first "conditioned" in 0.5% FBS without antibiotics for 24 h. (0.5% FBS was determined as an optimum by testing the effect of a range of FBS concentration on cell viability).
3. Next day, collect the mid-log phase bacteria-carrying siRNA-expression vector by centrifugation, resuspend the bacteria at 1 × 10^8 cfu/ml in IMDM without antibiotic, and coculture with PC-3M cells at 37°C for 30 min.
4. After exposure, wash the cell monolayers twice with serum-free IMDM containing 50 µg/ml gentamicin to kill all extracellular bacteria.
5. Add IMDM containing 10% FBS to the cells and then incubate at 37°C for 4 h.

6. Next, add tetracycline (10 μg/ml) to the cells and incubate at 37°C for 72 h in order to prevent intracellular bacterial multiplication. Stable PC-3M cell clones containing the target plasmids are selected with G418 (800 μg/ml, *see* **Note 4**) or screened for GFP expression using a fluorescence microscope (when the recombinant vector carried the GFP gene). The cells were maintained by treating with 200 μg/ml G418.
7. Use the transfected cells to assess for knockdown effects on the target gene or shRNA expression levels.

3.3. Animal Model: Nude Mice Tumor Orthotopic Implantation

1. Male BALB/c nude mice are inoculated with 2×10^6 PC-3M cells subcutaneously (s.c.) into the right flank of mice.
2. After the formation of palpable tumors (~5 mm by day 12), mice are killed by cervical dislocation.
3. Tumor tissues are excised, and cut into fragments (1 mm^3), and then implanted by SOI (surgical orthotopic implantation) *(30)* onto the prostate in nude mice. After proper exposure of the prostate to be implanted, 8–0 surgical sutures are used to penetrate the tumor pieces and attach them to the appropriate orthotopic organ. The incision in the skin was closed with a 7–0 surgical suture in one layer. 4. The animals are kept under isoflurane anesthesia during surgery. All procedures of the operation described above are performed with a 7 X magnification microscope.

3.4. Administration with Bacteria Carrying shRNA-Expressing Vector(*see* Note 5)

3.4.1. Oral Administration

1. Mice are fasted for 3–4 h without food and water.
2. Mice are pretreated with 100 μl of 3% sodium bicarbonate in PBS orally.
3. Immediately following buffer administration, bacteria are administered orally using a stainless steel gavage tube,100 μl (2×10^9 cfu) of bacteria carrying the siRNA plasmid into the lower esophagus.
4. After treatment, mice are given food and drinking water ad lib.

3.4.2. Intravenous Inoculation

1. Prep tail with alcohol swab.
2. Needle placement should be no closer to the body than half the length of the tail.
3. With the tail under tension, insert needle approximately parallel to the vein.
4. Insure proper placement by inserting needle at least 3 mm into the lumen of vein.
5. Administer 100 μl bacterial suspension (i.e., 2×10^7 cfu) in PBS into the tail vein using a tuberculin syringe fitted with a 25-gauge needle in a slow, fluid motion to avoid rupture of vessel.

3.4.3. Intraperitoneal Inoculation

1. Restrain mouse by grasping near base of tail; grasp nape of neck with the opposite hand.
2. Place tail between fingers to secure and control animal; prep the local area with alcohol swab.
3. Inoculate 0.5 ml bacterial suspension in PBS containing 5×10^5–10^6 cfu, using a 1 ml syringe.

3.5. Bacterial Titration and Bacterial Distribution in a Tumor Mouse Model

3.5.1. Direct CFU (Colony Forming Unit) Count

1. Inoculate a single colony of *S. typhimurium* carrying plasmid vectors into 5 ml LB with 50 μg/ml ampicillin and incubate at 37°C overnight.
2. Take 1 ml of the overnight culture and dilute 1:50 in fresh LB broth and restart incubation. Harvest bacteria when an $A_{600} = 0.6$ is reached.
3. Collect bacteria by centrifugation for 5 min at 4,000 rpm, and wash pellet twice in 2 ml cold PBS.
4. The washed bacteria were diluted to a final concentration of 10^8 cfu/100 μl cold PBS (an $OD_{600} = 1$ corresponds to 1×10^9 cfu/ml). Use diluted cells as soon as possible to avoid bacterial viability loss.
5. Inject 10^7 cfu/100 μl of bacteria via i.v. (tail vein) into tumor-bearing mice.
6. The mice were sacrificed at different times after the injection.
7. Tumors and relative normal tissues or organs such as liver, spleen, and kidney were removed, weighed aseptically, and homogenized in five volumes of ice-cold, sterile PBS.
8. Larger organs are first chopped in a sterile warring blender on ice bath (Fisher Scientific) before homogenization.
9. The homogenates are serially diluted in BHI broth and plated onto BHI agar with ampicillin (50 μg/ml) in triplicate, and incubated for 24 h at 37°C.
10. Colony-forming units per gram of tissue were assessed and averaged.

3.5.2. Observation Under a Fluorescence Microscope to Determine the Extent of Bacterial Infection (Cryostat Sectioning)

1. Add Tissue-Tek OCT compound to infected host cells; allow to equilibrate for at least 30 min.
2. Prepare a block using Tissue-Tek.
3. Place the tissue block from the GFP control group of mice into the cryostat and allow it to equilibrate to the cutting temperature (i.e., –17°C). Adjust the positioning of the block to align the block with the knife blade.
4. Prepare 12–14 μm thick sections using a microtome and collect onto frosted glass slides. Observe immediately under a fluorescence microscope (*see* **Fig. 8.2**)

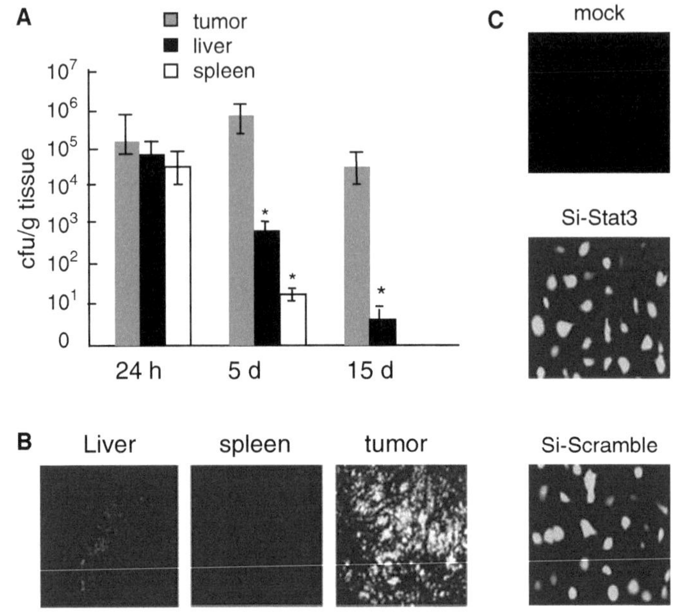

Fig. 8.2. GFP-based detection for quantitation and distribution of 57 vector carrying different recombinant constructs based upon the pGCsi-U6/Neo/GFP vector in different samples. (**a**) Quantification of bacterial foci at specified times (hours or days) in the various tissues of tumor bearing mice ($n = 3$). (**b**) Localization of attenuated *S. typhimurium* in C57BL6 mice bearing RM-1 prostate tumor by using GFP expression as a marker. (**c**) Expression of GFP from the psi-Stat3-GFP and psi-Scrambled vector is shown in stable infected RM-1 cells versus mock uninfected cells (magnification 400 X).

3.5.3. Detection of Transfected Cells by Fluorescence-Activated Cell Sorting (FACS)

1. Homogenize the treated tumor tissues and dilute with cold PBS (five times volume/sample weight, (w/v)).
2. Adjust to 1×10^7 host cells/ml with PBS after lysis of the red blood cells with 0.85%$NHCl_4$.
3. GFP-expressing cells are detected by flow cytometry.

3.6. Northern Blot Assays for Assessing RNAi effect at the mRNA level

3.6.1. Isolation of Total RNA

The RNA extraction protocol is designed to isolate total RNA from mammalian cells or tissue samples (e.g., frozen tissue specimens) from tumors.

1. If the samples are from cells grown in a monolayer, aspirate and discard the culture medium, wash the cells with PBS, place the culture dish on ice, and add directly 1 ml TRIzol reagent to a 3.5 cm diameter dish, and mix well.
2. If the samples are fresh tissues or frozen specimens from animals, perfuse the tissue with cold PBS and rinse one time to eliminate red blood cells. If necessary, quickly cut the tissue into small pieces, weigh the sample, add 1 ml of TRIzol reagent per 50–100 mg of tissue immediately, and homogenize

on ice. If you need to store the tissues, freeze the samples in liquid nitrogen first. When the liquid nitrogen stops churning, remove the samples from the liquid nitrogen and store them at –70°C.

3. Incubate the cell lysate or homogenate for 5 min at room temperature.
4. Add 0.2 ml chloroform per ml of TRIzol.
5. Mix by shaking the tubes vigorously by hand for 15 s.
6. Incubate at room temperature for 2–3 min.
7. Centrifuge the sample at 12,000× g for 15 min at 4°C.
8. Transfer the colorless aqueous phase to a fresh tube.
9. Add 0.5 ml isopropyl alcohol per 1 ml TRIzol, and vortex.
10. Incubate at room temperature for 10 min.
11. Centrifuge the sample at 12,000× g for 10 min at 4°C.
12. Remove supernatant carefully without disturbing the RNA pellet.
13. Wash the RNA pellet once with 75% ethanol, adding at least 1 ml of ethanol per ml TRIzol, mix well by vortexing.
14. Air-dry the RNA pellet for 5–10 min.
15. Dissolve RNA in RNase-free water.
16. Read the OD_{260} of 1/500 dilution: 1 OD_{260} = 35 μg/ml.
17. Store immediately at –80°C.

3.6.2. Northern Blotting

1. Prepare a 1.2% agarose gel. Mix 1.2 g of agarose with 72 ml of water, and heat until boiling. Cool to 65°C and add 10 ml of 10 X running buffer and 18 ml of 37% formaldehyde. Mix and pour into gel box.
2. Prepare the RNA samples as follows:

   Mix the following to each RNA sample
   a. RNA (up to 30 μg) 11 μl
   b. 10 X MOPS running buffer 5 μl
   c. Formamide 25 μl
   d. Formaldehyde (12.3 M) 9 μl
3. Denature the RNA by heating at 65°C for 10 min.
4. After chilling on ice, add 10 μl formaldehyde loading dye.
5. Load the samples, and electrophorese in a fume hood at 100 V until the blue dye has migrated to two-thirds of the length of the gel.
6. Visualize the gel on a UV transilluminator and photograph the gel prior to transfer
7. Measure the migration of the markers from the origin.

8. Transfer the RNA onto Hybond-N + membrane (or other positively charged nylon membrane). To remove formaldehyde, the gel must be rinsed several times in RNase-free water.
9. The denatured RNA may be transferred immediately to the nylon membrane by capillary elution or electroblotting. Capillary elution is conducted following the description by Sambrook et al. (*33*) and electroblotting should be performed according to the instructions of the apparatus manufacturer. Usually, transfer for a typical 6 mm thick gel should be completed in 1.5–2 h, and does not exceed 4 h to avoid the hydrolysis of small RNAs. UV-crosslink the RNA to the membrane.
10. Prehybridization: treat the membrane for 1–2 h at 42°C in 50% formamide, 5 X SSPE, 2 X Denhardt's reagent, and 0.1%SDS.
11. Probe design and labeling: the oligonucleotide probe should be designed to be homologous to the antisense strand of the cDNA sequence. Radiolabel the oligonucleotide with 6000 Ci/mmol of α-^{32}P dATP according to the manufacturer's instructions.
12. Hybridization: add the denatured radiolabeled probe directly to the prehybridization fluid, and incubate for 16–24 h at 42°C.
13. Wash the filter for 30 min at room temperature in 1 X SSC, 0.1% SDS, followed by three washes of 30 min at 68°C in 0.2 X SSC and 0.1% SDS.
14. Expose the filter to x-ray film or a phosphorimager. Quantify signals with the phosphorimager software.

3.7. Detection of shRNAsby Northern Blot(*see* Note 6)

Determination of the siRNA expression levels is critical in the assessment of stability or half-life of siRNA after transfection in mammalian cells. A DNA (or RNA) oligonucleotide complementary to the antisense of the siRNA or target-mRNA sequence can be synthesized and used to probe the expression of siRNA. Usually, small RNAs are separated by 12.5–15% polyacrylamide gels. Northern blots can also be employed to detect the larger precursor of siRNAs *(28, 29)*.

3.7.1. Isolation of Enriched Small RNAs

The procedure for isolation of small RNAs is adapted from the mirVana miRNA Isolation Kit manual. This procedure is designed for small scale RNA isolation from animal tissues or cultured cells. It can be also used with fresh or frozen cultured cells or animal tissues. The small RNA fraction can be enriched from the total RNA preparation.

1. Mix 50–100 μg of total RNA extracted using TRIzol reagent, with 5 volumes of lysis/binding buffer.
2. Add 1/10 volume of miRNA homogenate additive to RNA mixture, mix well by vortexing, and leave the mixture on ice for 10 min.
3. Add 1/3 volume of 100% ethanol to the RNA mixture and mix by vortexing.
4. For each sample, place a filter cartridge into one of the collection tubes supplied by the Ambion's kit. Pipet the RNA/ethanol mixture onto the filter cartridge. Up to 700 μl can be applied to a filter cartridge at a time.
5. Centrifuge for 1 min at 5,000× g to pass the mixture through the filter.
6. Collect the filtrate, add 2/3 volume of room temperature 100% ethanol to the filtrate, and mix well.
7. Pipet the filtrate/ethanol mixture onto a second filter cartridge.
8. Centrifuge for 1 min (at 5,000× g), and discard the flow-through. Reuse the collection tube for the washing steps.
9. Add 700 μl miRNA wash solution 1 to the filter cartridge and centrifuge for ~1 min at 5,000× g.
10. Discard the flow-through from the collection tube, and place the filter cartridge in the same collection tube.
11. Wash twice with 500 μl wash solution.
12. Discard the flow-through, replace the filter cartridge in the same collection tube, and spin for 1 min at 10,000× g to remove residual fluid from the filter.
13. Transfer the filter cartridge into a fresh collection tube, and apply 50 μl of 95°C elution solution, and close the tube.
14. Incubate at room temperature for 2 min and then spin for 1 min at 10,000× g to recover the RNA. Collect the eluate, which contains the small RNA-enriched fraction, and store it at –20°C. Repeat once the elution step in order to increase the RNA yield.

3.7.2. Separation on 15% Polyacrylamide Gel

1. Prepare denaturing 15% polyacrylamide gel as follows. Prepare 15 ml of 15% polyacrylamide gel mix with 8 M urea for use in the Bio-Rad Protean II minigel system. Use Ambion's reagents for preparation of the gel as follows: Mix following reagents first: 7.2 g Urea, 1.5 ml of 10 X TBE, 5.6 ml 40% acrylamide (acryl:bis:acryl = 19.1), and add nuclease-free water to 15 ml. Stir to mix. Then add 75 μl 10% ammonium persulfate, and 15 μl TEMED. Mix briefly and pour the gel immediately.

2. Mix 1–2 μg RNA with equal volume of gel loading buffer II.
3. Heat the sample for 2–5 min at 95–100°C, load the sample on the gel, and run at 30–45 mA.
4. Stop running when the bromophenol blue dye front has migrated to the bottom of the gel.
5. Soak the gel for 5 min in 0.5 μg/ml of EB solution in 1 X TBE. Wash the gel for 2–5 min in 1 X TBE buffer.
6. Visualize the RNA band using a UV transilluminator and collect the image on film or computer.

3.7.3. Detection of shRNA by Northern Blot

1. Run sample on gel (*see* **Sect. 3.7.2**)
2. After staining, transfer the RNA to a nylon membrane by electroblotting at 200 mA. Keep the membrane damp after blotting.
3. UV-crosslink the RNA to the membrane.
4. Specific DNA probe containing antisense shRNA target sequence should be designed and synthesized prior to the experiments. Use Ambion's *mir*Vana probe and marker kit for the 5′ end labeling (www.ambion.com).
5. Prehybridize for at least 1 h at 65°C.
6. Hybridize 8–24 h at room temperature.
7. Wash three times with wash solution at room temperature and once at 42°C.
8. Expose to x-ray film or a phosphorimager according to the manufacturer's instructions.

3.8. RT-PCR for Quantification of Specific Target RNA(*see* Note 2)

This procedure is adapted from the Invitrogen SuperScript First-Strand Synthesis System for the RT-PCR manual. The procedures of RT-PCR involve (1) the isolation of total RNA or mRNA from the cultured cells or animal tissue sample, (2) first strand cDNA synthesis using poly dT, random hexamer, or gene-specific primers, (3) removal of complementary template RNA, and (4) amplification of target gene cDNA by PCR using gene-specific primers. For efficient measurement of gene silencing, the positions of the designed primers should flank the sites of cleavage of target mRNA.

3.8.1. Reverse Transcription

1. Prepare RNA/primer mixture for first-strand cDNA synthesis: 3 μg/(1–4 μl) of total RNA (isolated from the samples-treated with your siRNA including control), 1 μl of random hexamers (50 ng/μl), 1 μl of 10 mM dNTP mix, and DEPC-treated water to 10 μl. Mix and spin the tube briefly. Incubate at 65°C for 5 min.

2. Add the following: 2 μl of 10× RT buffer, 4 μl of 25 mM $MgCl_2$, 2 μl of 0.1 M DTT, and 1 μl of RNaseout recombinant ribonuclease inhibitor.
3. Mix and spin briefly.
4. Incubate at 25°C for 2 min.
5. Add 1 μl (50 units) of SuperScript II RT.
6. Incubate at 25°C for 10 min.
7. Transfer the tube to 42°C and incubate for 50 min.
8. Stop reaction at 70°C for 15 min. Chill on ice.
9. Spin briefly and add 1 μl of RNase H and incubate at 37°C for 20 min.

3.8.2. PCR Amplification of the Target cDNA

1. Prepare PCR master mixture on ice: 5 μl of 10 X PCR buffer, 1.5 μl of 50 mM $MgCl_2$, 1 μl of 10mM dNTP mix, 1 μl of 10 μM sense primer, 1 μl of 10 μM of antisense primer, 0.5 μl of Taq DNA polymerase (5 units/μl), 2 μl of cDNA from the first-strand reaction, 38.1 μl of autoclaved, and distilled water to 50 μl of final volume.
2. Run the following program: 94°C for 2 min, then perform 20–35 cycles of PCR with optimized conditions for the sample.
3. Analyze 10 μl of the amplified sample by agarose gel.
4. The bands can be quantified with ImageQuant 5.0 software (Molecular Dynamics).

3.9. Western Blot

This procedure is adapted for minigels using the Bio-Rad Mini Trans-Blot cell for electrophoresis and electroblotting steps.

3.9.1. Lysis of Sample Containing Target Protein from Cultured Cells

1. Remove the tissue culture medium from the siRNA-treated cells cultured in a 24-well plate.
2. Rinse the cells once with 200 μl PBS, add 200 μl trypsin–EDTA, and incubate for 1 min at 37°C. Subsequently, add 800 μl DMEM medium to inactivate the trypsin.
3. Transfer the suspended cells to a chilled 1.5 ml microfuge tube and collect the cells by centrifugation at 3,000 rpm (700× g) for 4 min at 4°C.
4. Resuspend the cell pellet in ice-cold PBS and centrifuge again.
5. Remove the supernatant and add 25 μl of hot (90°C) 2 X concentrated Laemmli sample buffer to each cell pellet obtained from one well of a 24-well plate.
6. Incubate the samples for 3 min in a boiling water bath and vortex.

3.9.2. Extracting Proteins from Animal Tissues

1. Remove tissues from target organs; weigh and homogenize at 4°C in lysis buffer containing protein inhibitors.

2. Centrifuge lysate at 15,000× g for 30 min at 4°C.
3. Determine the protein content using the Bradford reagent (Bio-Rad).

3.9.3. Protein Analysis by SDS-PAGE Gels

1. Aliquot 15 µl of each protein lysate into a microcentrifuge tube, add 15 µl of 2 X SDS sample buffer, heat at 100°C for 3 min, and chill and load 10–15 µl into 12.5–15% SDS-PAGE.
2. Electrophorese at 15 to 20 mA until the tracking dye reaches the end of the gel (or 3 to 5 mA overnight).
3. Cut the membrane to same size as the gel plus extra 1–2 mm on each edge.
4. For PVDF membranes, first incubate the membrane for 1–2 s in 100% methanol, then equilibrate 10–15 min with the transfer buffer.
5. Electrophoretically transfer the proteins to the membrane for 30–60 min at 100 V at 4°C (or overnight at 14 V).
6. Place the membrane in heat-sealable plastic bag with 5 ml blocking buffer and seal bag. Incubate 30–60 min at room temperature with agitation.
7. Incubate the membrane with diluted primary antibody in blocking buffer (usually 1: 50–1000) for 30 min at room temperature with constant agitation.
8. Wash the membrane four times by agitating with 200 ml TBST, 10–15 min each wash.
9. Incubate the membrane with HRP- or AP-conjugated secondary antibody diluted in blocking buffer (1:10,000), and incubate for 30–60 min at room temperature with constant agitation.
10. Remove the membrane from bag and wash as in Step 8.
11. Develop the membrane with ECL according to the manufacturer's instructions.

4. Notes

1. PC-3M cell line was isolated from liver metastases produced in nude mice subsequent to intrasplenic injection of the androgen-insensitive PC-3 human prostate carcinoma cell line. It has enhanced tumorigenicity and produces a high incidence of well disseminated metastases *(29)*.

2. RT-PCR (reverse transcription-polymerase chain reaction) is the most sensitive technique for mRNA detection and quantitation currently available. RT-PCR can be used with as little as 1 ng or as much as 5 μg of total RNA. RT-PCR is more sensitive compared to Northern technique when studying the efficacy siRNAs.
3. pGCsiU6/Neo/GFP vector contains U6 promoter, polycloning sites (*Bam*HI and *Hin*dIII, CMV promoter-driven GFP protein expression, which can be used to assess transformed bacteria.
4. G418 is used for the selection and maintenance of eukaryotic cells expressing the *neoR* gene. G418 is an aminoglycoside antibiotic similar in structure to gentamicin B1, produced by *Micromonospora rhodorangea*. Unlike gentamicin, G418 blocks polypeptide synthesis in eukaryotic cells by binding irreversibly to 80S ribosomes. Resistance to G418 is conferred by the *neoR* gene from transposon Tn5 encoding an aminoglycoside 3″-phosphotransferase, APH 3′ II. This protein inactivates G418 by covalently modifying its amino or hydroxyl function therefore inhibiting the antibiotic-ribosome interaction *(31, 32)*. The concentration of G418 used for the selection varies with the cell type. Thus, working concentration should be determined for each cell type. G418 is a hazardous compound. Avoid contact with eyes, skin and clothes, harmful if swallowed.
5. Bacterial dose administered to animals varies and depends on the mouse type, age and weight, bacterial strain, and method of delivery.
6. Currently, two methods are available for shRNA detection: Northern blot on solid support using denaturing 15% polyacrylamide gel or ribonuclease protection assay (RPA). Ambion recommends the RPA using *mir*Vana miRNA Detection kit, and indicates that RPA is 100–500 times more sensitive than Northern analysis. Please refer to Ambion Catalog 2005–2006, page 105–109, and the *mir*Vana miRNA Isolation Kit, and the mirVana Probe and Marker kit. The *mir*Vana miRNA Detection Kit provides a fast and sensitive method for detecting small RNAs.

References

1. Hannon, G.J. (2002). RNA interference. *Nature* 418, 244–251.
2. Antoszczyk, S., Taira, K., and Kato, Y. (2006). Correlation of structure and activity of short hairpin RNA. *Nucleic Acids Symp Ser (Oxf)* 50, 295–296.
3. Hiroaki, K.H. and Taira, K. (2003). Short hairpin type of dsRNAs that are controlled by $tRNA^{Val}$ promoter significantly induce RNAi-mediated gene silencing in the cytoplasm of human cells. *Nucleic Acids Res.* 31, 700–707.

4. Zhang, L., Gao, L., Guo, B., et al. (2007). Intratumoral delivery and suppression of prostate tumor growth by attenuated *Salmonella enterica* serovar Typhimurium carrying plasmid- based siRNAs. *Cancer Res.* 67, 5859–5864.
5. Bermudes, D., Zheng, L.M., and King, L.C. (2002). Live bacteria as anticancer agents and tumor-selective protein deliver vectors. *Curr. Opin. Drug. Discov. Devel.* 5, 194–199.
6. Tjuvajev, J., Blasberg, R., Luo, X., et al. (2001). *Salmonella*-based tumor-targeted cancer therapy: tumor amplified protein expression therapy (TAPET) for diagnostic imaging. *J. Control Release* 74, 313–315.
7. Zheng, L., Luo, X., Feng, M., et al. (2000). Tumor amplified protein expression therapy: Salmonella as a tumor-selective protein delivery vector. *Oncol. Res.* 12, 127–135.
8. Forbes, N.S., Munn, L.L., Fukumura, D., et al. (2003). Sparse initial entrapment of systemically injected *Salmonella typhimurium* leads to heterogeneous accumulation within tumors. *Cancer Res.* 63, 5188–5193.
9. Pawelek, J.M., Low, K.B., and Bermudes, D. (2003). Bacteria as tumour-targeting vectors. *Lancet Oncol.* 4, 548–556.
10. Grosshans, H. and Slack, F.J. (2002). Micro-RNAs: small is plentiful. *J. Cell Biol.* 156, 17–21.
11. Amarzguioui, M. and Prydz, H. (2004). An algorithm for selection of functional siRNA sequences. *Biochem. Biophys. Res. Commun.* 316, 1050–1058.
12. Paul, C.P., Good, P.D., Winer, I., et al. (2002). Effective expression of small interfering RNA in human cells. *Nat. Biotechnol.* 20, 505–508.
13. Brummelkamp, T.R., Bernards, R., and Agami, R. (2002). A system for stable expression of short interfering RNAs in mammalian cells. *Science* 296, 550–553.
14. Brummelkamp, T.R., Bernards, R., and Agami, R. (2002). Stable suppression of tumorigenicity by virus-mediated RNA interference. *Cancer Cell* 2, 243–247.
15. Lambin, P., Theys, J., Landuyt, W., et al. (1998). Colonization of Clostridium in the body is restricted to hypoxic and necrotic areas of tumours. *Anaerobe* 4, 183–188.
16. Luo, X., Li, Z., Lin, S., et al. (2000). Antitumor effect of VNP20009, an attenuated Salmonella in murine tumor models. *Ocol. Res.* 12, 501–508.
17. Jazowiecka-Rakus, J. and Szala, S. (2004). Antitumour activity of *Salmonella typhimurium* VNP20047 in B16(F10) murine melanoma model. *Acta Biochim. Pol.* 51, 851–856.
18. Khan, S.A., Everest, P., and Servos, S. (1998). A lethal role for lipid A in Salmonella infections. *Mol. Microbiol.* 29, 571–579.
19. Low, K.B., Ittensohn, M., Le, T., et al. (1999). Lipid A mutant salmonella with suppressed virulence and TNF-α induction retain tumor-targeting in vivo. *Nat. Biotech.* 17, 3–41.
20. Miller, S.I., Kukral, A.M., and Mekalanos, J.J. (1989). A two-component regulatory system (phoP phoQ) controls Salmonella-typhimurium virulence. *Proc. Natl. Acad. Sci. USA* 86, 5054–5058.
21. Ronson, C.W., Nixon, B.T., and Ausubel, F.M. (1987). Conserved domains in Bacterial regulatory proteins that respond to environmental stimuli. *Cell* 49, 579–581.
22. Ernst, R.K., Guina, T., and Miller, S.I. (1999). How intracellular bacteria, survive: surface modifications that promote resistance to host innate immune responses. *J. Infect. Dis.* 179, S326–S330.
23. VanCott, J.L., Chatfield, S.N., Roberts, M., et al. (1998). Regulation of host immune responses by modification of Salmonella virulence genes. *Nat. Med.* 4, 1247–1252.
24. Guo, L., Lim, K.B., Gunn, J.S., et al. (1997). Regulation of lipid A modifications by *Salmonella typhimurium* virulence genes phoP-phoQ. *Science* 276, 250–253.
25. Dang, L.H., Bettegowda, C., Huso, D.L., et al. (2001). Combination bacteriolytic therapy for the treatment of experimental tumors. *Proc. Natl. Acad. Sci. U S A* 98, 15155–15160.
26 Low, K.B., Ittensohn, M., Luo, X., et al. (2004). Construction of VNP20009: A novel, genetically stable antibiotic-sensitive strain of tumor-targeting *Salmonella* for parenteral administation in humans. *Methods Mol. Med.* 90, 47–60.
27. Hohmann, E.L., Oletta, C.A., Killeen, K.P., et al. (1996). phoP/phoQ-deleted *Salmonella typhi* (Ty800) is a safe and immunogenic single-dose typhoid fevr vaccine in volunteers. *J. Infect. Dis.* 173, 1948–1014.
28. Hohmann, E.L., Oletta, C.A., and Miller, S.I. (1996). Evaluation of a phoP/phoQ-deleted, aroA-deleted live oral Salmonella typhi vaccine strain in human volunteers. *Vaccine* 14, 19–24.
29. Koslowski, J.M., Fidler, I.J., Campbell, D., et al. (1984). Metastatic behavior of human tumor cell lines grown in the nude mice. *Cancer Res.* 44, 3522–3529.

30. Hoffman, R.M. (1999). Orthotopic metastatic mouse models for anticancer discovery and evaluation: a bridge to the clinic. *Invest. New Drugs* 17, 343–359.
31. Davies, J. and Jimenez, A. (1980). A new selective agent for eukayotic cloning vector. *Am. J. Trop. Med. Hyg.* 29(5 Suppl.), 1089–1092.
32. Bar-Nun, S., et al. (1983). G-418, an enlongation inhibitor of 80S ribosomes. *Biochem. Biophys. Acta* 741, 123–127.

Chapter 9

The Therapeutic Potential of LNA-Modified siRNAs: Reduction of Off-Target Effects by Chemical Modification of the siRNA Sequence

Kees Fluiter, Olaf R. F. Mook, and Frank Baas

Abstract

Post-transcriptional gene silencing mediated by double-stranded RNA represents an evolutionarily conserved cellular mechanism. Small dsRNAs, such as microRNAs (miRNAs), are part of the main regulatory mechanisms of gene expression in cells. The possibilities of harnessing this intrinsic natural mechanism of gene silencing for therapeutic applications was opened up by the discovery by Tom Tuschl's team a few years ago that chemically synthesized small 21-mers of double-stranded RNA (small interfering RNA, siRNA) could inhibit gene expression without induction of cellular antiviral-like responses. siRNAs are especially of interest for cancer therapeutics because they allow specific inhibition of mutated oncogenes and other genes that aid and abet the growth of cancer cells. However, recent insights make it clear that siRNA faces some major hurdles before it can be used as a drug. Some of these problems are similar to those associated with classic antisense approaches, such as lack of delivery to specific tissues (other than the liver) or tumors, while other problems are more specific for siRNA, such as stability of the RNA molecules in circulation, off-target effects, interference with the endogenous miRNA machinery, and immune responses toward dsRNA. Chemical modifications of siRNA may help prevent these unwanted side effects. Initial studies show that minimal modifications with locked nucleic acids (LNA) help to reduce most of the unwanted side effects. In this chapter we will explore the limitations and possibilities of LNA-modified siRNA that may be used in future therapeutic applications.

Key words: RNA interference, siRNA, locked nucleic acid, off-target effects.

1. Introduction

The discovery of gene silencing by double-stranded RNA, known as RNA interference (RNAi), is one of the biggest breakthroughs of the last decade in the life sciences. This is reflected by the

M. Sioud (ed.), *Methods in Molecular Biology, siRNA and miRNA Gene Silencing, vol. 487*

DOI: 10.1007/978-1-60327-547-7_9

recent Nobel Prize awarded to Andrew Fire and Craig Mello, the two scientists who discovered the basic mechanism of RNAi in *C. elegans (1)*.

As can now be found in any biology textbook, double-stranded RNA binds to a protein complex, Dicer, which cleaves it into fragments. These fragmented double RNA strands are then bound to another protein complex, RISC. The two RNA strands are then separated and one of the RNA strands is eliminated while the other remains bound to the RISC complex and serves as a probe to detect mRNA molecules. When an mRNA molecule can pair with the RNA fragment on RISC, it is bound to the RISC complex, and then RISC directs either RNA cleavage, mediate translational repression, or induce chromatin modification.

RNA interference is important in the defense against viruses and limits the effects of transposons. In both cases, double-stranded RNA from a virus or a transposon is immediately cleaved by Dicer, RISC is activated, and the viral or transposon RNA is degraded *(2)*. However, small noncoding dsRNA is now also recognized as one of the key regulators of gene expression in general. Even before the advent of whole genome sequencing we have known that the majority of the transcribed genome is actually not coding for proteins. Most of us assumed that all the noncoding RNA was just evolutionary junk piled up in the genomes. Now we start to realize that many of these junk RNAs are actually involved in the RNAi pathway to regulate gene expression *(3)*. A conservative estimate is that thousands (and maybe many more) of noncoding microRNAs (miRNA) use the RNAi machinery to regulate gene expression, allowing very specific local regulation of mRNA levels, and thus the evolution of highly organized tissues within organisms such as the brain *(4)*.

After the discovery by Tuschl in 2001 *(5)* that synthetically synthesized double-stranded RNA 21-mers (siRNAs) are very effective in mediating RNAi and can be used as a research tool to study gene function in mammalian cells, siRNA technology is nowadays utilized as a “standard tool” in most labs for gene-function analysis, drug-target discovery and validation. But one of the more attractive features of siRNA is that this research tool in theory can also be used directly as a therapeutic approach. Any disease-causing gene can potentially be targeted. However, harnessing RNAi for therapy is not so straight forward as many initially have thought. There remains important obstacles for effective therapeutic use of siRNA: stability, efficacy, and specificity (*see* Chaps. 1 and 2). However, there is also progress and pilot siRNA clinical studies were started just 3 years after the initial discovery that RNAi can be used in mammalian cells for specific gene knockdown.

2. siRNA and Antisense Oligonucleotides: Facing the Same Hurdles

History tends to repeat itself over and over again. Also, in this new field of RNA interference we are facing exactly the same issues as antisense oligonucleotides were facing 30 years ago. Luckily, we have learned from the past and can use many of the solutions scientists working in the antisense field have found over the last three decades. Antisense oligonucleotides are normally defined as short, single-stranded DNA (-like) molecules that bind to mRNA through Watson–Crick base paring and then inhibit translation either through a steric block of the translation machinery or more effectively by recruiting an intracellular enzyme called RNase H, which recognizes DNA–RNA hybrids and will cleave the RNA in such a DNA–RNA hybrid (reviewed in Ref. *6*). This is very homologous with RNAi since both processes use short nucleotide sequences, use Watson–Crick base pairing to capture their target mRNAs, and both use intrinsic RNase activity in the cell to degrade the target mRNAs. Also homologous is the first hurdle both approaches face when used as a therapeutic molecule: *stability*. When the first antisense experiments were performed over three decades ago it quickly became clear that nature was equipped with formidable barriers that prevent foreign strands of DNA (or RNA) to muck about with gene expression within a cell. Not only it proved quite difficult to get antisense DNA inside cells, for which all kinds of transfection technologies were invented *(7)*, but even more problematic for in vivo usage were the wide range of nucleases that just destroyed the antisense strands in circulation before they could get to the intended tissues and could find a target mRNA. Progress in DNA amadite synthesis technology provided a real breakthrough in this field allowing the synthesis of chemically modified DNA that could resist nuclease activity.

However, for effective use in siRNA the chemical modifications must not hinder incorporation into RISC and the activity of RISC. So the chemistry used must mimic the function of RNA but should add increased enzymatic and chemical stability. The most promising generation of modifications which are of interest for siRNA are those involving substitutions of the 2′ position of the ribosyl ring in RNA analogues. These modifications of the sugar moiety are in general extremely resistant against nucleases and can be incorporated into RNA strands without altering the backbone conformation of the RNA, making them prime candidates for use in siRNAs. In vitro studies showed that several of these modifications are allowed in functional siRNAs *(8, 9, 10, 11)*. Phosphorothioates (PS) *(9, 11)* 2′-*O*-methyl (2′-*O*-Me) *(8, 10)*, 2′-*O*-allyl *(8)*, and 2′-deoxy-fluorouridine *(9, 11)* modifications have been examined for potential in vivo use. Some of the modified siRNAs were found to exhibit enhanced serum stability *(10)* and longer duration of action *(8)*. Modification of the 5′ end of the antisense strand with 2′-*O*-allyl *(8)* or chemical bloc-king of the 5′-hydroxyl group *(10)* resulted in a dramatic loss in activity, consistent with the proposed in vivo requirement for 5′ end phosphorylation. Also, more substantial modifications, such as total modification by 2′-*O*-Me *(9)* or PS modifications of every second or all internucleoside linkages *(9, 11)* increased cytotoxic effects and resulted in a significant decrease or complete loss of activity.

3. LNA-Modified siRNA

Modifications with conformational restricted ribosyl groups such as bridged- and locked nucleic acids may be even more promising for use in siRNAs. In locked nucleic acid (LNA) analogs, the ribose is locked with an extra methylene bridge that fixes the ribose moiety either in the C3′-endo (beta-d-LNA) or C2′-endo (alpha-l-LNA) conformation (**Fig. 9.1**). The beta-d-LNA modification results in significant increases in melting temperature T_m of up to several degrees per LNA residue and a very high resistance against nuclease activity. Introduction of LNA into classical antisense oligos has been shown to increase its serum stability *(12)*. In analogy, Braasch et al showed that LNA can be used to stabilize siRNAs without losing their function *(9)*. Recently, a systematic study on LNA-containing siRNAs has identified the number and positions of LNA molecules within the siRNA which still allow a functional siRNA *(13)*. Incorporation of LNA molecules into siRNA significantly increased its serum stability which

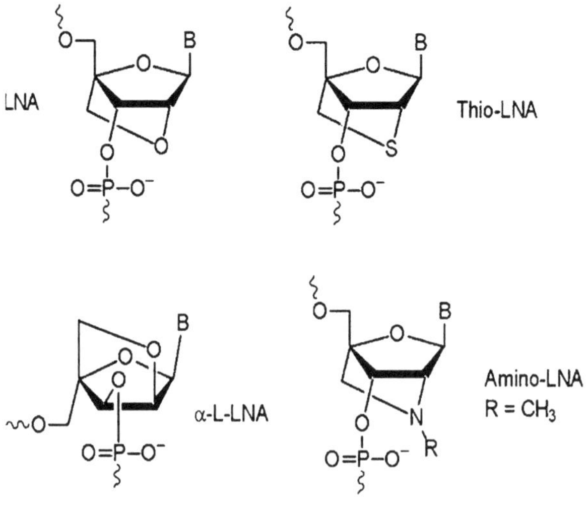

Fig. 9.1. Chemical structures of the four main LNA derivatives.

potentially favors successful in vivo applications *(13)*. Some LNA-stabilized siRNAs were still compatible with the intracellular siRNA machinery.

We have recently demonstrated that LNA-modified siRNA can mediate specific target knockdown with comparable efficacy as unmodified siRNA in vivo *(14)*. We designed LNA-modified siRNA (**Fig. 9.2**) and compared their in vitro and in vivo properties with unmodified siRNA in a mouse model system using GFP as target. Our studies on LNA-modified siRNA have shown that LNA offers the means to improve the serum half-life of siRNAs significantly. The introduction of only a few LNA moieties into siRNA is enough to protect siRNA against degradation (**Fig. 9.2**).

However, introduction of LNA molecules in siRNA can lower the efficacy in target knockdown. In general, like most other chemical modifications of siRNA LNA modifications of the antisense strand should be minimized to the 3′ overhangs. The sense strand can tolerate more modifications but there are also limitations. We found that heavily modified siRNA was not effective in GFP knockdown. Importantly, it is needed to leave the RISC cleavage site of the antisense strand unmodified *(15)*. Although the additional modifications in the heavily modified siRNA are only present in the sense strand we still see loss of efficacy *(14)*. Minimal modified siRNA retained its efficacy (albeit slightly lower than unmodified). Recent findings have demonstrated that RISC incorporates dsRNA and becomes active after passenger (sense) strand degradation in the RISC complex *(16)*. This may explain why heavily modified siRNA was inactive as possibly passenger strand degradation and, therefore, RISC activation could not occur. Nevertheless, using only a few LNA modifications as in our end-modified siRNA dramatically increased its half-life in serum while its efficacy in vivo was about equal as compared with nonmodified siRNA (**Fig. 9.3**). This suggests that lowered efficacy as observed with in vitro assays is compensated for

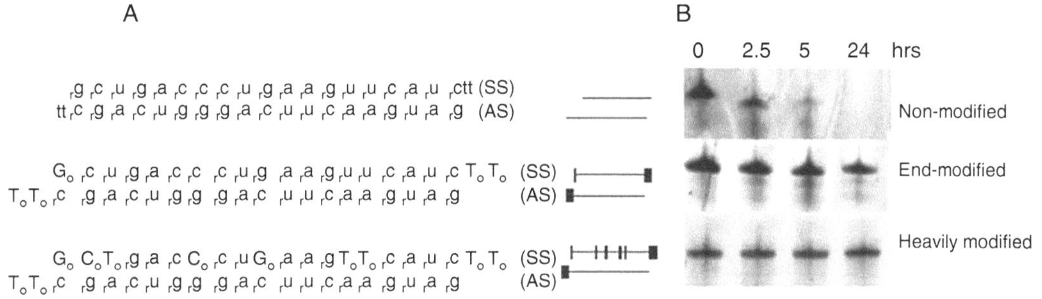

Fig. 9.2. (**a**) The three siRNA designs and DNA nucleotides are depicted in italic. LNA molecules are depicted in capitals. SS = sense sequence; AS = antisense sequence. (**b**) stability of the siRNA designs in mouse serum: unmodified siGFP, end-modified siGFP, and heavily modified siGFP were incubated in 100% fresh mouse serum and the degradation was monitored on a PAGE gel.

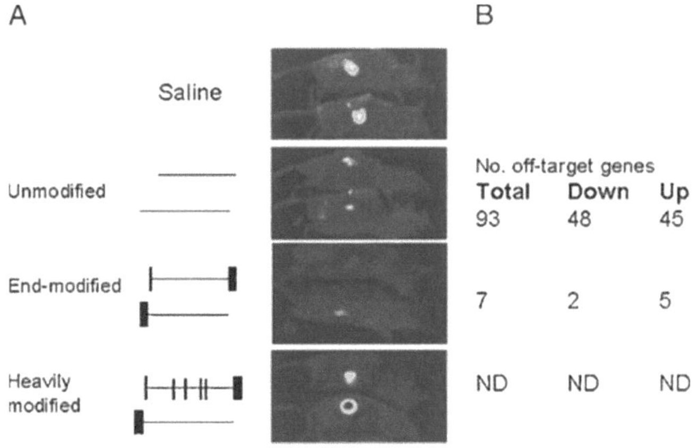

Fig. 9.3. (**a**) In vivo efficacy of the siRNA: unmodified siGFP, end-modified siGFP, and heavily modified siGFP were administered at a dosage of 0.25 mg/kg/day via osmotic minipumps. Seven days after implantation of the pumps, eGFP fluorescence in the tumors was analyzed using whole body imaging. (**b**) Using Affymetrix human whole genome arrays (U133 plus 2), we examined the number of differentially regulated genes in these tumor xenografts (**P** < 0.05 and over 3-fold difference). The complete list of genes can be found in Mook et al. *(14)*.

by increased in vivo stability. These characteristics potentially improve this molecule for therapeutic application but maybe more importantly we noticed that off-target effects in vivo were reduced with LNA modified siRNA in analogy with studies presented by others using 2′*O*-Me-modified siRNA *(17)*.

4. Nonspecific Effects of siRNA and How Chemical Modifications may Help Prevent These?

Now that siRNA has become a common technique in the life sciences we are becoming aware that siRNA can cause several types of nonspecific effects which will hinder its application in the clinic. The best known nonspecific effect associated with RNA interference is the interferon response. When a mammalian cell encounters a double-stranded RNA it is recognized as a viral by-product and an immune response is mounted, causing widespread regulation of gene transcription. The short length of the siRNA designs as reported first by the Tuschl group prevented the activation of the interferon response in mammalian cells *(5)*. Many thought that with the siRNA design the nonspecific responses towards RNAi were solved. When first reported, siRNAs seemed highly specific, as a single mutation in the target site could be demonstrated to completely abolish silencing. However, it was soon demonstrated that mutations in the target region of several siRNAs did not always inhibited knockdown *(18)*. Similar findings were reported in studies that indicated tolerance to wobble base pairing *(19, 20)*. Jackson et al. *(21)* reported severe nonspecific effects for genes with only a stretch of 11 nucleotides of sequence similarity between the target and the guide strand of the siRNA and it became clear that there were more problems with siRNA specificity. These effects were not related to the interferon effect, as was observed for both siRNAs *(22)* and short hairpin RNAs *(23)*, but in many cases were dependent on the sequence of the siRNA.

In the last 2 years there is a growing understanding of the mechanisms causing the sometimes widespread differential regulation of seemingly unrelated genes. Nonspecific effects of siRNA in an animal model or patient can be caused by:

1. Sequence dependent off-target effects by either accidental incorporation of the sense (passenger) strand into RISC or partial seed region homology of the siRNA towards a miRNA target.
2. Sequence independent side-effects such as interference with the miRNA machinery itself and inadvertent activation of the immune system which in vivo is quite sensitive in detecting dsRNA through RNA-sensing immunoreceptors such as Toll-like receptors. Judicious use of chemical modifications of the siRNA might help to minimize these nonspecific effects as is discussed below.

5. Sequence-Dependent Off-Target Effects

Microarray studies by Jackson et al. *(21, 24)* showed substantial off-target effects. They tested 24 siRNAs against two genes in their initial study, followed by a second study involving 6 other target genes. Importantly, they found that many of the genes that were downregulated showed sequence complementarity to the siRNAs in the 3′ UTRs of the transcripts. Moreover, the most enriched hexamers in the 3′ UTRs were complementary to nucleotides 2–7, the seed region shown to be important for microRNA (miRNA) targeting *(25, 26)*. They also noted that similarity to hexamers in positions 1–6 and 3–8 was also highly enriched *(21)*; this means that complementarity to the first 8 nucleotides of the 5′ end of siRNAs is most important for off-targeting. These results confirmed other reports in the literature *(27–30)*. Therefore, it is now clear that only limited complementarity between the target and the seed region of a miRNA (nucleotides 2–7/8) is sufficient to effect downregulation of the target sequence in itself. And additional binding of sequences in the miRNA 3′ end can increase the probability of silencing, determining target specificity within miRNA families *(31)*.

Chemical modifications may provide a solution to circumvent off-targeting. 2′*O*-Me modifications of the second base of the guide strand of the siRNA were reported to reduce off-targeting *(17)*. Importantly, this common RNA oligonucleotide modification does not affect the degree of silencing of the intended target, as could also be expected from previous reports *(32)*. Interestingly, the 2′-*O*-methyl modification was much more effective at reducing the levels of off-targeting when it was added to position 2 than when it was added to position 1 of the siRNA. We have found that LNA modifications also limit the off-target effects of siRNA. But in contrast, in our design we placed the LNA only in the overhangs and in the passenger strand. Initially we were afraid that the altered T_m of LNA-modified siRNA might affect the specificity for the target sequence, i.e., that the increase in T_m conferred by the LNA moieties would increase off-target effects. To test this assumption we performed whole genome expression profiling of treated tumor xenografts in vivo. The mice that had GFP-expressing tumor xenografts were systemically treated with unmodified siRNA and minimal LNA-modified siRNA against GFP and following GFP knockdown in the tumors the expression of the whole genome was analyzed using Affymetrix microarrays. When nonmodified siRNA was administered we found that in the tumors 93 genes were differentially regulated (**Fig. 9.3**). In contrast when LNA end-modified siRNA was administered (which resulted in similar GFP knockdown in the tumors) only 7 genes were found to be differentially regulated and there was no overlay between the differentially regulated genes in both experiments *(14)*. Therefore, cautious LNA modifications of the siRNA ends are sufficient to increase the biostability of the siRNA and LNA modifications limit the extent of off-target effects without affecting the on-target efficacy in vivo. Elmen et al. *(13)* also showed that LNA placement at the 5′ end of the passenger strand limited the off-target preference in an in vitro assay.

The reduction in off-target effects using chemically modified siRNA may be explained by several mechanisms. In the study by Jackson et al. *(17)* where they incorporated the 2′-*O*-methyl modification at the 2 position of the guide strand, it was suggested that there is some sort of steric hindering effect of the methyl group because position 2 has limited space in the RISC complex in which to accommodate a methyl group, forcing a conformational adjustment of the RISC complex, which limits the tolerance for mismatches in the seed region.

The other mechanism that may explain less off-target effects in our LNA modified siRNA is the limitation of the chance of effective incorporation of the sense strand into RISC. The placement of a chemical modification at the 5′ end of the sense strand forces preferential loading of the antisense strand in to RISC. And, if despite this guided RISC loading, the sense strand is incorporated into RISC, then 5′ phosphorylation, which is needed for RISC activation of the sense strand, may be prevented by the chemical modifications rendering RISC inactive.

6. sisiRNA

LNA enables a novel siRNA design invented by Jesper Wengel and Jorgen Kjems *(33)* that not only limits the chance of sense strand incorporation into RISC but may also allow more flexibility for the incorporation of chemical modifications into the siRNA. Small internally segmented interfering RNA (sisiRNA) entails an siRNA design where the passenger strand is nicked into two short separate strands and these are hybridized to the antisense strand using the high T_m inferred to by LNA modifications (**Fig. 9.4**). This "nicked" design seems equally effective in target knockdown as do conventional siRNA designs *(33)*. Interestingly, sisiRNA is reported to allow more chemically modified antisense strands, which are known to be nonfunctional with standard siRNA designs. Even quite bulky chemical modifications are tolerated *(33)*, which will open up new avenues for pharmacological applications. The nicked passenger strand has two advantages. First, the nicked strand is forced to become the passenger strand; it can never function as a functional guiding strand, thus the risk of passenger strand mediated off-target effects is reduced. Second, the passenger strand cleavage is not a rate-limiting step anymore limiting the risk of swamping the RISC machinery with a chemically modified passenger strand which is very stable but which may clog up the RISC complex. For siRNA to function the passenger

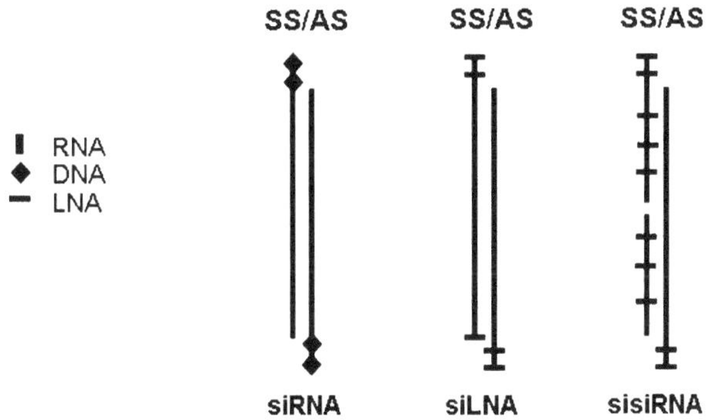

Fig. 9.4. Schematic representation of an siRNA, siLNA, and sisiRNA design. The positions of the DNA or LNA moieties are indicated.

strand must be cleaved *(34)*. Grimm et al. *(35)* demonstrated in vivo with short hairpin RNA (shRNA) constructs that high levels of shRNA resulted in dose-dependent liver injury and high morbidity. This was associated with the downregulation of liver-derived microRNAs (miRNAs), indicating possible competition of the latter with shRNAs for limiting cellular factors required for the processing of small RNAs. Since siRNA and miRNA essentially use the same molecular pathways there is a distinct possibility (but not yet proven) that a very stable chemically modified siRNA can cause problems similar to "throwing a spanner in the works." However, this fear is so far mainly theoretical. Indeed we ourselves have observed that a high level of modification of an siRNA is only detrimental for its efficacy, but in general we do not observe toxicity issues in cell lines with highly modified siRNAs. But the sisiRNA design might allow for more chemical modifications to be used and this may help to further open up therapeutic possibilities.

7. Preventing Inadvertent Stimulation of the Immune System

Negating off-target effects might be done using a clever strategy of chemical modifications and siRNA design. However, this is not the last concern that threatens the therapeutic use of siRNA. Nonspecific and toxic effects can also be caused by inadvertent stimulation of immune responses. The now classic 21-mer siRNA design as proposed by Tuschl *(5)* prevented the most prominent dsRNA-dependent protein kinase (PKR) interferon response in mammalian cells. However, it is becoming clear that in vivo the innate mammalian immune system is very adapt in recognizing nucleic acid species as signatures of potential pathogens, including the short siRNAs. A number of recent studies have pointed to immunological effects of siRNAs, including the induction of pro-inflammatory cytokines and type I interferon (as reviewed in Ref. *36)*. These immunological responses are mainly observed when delivery vehicles are used *(37)*. Immuno-recognition of RNA depends on certain molecular features such as length, double-strand configuration, sequence motifs, and nucleotide modifications. Judge et al. *(37)* described immunostimulatory motifs in siRNA. Avoiding these sequences in the design of siRNAs might yield molecules that induce less immune activation. RNA-sensing immunoreceptors include three members of the Toll-like receptor (TLR) family (TLR3, TLR7, TLR8) and cytosolic RNA-binding proteins like PKR and the helicases RIG-I and Mda5 *(38–40)*.

Detection of RNA molecules normally occurs during viral infection and triggers antiviral innate defense mechanisms including the induction of type I interferons and downregulation of gene expression. Type I interferon induction by synthetic siRNAs requires TLR7 and is sequence dependent, similar to the detection of CpG motifs in DNA by TLR9 *(36, 41)*. Identification of the exact molecular mechanisms of immune-recognition of RNA will allow the development of methods to avoid immune stimulation by siRNA. Importantly, Hornung et al. *(41)* demonstrated in a mouse model that TLR7 stimulation by siRNA could be inhibited by modification with LNA. Therefore LNA modifications in siRNA can be used to limit the inflammatory response towards the siRNA. Similar finds were reported for 2′-*O*-Me *(42)* and a recent report by the same group demonstrated that 2′-*O*-Me-modified RNA may have a general utility as a potent inhibitor of TLR7 and can function as an antagonist of immunostimulatory RNA, even when only a very limited percentage of positions are modified (<10%) (see Chap. 2) *(43)*.

The ability to limit the immune response using chemical modifications of the siRNA is likely to be of great value since these responses are especially aggravated when formulations of delivery vehicles and siRNA are used. And the strategy to use delivery agents to slow down clearance from the body and improve tissue penetration or even targeting is to become very important for future therapeutic applications.

8. The Future: Problems to Be Tackled

Pilot siRNA clinical studies began just 3 years after the initial discovery that RNAi works in mammalian cells. But we expect that the original unmodified siRNA designs will be of limited use as a therapeutic entity. This can be learned from the many experiences classic single-stranded antisense oligonucleotides have endured in clinical development. Luckily the "antisense field" has come up with a suite of chemical solutions that can be applied to siRNA to prevent rapid breakdown in circulation, prevent stimulation of the innate immune system, and prevent sequence dependent off-target effects. But there is still much room for optimization, especially improving circulation time and tissue uptake. Our own pharmacokinetic studies (unpublished data) reveal that siRNA molecules, also those modified with the LNA, are rapidly (within minutes) cleared from the blood with a main secretion route being the kidney–urine route just like other oligonucleotides *(44)*. This rapid renal clearance is also found by others *(45)* who show that 1 h after injection the amount of siRNA present in both kidneys was on average 40 times higher than in the other tissues. Despite the rapid clearance there is still some siRNA detectable in tissues 24 h postsystemic administration *(14, 46)* and these siRNAs can still silence the intended genes *(14)*. However, the fast clearance rate suggests that the effects we have observed until now with “naked” siRNA are suboptimal. Delaying the rapid clearance by either conjugating or complexing the siRNA to more bulky groups/particles, or by allowing association with large blood proteins, in order to prevent renal filtration might be a solution to this problem. In principle, specific tissue uptake can be improved by complexation of siRNA with carrier molecules like cationic liposomes, PEI or RGD-PEG-PEI *(41, 44, 47)*. However, the use of cationic liposomes has been associated with the induction of inflammatory responses, a potential problem for the in vivo application of siRNAs *(41, 48)*. One alternative might be the use of chitosan, a naturally occurring cationic polysaccharide that has been widely used in drug delivery systems and that also seems to work for siRNA *(49)*. The positively charged amines of chitosan allow electrostatic interaction with phosphate-bearing nucleic acids to form polyelectrolyte complexes while the protonated amine groups allow transport across cellular membranes. Another feasible option is conjugating the siRNA to a lipophilic group, such as conjugation to cholesterol which provided a clinical relevant way to improve upon the efficacy of the siRNA; however, the dosages needed were reported to be up to 50–80 mg/kg *(50, 51)* An attractive alternative to improve specific tissue delivery was reported by Song et al. *(52)* who showed that siRNA conjugated to Fab fragments specifically delivered siRNA into the targeted cells. We have recently shown that gapmer antisense containing amino LNA has comparable in vitro and in vivo properties as compared to LNA *(33, 53)*.This allows the use of “functionalized” LNA such as N-acylated and N-alkylated 2′-amino-LNA monomers *(54)*. The reactive amino group in gly-amino LNA might also be used to conjugate proteins of interest, and therefore enables to combine increased stability with the possibility to conjugate tumor-specific Fab fragments for targeted delivery of siRNA. If this results in improved in vivo efficacy of siRNA needs to be determined in future experiments.

Finally, the route of administration is also an important consideration for any drug. Also for siRNA it has been shown that different types of administration result in differences in biodistribution *(55)* and different dosage regimens will also likely contribute to differences. Therefore, it will be important for future therapeutic applications to consider all aspects: siRNA design, chemistry, delivery vehicle, administration route, and dosaging.

In conclusion, modification of siRNA with LNA molecules greatly enhances its stability in serum. However, it is clear that for classic siRNA designs LNA modifications should be kept at a minimum to allow compatibility with the RNAi machinery. Similar levels of target knockdown can be achieved in vivo by very low unmodified siRNA and end-modified siRNA concentration, suggesting that lowered efficacy was compensated for by increased stability. LNA modifications can significantly reduce off-target effects in vivo and initial observations indicate that LNA modifications can prevent unwanted attention from that part of the immune system that is prone to recognize nucleic acid strands. Because of its special high T_m LNA allows the design of new types of siRNA molecules with a "broken or nicked" sense-strand design (sisiRNA) that may prevent off-target effects caused by the passenger strand. In addition sisiRNA designs seem to tolerate more chemical modifications and this may be an exiting option to further improve the pharmacokinetic properties. There are still issues to be resolved for these molecules will effectively enter the clinic, but it will be very interesting to see how these new LNA-modified siRNA molecules will hold up in subsequent studies and whether they will keep their promising characteristics in future clinical trials.

References

1. Fire A., Xu S., Montgomery M.K., Kostas S.A., Driver S.E., and Mello C.C. (1998) Potent and specific genetic interference by double-stranded RNA in *Caenorhabditis elegans*. *Nature* 391, 806–811.
2. Meister G. and Tuschl T. (2004) Mechanisms of gene silencing by double-stranded RNA. *Nature* 431, 343–349.
3. Jackson R.J. and Standart N. (2007) How do microRNAs regulate gene expression? *Sci. Stke.* 367, rel; DOI: 10.1126/stke.3672007rel.
4. Berezikov E., Thuemmler F., van Laake L.W., Kondova I., Bontrop R., Cuppen E., and Plasterk R.H. (2006) Diversity of microRNAs in human and chimpanzee brain. *Nat. Genet.* 38, 1375–1377.
5. Elbashir S.M., Harborth J., Lendeckel W., Yalcin A., Weber K., and Tuschl T. (2001) Duplexes of 21-nucleotide RNAs mediate RNA interference in cultured mammalian cells. *Nature* 411, 494–498.
6. Crooke S.T. (1999) Molecular mechanisms of action of antisense drugs. *Biochim. Biophys. Acta* 1489, 31–44.
7. Mahato R.I., Cheng K., and Guntaka R.V. (2005) Modulation of gene expression by antisense and antigene oligodeoxynucleotides and small interfering RNA. *Expert Opin. Drug Deliv.* 2, 3–28.
8. Amarzguioui M., Holen T., Babaie E., and Prydz H. (2003) Tolerance for mutations and chemical modifications in a siRNA. *Nucleic Acids Res.* 31, 589–595.
9. Braasch D.A., Jensen S., Liu Y., Kaur K., Arar K., White M.A., and Corey D.R. (2003) RNA interference in mammalian cells by chemically-modified RNA. *Biochemistry* 42, 7967–7975.
10. Czauderna F., Fechtner M., Dames S., Aygun H., Klippel A., Pronk G.J., Giese K., and Kaufmann J. (2003) Structural variations and stabilising modifications of synthetic siRNAs in mammalian cells. *Nucleic Acids Res.* 31, 2705–2716.
11. Harborth J., Elbashir S.M., Vandenburgh K., Manninga H., Scaringe S.A., Weber K., and Tuschl T. (2003) Sequence, chemical, and structural variation of small interfering RNAs and short hairpin RNAs and the effect on mammalian gene silencing. *Antisense Nucleic Acid Drug Dev.* 13, 83–105.
12. Fluiter K., ten Asbroek A.L., de Wissel M.B., Jakobs M.E., Wissenbach M., Olsson H., Olsen O., Oerum H., and Baas F. (2003) In vivo tumor growth inhibition and biodistribution studies of locked nucleic acid (LNA) antisense oligonucleotides. *Nucleic Acids Res.* 31, 953–962.
13. Elmen J., Thonberg H., Ljungberg K., Frieden M., Westergaard M., Xu Y., Wahren B., Liang Z., Orum H., Koch T., and Wahlestedt C. (2005) Locked nucleic acid (LNA) mediated improvements in siRNA stability and functionality. *Nucleic Acids Res.* 33, 439–447.
14. Mook O.R., Baas F., de Wissel M.B., and Fluiter K. (2007) Evaluation of locked

nucleic acid-modified small interfering RNA in vitro and in vivo. *Mol. Cancer Ther.* 6, 833–843.

15. Martinez J. and Tuschl T. (2004) RISC is a 5′ phosphomonoester-producing RNA endonuclease. *Genes Dev.* 18, 975–980.
16. Leuschner P.J., Ameres S.L., Kueng S., and Martinez J. (2006) Cleavage of the siRNA passenger strand during RISC assembly in human cells. *EMBO Rep.* 7, 314–320.
17. Jackson A.L., Burchard J., Leake D., Reynolds A., Schelter J., Guo J., Johnson J.M., Lim L., Karpilow J., Nichols K., Marshall W., Khvorova A., and Linsley P.S. (2006) Position-specific chemical modification of siRNAs reduces "off-target" transcript silencing. *RNA* 12, 1197–205.
18. Boutla A., Delidakis C., Livadaras I., Tsagris M., and Tabler M. (2001) Short 5′-phosphorylated double-stranded RNAs induce RNA interference in Drosophila. *Curr. Biol.* 11, 1776–1780.
19. Harborth J., Elbashir S.M., Vandenburgh K., Manninga H., Scaringe S.A., Weber K., and Tuschl T. (2003) Sequence, chemical, and structural variation of small interfering RNAs and short hairpin RNAs and the effect on mammalian gene silencing. *Antisense Nucleic Acid Drug Dev.* 13, 83–105.
20. Saxena S., Jonsson Z.O., and Dutta A. (2003) Small RNAs with imperfect match to endogenous mRNA repress translation. Implications for off-target activity of small inhibitory RNA in mammalian cells. *J. Biol. Chem.* 278, 44312–44319.
21. Jackson A.L., Bartz S.R., Schelter J., Kobayashi S.V., Burchard J., Mao M., Li B., Cavet G., and Linsley P.S. (2003) Expression profiling reveals off target gene regulation by RNAi. *Nat. Biotechnol.* 21, 635–637.
22. Sledz C.A., Holko M., de Veer M.J., Silverman R.H., and Williams B.R. (2003) Activation of the interferon system by short-interfering RNAs. *Nat. Cell Biol.* 5, 834–839.
23. Bridge A.J., Pebernard S., Ducraux A., Nicoulaz A.L., and Iggo R. (2003) Induction of an interferon response by RNAi vectors in mammalian cells. *Nat. Genet.* 34, 263–264.
24. Jackson A.L., Burchard J., Schelter J., Chau B.N., Cleary M., Lim L., and Linsley P.S. (2006) Widespread siRNA "off-target" transcript silencing mediated by seed region sequence complementarity. *RNA* 12, 1179–1187.
25. Lewis B.P., Burge C.B., and Bartel D.P. (2005) Conserved seed pairing, often flanked by adenosines, indicates that thousands of human genes are microRNA targets. *Cell* 120, 15–20.
26. Lewis B.P., Shih I.H., Jones-Rhoades M.W., Bartel D.P., and Burge C.B. (2003) Prediction of mammalian microRNA targets. *Cell* 115, 787–798.
27. Birmingham A., Anderson E.M., Reynolds A., Ilsley-Tyree D., Leake D., Fedorov Y., Baskerville S., Maksimova E., Robinson K., Karpilow J., Marshall W.S., and Khvorova A. (2006) 3′ UTR seed matches, but not overall identity, are associated with RNAi off-targets. *Nat. Methods* 3, 199–204.
28. Lim L.P., Lau N.C., Garrett-Engele P., Grimson A., Schelter J.M., Castle J., Bartel D.P., Linsley P.S., and Johnson J.M. (2005) Microarray analysis shows that some microRNAs downregulate large numbers of target mRNAs. *Nature* 433, 769–773.
29. Lin X., Ruan X., Anderson M.G., McDowell J.A., Kroeger P.E., Fesik S.W., and Shen Y. (2005) siRNA-mediated off-target gene silencing triggered by a 7 nt complementation. *Nucleic Acids Res.* 33, 4527–4535.
30. Xie X., Lu J., Kulbokas E.J., Golub T.R., Mootha V., Lindblad-Toh K., Lander E.S., and Kellis M. (2005) Systematic discovery of regulatory motifs in human promoters and 3′ UTRs by comparison of several mammals. *Nature* 434, 338–345.
31. Brennecke J., Stark A., Russell R.B., and Cohen S.M. (2005) Principles of microRNA-target recognition. *PLoS Biol.* 3, e85.
32. Amarzguioui M., Holen T., Babaie E., and Prydz H. (2003) Tolerance for mutations and chemical modifications in a siRNA. *Nucleic Acids Res.* 31, 589–595.
33. Bramsen J.B., Laursen M.B., Damgaard C.K., Lena S.W., Babu B.R., Wengel J., and Kjems J. (2007) Improved silencing properties using small internally segmented interfering RNAs. *Nucleic Acids Res.* 35, 5886–5897
34. Matranga C., Tomari Y., Shin C., Bartel D.P., and Zamore P.D. (2005) Passenger-strand cleavage facilitates assembly of siRNA into Ago2-containing RNAi enzyme complexes. *Cell* 123, 607–620.
35. Grimm D., Streetz K.L., Jopling C.L., Storm T.A., Pandey K., Davis C.R., Marion P., Salazar F., and Kay M.A. (2006) Fatality in mice due to oversaturation of cellular microRNA/short hairpin RNA pathways. *Nature* 441, 537–541.
36. Schlee M., Hornung V., and Hartmann G. (2006) siRNA and isRNA: Two edges of one sword. *Mol. Ther.* 14, 463–470.

37. Judge A.D., Sood V., Shaw J.R., Fang D., McClintock K., and MacLachlan I. (2005) Sequence-dependent stimulation of the mammalian innate immune response by synthetic siRNA. *Nat. Biotechnol.* 23, 457–462.

38. Alexopoulou L., Holt A.C., Medzhitov R., and Flavell R.A. (2001) Recognition of double-stranded RNA and activation of NF-kappaB by Toll-like receptor 3. *Nature* 413, 732–738.

39. Sioud M. (2005) Induction of inflammatory cytokines and interferon responses by double-stranded and single-stranded siRNAs is sequence-dependent and requires endosomal localization. *J. Mol. Biol.* 348, 1079–1090.

40. Marques J.T. and Williams B.R. (2005) Activation of the mammalian immune system by siRNAs. *Nat. Biotechnol.* 23, 1399–1405.

41. Hornung V., Guenthner-Biller M., Bourquin C., Ablasser A., Schlee M., Uematsu S., Noronha A., Manoharan M., Akira S., de Fougerolles A., Endres S., and Hartmann G. (2005) Sequence-specific potent induction of IFN-alpha by short interfering RNA in plasmacytoid dendritic cells through TLR7. *Nat. Med.* 11, 263–270.

42. Judge A.D., Bola G., Lee A.C., and MacLachlan I. (2006) Design of noninflammatory synthetic siRNA mediating potent gene silencing in vivo. *Mol. Ther.* 13, 494–505.

43. Robbins M., Judge A., Liang L., McClintock K., Yaworski E., and Maclachlan I. (2007) 2′-O-methyl-modified RNAs Act as TLR7 Antagonists. *Mol. Ther.* 15, 1663–1669.

44. Urban-Klein B., Werth S., Abuharbeid S., Czubayko F., and Aigner A. (2005) RNAi-mediated gene-targeting through systemic application of polyethylenimine (PEI)-complexed siRNA in vivo. *Gene Ther.* 12, 461–466.

45. van de Water F.M., Boerman O.C., Wouterse A.C., Peters J.G., Russel F.G., and Masereeuw R. (2006) Intravenously administered short interfering RNA accumulates in the kidney and selectively suppresses gene function in renal proximal tubules. *Drug Metab. Dispos.* 34, 1393–1397.

46. Liua N., Dingb H., Vanderheydena J., Zhua Z., and Zhang Y. (2007) Radiolabeling small RNA with technetium-99m for visualizing cellular delivery and mouse biodistribution. *Surgery* 142, 262–269.

47. Schiffelers R.M., Ansari A., Xu J., Zhou Q., Tang Q., Storm G., Molema G., Lu P.Y., Scaria P.V., and Woodle M.C. (2004) Cancer siRNA therapy by tumor selective delivery with ligand-targeted sterically stabilized nanoparticle. *Nucleic Acids Res.* 32, e149.

48. Ma Z., Li J., He F., Wilson A., Pitt B., and Li S. (2005) Cationic lipids enhance siRNA-mediated interferon response in mice. *Biochem. Biophys. Res. Commun.* 330, 755–759.

49. Liu X., Howard K.A., Dong M., Andersen M.O., Rahbek U.L., Johnsen M.G., Hansen O.C., Besenbacher F., and Kjems J. (2007) The influence of polymeric properties on chitosan/siRNA nanoparticle formulation and gene silencing. *Biomaterials* 28, 1280–1288.

50. Soutschek J., Akinc A., Bramlage B., Charisse K., Constien R., Donoghue M., Elbashir S., Geick A., Hadwiger P., Harborth J., John M., Kesavan V., Lavine G., Pandey R.K., Racie T., Rajeev K.G., Rohl I., Toudjarska I., Wang G., Wuschko S., Bumcrot D., Koteliansky V., Limmer S., Manoharan M., and Vornlocher H.P. (2004) Therapeutic silencing of an endogenous gene by systemic administration of modified siRNAs. *Nature* 432, 173–178.

51. Krutzfeldt J., Rajewsky N., Braich R., Rajeev K.G., Tuschl T., Manoharan M., and Stoffel M. (2005) Silencing of microRNAs in vivo with 'antagomirs'. *Nature* 438, 685–689.

52. Song E., Zhu P., Lee S.K., Chowdhury D., Kussman S., Dykxhoorn D.M., Feng Y., Palliser D., Weiner D.B., Shankar P., Marasco W.A., and Lieberman J. (2005) Antibody mediated in vivo delivery of small interfering RNAs via cell-surface receptors. *Nat. Biotechnol.* 23, 709–717.

53. Fluiter K., Frieden M., Vreijling J., Rosenbohm C., De Wissel M.B., Christensen S.M., Koch T., Orum H., and Baas F. (2005) On the in vitro and in vivo properties of four locked nucleic acid nucleotides incorporated into an anti-H-Ras antisense oligonucleotide. *Chembiochem* 6, 1104–1109.

54. Sorensen M.D., Petersen M., and Wengel J. (2003) Functionalized LNA (locked nucleic acid): high-affinity hybridization of oligonucleotides containing N-acylated and N-alkylated 2′-amino-LNA monomers. *Chem. Commun. (Camb)* 17, 2130–2131.

55. Larson S.D., Jackson L.N., Chen L.A., Rychahou P.G., and Evers B.M. (2007) Effectiveness of siRNA uptake in target tissues by various delivery methods. *Surgery* 142, 262–269.

Chapter 10

pSM155 and pSM30 Vectors for miRNA and shRNA Expression

Junzhu Wu, Akua N. Bonsra, and Guangwei Du

Abstract

MicroRNAs (miRNAs) have key roles in diverse regulatory pathways, including control of developmental timing, cell differentiation, apoptosis, cell proliferation, and organ development. miRNAs regulate gene function through a process termed RNA interference (RNAi), which is a highly conserved intracellular mechanism that regulates posttranscriptional gene silencing. RNAi is triggered by double-stranded small interfering RNAs (siRNAs), which can be processed from small hairpin RNAs (shRNAs) generated from an expression vector. In some recently described vectors, the siRNAs are expressed from synthetic stem–loop precursors of microRNAs (miRNAs) driven by polymerase II promoters. We have reported new RNAi vectors, designated pSM155 and pSM30, that take into consideration miRNA processing and RNA splicing by placing the miRNA-based artificial miRNA expression cassettes inside of synthetic introns. Like the original miRNA vectors, these pSM155 and pSM30 constructs can efficiently downregulate the expression of their target genes. Moreover, the expression of a coexpressed fluorescent marker, EGFP, is substantially improved by this new design. The new vectors can also be used to express natural miRNAs and label cells expressing these miRNAs. These RNAi vectors thus provide new tools for gene suppression and miRNA expression. We describe in this chapter the protocols for selecting and cloning artificial and natural miRNAs (or shRNAs), evaluating their efficiency in downregulating gene expression, and also discuss the potential applications of these vectors.

Key words: microRNA, mRNA, small-hairpin RNA, RNA interference, RNA splicing, intron.

1. Introduction

MicroRNAs (miRNAs) are single-stranded RNAs (ssRNAs) of 19–25 nucleotides in length that are generated from endogenous hairpin-shaped transcripts (*see* Chaps. 18 and 19). They bind and guide cleavage of a variety of RNAs and/or their translational inhibition. Recent studies have revealed that miRNAs have key roles

M. Sioud (ed.), *Methods in Molecular Biology, siRNA and miRNA Gene Silencing, vol. 487*

DOI: 10.1007/978-1-60327-547-7_10

in diverse regulatory pathways, including control of developmental timing, cell differentiation, apoptosis, cell proliferation, and organ development. In animals, primary miRNAs (pri-miRNAs) are transcribed by RNA polymerase II, and contain 5′ CAP structures and 3′ poly(A) tails *(1, 2)*. The pri-miRNA is recognized and cleaved at a specific hairpin site by the nuclear microprocessor complex including an RNase III family enzyme, Drosha, to produce a miRNA precursor (pre-miRNA) of approximately 70–90 nucleotides with a 2 nucleotide 3′ overhang. This pre-miRNA is then transported to the cytoplasm by Exportin-5 *(1, 2)*, and then recognized by another RNase III, Dicer, and cleaved to produce a mature miRNA of ~22 nucleotides. miRNAs can be categorized into three groups according to their genomic context: exonic miRNA in noncoding transcripts, intronic miRNAs in noncoding transcripts, and intronic miRNAs in protein-coding transcripts.

The nature of inhibition of endogenous mRNA translation by miRNAs also promoted some groups to develop new RNA interference (RNAi) vectors based on natural miRNAs, i.e., miR30 and miR155 *(3– 5)*. The expression of siRNAs from the artificial miRNA is often driven by an RNA polymerase II promoter in this strategy, and has been shown to efficiently downregulate the expression of their target genes. This kind of new vector offers several advantages over the traditional RNAi vectors driven by RNA polymerase III promoters *(6, 7)*, including expression of several artificial miRNAs from a single transcript, and tissue-specific or regulated expression *(3, 5, 8)*.

Based on knowledge of miRNA biogenesis, we and others have reported the development of vectors in which the artificial miRNAs are expressed from artificial or natural introns *(3, 9, 10, 11)*. These vectors can be used to knockdown expression of reporter and endogenous genes and overexpress natural miRNAs to study their function, while providing a robust marker for the artificial miRNA-transfected cells *(9)*. In this chapter, we describe the application of two such vectors derived from miR155 and miR30 precursors, pSM155 and pSM30 *(9)*.

2. Materials

2.1. DNA Cloning and Plasmid Maps

1. Restriction enzymes, *Bsm*BI, *Msc*I, *Xho*I, and *Nhe*I (New England Biolabs, Ipswich, MA).
2. QIAEX II gel extraction kit (Qiagen, Valencia, CA).
3. Rapid ligation kit (Roche Diagnostics, Indianapolis, IN).
4. VectorNTI program (downloaded from Invitrogen, Carlsbad, CA, USA).

2.2. Cell Culture and Cell Lysis

1. Dulbecco's Modified Eagle Medium (DMEM) (Invitrogen) supplemented with 10% calf serum (CS, HyClone, Ogden, UT).
2. Opti-MEM-I, and LipofectAMINE Plus (Invitrogen).
3. Phosphate buffered saline (PBS): Prepare 10X stock with 1.37 M NaCl, 27 mM KCl, 100 mM Na_2HPO_4, and 18 mM KH_2PO_4 (adjust to pH 7.4 with HCl if necessary) and autoclave before storage at room temperature. Prepare working solution by dilution of one part with nine parts water.
4. Complete protease inhibitor cocktail tablets (Roche Diagnostics): Dissolve in PBS to make 25X stock solution.
5. Radioimmunoprecipitation assay (RIPA) buffer: 20 mM Tris-HCl, pH 7.4, 137 mM NaCl, 10% (v/v) glycerol, 0.1% (w/v) SDS, 0.5% (w/v) deoxycholate, 1% (v/v) Triton X-100, 1 mM EDTA, 1X Protease Inhibitor.

2.3. Western Blotting

1. Ponceau S Staining Solution: 0.1% (w/v) Ponceau S in 5% (v/v) acetic acid. Stain solution can be re-used several times.
2. Semi-dry transfer buffer: 48 mM Tris, pH 8.3, 39 mM glycine, 0.037% SDS, 20% methanol.
3. Tris-buffered saline (TBS): Prepare 10X stock with 1.37 M NaCl, 27 mM KCl, 250 mM Tris-HCl, pH 7.4, 1% Tween-20. Dilute 100 ml with 900 ml water for use.
4. Tris-buffered saline with Tween (TBS-T): Prepare 10X stock with 1.37 M NaCl, 27 mM KCl, 250 mM Tris-HCl, pH 7.4, 1% Tween-20. Dilute 100 ml with 900 ml water for use.
5. Blocking buffer: 1% (w/v) casein in TBS.
6. Antibody dilution buffer: 1% (w/v) casein in TBS-T.
7. Monoclonal anti-α-tubulin (Sigma-Aldrich, St Louis, MO).
8. Rabbit polyclonal GFP antibody (Abcam, Cambridge, MA).
9. Infrared IRDye-labeled secondary antibodies. Goat anti-mouse and anti-rabbit IgG conjugated to Alexa 680 (Invitrogen). Goat anti-mouse and anti-human IgG conjugated to IRDye 800 (Rockland Immunochemicals, Gilbertsville, PA).
10. Semi-dry blotting unit (Fisher Scientific, Pittsburg, PA).
11. Odyssey infrared imaging system (LI-COR Biosciences–Biotechnology, Lincoln, NE).

3. Methods

Some original artificial miRNA expression vectors were directly coupled to the coding sequence of a fluorescent protein to provide a

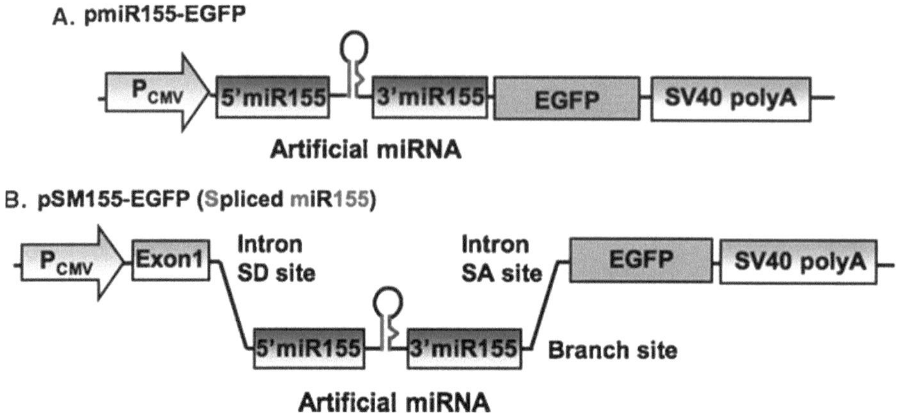

Fig. 10.1. Two strategies for coexpression of miRNA and EGFP. (**a**) In the traditional strategy, an EGFP open reading frame is directly linked to the miRNA expression cassette. (**b**) In the new strategy, the miRNA is located inside an intron.

way to identify transfected cells genuinely expressing the artificial miRNA *(5, 8)*. An example of this design is shown in **Fig. 10.1a**. The pri-miRNAs are processed in the nucleus by Drosha to set up transport of pre-miRNAs to the cytoplasm, where they are further processed to siRNA. However, this process simultaneously blocks the translation of the enhanced green fluorescent protein (EGFP) marker, because the resulting mRNA fragment lacks a 5′ CAP structure and is rapidly degraded. EGFP can be translated if the pri-miRNAs are exported to the cytoplasm before Drosha cleavage, but siRNAs are then not produced from these unprocessed pri-miRNAs *(9)*.

We hypothesized that both miRNA and EGFP expression could be accommodated if the pri-miRNAs were processed in nuclei without inactivating the EGFP component. To achieve this, we inserted the artificial miRNA-expressing cassettes into a chimeric intron composed of the 5′ donor site from the first intron of the human β-globin gene and the 3′ acceptor site from the intron of an immunoglobulin gene heavy chain variable region (derived from pCI-neo from Promega) (**Fig. 10.1b**) *(9)*. This design mimics the structure and processing of some natural miRNAs which consist of intronic miRNAs found in protein-coding sequences *(1, 2)*. Separation by splicing of the miRNA component from the 5′ CAP-exon-EGFP-3′ poly(A) component should facilitate Drosha processing of the pri-miRNA rather than cytoplasmic export. Also, the design should favor cytoplasmic export and translation of the EGFP RNA rather than intranuclear degradation. We have developed two vectors, pSM155 (Spliced miR155, **Fig. 10.2a**) and pSM30 (Spliced miR30, **Fig. 10.2b**) *(9)*, which have significantly improved the expression of the EGFP protein (**Fig. 10.3**) without affecting their ability of gene silencing (**Fig. 10.4**; only pSM155 is shown). These vectors and full sequences will be distributed upon request.

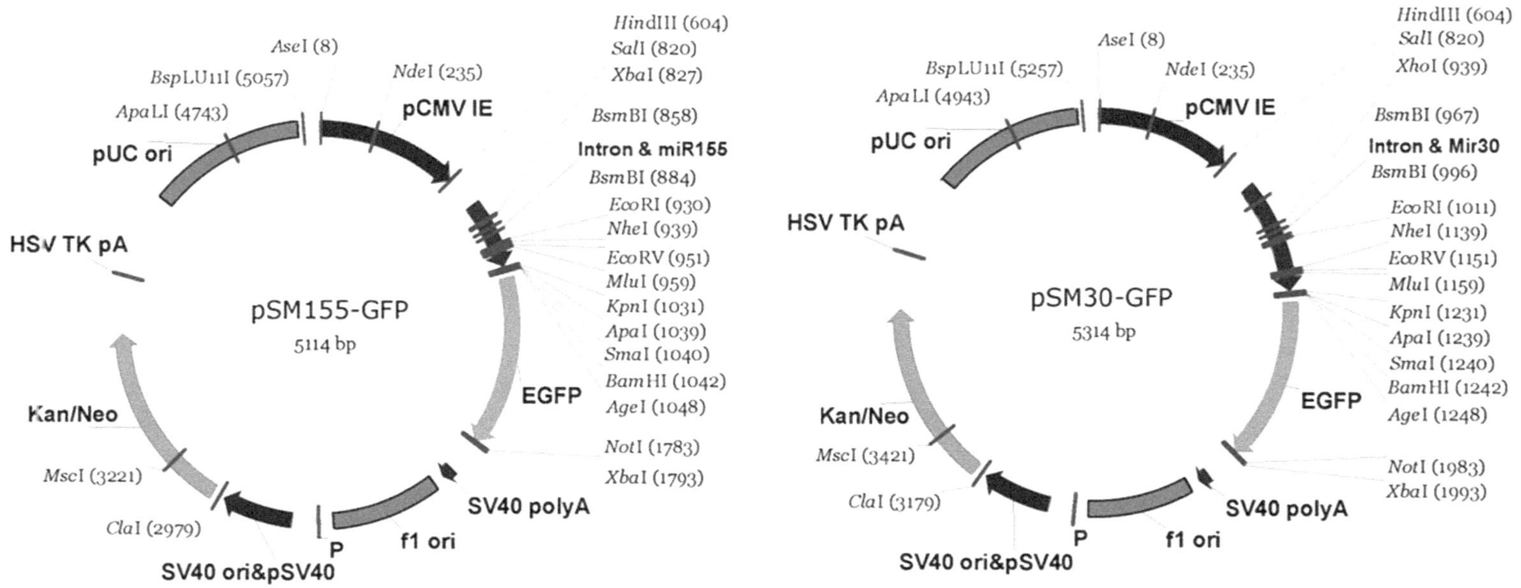

Fig. 10.2. Plasmid maps for pSM155 and pSM30. pSM155 (*left*) and pSM30 (*right*) are used to express artificial miRNAs for RNAi experiments by incorporating the sequence of the human miR155 miRNA precursor into a synthetic intron. The maps were generated using VectorNTI (full sequences and digital files are available upon request).

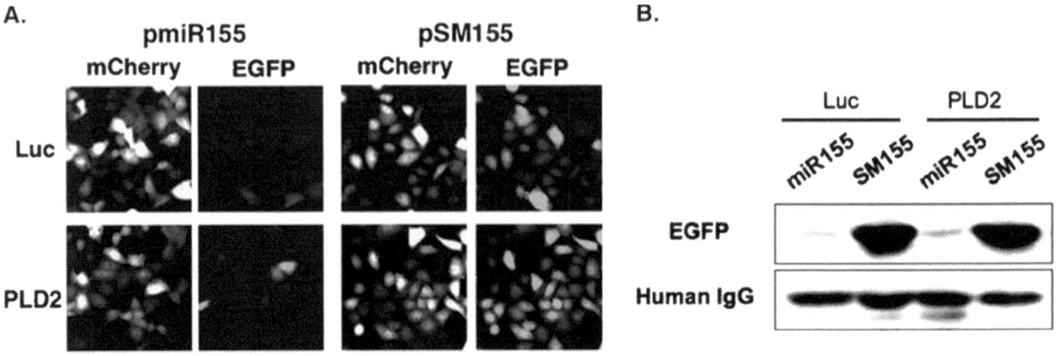

Fig. 10.3. The new pSM155 vector improves the expression of marker proteins in artificial miRNA-expressing cells. Both pmiR155 and pSM155 constructs contain an EGFP marker as illustrated in Fig. 10.1. A similar result was obtained with miR30-based vectors, pmiR30 and pSM30. (**a**) HeLa cells were transfected with pmiR155-luc, pmiR155-PLD2, pSM155-luc, or pSM155-PLD2, in combination with pcDNA3.1/mCherry, which encodes a red fluorescent protein. Expression of the EGFP marker protein is significantly improved in pSM155 for both constructs tested. As shown EGFP is expressed in all cells expressing mCherry when the pSM155 vector is used. However, it is only expressed in a few of the cells when pmiR155 is used. (**b**) HeLa cells were cotransfected with pmiR155-luc (lane 1), pSM155-luc (lane 2), pmiR155-PLD2 (lane 3), or pSM155-PLD2 (lane 4), and pRK-human IgG, which encodes a human IgG, was used as a transfection and loading control. The expression of EGFP and IgG was determined by the rabbit polyclonal GFP antibody/Alexa 680 goat anti-rabbit secondary IgG, and IRDye 800 goat anti-human IgG, respectively. (Reprinted by permission from Blackwell publishing: *FEBS J.* 273, 5421–5527 (2006)).

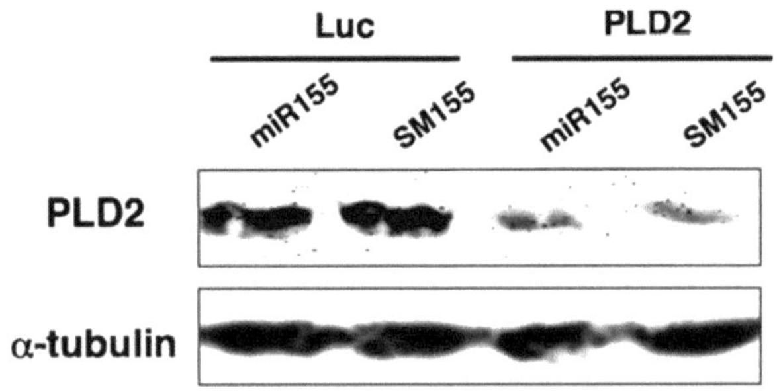

Fig. 10.4. The new pSM155 and its parental vector pmiR155 show similar efficiency in knocking down the expression of an endogenous gene, PLD2. HeLa cells were transfected with artificial miRNAs against luciferase (control) and PLD2 in pmiR155 and pSM155. Cell lysates were collected for Western blot analysis two days after transfection. PLD2 and α-tubulin were detected by a polyclonal antibody and a mouse monoclonal antibody, respectively, followed by goat anti-rabbit secondary IgG conjugated to Alexa 680 and goat anti-mouse antibody conjugated to IRDye 800. Fluorescence was recorded using an Odyssey infrared imaging system from LI-COR Bioscience-Biotechnology (Lincoln, NE, USA). (Reprinted by permission from Blackwell publishing: *FEBS J.* 273, 5421–5527 (2006)).

3.1. Selection of Target Sequences and Synthesis of Oligonucleotides

To select effective target sequences for artificial miRNA expression, we generally follow the guidelines for generation of effective small interfering RNA (siRNA) *(12, 13)*. Design of the oligo vectors can be also assisted by the algorithm on the Invitrogen Web site (www.invitrogen.com/rnai(for pSM155)) and Cold Spring Harbor Lab Web site (http://codex.cshl.edu(for pSM30)). An effective artificial miRNA must fit most but not necessarily all of the following rules *(12)*: overall low to medium G/C content

(30–50%), low internal stability at the 5′ antisense strand, high internal stability at the 5′ sense strand, absence of internal repeats or palindromes, A-form helix between miRNA and target mRNA, presence of an A at position 3 and 19 of sense strand, absence of a G or C at position 19 of sense strand, presence of a U at position 10 of sense strand, and absence of a G at position 13 of sense strand.

To enhance specificity, a miRNA also needs to have minimal homology with non-target RNAs and avoid low-stringency sequences. A pair of oligos includes cohesive ends, sense and antisense specific sequences matching the target mRNA, and a loop (64 nucleotides for the pSM155-based system and 67 nucleotides for the pSM30-based system). For each gene, we usually test four to five target sequences and choose at least two constructs for our experiments.

3.1.1. Synthesis of Oligos for pSM155 (An Example Is Shown in Fig. 10.5a) (see **Note 1***)*

1. To generate the top oligo sequence, combine these elements (from 5′ end to 3′ end): start with 5′ TGCTG, reverse complement of the 21-nt sense target sequence (this is the mature miRNA sequence), add GTTTTGGCCACTGACTGAC (loop), and add nucleotides 1–8 (5′-3′) of the sense target sequence and nucleotides 11–21 (5′-3′) of the sense target sequence.
2. To generate the bottom oligo sequence, perform the following steps: Remove 5′ TGCT from the top oligo sequence (new sequence starts with G), take the reverse complement of the sequence from step 1, and add CCTG to the 5′ end of the sequence from Step 2.

A. For pSM155

Top strand

Anti-sense target (21 nt) **Loop** Sense (nt 1-8 and 11-21)

TGCTGTATTCAGCCCATATCGTTTCA**GTTTTGGCCACTGACTGAC**AGAAACGATGGGCTGAATA

Bottom strand

CCTGTATTCAGCCCATCGTTTCAGTCAGTCAGTGGCCAAAACTGAAACGATATGGGCTGAATAC

B. For pSM30

Top strand

Sense target (22 nt) **Loop** Anti-sense (22 nt)

AGCGAGCTGCTGGTGCCAACCCTATT**TAGTGAAGCCACAGATGTA**AATAGGGTTGGCACCAGCAGCG

Bottom strand

GGCACGCTGCTGGTGCCAACCCTATTTACATCTGTGGCTTCACTAAATAGGGTTGGCACCAGCAGCT

Fig. 10.5. Examples of oligos used for cloning the artificial miRNAs into (**a**) pSM155 and (**b**) pSM30. A pair of oligos is required for cloning an artificial miRNA, which includes sequences for creating cohesive sites for the vector digested by *Bsm*BI, sense strand, miRNA loop, and anti-sense strand. Both pairs of oligos shown are derived from firefly luciferase and are intended to inhibit its expression.

3.1.2. Synthesis of Oligos for pSM30 (An Example Is Shown in Fig. 10.5b)

1. To generate the top oligo sequence, combine these elements (from 5′ end to 3′ end): start with 5′ AGCG, add 22-nt sense target sequence, change the first nucleotide to one which does not anneal to the last nucleotide in anti-sense, e.g., C to A, add TAGTGAAGCCACAGATGTA (loop), and add nucleotides of the anti-sense target sequence.
2. To generate the bottom oligo sequence, perform the following steps: Remove 5′ AGCG from the top oligo sequence (new sequence starts with G), take the reverse complement of the sequence from step 1 (this is the mature miRNA sequence), and add GGCA to the 5′ end of the sequence from Step 2.

3.2. Cloning of Artificial miRNAs

To simplify the cloning of artificial miRNAs without substantially altering the miRNA arm sequences, two inverted *Bsm*BI sites were placed internal to the arms of pSM155 and pSM30 (**Fig. 10.6**). *Bsm*BI is a class IIS restriction enzyme, and has a cleavage site different but next to its recognition site. A pair of oligos with appropriate four nucleotide overhangs generated above can be easily ligated to the cohesive sites of the vector generated by *Bsm*BI digestion. Incorporation of the *Bsm*BI sites greatly reduces the length of the oligos used for cloning, thus reducing the cost in purchasing oligos and avoiding the high error rate of long oligos.

3.2.1. Vector Preparation

1. Cut 1–2 μg of pSM155 or pSM30 in 15 μl New England Biolabs (NEB) restriction Buffer 3 at 55°C for 1 h.
2. Run the digested vectors on 0.8% agarose gel and purify them using standard commercially available gel purification kits according to the manufacturer's instruction.

3.2.2. Oligonucleotide Annealing

1. Dissolve oligos (25 μM final) with distilled water (100 μM stock concentration).
2. Take 5 μl from each oligo (top and bottom) and add 8 μl water and 2 μl 10X NEB restriction buffer (We usually use Buffer 3).
3. Boil the oligos for 4 min and leave the denatured oligos at room temperature for 15 min and then move them to 4°C for 10–15 min.
4. Dilute 2500-fold to get 10 nM double-stranded oligos (250-fold dilution with water, quickly followed by 10-fold with 1X NEB Buffer 3; or add 0.2 μl of annealed oligos to 500 μl 1X NEB Buffer 3) (*see* **Note 2**).

3.2.3. Ligation Using Rapid Ligation Kit

1. Mix 2 μl of the annealed oligos with 2 μl vector and H_2O (~5 ng or 1.5×10^{-3} pmol) (digested with *Bsm*BI) (oligos: vector, ~15:1).
2. Add 1 μl DNA dilution buffer, mix well, and then spin down briefly.

A. pSM155

```
                                                                SalI     XbaI
                                                              ~~~~~~~~ ~~~~~~~~
781 ACTGGGCTTG TCGAGACAGA GAAGACTCTT GCGTTTCGGT CGACTTCTAG AGCTCTGGAG
    TGACCCGAAC AGCTCTGTCT CTTCTGAGAA CGCAAAGCCA GCTGAAGATC TCGAGACCTC

                                 BsmBI              BsmBI
     miR155 5'                  ~~~~~~             ~~~~~~~~         miR155 3'
841 GCTTGCTGAA GGCTGTATGC TAGAGACGTA CAGATGCGTC TCACAGGACA CAAGGCCTGT
    CGAACGACTT CCGACATACG ATCTCTGCAT GTCTACGCAG AGTGTCCTGT GTTCCGGACA

                                        EcoRI        NheI        EcoRV      MluI
                                      ~~~~~~~~    ~~~~~~~~     ~~~~~~~~     ~~~
901 TACTAGCACT CACATGGAAC AAATGGCCGA ATTCGGCGCT AGCTGCTGAT ATCGCATACG
    ATGATCGTGA GTGTACCTTG TTTACCGGCT TAAGCCGCGA TCGACGACTA TAGCGTATGC
```

B. pSM30

```
                                                                SalI
                                                              ~~~~~~~~
781 ACTGGGCTTG TCGAGACAGA GAAGACTCTT GCGTTTCGGT CGACTAGGGA TAACAGGGTA
    TGACCCGAAC AGCTCTGTCT CTTCTGAGAA CGCAAAGCCA GCTGATCCCT ATTGTCCCAT

841 ATTGTTTGAA TGAGGCTTCA GTACTTTACA GAATCGTTGC CTGCACATCT TGGAAACACT
    TAACAAACTT ACTCCGAAGT CATGAAATGT CTTAGCAACG GACGTGTAGA ACCTTTGTGA

                                                               XhoI
              miR30 5' arm                                   ~~~~~~~~
901 TGCTGGGATT ACTTCTTCAG GTTAACCCAA CAGAAGGCTC GAGAAGGTAT ATTGCTGTTG
    ACGACCCTAA TGAAGAAGTC CAATTGGGTT GTCTTCCGAG CTCTTCCATA TAACGACAAC
               BsmBI               BsmBI                     EcoRI
              ~~~~~~              ~~~~~~~~                  ~~~~~~~~
961 ACAGTGAGCG AGAGACGGAA GCCACAGACG TCTCATGCCT ACTGCCTCGG AATTCAAGGG
    TGTCACTCGC TCTCTGCCTT CGGTGTCTGC AGAGTACGGA TGACGGAGCC TTAAGTTCCC

1021 GCTACTTTAG GAGCAATTAT CTTGTTTACT AAAACTGAAT ACCTTGCTAT CTCTTTGATA
     CGATGAAATC CTCGTTAATA GAACAAATGA TTTTGACTTA TGGAACGATA GAGAAACTAT

                                                                       NheI
       miR30 3' arm                                                     ~~~
1081 CATTTTTACA AAGCTGAATT AAAATGGTAT AAATTAAATC ACTTTTTTCA ATTCGGCGCT
     GTAAAAATGT TTCGACTTAA TTTTACCATA TTTAATTTAG TGAAAAAAGT TAAGCCGCGA

     NheI   EcoRV          MluI
     ~~~   ~~~~~~~~      ~~~~~~~~
1141 AGCTGCTGAT ATCGCATACG CGTGGACCAG ATAGGCACCT ATTGGTCTTA CTGACATCCA
     TCGACGACTA TAGCGTATGC GCACCTGGTC TATCCGTGGA TAACCAGAAT GACTGTAGGT
```

Fig. 10.6. Sequences around miRNA expression cassettes in (**a**) pSM155 and (**b**) pSM30. The arms of miR155 and miR30 are shaded. Cutting both vectors with *Bsm*BI removes its recognition sites and generates cohesive ends for miRNA cloning. The entire miR155 or miR30 cassettes in these vectors can also be removed and replaced with natural miRNAs for their overexpression.

3. Add 5 µl 2X ligation buffer and mix. Add 0.5 µl T4 DNA ligase, mix well, and spin down briefly.
4. Incubate at RT for 5 min.

3.2.4. Transformation

Any general *E. coli* strain used for cloning, e.g., DH5α, XL-Blue, is suitable. Add 3 μl of ligation mixture to 50 μl of competent cells and follow the regular transformation method.

3.2.5. Characterization of Recombinant Plasmids

The cloning efficiency of our current strategy is excellent and there is very low empty vector background. We often purify plasmids from two colonies for further characterization. For the pSM155 vector, the plasmid DNA is digested with *Msc*I. The linearized parental vector is 5114 bp, and the recombinant construct generates 2784-bp and 2368-bp fragments (*see* **Note 3**). For the pSM30 vector, the plasmid DNA is digested with *Xho*I/*Nhe*I (*see* **Note 3** and **4**). The empty pSM30 vector is used as control. The recombinant clone and the vector release 239-bp and 200-bp fragments, respectively. Finally, the candidate recombinant clones need to be confirmed by sequencing (*see* **Note 5**).

3.3. Cloning of Natural Endogenous miRNAs

Both pSM155 and pSM30 could also be used to express natural miRNAs, and thus label cells transfected by exogenous miRNAs. A full-length miRNA can be amplified by PCR using a top primer with a 5′ flanking *Sal*I restriction site and a bottom primer with a 5′ flanking *Nhe*I, *Eco*RV, or *Mlu*I restriction site. The PCR product digested with these restriction enzymes is then cloned into pSM155 or pSM30 cut with the same enzymes (**Fig. 10.6**).

3.4. Determination of Knockdown Efficiencies of pSM155 and pSM30 miRNA Constructs

The knockdown efficiency is often determined by Western blotting (*see* **Note 6**). An irrelevant artificial miRNA construct, such as that targeting luciferase, can be used as a control. In some cases, immunofluorescent microscopy can also be used. If an antibody for the gene of interest is not available, reverse transcription-PCR is then the preferred method. We will describe Western blot analysis using the Odyssey Infrared Imaging System from LI-COR Biosciences–Biotechnology.

3.4.1. Cell Culture and Transfection

1. Culture HeLa cells in DMEM containing 10% FCS.
2. One day prior to transfection, plate the cells to a 6-well plate, and grow overnight. The cells should be around 40–60% confluent at the time of transfection.
3. On the next day, dilute 0.5–1 μg DNA in 100 μl Opti-MEM in a microtube, mix, and then add 6 μl of Plus Reagent, and mix again.
4. Incubate at room temperature for 15 min.
5. Dilute 4 μl of Lipofectamine in 100 μl Opti-MEM in a different microtube and mix gently.
6. Combine Lipofectamine and DNA, and then mix.
7. Incubate the mixture for 15 min at room temperature.
8. During the two 15 min interval, replace the growth medium with Opti-MEM.

9. Add the transfection mixture dropwise to each well.
10. Incubate the plates at 37°C for 4–5 h.
11. Remove the transfection mixture
12. Add 2 ml growth media and incubate at 37°C
13. Collect the cells 2–3 days later (*see* **Note 7**).

3.4.2. Cell Lysates and SDS-PAGE

1. Remove culture medium from 6-well plates using vacuum.
2. Add 1 ml ice-cold PBS containing 5 mM EDTA to cells.
3. Leave cells on ice until they detach from the plate (*see* **Note 8**).
4. Transfer cells to a microtube.
5. Collect cells by briefly spinning for 20–30 s at top speed.
6. Remove supernatants and re-suspend pellets with 60 μl RIPA buffer containing 1X protease inhibitor cocktails, keep them on ice for 5 min.
7. Remove the nuclei by low speed spinning (about 2000–3000× *g*) at 4°C for 2–3 min.
8. Transfer the supernatant to a new microtube containing SDS loading buffer (to make the final concentration to 1X).
10. After separation by standard SDS-PAGE, transfer proteins to a nitrocellulose membrane by semi-dry transfer (*see* **Note 9**). A 9 × 6 mm mini-gel is transferred at 200 mA for 1 h. To check transfer efficiency, incubate the membrane for 1 min in Ponceau S staining solution with gentle agitation. Rinse the membrane in distilled water until the background is clean. The stain can be completely removed from the protein bands by continued washing.
11. Wet the membrane in TBS for several minutes and incubate the membrane in blocking buffer for 1 h (*see* **Note 10**). Be sure to use sufficient blocking buffer to cover the membrane.
12. Dilute primary antibody in blocking buffer. Optimum dilution depends on your antibody and should be determined empirically. An internal loading control such as α-tubulin or actin is recommended. To lower background, add 0.1–0.2% Tween-20 to the diluted antibody before incubation.
13. Incubate the membranes in primary antibody for 60 min or longer at room temperature with gentle shaking. Optimum incubation times vary for different primary antibodies. Use enough antibody solution to completely cover the membrane.
14. Wash membrane four times for 5 min each at room temperature in TBS-T with gentle shaking.
15. Dilute the fluorescently labeled secondary antibody in blocking buffer. Avoid prolonged exposure of antibody and membrane to light. Recommended dilution is 1:10,000 (suggested dilution range is 1:5000 to 1:25,000) (*see* **Note**

11). Add Tween-20 to the diluted antibody as you did for the primary antibody. Add SDS if desired (*see* **Note 12**). Incubate secondary antibody with membrane at room temperature for 60 min (*see* **Notes 13**).

16. Wash membrane three times for 5 min each at room temperature in TBS-T with gentle shaking.
17. Rinse membrane with TBS for 5 min to remove residual Tween-20.
18. Scan the membrane in the appropriate channels. Alexa 680 and IRdye 800-labeled secondary antibodies are scanned at 700 and 800 channels, respectively (*see* **Notes 14**).

3.5. Other Technical and Potential Therapeutic Applications

In addition to wide applications, knocking-down the expression of disease-related genes using small interfering RNAs (siRNAs) is increasingly used as a strategy to treat a variety of illnesses. Like chemically synthesized siRNAs and shRNAs-expressed from Pol III promoters, such as U6 or H1, the miRNA-expressed siRNAs also have great potential in treating human diseases. More recently we have described the design of pSM155 and pSM30 vectors that provide a reliable fluorescent marker for monitoring transfected cells. With some modification, the pSM155 and pSM30 could also be adapted for other applications in research and therapeutics.

1. pSM155 and pSM30 can be used to express two or more miRNAs from a single vector, which can be designed against either the same target mRNA or two different target mRNAs. To clone two different miRNAs into pSM155, the first miRNA cassette purified by cutting with *Xba*I/*Mlu*I can be cloned into the second miRNA expression vector digested with *Nhe*I/*Mlu*I restriction enzymes. The ability to effectively express two synthetic miRNAs from a single transcript has been demonstrated in a similar design *(3)*.
2. If other fluorescent proteins such as mCherry and dsRed are needed to label the transfected cells, the EGFP can be replaced by genes encoding these proteins in both vectors using *Not*I and *Kpn*I, *Apa*I, *Sma*I, *Bam*HI, or *Age*I (*see* maps in **Fig. 10.2**).
3. The EGFP marker can also be replaced by an antibiotic selection marker such as hygromycin in both pSM155 and pSM30 vectors. Expression of the antibiotic selection marker and miRNA from a single mRNA transcript allows to select "true clones" when stable cell lines expressing artificial miRNAs are generated.
4. RNAi and rescue experiments in the same vector—The cDNA containing wobble mutations at the sites where the artificial miRNA targets can be cloned into the same vector.

5. The utilization of Pol II promoters allows the new pSM155, pSM30, and other miRNA-based vectors to drive tissue-specific regulation or inducible expression. For example, introducing tissue-specific promoters into miRNA-based vectors will allow targeting to specific tissues. An inducible promoter controlled by small molecules can also be adapted to drive miRNAs expression in order to avoid chronic toxicity of RNAi.

Despite the encouraging results obtained with miRNA-based gene silencing, the technique may not be as efficient as synthetic siRNAs and the traditional Pol III-driven shRNAs in some cell types and tissues. The more complex structure of miRNAs requires additional processing steps, which may eventually lead to generation of less mature siRNAs. In this respect, it has been reported that many miRNA primary transcripts are present at high levels but are not processed by the enzyme Drosha in early mouse development and in human primary tumors *(14)*.

4. Notes

1. It is not necessary to use phosphorylated oligonucleotides. If nonphosphorylated oligos are used for cloning, make sure that the vector is not dephosphorylated.
2. Keep the oligos on ice, do not leave diluted oligonucleotides at temperatures higher than room temperature. Cooled samples can be stored at –20°C.
3. Since the cloning efficiency is very high, we often directly sequence the candidate plasmids without restriction enzyme characterization.
4. 1.5% agarose gel is used to separate the 239-bp fragment from the 200-bp fragment.
5. The mutation rate of the synthesized long oligos is usually high. Point mutation and deletion were often seen. It is necessary to confirm the candidate constructs by DNA sequencing.
6. One key to get reliable results is getting high transfection efficiency. We only perform Western blotting when more than 80–90% cells are transfected. If transfection is a problem, cotransfection of the gene of interest with an epi-tag such as flag or myc and the miRNA construct would be an alternative method in the initial screening experiments of efficient miRNA constructs. In this case, a primary antibody against the epi-tag can be used. This method is very fast and reliable, especially when many constructs need to be tested and genes of interest are expressed only in cells with low

transfection efficiency. Subsequent to selection, the silencing potential of candidate constructs should be tested using endogenous cognate genes.

7. The time required for effective knockdown can vary for different genes.
8. The time for detachment depends on the cell types. It usually takes about for 3–5 min for most cell types.
9. Nitrocellulose membrane is preferred for detection by fluorescently labeled secondary antibodies. If PVDF membrane has to be used, those with low fluorescence should be chosen.
10. It is critical to avoid a blocking solution containing Tween-20. Tween-20 increases fluorescent background.
11. For detection of small amounts of protein, try using more secondary antibody (1:5000–1:10,000).
12. Adding 0.01–0.02% SDS to the diluted secondary antibody (in addition to Tween-20) will substantially reduce membrane background, particularly when using PVDF. However, do not add SDS during blocking or to the diluted primary antibody.
13. Diluted secondary antibody can be saved and re-used. Store at 4°C and protect from light. However, for best sensitivity and performance, use freshly diluted antibody solution.
14. Protect the membrane from light until it has been scanned. Keep the membrane wet if you plan to strip and re-use it. Once a membrane has dried, stripping is ineffective. Blots can be allowed to dry before scanning if desired. Signal strength may be enhanced on a dry membrane. The fluorescent signal on the membrane will remain stable for several months, or longer, if protected from light. Membranes may be stored dry or in PBS buffer at 4°C.

References

1. Cullen, B. R. (2004) Transcription and processing of human microRNA precursors. *Mol Cell* 16, 861–865.
2. Kim, V. N. (2005) MicroRNA biogenesis: coordinated cropping and dicing. *Nat Rev Mol Cell Biol* 6, 376–385.
3. Chung, K. H., Hart, C. C., Al-Bassam, S., Avery, A., Taylor, J., Patel, P. D., Vojtek, A. B., and Turner, D. L. (2006) Polycistronic RNA polymerase II expression vectors for RNA interference based on BIC/miR-155. *Nucleic Acids Res* 34, e53.
4. Silva, J. M., Li, M. Z., Chang, K., Ge, W., Golding, M. C., Rickles, R. J., Siolas, D., Hu, G., Paddison, P. J., Schlabach, M. R., Sheth, N., Bradshaw, J., Burchard, J., Kulkarni, A., Cavet, G., Sachidanandam, R., McCombie, W. R., Cleary, M. A., Elledge, S. J., and Hannon, G. J. (2005) Second-generation shRNA libraries covering the mouse and human genomes. *Nat Genet* 37, 1281–1288.
5. Stegmeier, F., Hu, G., Rickles, R. J., Hannon, G. J., and Elledge, S. J. (2005) A lentiviral microRNA-based system for single-copy polymerase II-regulated RNA interference in mammalian cells. *Proc Natl Acad Sci USA* 102, 13212–13217.
6. Brummelkamp, T. R., Bernards, R., and Agami, R. (2002) A system for stable expression

of short interfering RNAs in mammalian cells. *Science* 296, 550–553.

7. Paddison, P. J., Caudy, A. A., Bernstein, E., Hannon, G. J., and Conklin, D. S. (2002) Short hairpin RNAs (shRNAs) induce sequence-specific silencing in mammalian cells. *Genes Dev* 16, 948–958.
8. Dickins, R. A., Hemann, M. T., Zilfou, J. T., Simpson, D. R., Ibarra, I., Hannon, G. J., and Lowe, S. W. (2005) Probing tumor phenotypes using stable and regulated synthetic microRNA precursors. *Nat Genet* 37, 1289–1295.
9. Du, G., Yonekubo, J., Zeng, Y., Osisami, M., and Frohman, M. A. (2006) Design of expression vectors for RNA interference based on miRNAs and RNA splicing. *FEBS J* 273, 5421–5427.
10. Lin, S. L. and Ying, S. Y. (2006) Gene silencing in vitro and in vivo using intronic microRNAs. *Methods Mol Biol* 342, 295–312.
11. Zhou, H., Xia, X. G., and Xu, Z. (2005) An RNA polymerase II construct synthesizes short-hairpin RNA with a quantitative indicator and mediates highly efficient RNAi. *Nucleic Acids Res* 33, e62.
12. Mittal, V. (2004) Improving the efficiency of RNA interference in mammals. *Nat Rev Genet* 5, 355–365.
13. Ui-Tei, K., Naito, Y., Takahashi, F., Haraguchi, T., Ohki-Hamazaki, H., Juni, A., Ueda, R., and Saigo, K. (2004) Guidelines for the selection of highly effective siRNA sequences for mammalian and chick RNA interference. *Nucleic Acids Res* 32, 936–948.
14. Thomson, J. M., Newman, M., Parker, J. S., Morin-Kensicki, E. M., Wright, T., and Hammond, S. M. (2006) Extensive post-transcriptional regulation of microRNAs and its implications for cancer. *Genes Dev* 20, 2202–2207.

Chapter 11

Targeting Oncogenes with siRNAs

Olaf Heidenreich

Abstract

Current cancer chemotherapies heavily rely on the unspecific inhibition of proliferating cells. This lack of tumour cell specificity results in severe toxic side effects and may only hardly affect quiescent cancer stem cells consequently leading to relapse. Since oncogenes are exclusively expressed in malignant and pre-malignant cells, they may provide unique, cancer cell specific targets for therapeutic strategies. However, their role in maintaining the malignant phenotype is frequently unknown. Furthermore, oncogenic transcription factors are generally considered to be "undruggable" with conventional small molecule approaches. Oncogene-specific RNA interference offers here new and exciting options to analyse oncogene functions directly in the malignant environment. Moreover, such approaches may permit the targeting of oncogenic transcription factors, thereby considerably extending the number of cancer-specific target structures. In this chapter, several rationales and practical aspects of oncogene targeting with siRNAs are discussed. Special emphasis is given to the application of RNA interference to haematopoietic cells, which are generally hard to transfect. In particular, solving the problem of systemic siRNA/shRNA delivery will greatly advance the inclusion of RNA interference strategies into more efficient and specific therapeutic strategies.

Key words: RNA interference, short interfering RNA, short hairpin RNA, siRNA delivery, oncogene, chromosomal rearrangements, point mutations.

1. Introduction

The long-term disease-free survival rates achieved by most cancer chemotherapies are still very disappointing. Most current cancer treatment strategies are profoundly hampered by severe toxic side effects caused by a dissatisfying low specificity for cancerous tissues. Moreover, cancer cells may lose their dependence on the drug target either by mutating the target or by activating pathways circumventing the drug target. Furthermore,

M. Sioud (ed.), *Methods in Molecular Biology, siRNA and miRNA Gene Silencing, vol. 487*

DOI: 10.1007/978-1-60327-547-7_11

chemotherapy may not affect cancer stem cells thereby failing to completely eradicate the tumour and ultimately leading to relapse. Therefore, there is a need for identifying new therapeutic targets and guiding therapies to cancer cells (*see* **Chap. 3**).

1.1. Oncogenes – Targets for Cancer Therapy?

Taking the limitations mentioned above into account, an ideal target for a therapeutic approach would be a gene (or its encoded products) with the following characteristics:

1. Which plays a very central and essential role in tumour biology, thus being less likely to become obsolete for tumour maintenance.
2. Which is essential for maintaining the cancer stem cell?.
3. Which is exclusively expressed in the tumour, but not in normal cells.
4. Which can be easily retargeted in case of mutations by modified or new drugs (drug tailoring).

Oncogenes derived from cellular proto-oncogenes carry genetic changes resulting either in changed expression levels and/or in RNAs or proteins with new properties. Alternatively, oncogenes can be introduced into cells by pathogenic vectors such as papilloma viruses or hepatitis B virus. The significance for tumour maintenance is for many oncogenes still not known, since their role in tumour biology and cancer stem cell persistence is insufficiently understood. This knowledge, however, is an essential requirement for the selection of promising therapeutic targets. The dependence of cancer stem cells (CSCs) on oncogenes is of particular interest, since they have been shown to maintain tumours such as breast cancer, glioma and leukaemias *(1–3)*, and may contribute to drug resistance, as recently shown for chronic myeloid leukaemia *(4)*. Thus, it is imperative to develop strategies for specifically targeting and eradicating CSCs in order to cure cancer including leukaemias.

An oncogene might be instrumental for initiating the malignant transformation, but might become dispensable during the course or tumourigenesis due to secondary events. Interfering with such an oncogene will be of limited therapeutic value, as only early malignant stages might be affected. Nevertheless, tumour cells may become dependent on a single oncogene and signalling pathways controlled by this oncogene. This dependence is frequently termed "oncogene addiction" *(5)*. Furthermore, an imbalance of pro-survival and pro-apoptotic signals caused by oncogene interference may result in an "oncogenic shock," which has been recently described as a consequence of inhibiting oncogenic kinases *(6, 7)*. In any case, a possible oncogene addiction is not always easy to examine; it may heavily depend on the cellular environment. For instance, ectopic expression of certain leukaemic fusion genes such as RUNX1/RUNX1T1 (AML1/MTG8, AML1/ETO) or MLL/AFF1 (MLL/AF4) results in antiproliferative effects in many cell types. Nevertheless, the corresponding leukaemic cells are strictly dependent on the continuous presence of these fusion genes *(8, 9)*. Obviously, the tumour-promoting effects of those oncogenes are strongly influenced by the cellular environment and are, thus, lineage-dependent *(10)*. Therefore, studying oncogene addiction may require the manipulation of the oncogene of interest in an appropriate context, i.e., in the tumour cell or in its untransformed precursor cell.

Based on their expression pattern, oncogenes can be divided into two major classes: those which are exclusively expressed in malignant and pre-malignant cells, and those which show elevated expression levels in these cells, but which are also expressed in normal tissues. For instance, chromosomal abnormalities including translocations frequently yield fusion genes, which are exclusively expressed in tumours *(11 – 13)*. Well known examples are *BCR/ABL1* in leukaemia *(14*, see Chap. 21), *EWS/FLI1* in Ewing sarcoma *(15)*, *PAX3/FKHR* in alveolar rhabdomyosarcoma *(16)* and *TMPRSS2/ERG* in prostate cancer*(17)*. Such fusion oncogenes fulfill the third requirement of an ideal target. Similarly, point mutations within the open reading frame, which are frequently found in *RAS* members or p53, also result in molecules uniquely expressed in malignant and pre-malignant cells *(18, 19)*. In contrast, mutations including chromosomal rearrangements not affecting the encoded transcript and protein may result in increased expression levels, but do not permit a qualitative discrimination against non-malignant tissues. For example, the translocations t(8;14) or t(14;18) juxtaposes an immunoglobulin enhancer to the *MYC* or *BCL2* gene, respectively *(20)*. In these examples, suppression of such an oncogene must result in a substantially more pronounced effect in the tumour cell compared to a normal (stem) cell, which expresses the corresponding proto-oncogene, to achieve a sufficiently specific anti-tumour effect.

Many oncogenes and the majority of the fusion genes are transcription factors, which, unlike kinases or G-proteins, are frequently considered to be "undruggable" by small molecule approaches. Thus, a substantial fraction of putatively very interesting oncogenes are currently not accessible to more commonly employed targeting strategies. Furthermore, several mechanisms including mutation of the binding site of small molecules may lead to resistance. For instance, mutations in the catalytic domain of BCR/ABL1 are responsible for final treatment failures with imatinib mesylate, an inhibitor of ABL1, KIT and PDGFR kinase activities *(21)*. For these reasons, new compounds are desirable, which permit targeting of a wide variety of oncogenes including transcription factors, and which may even be quickly adapted to possible "escape" mutations. Antisense compounds including small interfering RNAs offer here new possibilities (**Fig. 11.1**). Since siRNAs bind to the transcript, they can be easily targeted against almost any gene including the "undruggable" transcription factors. Furthermore, if binding of

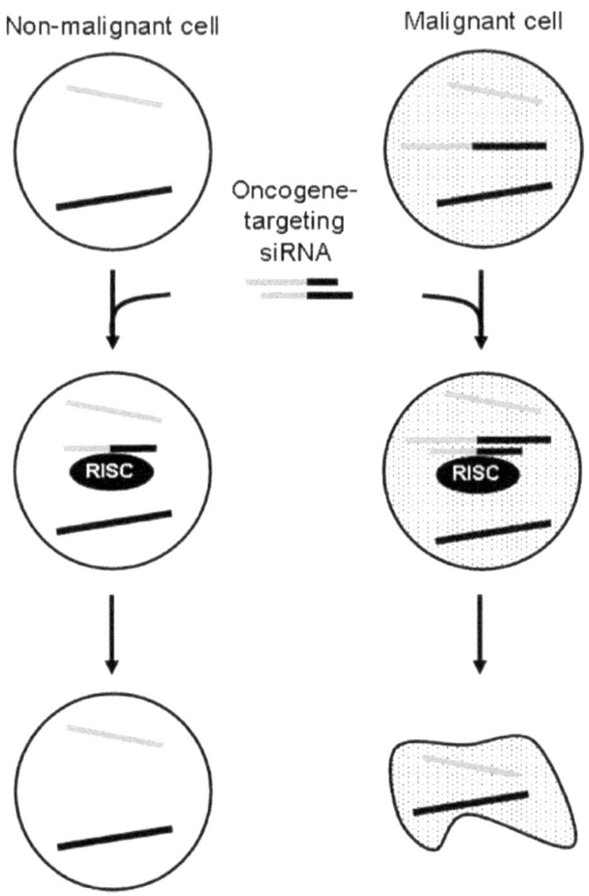

Fig. 11.1. RNAi-mediated oncogene suppression. Only malignant or pre-malignant cells express oncogenes. Targeting oncogenes with siRNAs or shRNAs specific for the mutated region will only cause RISC-mediated degradation of the oncogenic transcript, but not of the unmutated proto-oncogenic sequences. Consequently, only expansion of malignant cells will be affected, whereas the functions of normal cells should not be compromised.

the original siRNA is affected by mutations in its complementary sequence, the siRNA sequence can be adapted to the new, mutated target. However, these advantages are currently outweighed by several disadvantages, with the most challenging one being the in vivo delivery of siRNAs. In the following sections, I will discuss the possibilities and current limitations of RNA interference to study and to inhibit oncogene functions.

2. siRNA Design

2.1. General Considerations

Like other antisense-based approaches, a sequence-specific siRNA effect depends on the hybridisation to complementary sequence stretches within the target transcript. However, unlike more conventional antisense oligodeoxynucleotides (AS-ODN), which after binding to RNA recruit with RNaseH the degrading enzyme, siRNAs are already part of a ribonucleoprotein complex, the RNA-induced silencing complex (RISC). However, like AS-ODNs, siRNAs may also affect targets with imperfect homology. A short sequence stretch from nucleotide 2 to nucleotide 8, the so-called seed sequence, substantially contributes to productive binding probably by forming the initial base pairs with the target sequence *(22)*. The association between siRNA and target RNA may also be influenced by the secondary structure of the latter. Very stable double-stranded regions are less amenable to siRNA and RISC binding than open, unstructured regions *(23, 24)*. Furthermore, both AS-ODN and siRNAs may associate with other proteins including Toll-like receptors *(25)*. These different interactions may impact the target specificity of siRNAs and must be considered for its design. Moreover, siRNAs are double-stranded prior to becoming part of RISC. The 5′-terminal thermodynamic stability crucially influences the strand selection for RISC with the strand containing the less stable 5′-end being preferentially incorporated *(26)*.

For siRNA design, *see* Chap. 1. Also, an excellent open access resource for siRNA design is the siRNA selection programme of the Whitehead Institute (http://jura.wi.mit.edu/bioc/siRNAext/). This programme selects siRNAs according to Elbashir et al., Schwarz et al. and Khvorova et al. *(26 – 28)*. Furthermore, it offers identification of alternative siRNA target by different blast strategies. However, this software does not consider the effects of target RNA secondary and tertiary structure and it has some limitations as discussed in Chap. 1.

2.2. Design of Oncogene-Specific siRNAs

The choice of an oncogene-specific siRNA is strongly dependent on the sequence space available for the selection of appropriate siRNA target sites. In the case of viral oncogenes such as E6 or E7 of the oncogenic human papilloma viruses 16 or 18, the complete mRNA can be consulted for siRNA design *(29)*. The same holds true for oncogenes with mutations not being transcribed, although here the design of tumour-specific siRNAs is impossible. However, the situation is different for oncogenes with mutations being presented in the oncogenic transcript such as fusion transcripts or transcripts with point mutations. In these cases, the sequence space available for the design of cancer cell-specific siRNAs is very limited, as the siRNA binding site must include the corresponding mutation.

Several fusion sequences of oncogene transcripts such as *RUNX1/RUNX1T1* (*AML1/ETO, AML1/MTG8*) or the major isoforms of *BCR/ABL1* permit the design of siRNAs according to established rules such as proper strand selection by 5′ stability or avoidance of GU-rich regions to reduce interferon responses *(30, 31)*. However, such a "conventional" design is not always possible. For instance, the fusion site of *MLL/AFF1* (*MLL/AF4*) does not permit the design of siRNAs according to the 5′ stability rule *(9)*. Nevertheless, in such cases, a systematic scan of the fusion site with several siRNAs may identify "non-canonical" siRNAs with sufficient efficacy and specificity *(32)*. In the case of *MLL/AFF1*, we designed 14 different siRNAs with target sites moved by one nucleotide each. Using this approach, we obtained two non-canonical siRNAs, which caused a substantial decrease in fusion transcript levels. However, the one located more towards the 3′-side (i.e., the *AFF1* side) of the fusion did also affect the transcript levels of the non-fused *AFF1 (9)*.

Table 11.1 shows a collection of siRNAs targeting either fusion oncogenes or oncogenes carrying point mutations. Interestingly, of the 16 synthetic siRNAs, only 9 have a 3′-overhang on both sides, whereas 7 siRNAs contain an antisense strand, which is two nucleotides longer than the sense strand, and, thus, have one sticky and one blunt end *(9)*. It is currently unknown, if these siRNAs are further processed by DICER prior or during RISC assembly. Furthermore, of all 22 siRNAs and short hairpin RNAs (shRNAs) shown, more than a third do not comply with the "5′-stability" rule showing that these conventional design rules may be applied flexibly. In the case of the siRNAs/shRNAs targeting fusion transcripts, the fusion point is in 10 of 13 constructs either located in the centre or on the 5′-side of the siRNA guide (i.e., antisense) strand. Since RISC association seems to proceed via initial contacts between the seed sequence of the guide strand and the target RNA *(24)*, a fusion site located in the 3′-side of the target RNA binding sequence and, thus, on the 5′-half of the guide strand, should only allow efficient binding to the fusion transcript, but not to the non-fused wild type transcripts. In contrast, positioning the fusion site too far to its 3′-terminus may result in loss of fusion transcript specificity as shown for *MLL/AFF1 (9)*. Taken together, these results suggest that positioning the fusion site either in the centre or the 3′-site of the siRNA binding site is important for specificity. Moreover, diverging from the "5′-stability" rule may still result in active siRNAs.

The specific targeting of point mutations is significantly more demanding than that of fusion sites. With one exception, for all examples shown the mutated nucleotide locates to the central part of the siRNA binding site. This design achieved a decent, but, in most cases, not an absolute specificity for the mutated transcript. Notably, all mutations are located downstream of the seed sequence of the siRNA guide strand. Thus, the limited specificity seen may not be completely surprising. In this context, Saxena and colleagues have examined the influence of location of the point mutation on siRNA specificity and activity both on target RNA and protein levels *(33)*. They show that mismatches on the 5′-end of the siRNA guide strand affecting the seed sequence severely affect siRNA activity. It remains to be seen whether positioning oncogenic point mutations into the seed sequence will also result in an improved specificity.

3. siRNA Transfection

Despite many efforts, the cellular uptake of preformed siRNAs still constitutes the major obstacle for their possible therapeutic application. Unlike most small molecular drugs, the anionic siRNAs cannot penetrate cellular membranes. This inability limits their uptake both through the plasma membrane as well as by endocytosis. In cell culture, many different chemical approaches facilitate the endocytotic uptake of siRNAs, and several physical approaches permit direct entry through the plasmalemma. The first group includes many cationic compounds such as cationic lipids, peptides or polyethyleneimine (PEI) derivatives, whereas the most prominent physical approach is electroporation.

An important question to consider is how to monitor siRNA uptake. In general, two different approaches can be taken. The first approach is the co-transfection of a siRNA with a plasmid encoding a reporter such as green fluorescent protein. The underlying assumption is that cells that show GFP expression will also have taken up siRNA. Although this assumption is valid in most of the cases, it may substantially underestimate the siRNA transfection efficacy. In order to get expressed, the plasmid has to be taken up by endocytosis, followed by escape from the endosome and translocation into the nucleus, where it has to associate with the appropriate transcription machinery. In contrast, siRNAs have "only" to escape from the endosome in order to become part of the cytoplasmatically located RNA-induced silencing complex (RISC). Thus, as the nuclear uptake in particular provides a substantial obstacle to reporter gene expression, a substantially higher fraction of cells may have productively taken up siRNAs than indicated by the GFP-positive fraction.

An alternative is the direct labelling of siRNAs. In this case, a fluorescent dye such as Cy3 or Cy5 is covalently linked during the chemical synthesis to the 5′-end of the sense strand. The advantage here is that the siRNA can be directly monitored, whilst modification of the sense strand does not affect the silencing activity of the antisense strand. Uptake is frequently monitored by flow cytometry and/or fluorescence microscopy. Flow cytometry is more sensitive compared to fluorescence microscopy and allows easy quantitation of label-positive cells, but does not provide any information about intracellular siRNA distributions. Cells with siRNAs entrapped in the endosome will count positive, even if no siRNA has reached the cytoplasm. Fluorescence microscopy can provide this information, but, at least in our hands, this depends on the fluorescent label. So far, we have been unable to detect substantial amounts of fluorescein-labelled siRNAs in the cytoplasm of healthy cells, whereas this is no problem with Cy3- or Cy5-labelled siRNA *(30, 34)*. The reason for this difference is not known yet.

Both procedures have in common the use of a surrogate marker, i.e., reporter gene expression or fluorescent label. The more direct and more valid approach is the test for siRNA function. A 70% decrease in target transcript or target protein levels indicates a transfection efficacy of at least this value. Thus, the most appropriate approach to obtain a lower estimate for functional siRNA uptake is to measure directly the reduction in target RNA or protein. Ideally, an siRNA with an established activity should be employed to determine functional uptake in a new cell system.

Chemical transfection uses cationic lipid formulations such as DOTAP or Lipofectamine, cationic polymers such as PEI or chitosan, or peptides, which either form non-covalent complexes with or covalently linked to siRNAs. All these compounds facilitate endocytotic uptake and endosomal escape of siRNAs. These approaches often yield excellent siRNA transfection efficacies in adherent cell culture, but perform comparatively poorly with suspension cells. One likely reason for this difference is the much larger surface of adherent cells compared to suspension cells, which eases the adhesion and subsequent uptake of the siRNA complexes.

siRNA transfection by electroporation has been successfully applied to adherent and suspension cells. Electroporation of adherent cells requires their dissociation from the surface, a step which is obviously unnecessary for suspension cells. However, dependent on the electroporation method, siRNA delivery may be associated with quite substantial cell death. One reason for this is the uncritical adaptation of protocols optimized for plasmid transfection. Due to their smaller size, siRNAs can be efficiently electroporated under substantially milder conditions than plasmids. For instance, the electroporation protocol used by us for siRNA delivery to a wide range of haematopoietic cells including primary cells achieves high efficiencies (>80%) with negligible (<10%) cell death (**Table 11.1**).

The following protocol describes the electroporation of the leukaemic cell line Kasumi-1, which expresses the leukaemic fusion gene *RUNX1/RUNX1T1*, with siRNAs. The active siRNA targets the fusion site of the *RUNX1/RUNX1T1* transcript, whereas the control siRNA is homologous to the *MLL/AFF1* fusion site, which is not expressed in Kasumi-1.

Table 11.1
siRNAs targeting oncogenes

Oncogene	siRNA sequence	Comments	Ref.
Fusion genes			
BCR/ABL1	5′-GCAGAGUUCAAAAGCCCUUtt-3′ 3′-ttCGUCUCAAGUUUUCGGGAA-5′	Targets exon 14 – exon 2 variant	*(31)*
BCR/ABL1	5′-CAGAGUUCAAAAGCCCUUCAG-3′ 3′-UCGUCUCAAGUUUUCGGGAAGUC-5′	Targets exon 14 – exon 2 variant	*(40)*
BCR/ABL1	5′-AAUAAGGAAGAAGCCCUUCAG-3′ 3′-AGUUAUUCCUUCUUCGGGAAGUC-5′	Targets exon 13 – exon 2 variant	*(65)*
BCR/ABL1	5′-AUCAAUAAGGAAGAAGCCCUU-3′ 3′-GGUAGUUAUUCCUUCUUCGGGAA-5′	Targets exon 13 – exon 2 variant	*(65)*
BCR/ABL1	5′-GGAGACGCAGAAGCCCUUCAG-3′ 3′-UACCUCUGCGUCUUCGGGAAGUC-5′	Targets exon 1 – exon 2 variant	*(65)*
ETV6/PDGFRB (TEL/PDGFRB)	5′-AUGAAGAAGCCUUGCCCUUUU-3′ 3′-UUUACUUCUUCGGAACGGGAA-5′	shRNA	*(48)*
EWSR1/FLI1 (EWS/FLI1)	5′-GCAGCAGAACCCUUCUUAUtt-3′ 3′-ttCGUCGUCUUGGGAAGAAUA-5′		*(44)*
EWSR1/FLI1 (EWS/FLI1)	5′-GGCAGCAGAACCCUUCUUAcg-3′ 3′-gcCCGUCGUCUUGGGAAGAAU-5′	Also active as shRNA *(51)*	*(66)*
EWSR1/FLI1 (EWS/FLI1)	5′-GCAGAACCCUUCUUAUGACUU-3′ 3′-UUCGUCUUGGUCAGAAUACUG-5′	Shows activity in vivo	*(52)*

(continued)

Table 11.1 (continued)

Oncogene	siRNA sequence	Comments	Ref.
MLL/AFF1 (MLL/AF4)	5′-AAAAGCAGACCUACUCCAAUG-3′ 3′-UCUUUUCGUCUGGAUGAGGUUAC-5′	Targets exon 9 – exon 4 variant, affects AFF1	*(9)*
MLL/AFF1 (MLL/AF4)	5′-ACUUUAAGCAGACCUACUCCA-3′ 3′-CCUGAAAUUCGUCUGGAUGAGGU-5′	Targets exon 10 – exon 4 variant	*(9)*
MLL/AFF1 (MLL/AF4)	5′-AAGAAAAGCAGACCUACUCCA-3′ 3′-UUUUCUUUUCGUCUGGAUGAGGU-5′	Targets exon 9 – exon 4 variant	*(9)*
NPM1/ALK	5′-CACUUAGUAGUGUACCGCCtt-3′ 3′-ttGUGAAUCAUCACAUGGCGG-5′	Low activity reported (67)	*(68)*
NPM1/ALK	5′-CAGCACUUAGUAGUGUACCGC-3′ 3′-CUGUCGUGAAUCAUCACAUGG-5′		*(67)*
NPM1/ALK	5′-CUUAGUAGUGUACCGCCGCUU-3′ 3′-UUGAAUCAUCACAUGGCGGCG-5′	shRNA	*(48)*
RUNX1/RUNX1T1 (AML1/MTG8, AML1/ETO)	5′-CCUCGAAAUCGUACUGAGAAG-3′ 3′-UUGGAGCUUUAGCAUGACUCU-5′	Also active as shRNA *(43)*	*(30)*
Point mutations			
BRAF E599V	5′-GCUACAGAGAAAUCUCGAUUU-3′ 3′-UUCGAUGUCUCUUUAGAGCUA-5′	shRNA	*(69)*
HRAS G12V	5′-GGGCGCCGUCGGUGUGGGCAAG-3′ 3′-UUCCCGCGGCAGCCACACCCGUUC-5′	shRNA	*(70)*

KRAS G12C	5′-GUUGGAGCUUGUGGCGUAGUU-3′ 3′-UUCAACCUCGAACACCGCAUC-5′	shRNA	*(71)*
KRAS G12N	5′-GUUGGAGCUGAUGGCGUAGUU-3′ 3′-UUCAACCUCGACUACCGCAUC-5′	shRNA	*(71)*
KRAS G12V	5′-GUUGGAGCUGUUGGCGUAGUU-3′ 3′-UUCAACCUCGACAACCGCAUC-5′	shRNA	*(72)*
NRAS Q61R	5′-CGAGAAGAGUACAGUGCCAUGtt-3′ 3′-ttGCUCUUCUCAUGUCACGGUA-5′	Also active as shRNA	*(73)*
TP53 R248W	5′-GCAUGAACUGGAGGCCCAUtt-3′ 3′-ttCGUACUUGACCUCCGGGUA-5′		*(74)*

Mutations are underlined. Lower case letters indicate 2′-deoxynucleotides.

3.1. Required Materials

1. Kasumi-1 (German Collection of Microorganisms and Cell Culture No. ACC220)
2. Complete medium RPMI1640 + 2 mM glutamine + 10% foetal calf serum
3. Neubauer chamber
4. Table top centrifuge
5. 4 mm electroporation cuvettes (Eurogentec, Southampton, UK)
6. Square wave electroporator EPI2500 (Fischer, Heidelberg, Germany; www.electroporation.eu; *see* **Note 1**)
7. 20 μM RUNX1/RUNX1T1 siRNA in 25 mM HEPES pH 7.5, 100 mM NaCl (*see* **Note 2**)
8. 20 μM MLL/AFF1 siRNA in 25 mM HEPES pH 7.5, 100 mM NaCl
9. 24-well plate

3.2. An Experimental Protocol

1. Count cells in Neubauer chamber.
2. Spin 3×10^6 cells for 5 min at 300× g at room temperature.
3. Discard supernatant.
4. Resuspend cells in 300 μl prewarmed complete medium to a final concentration of 10^7 cells/ml (*see* **Note 3**).
5. Transfer aliquots of 100 μl into three electroporation cuvettes (*see* **Note 3**).
6. Immediately prior to electroporation, add 0.5 μl 20 μM RUNX1/RUNX1T1 siRNA to cuvette #2, and 0.5 μl 20 μM MLL/AFF1 siRNA to cuvette #3 followed by gentle swirling. Cuvette #1 represents the "Mock" control (*see* **Note 4**).
7. Electroporate at 330 V for 10 ms (*see* **Note 5**).
8. Incubate cuvettes for 15 min at room temperature.
9. Transfer each cell suspension to 2 ml complete medium in a well of a 24 well plate to achieve a final cell concentration of 5×10^6 cells/ml (*see* **Note 6**).
10. Incubate for 24–96 h in an incubator at 37°C, 5% CO_2 and 95% humidity (*see* **Note 7**).

The experimental protocol described in **Sect. 3.2** is also suitable for repetitive electroporations. This is of particular interest, if the target protein is very stable, or if long-lasting protein depletion is required to study cellular processes such as changes in DNA methylation.

An alternative to the exogenous delivery of preformed siRNAs is the endogenous siRNA delivery. Notably, several publications show that the same sequence shows similar efficiencies both as siRNAs and as shRNAs (**Table 11.1**). For endogenous delivery, cells are transfected with a plasmid or recombinant virus encoding an expression cassette for an siRNA precursor. Inside the cell, an RNA hairpin is transcribed, which is then further processed to the mature siRNA. Two different classes of expression cassettes can be distinguished. The first class uses RNA polymerase III-dependent promoters such as the U6 or the H1 promoter followed by a reasonably short sequence of less than 50 nucleotides encoding a short hairpin RNA (shRNA). Upon transcription, the shRNA is exported to the cytoplasm, processed by Dicer to the siRNA, whose guide strand finally becomes part of the RISC. The second class uses RNA polymerase II-dependent promoters. In this case, the guide strand sequence is embedded in a pre-micro RNA backbone such as pre-mir 30, which forms an approximately 100 nucleotides long hairpin structure. Since such hairpin structures can be incorporated into the 3′-untranslated region of reporter transcripts such as GFP or DS-RED, their expression may be easily monitored by such surrogate markers. Here, the pre-miRNA-like hairpin is first cut out by the RNase Drosha followed by nuclear export and further processing by Dicer. Nevertheless, no Pol II-dependent oncogene-specific RNAi has yet been published.

On the one hand, the stable expression of shRNAs or miRNAs may also allow the study of oncogene functions in vivo, where delivery of preformed siRNAs is still problematic. On the other, the application of shRNA or miRNA approaches in studying oncogene function may be limited by their inherent stable expression and permanent intracellular presence. An essential role of an oncogene of interest in cellular proliferation would result in a strong counter-selection, thus either eliminating siRNA-expressing cells, or selecting for cells with only modest oncogene suppression or for cells which do not depend on the oncogene anymore. For instance, high levels of BCR/ABL1 shRNA were not tolerated in *BCR/ABL1*-expressing K562 cells *(35)*. A way out of this possible dilemma is the inducible shRNA expression. Inducible options exist for both Pol II- and Pol III-based expression systems, with tetracycline (or its derivatives)-inducible systems being currently the most popular ones *(36–38)*. However, so far only Pol III-based systems have been described for the inducible expression of oncogene-specific shRNAs.

4. Functional Analyses of Oncogenes Using RNAi

As already stated above, an oncogene is only a promising therapeutic target, if the corresponding malignancy still depends on its function at diagnosis and later stages of disease. It is therefore not surprising that the majority of all studies targeting oncogenes by RNAi examine the consequences of oncogene depletion on the proliferation and apoptosis of tumour cells expressing the corresponding oncogene. Several studies dealt with the leukaemic fusion oncogenes such as *BCR/ABL1* or *RUNX1/RUNX1T1*. For instance, both siRNA and shRNA-mediated BCR/ABL1 depletion inhibited proliferation of t(9;22)-positive leukaemic cell lines and primary cells, and restored growth factor dependence in cells ectopically expressing *BCR/ABL1 (31, 35, 39 –41)*. Notably, down-modulation of imatinib mesylate-resistant BCR/ABL1 forms restored at least partially drug sensitivity in cells expressing the two BCR/ABL1 mutants *(40)*.

Unlike the translocation t(9;22), most of the leukaemic translocations affect transcription factors such as MLL or RUNX1 (AML1) *(42)*. We could show for the associated fusion genes *RUNX1/RUNX1T1* (*AML1/MTG8 or AML1/ETO*) and *MLL/AFF1* (*MLL/AF4*) a central function for the maintenance of the leukaemic phenotype in cell culture and in xenotransplantation studies *(8, 9, 30)*. Depletion of MLL/AFF1 triggered apoptosis in acute lymphoblastic leukaemia cell lines, whereas suppression of *RUNX1/RUNX1T1* not only facilitates myeloid differentiation and inhibits colony formation, but also induces cellular senescence *(8, 30, 43)*. This is all the more noteworthy, as these studies used cell lines which are already adapted to normal culture conditions probably by acquiring additional genetic changes, and which were for that reason less likely to depend on these fusion genes. Thus, even cell lines can still be addicted to a particular oncogene. Moreover, *RUNX1/RUNX1T1* not only shares this senescence-suppressing property with *EWSR1/FLI1 (EWS/FLI1)*, the fusion gene associated with Ewing sarcoma, but like the latter establishes senescence not via the more conventional CDKN2A (p16) or CDKN1A (p21) pathways, but via a more uncommon CDKN1B (p27)-dependent way *(8, 44)*. Thus, two oncogenes associated with very distinct malignancies use similar strategies to support the self-renewal of tumour cells. Furthermore, several fusion genes also control the expression of microRNAs *(45, 46, 47)*. Therefore, in addition to the transcriptional level, these oncogenes control gene expression by posttranscriptional, co- and post-translational mechanisms, whose studies require systematic interlinked transcriptome, and proteome and metabolome analyses.

A therapeutically very relevant question is the role of oncogenes in drug response and resistance. I have already mentioned the, at least partial, restoration of imatinib mesylate sensitivity after siRNA-mediated knock-down of either overexpressed or mutated BCR/ABL1 *(40)*. Imatinib mesylate, however, is not specific for BCR/ABL1 or ABL1, but also affects the kinase activities of KIT, Abl-related gene ARG and PDGF receptor. Therefore, besides BCR/ABL1, imatinib also inhibits the activity of the oncogenic fusion protein ETV6/PDGFRB (TEL/PDGFRB) found in chronic myelomonocytic leukaemia (CMML). However, as in the case of t(9:22)-associated leukaemias, imatinib is not curative as a single agent, and the emergence of point mutations may lead to imatinib resistance. Chen and colleagues showed that stable knockdown of ETV6/PDGFRB sensitises cells to imatinib and to rapamycin, an inhibitor of FRAP1 (mTOR) *(48)*. In an extension of this work, the authors demonstrated that depletion of NPM1/ALK, a fusion gene associated with anaplastic large-cell lymphoma (ALCL), correlates with a substantially increased rapamycin sensitivity *(48)*.

As already stated before, oncogenes encoding transcription factors are currently difficult to target with substances of low molecular weight. A possible way out of this dilemma might be the identification of drugs or substances which interfere with essential oncogene functions. Stegmaier and colleagues used a very intelligent approach to identify substances which mimic at least in part gene expression signatures associated with an oncogene knockdown *(49)*. For that, they suppressed *EWSR1/FLI1* in a Ewing sarcoma cell line using an siRNA targeting the 3′-UTR of the fusion gene. As this part is also present in the unmutated *FLI1* transcript, they also used an ectopically expressed fusion cDNA, which did not contain this target sequence to rescue any siRNA-related phenotype. Using this depletion/rescue system, the authors identified a gene expression signature consisting of 14 genes which was then used for a large scale screening of more than 1,000 small-molecule compounds. They found with arabino cytidine (AraC), a substance which triggered the degradation of EWSR1/FLI1, a profound inhibition of growth and expansion of Ewing carcinoma cells both in cell culture and in xenotransplanted nude mice. In a similar study, the same group identified several corticosteroids and the DNMT inhibitor 5-Aza-2′-deoxycytidine as compounds yielding a similar expression signature in a t(8;21)-positive AML cell line as the knockdown of *RUNX1/RUNX1T1* *(50)*. The application of knockdown-associated expression signatures to small molecule library screening may hold great promise for direct or indirect targeting of oncogenes.

Finally, RNAi-mediated oncogene inhibition in combination with gene expression profiling has been exploited to identify the cell of origin of Ewing sarcoma. Endothelial cells, neuroectodermal cells and mesenchymal stem cells (MSCs) have been suggested to give rise to Ewing sarcoma. A current study compared the global expression profiles of Ewing tumour cells with and without *EWSR1/FLI1* suppression with those of normal tissues *(51)*. Knockdown of the fusion gene caused a signature overlap with MSCs. Moreover, EWSR1/FLI1 depletion facilitated adipogenic and osteogenic differentiation, a typical feature of MSCs.

Thus, Ewing sarcomas are likely to originate within the MSC compartment, which is also in agreement with the primary localisation of the malignancy in the long bones of the extremities.

5. Outlook – The Next Frontiers

Up to now, RNAi has mainly been used for functional studies in the oncogene field. siRNA-mediated oncogene depletion yielded important insights into the role of oncogenes in the maintenance of the malignant phenotype and elucidated molecular mechanisms responsible for these functions. Hitherto, most of these studies are restricted to the ex vivo, i.e., cell culture situation. One important question to address next is a possible oncogene addiction in vivo using appropriate animal models. Such settings will also allow studying the role of oncogenes in the maintenance of cancer stem cells (CSCs). An essential role should result in an elimination of CSCs upon oncogene knockdown and, consequently, in a reduced mortality. So far, only a few studies have been published addressing this question *(9, 48, 52)*. One obvious limitation for such experiments is the in vivo delivery of siRNAs or shRNA-encoding vectors. Our group avoided this problem by transfecting leukaemic cells with MLL/AFF1 siRNA ex vivo prior to xenotransplantation *(9)*. A similar approach was taken by Chen et al., who transduced cells with a retroviral vector encoding an ETV6/PDGFRB siRNA followed by transplantation into murine hosts *(48)*. In contrast to the previous study, this group observed only a delayed disease onset. It is currently unclear, whether this indicates only a limited dependency of tumour initiating cells on ETV6/PDGFRB, or whether cells with a diminished or silenced shRNA expression finally caused the disease. Unfortunately, the authors did not examine whether the malignant cells still showed diminished ETV6/PDGFRB levels. Both ex vivo approaches do not permit analysis of oncogene functions in established tumours. Since the RNAi molecule is already present at the time of transplantation, changes in morbidity and mortality may simply be due to an impaired engraftment. Inducible shRNA expression systems offer new and exciting options. They combine the advantage of ex vivo manipulation with an absence of RNAi during the course of tumour engraftment and establishment.

Alternatively, preformed siRNAs may be systemically delivered. This route is more desirable than any viral approach, as it is not riddled to such an extent by immune responses, nor can it result in genomic changes due to integration possibly leading to insertion mutagenesis. Moreover, systemic siRNA delivery would be more amenable to augment existing chemotherapy protocols.

Notably, two reports describe such approaches for EWSR1/FLI1 using either a targeted cyclodextrin approach or nanocapsules *(52, 53)*. In each study, an inhibition of tumour growth could be observed. Alternative approaches may include polyplexes consisting of polyethyleneimine (PEI) and polyethylene glycol (PEG) *(54)*, chitosan *(55)*, atelocollagen *(56)*, liposomes with and without PEG shielding *(57, 58)* or chemically stabilised siRNA containing a cholesterol moiety *(59)*. Several reports describe a successful targeted siRNA delivery using polyplexes, aptamers, antibodies and peptides *(60–64)*. Interestingly, siRNA-peptide complexes are able to pass the blood-brain barrier indicating transendothelial delivery *(64)*. It remains to be seen how these encouraging developments translate into clinical application.

Another problem for any oncogene-related therapeutic siRNA approach lies in the specificity of siRNAs. For instance, siRNAs targeting different versions of the leukaemic fusion gene *MLL/AFF1* inhibit leukaemic growth only in those cells expressing this very variant, whereas cells expressing another variant are not affected *(9)*. This may be good proof for siRNA specificity. However (and fortunately), there is only a small absolute number of infant leukaemia patients, of which some 50% present a *MLL/AFF1*-associated ALL. Thus, the specific targeting of an oncogene (variant) causes a further fragmentation of the often already limited patient pool. This does not only limit the interest of the pharmaceutical industry in developing such very specific approaches, but it will also complicate the acquisition of sufficiently large patient numbers for clinical studies.

The ultimate goal of oncogene-targeting RNAi approaches is the selective elimination of tumour cells including cancer stem cells without damaging normal tissues. The examples and procedures discussed in this chapter show that RNAi already substantially contributed to our understanding of oncogene biology. The next steps must be the translation of this knowledge and of the RNAi technology into therapeutic approaches. Given the impressive progress made within the last 5 years, anti-oncogenic RNAi approaches may not be so far away from the bedside.

6. Notes

1. This electroporator uses square wave pulses. In contrast to the more common transient decay electroporators, which completely discharge via the cuvette, and where the rate of discharge depends on the volume in the cuvette, square wave electroporators keep a given voltage constant for a defined time independent of the cuvette volume. This feature allows

the optimisation of the electroporation with small volumes requiring low amounts of siRNAs. Scaling up to larger volumes of up to 800 μl does not require readjustment of the electroporation conditions.

2. siRNAs may be alternatively diluted in a buffer consisting of 30 mM HEPES pH 7.3, 1 mM $MgCl_2$, and 100 mM KCl.
3. Cell concentrations may be increased to up to 10^8 cells/ml. Volumes in the 4 mm electroporation cuvette may vary from 100 to 750 μl. Thus, up to 7.5×10^7 cells may be transfected with siRNAs in a single electroporation. Assuming a final siRNA concentration of 100 nM, up to 10^6 cells can be transfected per pmol siRNA. In comparison, chemical transfections using cationic lipids and an siRNA concentration of 5 nM in 2 ml medium and 10^6 cells achieve a ratio of about 10^5 cells/pmol siRNA.
4. Efficient siRNAs require concentrations of about 100 nM. siRNA concentrations may be increased to up to 1–2 μM. We prefer siRNAs with proven activity as controls. Good candidates are siRNAs targeting fusion genes not being expressed in the cell of interest.
5. The voltage may be varied between 300 and 400 V. Most cell types tested showed an optimum at 350 V for 10 ms.
6. The dilution should achieve optimal cell concentrations for culture. Most suspension cells have optimal cell densities in the range from 3×10^5/ml to 10^6/ml.
7. Effects on transcript levels are visible within 16 h. Effects on protein levels depend on the protein half-life.

Acknowledgements

I wish to thank Simon Bomken for carefully reading the manuscript. I gratefully acknowledge support from the Deutsche Krebshilfe, the Deutsche Jose Carreras Leukaemie-Stiftung, the Wilhelm Sander-Stiftung and the North of England Children's Cancer Research Fund.

References

1. Al-Hajj, M., Wicha, M. S., Benito-Hernandez, A., Morrison, S. J., and Clarke, M. F. (2003) Prospective identification of tumorigenic breast cancer cells. *Proc Natl Acad Sci U S A* 100, 3983–3988.
2. Singh, S. K., Hawkins, C., Clarke, I. D., Squire, J. A., Bayani, J., Hide, T., Henkelman, R. M., Cusimano, M. D., and Dirks, P. B. (2004) Identification of human brain tumour initiating cells. *Nature* 432, 396–401.
3. Pearce, D. J., Taussig, D., Zibara, K., Smith, L. L., Ridler, C. M., Preudhomme, C., Young, B. D., Rohatiner, A. Z., Lister, T. A., and Bonnet, D. (2006) AML engraftment in the NOD/SCID assay reflects the outcome of AML: implications for our understanding

of the heterogeneity of AML. *Blood* 107, 1166–1173.

4. Bhatia, R., Holtz, M., Niu, N., Gray, R., Snyder, D. S., Sawyers, C. L., Arber, D. A., Slovak, M. L., and Forman, S. J. (2003) Persistence of malignant hematopoietic progenitors in chronic myelogenous leukemia patients in complete cytogenetic remission following imatinib mesylate treatment. *Blood* 101, 4701–4707.
5. Weinstein, I. B. and Joe, A. K. (2006) Mechanisms of disease: Oncogene addiction--a rationale for molecular targeting in cancer therapy. *Nat Clin Pract Oncol* 3, 448–457.
6. Sharma, S. V., Gajowniczek, P., Way, I. P., Lee, D. Y., Jiang, J., Yuza, Y., Classon, M., Haber, D. A., and Settleman, J. (2006) A common signaling cascade may underlie "addiction" to the Src, BCR-ABL, and EGF receptor oncogenes. *Cancer Cell* 10, 425–435.
7. Sharma, S. V. and Settleman, J. (2006) Oncogenic shock: turning an activated kinase against the tumor cell. *Cell Cycle* 5, 2878–2880.
8. Martinez, N., Drescher, B., Riehle, H., Cullmann, C., Vornlocher, H. P., Ganser, A., Heil, G., Nordheim, A., Krauter, J., and Heidenreich, O. (2004) The oncogenic fusion protein RUNX1-CBFA2T1 supports proliferation and inhibits senescence in t(8;21)-positive leukaemic cells. *BMC Cancer* 4, 44.
9. Thomas, M., Gessner, A., Vornlocher, H. P., Hadwiger, P., Greil, J., and Heidenreich, O. (2005) Targeting MLL-AF4 with short interfering RNAs inhibits clonogenicity and engraftment of t(4;11)-positive human leukemic cells. *Blood* 106, 3559–3566.
10. Garraway, L. A. and Sellers, W. R. (2006) Lineage dependency and lineage-survival oncogenes in human cancer. *Nat Rev Cancer* 6, 593–602.
11. Rabbitts, T. H. (1994) Chromosomal translocations in human cancer. *Nature* 372, 143–149.
12. Greaves, M. F. and Wiemels, J. (2003) Origins of chromosome translocations in childhood leukaemia. *Nat Rev Cancer* 3, 639–649.
13. Borkhardt, A. and Heidenreich, O. (2004) RNA interference as a potential tool in the treatment of leukaemia. *Expert Opin Biol Ther* 4, 1921–1929.
14. Stam, K., Heisterkamp, N., Grosveld, G., de Klein, A., Verma, R. S., Coleman, M., Dosik, H., and Groffen, J. (1985) Evidence of a new chimeric bcr/c-abl mRNA in patients with chronic myelocytic leukemia and the Philadelphia chromosome. *N Engl J Med* 313, 1429–1433.
15. May, W. A., Gishizky, M. L., Lessnick, S. L., Lunsford, L. B., Lewis, B. C., Delattre, O., Zucman, J., Thomas, G., and Denny, C. T. (1993) Ewing sarcoma 11;22 translocation produces a chimeric transcription factor that requires the DNA-binding domain encoded by FLI1 for transformation. *Proc Natl Acad Sci U S A* 90, 5752–5756.
16. Galili, N., Davis, R. J., Fredericks, W. J., Mukhopadhyay, S., Rauscher, F. J., 3rd, Emanuel, B. S., Rovera, G., and Barr, F. G. (1993) Fusion of a fork head domain gene to PAX3 in the solid tumour alveolar rhabdomyosarcoma. *Nat Genet* 5, 230–235.
17. Tomlins, S. A., Rhodes, D. R., Perner, S., Dhanasekaran, S. M., Mehra, R., Sun, X. W., Varambally, S., Cao, X., Tchinda, J., Kuefer, R., Lee, C., Montie, J. E., Shah, R. B., Pienta, K. J., Rubin, M. A., and Chinnaiyan, A. M. (2005) Recurrent fusion of TMPRSS2 and ETS transcription factor genes in prostate cancer. *Science* 310, 644–648.
18. Schubbert, S., Shannon, K., and Bollag, G. (2007) Hyperactive Ras in developmental disorders and cancer. *Nat Rev Cancer* 7, 295–308.
19. Vousden, K. H. and Lu, X. (2002) Live or let die: the cell's response to p53. *Nat Rev Cancer* 2, 594–604.
20. Mitelman, F., Johansson, B., and Mertens, F. (2007) The impact of translocations and gene fusions on cancer causation. *Nat Rev Cancer* 7, 233–245.
21. Copland, M., Jorgensen, H. G., and Holyoake, T. L. (2005) Evolving molecular therapy for chronic myeloid leukaemia--are we on target? *Hematology* 10, 349–359.
22. Dykxhoorn, D. M., Novina, C. D., and Sharp, P. A. (2003) Killing the messenger: short rnas that silence gene expression. *Nat Rev Mol Cell Biol* 4, 457–467.
23. Kretschmer-Kazemi Far, R. and Sczakiel, G. (2003) The activity of siRNA in mammalian cells is related to structural target accessibility: a comparison with antisense oligonucleotides. *Nucleic Acids Res* 31, 4417–4424.
24. Ameres, S. L., Martinez, J., and Schroeder, R. (2007) Molecular basis for target RNA recognition and cleavage by human RISC. *Cell* 130, 101–112.
25. Hornung, V., Guenthner-Biller, M., Bourquin, C., Ablasser, A., Schlee, M., Uematsu, S., Noronha, A., Manoharan, M., Akira, S., de Fougerolles, A., Endres, S., and Hartmann, G. (2005) Sequence-specific potent induction of IFN alpha by short interfering RNA in plasmacytoid dendritic cells through TLR7. *Nat Med* 11, 263–270.

26. Khvorova, A., Reynolds, A., and Jayasena, S. D. (2003) Functional siRNAs and miRNAs exhibit strand bias. *Cell* 115, 209–216.
27. Elbashir, S. M., Harborth, J., Lendeckel, W., Yalcin, A., Weber, K., and Tuschl, T. (2001) Duplexes of 21-nucleotide RNAs mediate RNA interference in cultured mammalian cells. *Nature* 411, 494–498.
28. Schwarz, D. S., Hutvagner, G., Du, T., Xu, Z., Aronin, N., and Zamore, P. D. (2003) Asymmetry in the assembly of the RNAi enzyme complex. *Cell* 115, 199–208.
29. Jiang, M., and Milner, J. (2002) Selective silencing of viral gene expression in HPV-positive human cervical carcinoma cells treated with siRNA, a primer of RNA interference. *Oncogene* 21, 6041–6048.
30. Heidenreich, O., Krauter, J., Riehle, H., Hadwiger, P., John, M., Heil, G., Vornlocher, H. P., and Nordheim, A. (2003) AML1/MTG8 oncogene suppression by small interfering RNAs supports myeloid differentiation of t(8;21)-positive leukemic cells. *Blood* 101, 3157–3163.
31. Scherr, M., Battmer, K., Winkler, T., Heidenreich, O., Ganser, A., and Eder, M. (2003) Specific inhibition of bcr-abl gene expression by small interfering RNA. *Blood* 101, 1566–1569.
32. Thomas, M., Greil, J., and Heidenreich, O. (2006) Targeting leukemic fusion proteins with small interfering RNAs: recent advances and therapeutic potentials. *Acta Pharmacol Sin* 27, 273–281.
33. Saxena, S., Jonsson, Z. O., and Dutta, A. (2003) Small RNAs with imperfect match to endogenous mRNA repress translation. Implications for off-target activity of small inhibitory RNA in mammalian cells. *J Biol Chem* 278, 44312–44319.
34. Dunne, J., Drescher, B., Riehle, H., Hadwiger, P., Young, B. D., Krauter, J., and Heidenreich, O. (2003) The apparent uptake of fluorescently labeled siRNAs by electroporated cells depends on the fluorochrome. *Oligonucleotides* 13, 375–380.
35. Scherr, M., Battmer, K., Schultheis, B., Ganser, A., and Eder, M. (2005) Stable RNA interference (RNAi) as an option for anti-bcr-abl therapy. *Gene Ther* 12, 12–21.
36. Matsukura, S., Jones, P. A., and Takai, D. (2003) Establishment of conditional vectors for hairpin siRNA knockdowns. *Nucleic Acids Res* 31, e77.
37. van de Wetering, M., Oving, I., Muncan, V., Pon Fong, M. T., Brantjes, H., van Leenen, D., Holstege, F. C., Brummelkamp, T. R., Agami, R., and Clevers, H. (2003) Specific inhibition of gene expression using a stably integrated, inducible small-interfering-RNA vector. *EMBO Rep* 4, 609–615.
38. Stegmeier, F., Hu, G., Rickles, R. J., Hannon, G. J., and Elledge, S. J. (2005) A lentiviral microRNA-based system for single-copy polymerase II-regulated RNA interference in mammalian cells. *Proc Natl Acad Sci U S A* 102, 13212–13217.
39. Wilda, M., Fuchs, U., Wossmann, W., and Borkhardt, A. (2002) Killing of leukemic cells with a BCR/ABL fusion gene by RNA interference (RNAi). *Oncogene* 21, 5716–5724.
40. Wohlbold, L., Van Der Kuip, H., Miething, C., Vornlocher, H. P., Knabbe, C., Duyster, J., and Aulitzky, W. E. (2003) Inhibition of bcr-abl gene expression by small interfering RNA sensitizes for imatinib mesylate (STI571). *Blood* 102, 2236–2239.
41. Withey, J. M., Marley, S. B., Kaeda, J., Harvey, A. J., Crompton, M. R., and Gordon, M. Y. (2005) Targeting primary human leukaemia cells with RNA interference: Bcr-Abl targeting inhibits myeloid progenitor self-renewal in chronic myeloid leukaemia cells. *Br J Haematol* 129, 377–380.
42. Look, A. T. (1997) Oncogenic transcription factors in the human acute leukemias. *Science* 278, 1059–1064.
43. Fazi, F., Zardo, G., Gelmetti, V., Travaglini, L., Ciolfi, A., Di Croce, L., Rosa, A., Bozzoni, I., Grignani, F., Lo-Coco, F., Pelicci, P. G., and Nervi, C. (2007) Heterochromatic gene repression of the retinoic acid pathway in acute myeloid leukemia. *Blood* 109, 4432–4440.
44. Matsunobu, T., Tanaka, K., Nakamura, T., Nakatani, F., Sakimura, R., Hanada, M., Li, X., Okada, T., Oda, Y., Tsuneyoshi, M., and Iwamoto, Y. (2006) The possible role of EWS-Fli1 in evasion of senescence in Ewing family tumors. *Cancer Res* 66, 803–811.
45. Scherr, M., and Eder, M. (2007) Gene silencing by small regulatory RNAs in mammalian cells. *Cell Cycle* 6, 444–449.
46. Fazi, F., Racanicchi, S., Zardo, G., Starnes, L. M., Mancini, M., Travaglini, L., Diverio, D., Ammatuna, E., Cimino, G., Lo-Coco, F., Grignani, F., and Nervi, C. (2007) Epigenetic silencing of the myelopoiesis regulator microRNA-223 by the AML1/ETO oncoprotein. *Cancer Cell* 12, 457–466.
47. Nakamura, T., Canaani, E., and Croce, C. M. (2007) Oncogenic All1 fusion proteins target Drosha-mediated microRNA processing. *Proc Natl Acad Sci U S A* 104, 10980–10985.

48. Chen, J., Wall, N. R., Kocher, K., Duclos, N., Fabbro, D., Neuberg, D., Griffin, J. D., Shi, Y., and Gilliland, D. G. (2004) Stable expression of small interfering RNA sensitizes TEL-PDGFbetaR to inhibition with imatinib or rapamycin. *J Clin Invest* 113, 1784–1791.
49. Stegmaier, K., Wong, J. S., Ross, K. N., Chow, K. T., Peck, D., Wright, R. D., Lessnick, S. L., Kung, A. L., and Golub, T. R. (2007) Signature-based small molecule screening identifies cytosine arabinoside as an EWS/FLI modulator in Ewing sarcoma. *PLoS Med* 4, e122.
50. Stegmaier, K., Corsello, S. M., Ross, K. N., and Golub, T. R. (2006) ASH annual meeting abstracts, *Blood* 108, 729.
51. Tirode, F., Laud-Duval, K., Prieur, A., Delorme, B., Charbord, P., and Delattre, O. (2007) Mesenchymal stem cell features of Ewing tumors. *Cancer Cell* 11, 421–429.
52. Hu-Lieskovan, S., Heidel, J. D., Bartlett, D. W., Davis, M. E., and Triche, T. J. (2005) Sequence-specific knockdown of EWS-FLI1 by targeted, nonviral delivery of small interfering RNA inhibits tumor growth in a murine model of metastatic Ewing's sarcoma. *Cancer Res* 65, 8984–8992.
53. Toub, N., Bertrand, J. R., Tamaddon, A., Elhamess, H., Hillaireau, H., Maksimenko, A., Maccario, J., Malvy, C., Fattal, E., and Couvreur, P. (2006) Efficacy of siRNA nanocapsules targeted against the EWS-Fli1 oncogene in Ewing Sarcoma. *Pharm Res* 23, 892–900.
54. Mao, S., Neu, M., Germershaus, O., Merkel, O., Sitterberg, J., Bakowsky, U., and Kissel, T. (2006) Influence of polyethylene glycol chain length on the physicochemical and biological properties of poly(ethylene imine)-graft-poly(ethylene glycol) block copolymer/SiRNA polyplexes. *Bioconjug Chem* 17, 1209–1218.
55. Howard, K. A., Rahbek, U. L., Liu, X., Damgaard, C. K., Glud, S. Z., Andersen, M. O., Hovgaard, M. B., Schmitz, A., Nyengaard, J. R., Besenbacher, F., and Kjems, J. (2006) RNA interference in vitro and in vivo using a novel chitosan/siRNA nanoparticle system. *Mol Ther* 14, 476–484.
56. Minakuchi, Y., Takeshita, F., Kosaka, N., Sasaki, H., Yamamoto, Y., Kouno, M., Honma, K., Nagahara, S., Hanai, K., Sano, A., Kato, T., Terada, M., and Ochiya, T. (2004) Atelocollagen-mediated synthetic small interfering RNA delivery for effective gene silencing in vitro and in vivo. *Nucleic Acids Res* 32, e109.
57. Pirollo, K. F., Rait, A., Zhou, Q., Hwang, S. H., Dagata, J. A., Zon, G., Hogrefe, R. I., Palchik, G., and Chang, E. H. (2007) Materializing the potential of small interfering RNA via a tumor-targeting nanodelivery system. *Cancer Res* 67, 2938–2943.
58. Zimmermann, T. S., Lee, A. C., Akinc, A., Bramlage, B., Bumcrot, D., Fedoruk, M. N., Harborth, J., Heyes, J. A., Jeffs, L. B., John, M., Judge, A. D., Lam, K., McClintock, K., Nechev, L. V., Palmer, L. R., Racie, T., Rohl, I., Seiffert, S., Shanmugam, S., Sood, V., Soutschek, J., Toudjarska, I., Wheat, A. J., Yaworski, E., Zedalis, W., Koteliansky, V., Manoharan, M., Vornlocher, H. P., and MacLachlan, I. (2006) RNAi-mediated gene silencing in non-human primates. *Nature* 441, 111–114.
59. Soutschek, J., Akinc, A., Bramlage, B., Charisse, K., Constien, R., Donoghue, M., Elbashir, S., Geick, A., Hadwiger, P., Harborth, J., John, M., Kesavan, V., Lavine, G., Pandey, R. K., Racie, T., Rajeev, K. G., Rohl, I., Toudjarska, I., Wang, G., Wuschko, S., Bumcrot, D., Koteliansky, V., Limmer, S., Manoharan, M., and Vornlocher, H. P. (2004) Therapeutic silencing of an endogenous gene by systemic administration of modified siRNAs. *Nature* 432, 173–178.
60. Schiffelers, R. M., Ansari, A., Xu, J., Zhou, Q., Tang, Q., Storm, G., Molema, G., Lu, P. Y., Scaria, P. V., and Woodle, M. C. (2004) Cancer siRNA therapy by tumor selective delivery with ligand-targeted sterically stabilized nanoparticle. *Nucleic Acids Res* 32, e149.
61. Chu, T. C., Twu, K. Y., Ellington, A. D., and Levy, M. (2006) Aptamer mediated siRNA delivery. *Nucleic Acids Res* 34, e73.
62. McNamara, J. O., 2nd, Andrechek, E. R., Wang, Y., Viles, K. D., Rempel, R. E., Gilboa, E., Sullenger, B. A., and Giangrande, P. H. (2006) Cell type-specific delivery of siRNAs with aptamer-siRNA chimeras. *Nat Biotechnol* 24, 1005–1015.
63. Song, E., Zhu, P., Lee, S. K., Chowdhury, D., Kussman, S., Dykxhoorn, D. M., Feng, Y., Palliser, D., Weiner, D. B., Shankar, P., Marasco, W. A., and Lieberman, J. (2005) Antibody mediated in vivo delivery of small interfering RNAs via cell-surface receptors. *Nat Biotechnol* 23, 709–717.
64. Kumar, P., Wu, H., McBride, J. L., Jung, K. E., Kim, M. H., Davidson, B. L., Lee, S. K., Shankar, P., and Manjunath, N. (2007) Transvascular delivery of small interfering RNA to the central nervous system. *Nature* 448, 39–43.
65. Wohlbold, L., van der Kuip, H., Moehring, A., Granot, G., Oren, M., Vornlocher, H. P.,

and Aulitzky, W. E. (2004) All common p210 and p190 Bcr-abl variants can be targeted by RNA interference. *Leukemia* 19, 290–292.

66. Prieur, A., Tirode, F., Cohen, P., and Delattre, O. (2004) EWS/FLI-1 silencing and gene profiling of Ewing cells reveal downstream oncogenic pathways and a crucial role for repression of insulin-like growth factor binding protein 3. *Mol Cell Biol* 24, 7275–7283.
67. Hsu, F. Y., Zhao, Y., Anderson, W. F., and Johnston, P. B. (2007) Downregulation of NPM-ALK by siRNA causes anaplastic large cell lymphoma cell growth inhibition and augments the anti cancer effects of chemotherapy in vitro. *Cancer Invest* 25, 240–248.
68. Ritter, U., Damm-Welk, C., Fuchs, U., Bohle, R. M., Borkhardt, A., and Woessmann, W. (2003) Design and evaluation of chemically synthesized siRNA targeting the NPM-ALK fusion site in anaplastic large cell lymphoma (ALCL). *Oligonucleotides* 13, 365–373.
69. Hingorani, S. R., Jacobetz, M. A., Robertson, G. P., Herlyn, M., and Tuveson, D. A. (2003) Suppression of BRAF(V599E) in human melanoma abrogates transformation. *Cancer Res* 63, 5198–5202.
70. Yang, G., Thompson, J. A., Fang, B., and Liu, J. (2003) Silencing of H-ras gene expression by retrovirus-mediated siRNA decreases transformation efficiency and tumorgrowth in a model of human ovarian cancer. *Oncogene* 22, 5694–5701.
71. Fleming, J. B., Shen, G. L., Holloway, S. E., Davis, M., and Brekken, R. A. (2005) Molecular consequences of silencing mutant K-ras in pancreatic cancer cells: justification for K-ras-directed therapy. *Mol Cancer Res* 3, 413–423.
72. Brummelkamp, T. R., Bernards, R., and Agami, R. (2002) Stable suppression of tumorigenicity by virus-mediated RNA interference. *Cancer Cell* 2, 243–247.
73. Eskandarpour, M., Kiaii, S., Zhu, C., Castro, J., Sakko, A. J., and Hansson, J. (2005) Suppression of oncogenic NRAS by RNA interference induces apoptosis of human melanoma cells. *Int J Cancer* 115, 65–73.
74. Martinez, L. A., Naguibneva, I., Lehrmann, H., Vervisch, A., Tchenio, T., Lozano, G., and Harel-Bellan, A. (2002) Synthetic small inhibiting RNAs: efficient tools to inactivate oncogenic mutations and restore p53 pathways. *Proc Natl Acad Sci U S A* 99, 14849–14854.

Chapter 12

Targeting Stromal–Cancer Cell Interactions with siRNAs

Seyedhossein Aharinejad, Mouldy Sioud, Trevor Lucas, and Dietmar Abraham

Abstract

Tumors are composed of both malignant and normal cells, including fibroblasts, endothelial cells, mesenchymal stem cells, and inflammatory immune cells such as macrophages. These various stromal components interact with cancer cells to promote growth and metastasis. For example, macrophages, attracted by colony-stimulating factor-1 (CSF-1) produced by tumor cells, in turn produce various growth factors such as vascular endothelial growth factor, which supports the growth of tumor cells and their interaction with blood vessels leading to enhanced tumor cell spreading. The activation of autocrine and paracrine oncogenic signaling pathways by stroma-derived growth factors and cytokines has been implicated in promoting tumor cell proliferation and metastasis. Furthermore, matrix metalloproteinases (MMPs) derived from both tumor cells and the stromal compartment are regarded as major players assisting tumor cells during metastasis. Collectively, these recent findings indicate that targeting tumor–stroma interactions is a promising strategy in the search for novel treatment modalities in human cancer. This chapter summarizes our current understanding of the tumor microenvironment and highlights some potential targets for therapeutic intervention with small interfering RNAs.

Key words: Stroma, matrix metalloproteinases, extracellular matrix, colony-stimulating factor, vascular endothelial growth factor, tumor macrophages, angiogenesis, metastasis, RNA interference, small interfering RNAs.

1. Tumor–Host Interactions and the Tumor Microenvironment

1.1. The Cellular Microenvironment of Cancer Cells

The development of cancer is a complex, multistage process during which a normal cell undergoes genetic changes that result in phenotypic alterations and acquisition of the ability to invade and colonize distant sites *(1, 2)*. Solid tumors are composed of both malignant and normal cells. Targeting the complex interaction between genetically instable neoplastic cells, the surrounding extracellular matrix (ECM), and stromal cells such as fibroblasts,

M. Sioud (ed.), *Methods in Molecular Biology, siRNA and miRNA Gene Silencing, vol. 487*

DOI: 10.1007/978-1-60327-547-7_12

inflammatory cells, and endothelial lineage cells is a promising strategy in the search for novel treatment modalities in human cancers *(1, 2)*. Changes in the tumor microenvironment can lead to ECM modification, infiltration of inflammatory cells, and alter the activity of matrix metalloproteinases (MMPs), which are essential regulatory factors in tumor growth and invasion *(3)*. In addition, mediators released from both the stroma and tumor cells can lead to the induction of angiogenesis by shifting the balance between factors that promote and inhibit angiogenesis *(4–7)*, allowing the growth of tumors to macroscopic levels *(8– 12)*.

1.2. Macrophages, Angiogenesis, and the Extracellular Matrix ECM)

Macrophages are common components of the tumor stroma *(9)* that modify the ECM and influence new capillary growth by several different mechanisms *(12, 13)*. Macrophages can produce growth factors, cytokines, proteolytic enzymes, and matrix molecules that act directly to stimulate vascularization by stimulating endothelial cell proliferation, migration and differentiation in vitro, and angiogenesis in vivo *(9, 12)*. Macrophages can also modify the ECM either through the direct production of ECM components or the production of proteases that alter ECM structure and composition *(13)*. The composition of the ECM dramatically influences endothelial cell shape and morphology and profoundly influences capillary growth *(8)*. Importantly, recruitment of macrophages to tumors can significantly increase metastatic progression *(14)*. Macrophages can also secrete cytokines that stimulate other cells to synthesize or degrade ECM molecules. Remodeling of the ECM is crucial to both angiogenesis and tumorigenesis and primarily involves the MMP family of proteolytic enzymes. MMPs degrade the ECM including the basement membrane, and in conjunction with soluble growth factors foster the migration and proliferation of endothelial cells. This process promotes angiogenesis and also allows tumors to spread locally and distantly *(3)*. Strict regulation of MMP expression is critical for maintenance of proper ECM homeostasis; however, in malignancies high levels of MMPs are often synthesized not only by cancer cells but also by adjacent and intervening stromal cells *(15)*.

1.3. Fibroblasts in the ECM

Fibroblasts in the tumor stroma have been termed carcinoma-associated fibroblasts (CAFs), myofibroblasts, or tumor-associated fibroblasts and are important promoters of tumor growth and progression *(16, 17)*. CAFs are commonly identified by expression of α-smooth-muscle actin *(18)*. CAFs secrete and deposit laminin and type IV collagen to produce basal membranes and the ECM components—collagen types I, III, and IV and fibronectin *(19)*. Furthermore, CAFs facilitate the invasiveness of noninvasive cancer cells when coinjected into mice *(21)*. CAFs also remodel the tumor–stromal ECM and are recognized as a source of paracrine (cell to cell) growth factors that influence the growth of carcinoma cells *(20, 22)*. CAFs can secrete stromal derived factor-1 (SDF-1, CXCL-12), which induces angiogenesis and promotes the proliferation of CXCR4-expressing tumor cells *(23)*. CAF expression of fibroblast activation protein (FAP), which displays dipeptidyl peptidase and collagenase activity, promotes tumor growth and has been targeted with monoclonal antibodies in clinical trials *(24)*. Conversely, CAF transforming growth factor-beta (TGF-β) signaling suppresses epithelial transformation by paracrine hepatocyte growth factor (HGF) activation of the phosphorylated HGF receptor c-met and c-myc *(25)*. Moreover, fibroblasts are a major source of vascular endothelial growth factor (VEGF), which plays a central role in the formation of new blood vessels in the developing tumor mass *(26)*.

1.4. The Role of Endothelial Progenitor Cells (EPCs) in the ECM

Microvessels may develop either from existing capillary networks by neoangiogenesis and arteriogenesis or bone marrow derived angioblasts or circulating EPCs in a process known as vasculogenesis *(27)*. Originating in the bone marrow, EPCs can be mobilized during tumor development and through the release of trophic factors, cancer cells and host cells are able to effectively induce the homing of EPCs to sites of vascular growth within the tumor stroma *(28)*. EPCs have the ability to form endothelial colonies in vitro and may be recruited from the bone marrow after tumor growth *(29)*. Chemotactic agents that are responsible for this process may include VEGF *(30)* and stromal-derived factor-1 (SDF-1) *(31)*. SDF-1 recruits vascular progenitors *(32)* through interaction with the CXCR4 (chemokine, CXC-motif, receptor 4) receptor, which is expressed on EPCs *(31)*.

2. Influencing Tumor–Host Interactions by Targeting Tumor-Associated Macrophages (TAMs)

2.1. Colony-Stimulating Factor-1 (CSF-1) and Macrophages

The production of macrophages is regulated by CSF-1, also called macrophage-CSF (M-CSF) *(33)*. CSF-1 is produced by a variety of cell types such as fibroblasts or macrophages and prevents the death of monocytes and promotes their differentiation into macrophages *(34, 35)*. CSF-1 also induces or augments the production of a variety of cytokines by macrophages such as TNF-α *(36)*. Macrophages most likely enhance tumor progression through paracrine circuits involving the production of CSF-1 by tumor cells *(9)* or other host-derived stromal cells and by ECM-modulating functions mediated by MMPs *(12)* to accelerate angiogenesis in vivo *(36)*. Consistent with a proangiogenic effect, recent work suggests that CSF-1 also stimulates monocytes to secrete biologically active VEGF *(37)*. VEGF is a key factor in tumor angiogenesis and is upregulated in numerous malignant tumors. The biological effects of VEGF are mediated by VEGF-receptor 1 (VEGF-R1, Flt-1) and VEGF-R2 (KDR/Flk-1).

2.2. CSF-1 Signaling Pathways

CSF-1 is a disulfide-linked homodimeric growth factor that binds the integral membrane receptor tyrosine kinase (CSF-1R) product of the c-*fms* proto-oncogene *(38)*. Similar to other tyrosine kinase receptors, ligand binding stabilizes CSF-1R dimerization to activate the receptor through autophosphorylation in trans, thereby initiating a series of membrane-proximal tyrosine phosphorylation cascades leading to rapid stimulation of cytoskeletal remodeling, gene transcription, and protein translation *(39)*. Many of the downstream tyrosine-phosphorylated proteins, such as the p85 regulatory subunit of phosphatidylinositol 3-kinase (PI3K), Cbl, and Gab3, have been shown to be important in regulating macrophage survival, differentiation, and motility *(40)*.

2.3. CSF-1 and Solid Tumors

CSF-1 is widely overexpressed in tumors of the reproductive system. In breast cancer, CSF-1 expression has been shown to correlate with high grade and poor prognosis associated with dense leucocytic infiltration *(41)*. High levels of CSF-1R mRNA have been observed in ovarian and endometrial cancers and elevated levels correlated with high histological grade and advanced clinical presentation *(42)*. Over half of invasive ovarian adenocarcinomas and endometrial cancers co-express CSF-1 and CSF-1R *(43)*. Constitutive production of CSF-1 has been reported in normal ovarian epithelial cultures at levels comparable with ovarian cancer cell lines *(44)*. However, the co-expression of CSF-1 and CSF-1R may establish an autocrine loop that plays a role in metastatic progression. Serum levels of CSF-1 are markedly elevated in patients with endometrial cancer associated with active or recurrent disease *(45)*. Increased serum CSF-1 levels also characterize most clinical cases of epithelial ovarian cancers *(46)* and CSF-1 is considered a tumor marker for ovarian germ cell tumors *(47)*.

2.4. CSF-1 and Tumor Cell Invasion

Osteopetrotic CSF-1 (*op/op*) mice, which have a CSF-1 gene defect and a profound macrophage deficiency *(48)*, have been used as a model to examine tumor growth. These mice show an impaired tumor development (Lewis lung carcinoma) when compared to normal littermates that is reversed by CSF-1 treatment *(49)*. Crossing CSF-1 (*op/op*) mice with a transgenic mouse susceptible to mammary cancer prevented macrophage accumulation in mammary tumors. In the macrophage-deficient mice, the incidence and initial rates of growth of primary tumors were not different from those seen in normal mice, but the rate of tumor progression was slowed and metastatic ability was almost completely abrogated when compared with mice that contained normal numbers of macrophages. Overexpression of CSF-1 in wild-type mice also accelerated tumor progression and increased rates of metastasis *(14)*. Another study has shown that CSF-1 promotes tissue invasion by enhancing ECM-degrading proteinase MMP-2 production by lung cancer cells *(50)*. In some instances, malignant cells co-express

CSF-1 and CSF-1R, raising the possibility of autocrine growth control by CSF-1 in the development of these malignancies *(51)*.

2.5. CSF-1 in Breast Cancer

The mechanism by which mammary epithelial cells undergo genetic changes that result in acquisition of the ability to invade and colonize distant sites is complex *(52, 53)*. Normal and malignant mammary epithelium and the surrounding stromal cells produce and respond to various growth factors. Among the stromal cells, macrophages play a unique role because they are recruited to mammary gland carcinomas *(7, 53)*. The fact that in the absence of such tumor-associated macrophages, metastatic progression of mammary gland tumors is profoundly reduced *(14)* as well as the fact that CSF-1 blockade suppresses tumor growth, MMP production, and macrophage recruitment in embryonic tumors and colon cancer *(54)*, support the paradigm that CSF-1 enhances progression of malignancies through effects on the recruitment and control of macrophages that regulate tumor cell growth, angiogenesis, and the ECM. The recent discovery of highly specific, small interfering (si)RNA molecules as promising candidate therapeutics to specifically and potently modify gene expression led us to hypothesize that blocking CSF-1 using this approach could efficiently suppress breast cancer development. It should be noted that human MCF-7 mammary carcinoma cells express both mRNA and protein for CSF-1 and CSF-1R in vitro. When MCF-7 cells were xenografted to immunodeficient nude mice, cancer cell CSF-1 expression was lost but host (mouse) cells were stimulated to overexpress CSF-1.

2.6. CSF-1 siRNAs Against CSF-1 and c-fms Downregulate Target Proteins and Suppress Mammary Tumor Growth

CSF-1 and CSF-1R siRNAs suppress target gene expression in a sequence- and dose-dependent manner in vitro. Mice bearing human MCF-7 mammary carcinoma xenografts were treated with five intratumoral injections of CSF-1 siRNA, CSF1-R siRNA, scrambled control siRNA, or Ringer's solution (control). siRNA treatment was well tolerated and no significant changes in the cellular blood count of treated mice was observed. Anti-CSF-1 and CSF-1R siRNAs suppressed mammary tumor growth by 45 and 40%, respectively, and selectively downregulated target protein expression in tumor lysates.

2.7. CSF-1 and CSF-1R Blockade Downregulates Mouse MMP-2 and MMP-12 Expression and Decreases Angiogenic Activity in MCF-7 Mammary Tumor Xenografts

After human MCF-7 cell xenografting in mice, macrophage invasion in the tumor xenografts was observed. In association with this, host (mouse) MMP-2 and MMP-12 (a macrophage-specific protease involved in ECM remodeling) were strongly expressed during tumor progression in control animals. Treatment with CSF-1 siRNA or CSF-1R siRNA reduced macrophage recruitment to the tumor and intratumoral levels of both MMP-2 and MMP-12.

Histomorphometrical analysis of mammary tumors showed an increased density of proliferating endothelial cells with tumor progression that was decreased after CSF-1 and CSF-1R siRNA blockade. In addition, VEGF-A mRNA levels increased with tumor progression and were reduced in CSF-1 and CSF-1R siRNA-treated mice. CSF-1 and CSF-1R blockade, however, did not significantly affect tissue mRNA expression of the VEGF-A receptors Flt-1 and KDR. These data indicated that blocking CSF-1 or CSF-1R is associated with decreased VEGF-A expression and reduced angiogenic activity in mammary tumor xenografts. The median survival of animals in the control group was 62 days, which was significantly increased in mice after treatment with CSF-1 siRNA (103 days) and slightly (but not significantly) increased after treatment with CSF-1R siRNA (76 days).

2.8. The Role of TNF-α

TNF-α is a pivotal cytokine in tumor growth and is produced by both macrophages and malignant cells *(55)*. TNF-α can induce tumor cell apoptosis in neoplastic tissues *(56)* but may promote tumor growth at lower levels *(6)* and has been referred to as a tumor-promoting factor *(57)*. TNF-α induces macrophage expression of the type-IV collagenase MMP-9 by macrophages which degrades ECM collagen *(58)*. TNF-α has also been identified as a key inducer of CSF-1 production *(59)*. Utilizing siRNA to explore the mechanisms behind colon cancer upregulation of host factors to promote tumorigenesis, mice bearing human colon cancer xenografts were treated with siRNA. From a methodological perspective, xenotransplanted flank tumors in *nu/nu* mice received 10 μg of human scrambled siRNA control, TNF-α siRNA, mouse CSF-1 siRNA, or a combination of human TNF-α and mouse CSF-1 siRNA treatment cycled on days 8, 12, 14, 17, and 20 following tumor cell injection. On day 22, the animals were sacrificed. Treatment with human TNF-α siRNA, mouse CSF-1 siRNA, or combined human TNF-α/mouse CSF-1 siRNA suppressed tumor growth by 34, 47, and 50%, respectively. Analysis of the isolated tumors showed that, in addition to CSF-1, levels of host and cancer cell-derived TNF-α mRNA expression increase dramatically during tumor development *(60)*. Tumor cell-derived TNF-α can also influence the migration of macrophages in vitro and cellular proliferation as well as TAM infiltration are reduced following TNF-α siRNA treatment of colon cancer in vivo *(60)*. These findings support other reports of high CSF-1 serum levels correlating to poor prognosis in colon carcinoma patients *(61)* and increased soluble TNF-R1 in colorectal cancer patients *(62)*. Since TNF-α regulates CSF-1 expression in macrophages, it was then shown by co-culturing colon cancer cells with macrophages that TNF-α expression by SW620 cells leads to the induction of macrophage TNF-α and CSF-1, and that CSF-1 in turn increases macrophage VEGF-A and MMP-2 expression *(60)*. In breast cancer, TNF-α is known to regulate MMP production in macrophages which is associated with increased invasiveness *(63)*.

2.9. The Role of EMMPRIN in Tumor–Stroma Interactions

Tumor cells stimulate MMP production by stromal cells via soluble cytokines *(64, 65)* or through intercellular interactions mediated by adhesion molecules such as the ECM MMP inducer (EMMPRIN) *(66)*. EMMPRIN (also known as the M6 leukocyte activation antigen, CD147, or murine basigin) is a differentially glycosylated (32–60 kDa) member of the immunoglobulin superfamily *(67)* containing two extracellular immunoglobulin domains, a transmembrane domain, and a 39-amino acid cytoplasmic domain *(66)*. CD147 forms homo-oligomers in the plasma membrane similar to other members of the Ig superfamily *(67)*. Accumulating evidence suggests a prominent role for CD147 in mediating interactions both between tumor cells and between tumor cells and host stromal cells to promote a number of events during cancer progression including MMP production *(66)*. CD147 is highly expressed on the surface of tumor cells and stimulates the production of MMPs by adjacent stromal fibroblasts *(14, 68)*. To analyze the role of CD147 colon cancer growth in vivo, we treated mice bearing human colon cancer xenografts with intratumoral injections (10 μg) of siRNA against human CD147/EMMPRIN, mouse CD147/Basigin, combined human CD147/EMMPRIN, and mouse CD147/Basigin or scrambled control siRNAs. The treatment was cycled on days 12, 14, 17, and 20. On day 22, animals were sacrificed. Mouse CD147/Basigin siRNA (993 ± 83 mg; $P < 0.05$) treatment markedly reduced tumor weights by 26% compared to untreated controls (1343 ± 139 mg) whereas tumor weights were not influenced by human CD147/EMMPRIN siRNA. Although cancer cell CD147/EMMPRIN tissue mRNA significantly declined, host (mouse) CD147/Basigin and MMP-2 tissue expression increased following human CD147/EMMPRIN blockade. In contrast, treatment with mouse CD147/Basigin or combined EMMPRIN/Basigin siRNA downregulated host (mouse) CD147/Basigin and MMP-2 tissue mRNA expression. These data showed that specific tumor–stromal cell interactions mediated by CD147 downregulate fibroblast MMP-2 production and promote colon cancer growth in vivo *(69)*.

3. Targeting the Interplay Between Stroma and Tumor Cells

The tumor microenvironment is an integral part of its anatomy and physiology and interactions between neoplastic cells and the surrounding microenvironment are crucial to each step of tumorigenesis. Cancer cells themselves are able to generate a supportive microenvironment by producing stroma-modulating growth factors such as members of the fibroblast growth factor (FGF) and the vascular endothelial growth factor (VEGF) families, platelet-derived growth factor (PDGF), epidermal growth factor (EGF), CSFs, and others *(17)*.

Secretion of these factors disrupts normal tissue homeostasis and induces remodeling such as angiogenic *(70)* and inflammatory responses *(71)*. These factors also activate surrounding stromal cells such as fibroblasts and recruit and activate macrophages, which in turn leads to the production and secretion of additional growth factors and proteases, emphasizing the importance of crosstalk between cancer cells and stromal cells in tumor development *(17)*.

One major advantage of therapies targeting cells in the microenvironment is that these are genetically stable in contrast to tumor cells in which drug resistance can develop through the acquisition of compensatory mutations. However, there are some reports indicating that stromal regions from cancers exhibit genetic alterations *(71, 72, 73)* and it is still an unanswered question whether cancer-associated fibroblasts (CAFs) harbor genetic and/or only epigenetic alterations that lead to activated, tumor-enhancing phenotypes *(74)*.

Another limitation of targeting the cells in the tumor microenvironment is that a delicate balance exists between their tumor-inhibitory and tumor-promoting functions. Thus, understanding the complex cellular and molecular networks affecting stromal cells in the tumor mass and identification of the key molecular differences between these cells under normal tissue homeostasis compared to when they have been altered by the tumor microenvironment will help to provide information for targeting the tumor microenvironment *(75)*.

3.1. Targeting TAMs

Macrophages recruited by tumors may be classified as inflammatory type 1 macrophages having anti-tumor activity or more commonly to type 2 (M2) macrophages that are pro-angiogenic and stimulate tumor growth *(76)*. By producing growth factors and cytokines, including epidermal growth factor (EGF), members of the fibroblast growth factor (FGF) family, transforming growth factor (TGF)-β, TNF-α, and some chemokines, TAMs affect tumor cell proliferation *(66, 67)*. These macrophages also can produce enzymes and inhibitors which regulate the digestion of the ECM within the tumor stroma, such as MMPs, plasmin, and urokinase-type plasminogen activator (uPA) and its receptor uPAR, that not only remodel the ECM, but also generate reactive cleavage products of ECM molecules and release and activate pro-angiogenic factors such as VEGF and others *(76)*. Additionally, macrophages can also express the angiogenic growth factor VEGF *(37)*. Thus, tumors may therefore recruit macrophages and create a microenvironment that causes macrophages to suppress immune function and instead modify the ECM and influence capillary growth by different mechanisms to promote and sustain tumor angiogenesis, which is one of the contributions of the stroma to tumor progression *(12)*.

Macrophages are produced mainly under the influence of the growth factor CSF-1 *(35)*. CSF-1 prevents the death of mononuclear cells in culture, promotes their differentiation into macrophages, and induces or augments the production of a variety of cytokines by macrophages *(35)*. Accumulating reports suggest that CSF-1 and its receptor play an important role in malignancies. For example, high levels of CSF-1 have been observed in patients with tumors of the head and neck, in men with metastatic prostate cancer, and in women with metastatic breast cancer *(77)*. CSF-1 promotes the metastatic potential of mammary carcinoma xenografts in mice by regulating the infiltration and function of TAMs *(14)*.

Inhibition of upregulated host CSF-1 in human embryonic, colon, and breast cancer xenografts in mice suppresses tumor growth and increases mouse survival associated with decreased tumor vascularity, reduced expression of angiogenic factors and MMPs, and decreased macrophage recruitment to tumors *(52, 54)*. In the colon cancer model, upregulated host macrophage CSF-1 was mediated by TNF-α secreted by colon cancer cells, and TNF-α successfully stimulated macrophage migration and proliferation *(60)*. Moreover, some cancer cells produce CSF-1 and directly influence macrophages *(78)*. Thus, interaction between cancer cells and the surrounding tumor microenvironment leads to upregulation of CSF-1 which in turn leads to macrophage recruitment. The tumor microenvironment educates these tumor-associated macrophages to perform supportive roles that promote tumor progression and metastasis *(79)*. Of interest, these molecular and phenotypic changes could be exploited therapeutically and should enhance the antitumor selectivity of targeting macrophages associated with tumors.

3.2. Tumor-Activated Fibroblasts

Carcinoma-activated fibroblasts can promote tumorigenesis through the expression of CXCL12/stromal derived factor-1 *(23, 80)*, a chemokine ligand for the widely distributed chemokine receptor CXCR4 which promotes angiogenesis by recruiting marrow-derived precursors that contribute to vessel development. Molecules produced in activated fibroblasts, such as the fibroblast activation protein (FAP) *(80, 81)*, cathepsin, MMP1 and MMP3, which degrade and remodel the ECM *(82)*, urokinase, plasminogen activator (uPA), which is required for the activation of plasminogen to plasmin, and the secreted pro-migratory ECM component tenascin *(83)* could provide promising selective targets in the tumor stroma (see chapter 13). Additionally, CAFs express a range of growth factors and cytokines such as insulin-like growth factor 1 (IGF1) and HGF *(22)*, which promote tumor cell survival *(84)* as well as tumor cell migration and invasion *(83, 85)*, and VEGF or monocyte chemoattractant protein-1 (MCP-1) *(86)*, which further activate the tumor stroma through the stimulation of angiogenesis and the recruitment of inflammatory cells. Similar to tumor macrophages, CAFs have a distinct phenotype compared with normal fibroblasts *(20)*, which neither produce CXCL12 nor induce angiogenesis, and only marginally contribute to tumor growth *(23)*. Importantly, fibroblasts are a major cell type in the tumor stroma and various types of invasive human carcinomas usually include large numbers of these cells, suggesting that stromal fibroblasts represent an attractive target for therapeutic intervention *(87)*.

3.3. Endothelial Cells (ECs), EPCs, and Lymphatic Endothelial Cells (LECs)

Tumor growth beyond a certain size is accompanied by increased angiogenesis, and is considered as a necessary event for the transition of a benign tumor into a large tumor with the ability to spread and metastasize. The tumor vasculature is derived by angiogenesis, new blood vessel growth from pre-existing vessels, and vasculogenesis, the recruitment of circulating endothelial progenitor cells *(88, 89)*. A plethora of anti-angiogenic agents inhibiting either angiogenic growth factors or their receptors have been developed and tested in preclinical and clinical studies. Amongst them the VEGF axis represents an important target for antitumor therapy, which includes anti-VEGF agents and inhibitors to reduce VEGF receptor kinase activity. The VEGF family consists of five members (VEGF-A, -B, C- D-, and PlGF) and three VEGF tyrosine kinase receptors (VEGFR-1, -2, -3) and several co-receptors, such as heparansulfate proteoglycans and neuropilins *(90)*. VEGFR-1 transduces weak signals for endothelial cell and pericyte growth and survival and a synergy exists between VEGFR-1 and R-2 specific ligands allowing the modulation of a variety of VEGFR-dependent signals. The principal function of the VEGFR-2 is the stimulation of vascular endothelial cell survival and growth and promotion of angiogenesis *(91)*. Both VEGFR-1 and VEGFR-2 are expressed in vascular endothelial cells and also in some tumor cells. In addition, VEGFR-1 is expressed by monocytes and macrophages, whereas VEGFR-2 is also expressed on hematopoietic stem cells *(92)*. VEGFR-3, a receptor for the lymphatic growth factors VEGF-C and VEGF-D, but not for VEGF-A, regulates normal and tumor lymphangiogenesis allowing increased spread of tumor metastases through the lymphatics *(91)*. PDGF and its receptors have been suggested as another system involved in tumor lymphangiogenesis *(92)*, thus representing an additional therapeutic target for inhibition of lymphatic metastasis. Clearly, metastasis either through angiogenesis or lymphangiogenesis always needs to be considered together with the environmental conditions in which metastatic tumors develop. Metastatic cells need an appropriate microenvironment in which they can survive and proliferate *(93)*.

Genes specifically expressed in endothelial cells may serve as additional attractive targets for an RNAi-based approach to interfere with angiogenesis. CD31/platelet endothelial cell adhesion molecule 1 (PECAM-1) is one candidate target with vasculature-specific expression including endothelial cells, platelets, monocytes, neutrophils, and selected T cells and CD-31 possibly contribute to neovascularization during tumor growth *(94)*. Another way to block blood supply to the developing tumor is to prevent homing of EPCs to the tumor site. The SDF-1-CXCR-4 signaling axis is used to recruit EPCs into tumors, and high levels of CXCR-4 expression by various types of human carcinoma cells are clinically associated with a poor prognosis *(95, 96)*, thus representing a promising target for blocking EPC homing. However, there is considerable controversy as to the exact contribution of EPCs to the tumor endothelium, leaving it an open question whether blocking EPC recruitment to the tumor could be a potential anti-angiogenic and anti-tumorigenic therapeutic approach.

3.4. The ECM

The noncellular compartment consists of the various components of the ECM including collagen, laminin, fibronectin, and heparan sulfate proteoglycans, whose composition directly and indirectly influences the phenotype of the cellular compartment *(97, 98)*. Hyaluronan is produced by hyaluronan synthases and the main source of hyaluronan in many human cancers is the stroma. There is also increasing evidence that the signaling functions of hyaluronan are important for promoting tumor cell growth, invasion, and metastasis *(99)*.

One specialized ECM component is the basement membrane (BM), which separates the epithelium from the stroma and underlying endothelial cells, pericytes, and other cell types. Maintaining organ homeostasis can prevent neoplastic transformation in normal tissues by ensuring firm cell–cell contacts, mediated by tight-junction proteins and cell adhesion molecules, such as β1 integrins and E-cadherin *(100)*.

In normal tissue homeostasis, the interacting network of proteases and their natural inhibitors maintain a proteolytic balance. During cancer progression, this balance is disturbed by overexpression of proteases including proteases of the cysteine, serine, and MMP classes as well as endoglycosidases such as heparanase *(88)*. uPA is a specific serine protease that converts plasminogen into its active form, plasmin, which is a broad-spectrum protease involved in the degradation of the ECM and activation of latent MMPs *(101)*. Besides its proteolytic function, the uPA-uPAR system promotes various cell responses such as tumor cell migration, adhesion, and proliferation due to its interaction with various integrins *(101)*. The MMPs form a family of structurally and functionally related zinc endopeptidases which collectively can degrade virtually all ECM components *(98)*. Tumor-derived MMPs and host/stromal-derived MMPs also have been regarded as major molecules assisting tumor cells during metastasis. Importantly, MMPs can exhibit pro-metastatic as well as anti-metastatic roles *(102)*.

Controlled BM/ECM degradation is necessary for angiogenesis and invasion of tumor cells into both the surrounding normal tissue and the blood and lymphatic systems *(17)*. Proteases can be pro-angiogenic, by releasing reactive ECM fragments and activating ECM-bound angiogenic growth factors such as VEGF, bFGF and others. At the same time, unbalanced protease expression can also provide anti-angiogenic signals, such as by generating endogenous angiogenesis inhibitors through the proteolytic modification of ECM components in the matrix. In many cancers, matrix-degrading enzymes are provided by infiltrating innate immune cells *(15, 79)* and inhibition of certain enzymes, such as cysteine cathepsins, or heparanase, offers the potential to block multiple nodes in the tumor microenvironment. The proteolytic activity of cathepsin B not only involves the direct degradation of ECM proteins, but also indirectly activates MMPs and receptor-bound uPA *(103)*. In addition, cathepsin B has been suggested to increase MMP activity by inactivating tissue inhibitors of MMPs (TIMPs). However, broad-spectrum MMP inhibitors (MMP-I) have not shown promising results in clinical trials, probably due to the fact that ECM degradation is a delicate balance and that MMPs can also release anti-angiogenic proteins such as endostatin and angiostatin *(97)*, which inhibit tumorigenesis. Experimental data clearly point to the complexity of MMP functions in vivo, rather than indicating that MMPs are ineffective targets for cancer therapy. Thus, the design of MMP inhibitors, not only targeting a specific MMP, but also targeting MMPs at a specific site or tissue, especially during early stages of metastatic disease, is of critical importance *(102)*.

4. Strategies for the In Vivo Application of siRNAs

Several delivery methods for siRNAs are currently under investigation (*see* Chaps. 2–6) However, their ultimate success will depend on many relevant parameters including stability, transfection efficacy, as well as the ability of the transfer vectors to overcome biological barriers after systemic or local administration to reach target cells or tissues *(104)*. Over the past few years, chemically synthesized siRNA has become the standard tool for gene expression silencing in vitro. However, some serum nucleases can

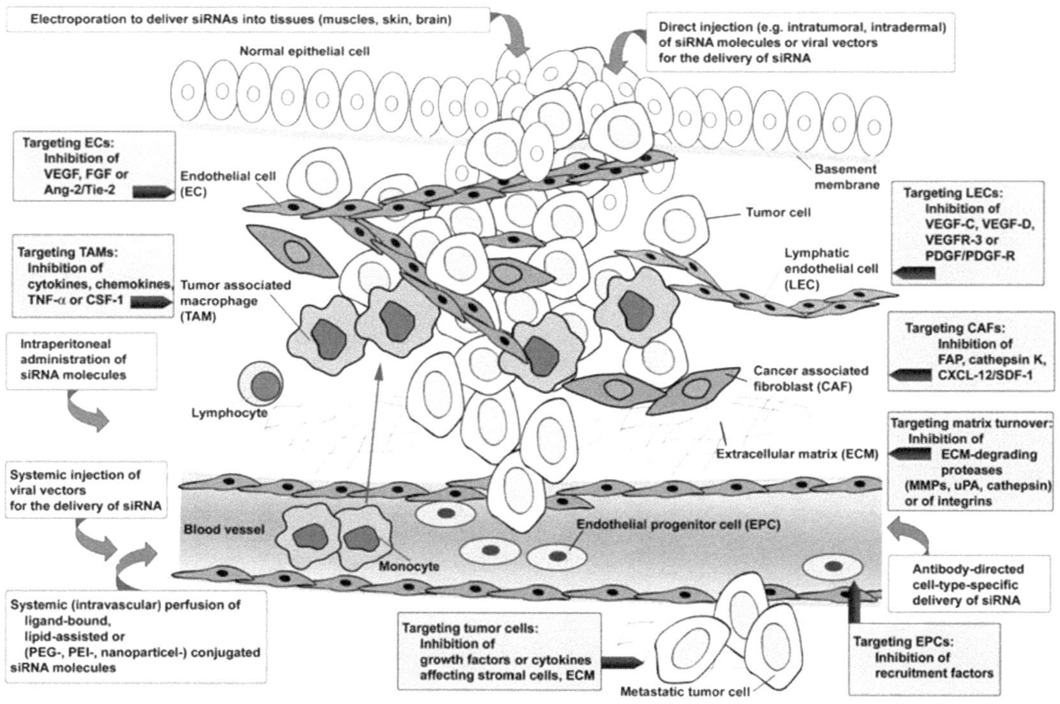

Fig. 12 1. Potential therapeutic strategies for the delivery of RNAi molecules targeting different cells in the tumor microenvironment. Tumor cells as well as stromal cells secrete growth factors, chemokines, and proteases, which can act in autocrine and paracrine manners. In addition, secretion of extracellular matrix (ECM) components is initiated. The degradation of ECM components together with tumor- and host-cell derived growth factors and cytokines then induce angiogenesis as well as recruitment and activation of inflammatory cells, endothelial progenitor cells and fibroblasts, which secrete further pro-tumorigenic factors to amplify these signals thereby promoting (lymph)angiogenesis, tumor growth, and metastasis. RNA interference based strategies for gene-specific therapeutics targeting the tumor microenvironment include the local or systemic delivery of synthetic double-stranded siRNAs or by plasmid and viral vector systems that express double-stranded short hairpin RNAs (shRNAs) that are subsequently processed to siRNAs by the cellular machinery. Among the targets to inhibit in stromal cells (ECs, TAMs, CAFs, EPCs, and LECs) are growth factors and their receptors (e.g., VEGF/VEGF-R family, FGF/FGF-R family, Ang/Tie-2, CSF-1/CSF-1R, PDGF/PDGFR family), cytokines (e.g., TNF-α), chemokines (e.g., SDF-1/CXCR-4 axis), basement membrane- and ECM-degrading enzymes (MMPs, uPA, cathepsin), ECM-contact factors mediating signaling (integrins), and intracellular components (e.g., factors of the NF-κB pathway). In addition tumor cell-derived factors affecting surrounding stromal cells and the ECM can be targeted. Abbreviations: Ang-2, angiopoietin-2; CAF, cancer associated fibroblast; CSF-1, colony-stimulating factor 1; CSF-1R, CSF-1 receptor; CXCL-12, CXC-motive ligand-12 EC, endothelial cell; ECM, extracellular matrix; EGF, epidermal growth factor; EMMPRIN, extracellular matrix metalloproteinase inducer; EPC, endothelial progenitor cell; FAP, fibroblast activated protein; FGF, fibroblast growth factor; IKK, IκB kinase; LEC, lymphatic endothelial cell; MMP, matrix metalloproteinases; NF-κB, nuclear factor of κB; PDGF, platelet-derived growth factor; PDGF-R, PDGF receptor; PEG, polyethylene glycol; PEI, polyethyleneimine; SDF-1, stromal cell-derived factor 1; TAM, tumor associated macrophage; TNF-α tumor necrosis factor- α; uPA, urokinase plasminogen activator; VEGF, vascular endothelial growth factor; VEGF-R, VEGF receptor.

degrade siRNAs, and thus chemical modifications that improve stability and protect against nuclease degradation have been developed *(105)*. Alternatively, complexing siRNAs with polyethyleneimine (PEI) *(106, 107)*, atelocollagen *(108)*, or cholesterol *(109)* has improved stability and delivery in vivo (*see* Chap. 4). The most difficult task in transferring this technology to an in vivo setting is to develop appropriate delivery strategies. A number of different approaches have been developed for the in vivo delivery of siRNA (**Fig. 12.1**).

Approaches for the delivery of siRNA in vivo include intravascular, intraperitoneal, and intratumoral administration of siRNAs, attractive routes because of their inherent simplicity, or by electroporation of synthetic siRNAs directly into target tissues *(120, 121)*. Intradermal administration has also been used to deliver siRNA in vivo *(122)*. In an effort to improve methods for targeted, tissue-specific delivery, lipid-based strategies for in vivo delivery of RNAi-based applications *(123)* or formulations of siRNA in ligand-targeted nanoparticles with PEGylated PEI and a peptide ligand attached to polyethylene glycol (PEG) to target the tumor neovasculature *(124)* have been developed. Of importance, siRNA molecules transfected into dividing cells become diluted. To tackle this problem, vector-based systems for the stable expression of siRNAs in target cells have been developed (*see* Chap. 3), although their use is limited by low transfection efficiencies. Consequently, viral vector systems for short hairpin (sh)RNA delivery have been developed *(125)*, although there are concerns over the safety of using viral vectors for therapy. Cell-specific targeting of siRNA expression is a key issue when considering RNAi for therapeutic applications in vivo (*see* Chap. 2). The recent development of antibody-directed delivery of siRNA therapeutics by combining the nucleic-acid-binding properties of the small basic protein protamine with antibodies to achieve cell-type-specific siRNA delivery in vivo constitutes a promising approach for clinical application of siRNA therapeutics *(126)*.

In conclusion, siRNA-mediated gene silencing is a powerful tool to enhance understanding of the tumor microenvironment, allowing identification of the key players involved in tumor growth, invasion, and metastasis. Ideally, targeting stromal cells that have been modified by the tumor microenvironment to gain tumor-promoting properties should spare their unmodified counterparts in normal tissues, which is a prerequisite to translate potential therapeutic RNAi strategies targeting the tumor microenvironment (**Table 12.1**) into efficient therapies in the clinic.

Table 12.1
Potential therapeutic targets involved in tumor–stromal cell interactions

Targeted stromal element	Alteration induced by siRNA	siRNA studies in vivo	General preclinical and clinical results
Stromal cells			
Endothelial cells	Blocking endothelial cell proliferation, survival and vessel formation	Blockade of VEGFR-2 inhibits tumor angiogenesis *(124)* Downregulation of CD-31 inhibits tumor growth and reduces metastasis formation *(127)*	Various therapeutic strategies to inhibit VEGF signaling *(90)*
EPCs	Inhibition of EPC recruitment, such as by blocking CXCR4, the receptor for SDF-1		SDF-1/CXCR4 signaling recruits EPCs to tumors and directly enhances tumor cell growth *(23)* Inhibition of the SDF-1/CXCR4 axis decreases the growth of gastrointestinal tumors through the suppression of tumor neoangiogenesis in mice *(122)*
Tumor-associated fibroblasts	Inhibition of SDF-1, cathepsin K *(80)*, or FAP *(81)* expression		FAP potentiates tumor growth in an animal model *(123)* Phase I clinical trials using an anti-FAP antibody were conducted in patients *(24)*
Macrophages	Inhibiting macrophage recruitment, transcriptional activation, or survival factors *(76, 79)*	Blockade of CSF-1 or the CSF-1R suppresses mammary tumor growth and increases survival *(52)*	Reduces tumor malignancy in preclinical models of breast cancer *(14, 79)*
ECM molecules			
Tenascin	Blockade of tenascin synthesis to inhibit tumor growth	Treatment of glioma patients with siRNA against tenascin-C following brain surgery *(120)*	Tenascin C regulates angiogenesis and tumor cell migration in preclinical models of melanoma and breast cancer *(124, 125)*

(continued)

Table 12.1 (continued)

Targeted stromal element	Alteration induced by siRNA	siRNA studies in vivo	General preclinical and clinical results
Hyaluronan	Inhibition of the hyaluronan receptor CD44 *(128)* or the hyaluronan synthases (HAS) in stromal cells and tumor cells *(129)*		Promotes tumor growth in preclinical models of prostate and breast cancer *(129, 126)*
Matrix-degrading proteases			
MMPs	Inhibition in different cell types (stromal cells and cancer cells)	Downregulation of MMP-9 and uPAR decreases breast cancer growth *(121)*	MMPs are key regulators of tumor–host cell interactions *(98)*. MMPs can both promote and inhibit angiogenesis and metastasis and can generate reactive ECM fragments. Negative results in clinical trials involving different treatment combinations *(102)*
EMMPRIN/CD147	Inhibition of expression	Blockade of host CD147 expression suppresses tumor growth in human colon cancer xenografts *(69)*	CD147 overexpressing breast cancer cells exhibit accelerated growth and increased invasiveness *(127)*
uPA and cathepsin B	Altering expression levels of uPA, its receptor uPAR, or cathepsin B	Cathepsin B and uPAR siRNAs suppress intracranial tumor growth in gliomas *(128)*	Inhibition of uPAR expression impedes tumor growth and metastasis in murine tumor models *(129)* Cathepsin B has been implicated in progression of various human tumors *(130)*

PAI-1	Altering expression levels	Development of in vivo technology *(131)*	Host-produced PAI is involved in cancer cell invasion and angiogenesis *(132)* PAI-1 expression by stromal fibroblasts and ECs promotes tumor growth and spread *(133)*
Regulatory molecules			
Growth factors and cytokines produced by tumor cells and/or stromal cells (VEGF-A, VEGF-C, VEGF-D, PDGF, EGF, CSF-1, HGF, IGF-1, TNF-a, and others) and their receptors	Altering expression levels resulting in an inhibition of signaling	VEGF inhibition suppresses tumor angiogenesis and tumor growth *(134, 135)* Blockade of TNF-a and CSF-1 expression suppresses tumor growth in human cancer xenografts *(52, 60)*	Various therapeutic strategies to inhibit growth factor signaling *(79, 90,136)*
HIF-1a	Altering expression levels	Silencing of HIF-1 attenuates glioma growth *(137)*	Inhibition of HIF-1a expression suppresses growth of glioma xenografts *(138)*
Chemokines and chemokine receptors	Altering expression levels	Silencing of CXCR4 blocks breast cancer metastasis *(139)*	Chemokine/chemokine receptor interactions in the stromal compartment are involved in angiogenesis, cellular invasion, and metastasis *(140)*

Abbreviations: CSF-1, colony-stimulating factor-1; CSF-1R, CSF-1 receptor; CXCR-4, CXC-motive receptor 4; ECM, extracellular matrix; EGF, epidermal growth factor; EMMPRIN, extracellular matrix metalloproteinase inducer; EPC, endothelial progenitor cell; FAP, fibroblast activated protein; HGF, hepatocyte growth factor; HIF-1 α hypoxia inducible factor-1α; IGF-1, insulin-like growth factor 1; MMP, matrix metalloproteinases; PAI-1, plasminogen activator inhibitor 1; PDGF, platelet-derived growth factor; SDF-1, stromal cell-derived factor 1; TNF-α, tumor necrosis factor-α; uPA, urokinase plasminogen activator; uPAR, uPA receptor; VEGF, vascular endothelial growth factor; VEGF-R2, VEGF receptor 2.

References

1. Farber, E. (1984) The multistep nature of cancer development. *Cancer Res*, 44, 4217–4223.
2. Weinberg, R.A. (1989) Oncogenes, antioncogenes, and the molecular bases of multistep carcinogenesis. *Cancer Res*, 49, 3713–3721.
3. Stetler-Stevenson, W.G., and Yu, A.E. (2001) Proteases in invasion: matrix metalloproteinases. *Semin Cancer Biol*, 12, 143–152.
4. Fidler, I.J. (2003) The pathogenesis of cancer metastasis: the 'seed and soil' hypothesis revisited. *Nat Rev Cancer*, 3, 453–458.
5. Folkman, J. (1995) Angiogenesis in cancer, vascular, rheumatoid and other disease. *Nat Med*, 1, 27–31.
6. Folkman, J. (1995) Angiogenesis inhibitors generated by tumors. *Mol Med*, 1, 120–122.
7. Liotta, L.A., and Kohn, E.C. (2001) The microenvironment of the tumour-host interface. *Nature*, 412, 375–379.
8. Folkman, J., Watson, K., Ingber, D., and Hanahan, D. (1989) Induction of angiogenesis during the transition from hyperplasia to neoplasia. *Nature*, 339, 58–61.
9. Mantovani, A., Bottazzi, B., Colotta, F., Sozzani, S., and Ruco, L. (1992) The origin and function of tumor-associated macrophages. *Immunol Today*, 13, 265–270.
10. Weidner, N., Semple, J.P., Welch, W.R., and Folkman, J. (1991) Tumor angiogenesis and metastasis--correlation in invasive breast carcinoma. *N Engl J Med*, 324, 1–8.
12. Weidner, N., Carroll, P.R., Flax, J., Blumenfeld, W., and Folkman, J. (1993) Tumor angiogenesis correlates with metastasis in invasive prostate carcinoma. *Am J Pathol*, 143, 401–409.
12. Polverini, P.J., Cotran, P.S., Gimbrone, M.A., Jr., and Unanue, E.R. (1977) Activated macrophages induce vascular proliferation. *Nature*, 269, 804–806.
13. Sunderkotter, C., Steinbrink, K., Goebeler, M., Bhardwaj, R., and Sorg, C. (1994) Macrophages and angiogenesis. *J Leukoc Biol*, 55, 410–422.
14. Lin, E.Y., Nguyen, A.V., Russell, R.G., and Pollard, J.W. (2001) Colony-stimulating factor 1 promotes progression of mammary tumors to malignancy. *J Exp Med*, 193, 727–740.
15. Coussens, L.M., and Werb, Z. (1996) Matrix metalloproteinases and the development of cancer. *Chem Biol*, 3, 895–904.
16. Baglole, C.J., Ray, D.M., Bernstein, S.H., Feldon, S.E., Smith, T.J., Sime, P.J., and Phipps, R.P. (2006) More than structural cells, fibroblasts create and orchestrate the tumor microenvironment. *Immunol Invest*, 35, 297–325.
17. Mueller, M.M., and Fusenig, N.E. (2004) Friends or foes - bipolar effects of the tumour stroma in cancer. *Nat Rev Cancer*, 4, 839–849.
18. Gabbiani, G. (2003) The myofibroblast in wound healing and fibrocontractive diseases. *J Pathol*, 200, 500–503.
19. Chang, H.Y., Chi, J.T., Dudoit, S., Bondre, C., van de Rijn, M., Botstein, D., and Brown, P.O. (2002) Diversity, topographic differentiation, and positional memory in human fibroblasts. *Proc Natl Acad Sci U S A*, 99, 12877–12882.
20. Kalluri, R., and Zeisberg, M. (2006) Fibroblasts in cancer. *Nat Rev Cancer*, 6, 392–401.
21. Dimanche-Boitrel, M.T., Vakaet, L., Jr., Pujuguet, P., Chauffert, B., Martin, M.S., Hammann, A., Van Roy, F., Mareel, M., and Martin, F. (1994) In vivo and in vitro invasiveness of a rat colon-cancer cell line maintaining E-cadherin expression: an enhancing role of tumor-associated myofibroblasts. *Int J Cancer*, 56, 512–521.
22. Bhowmick, N.A., Chytil, A., Plieth, D., Gorska, A.E., Dumont, N., Shappell, S., Washington, M.K., Neilson, E.G., and Moses, H.L. (2004a) TGF-beta signaling in fibroblasts modulates the oncogenic potential of adjacent epithelia. *Science*, 303, 848–851.
23. Orimo, A., Gupta, P.B., Sgroi, D.C., Arenzana-Seisdedos, F., Delaunay, T., Naeem, R., Carey, V.J., Richardson, A.L., and Weinberg, R.A. (2005) Stromal fibroblasts present in invasive human breast carcinomas promote tumor growth and angiogenesis through elevated SDF-1/CXCL12 secretion. *Cell*, 121, 335–348.
24. Scott, A.M., Wiseman, G., Welt, S., Adjei, A., Lee, F.T., Hopkins, W., Divgi, C.R., Hanson, L.H., Mitchell, P., Gansen, D.N., Larson, S.M., Ingle, J.N., Hoffman, E.W., Tanswell, P., Ritter, G., Cohen, L.S., Bette, P., Arvay, L., Amelsberg, A., Vlock, D., Rettig, W.J., and Old, L.J. (2003) A Phase I dose-escalation study of sibrotuzumab in patients with advanced or metastatic fibroblast activation protein-positive cancer. *Clin Cancer Res*, 9, 1639–1647.
25. Bhowmick, N.A., Neilson, E.G., and Moses, H.L. (2004) Stromal fibroblasts in cancer initiation and progression. *Nature*, 432, 332–337.

26. Fukumura, D., Xavier, R., Sugiura, T., Chen, Y., Park, E.C., Lu, N., Selig, M., Nielsen, G., Taksir, T., Jain, R.K., and Seed, B. (1998) Tumor induction of VEGF promoter activity in stromal cells. *Cell*, 94, 715–725.
27. Carmeliet, P. (2000) Mechanisms of angiogenesis and arteriogenesis. *Nat Med*, 6, 389–395.
28. Goon, P.K., Lip, G.Y., Boos, C.J., Stonelake, P.S., and Blann, A.D. (2006) Circulating endothelial cells, endothelial progenitor cells, and endothelial microparticles in cancer. *Neoplasia*, 8, 79–88.
29. Lyden, D., Hattori, K., Dias, S., Costa, C., Blaikie, P., Butros, L., Chadburn, A., Heissig, B., Marks, W., Witte, L., Wu, Y., Hicklin, D., Zhu, Z., Hackett, N.R., Crystal, R.G., Moore, M.A., Hajjar, K.A., Manova, K., Benezra, R., and Rafii, S. (2001) Impaired recruitment of bone-marrow-derived endothelial and hematopoietic precursor cells blocks tumor angiogenesis and growth. *Nat Med*, 7, 1294–1201.
30. Rafii, S., Meeus, S., Dias, S., Hattori, K., Heissig, B., Shmelkov, S., Rafii, D., and Lyden, D. (2002) Contribution of marrow-derived progenitors to vascular and cardiac regeneration. *Semin Cell Dev Biol*, 13, 61–67.
31. Yamaguchi, J., Kusano, K.F., Masuo, O., Kawamoto, A., Silver, M., Murasawa, S., Bosch-Marce, M., Masuda, H., Losordo, D.W., Isner, J.M., and Asahara, T. (2003) Stromal cell-derived factor-1 effects on ex vivo expanded endothelial progenitor cell recruitment for ischemic neovascularization. *Circulation*, 107, 1322–1328.
32. Aghi, M., Cohen, K.S., Klein, R.J., Scadden, D.T., and Chiocca, E.A. (2006) Tumor stromal-derived factor-1 recruits vascular progenitors to mitotic neovasculature, where microenvironment influences their differentiated phenotypes. *Cancer Res*, 66, 9054–9064.
33. Sherr, C.J., Rettenmier, C.W., Sacca, R., Roussel, M.F., Look, A.T., and Stanley, E.R. (1985) The c-fms proto-oncogene product is related to the receptor for the mononuclear phagocyte growth factor, CSF-1. *Cell*, 41, 665–676.
34. Lopez, M., Martinache, C., Canepa, S., Chokri, M., Scotto, F., and Bartholeyns, J. (1993) Autologous lymphocytes prevent the death of monocytes in culture and promote, as do GM-CSF, IL-3 and M-CSF, their differentiation into macrophages. *J Immunol Methods*, 159, 29–38.
35. Stanley, E.R. (2000) CSF-1. In Oppenheim, J., and Feldmann, M. (eds.), *Cytokine Reference: A Compendium of Cytokines and other Mediators of Host Defence*. Academic press, London, pp. 912–934.
36. James, S.L., Cook, K.W., and Lazdins, J.K. (1990) Activation of human monocyte-derived macrophages to kill schistosomula of *Schistosoma mansoni* in vitro. *J Immunol*, 145, 2686–2690.
37. Eubank, T.D., Galloway, M., Montague, C.M., Waldman, W.J., and Marsh, C.B. (2003) M-CSF induces vascular endothelial growth factor production and angiogenic activity from human monocytes. *J Immunol*, 171, 2637–2643.
38. Roth, P., and Stanley, E.R. (1992) The biology of CSF-1 and its receptor. *Curr Top Microbiol Immunol*, 181, 141–167.
39. Yeung, Y.G., and Stanley, E.R. (2003) Proteomic Approaches to the Analysis of Early Events in Colony-stimulating Factor-1 Signal Transduction. *Mol Cell Proteomics*, 2, 1243–1255.
40. Pixley, F.J., and Stanley, E.R. (2004) CSF-1 regulation of the wandering macrophage: complexity in action. *Trends Cell Biol*, 14, 628–638.
41. Lin, E.Y., Gouon-Evans, V., Nguyen, A.V., and Pollard, J.W. (2002) The macrophage growth factor CSF-1 in mammary gland development and tumor progression. *J Mammary Gland Biol Neoplasia*, 7, 147–162.
42. Bast, R.C., Jr., Boyer, C.M., Jacobs, I., Xu, F.J., Wu, S., Wiener, J., Kohler, M., and Berchuck, A. (1993) Cell growth regulation in epithelial ovarian cancer. *Cancer*, 71, 1597–1601.
43. Baiocchi, G., Kavanagh, J.J., Talpaz, M., Wharton, J.T., Gutterman, J.U., and Kurzrock, R. (1991) Expression of the macrophage colony-stimulating factor and its receptor in gynecologic malignancies. *Cancer*, 67, 990–996.
44. Lidor, Y.J., Xu, F.J., Martinez-Maza, O., Olt, G.J., Marks, J.R., Berchuck, A., Ramakrishnan, S., Berek, J.S., and Bast, R.C., Jr. (1993) Constitutive production of macrophage colony-stimulating factor and interleukin-6 by human ovarian surface epithelial cells. *Exp Cell Res*, 207, 332–339.
45. Kacinski, B.M., Chambers, S.K., Stanley, E.R., Carter, D., Tseng, P., Scata, K.A., Chang, D.H., Pirro, M.H., Nguyen, J.T., Ariza, A.et-al. and (1990) The cytokine CSF-1 (M-CSF) expressed by endometrial carcinomas in vivo and in vitro, may also be a circulating tumor marker of neoplastic disease activity in endometrial carcinoma

patients. *Int J Radiat Oncol Biol Phys*, 19, 619–626.
46. Suzuki, M., Ohwada, M., Aida, I., Tamada, T., Hanamura, T. and Nagatomo, M. (1993) Macrophage colony-stimulating factor as a tumor marker for epithelial ovarian cancer. *Obstet Gynecol*, 82, 946–950.
47. Suzuki, M., Kobayashi, H., Ohwada, M., Terao, T., and Sato, I. (1998) Macrophage colony-stimulating factor as a marker for malignant germ cell tumors of the ovary. *Gynecol Oncol*, 68, 35–37.
48. Yoshida, H., Hayashi, S., Kunisada, T., Ogawa, M., Nishikawa, S., Okamura, H., Sudo, T., and Shultz, L.D. (1990) The murine mutation osteopetrosis is in the coding region of the macrophage colony stimulating factor gene. *Nature*, 345, 442–444.
49. Nowicki, A., Szenajch, J., Ostrowska, G., Wojtowicz, A., Wojtowicz, K., Kruszewski, A.A., Maruszynski, M., Aukerman, S.L., and Wiktor-Jedrzejczak, W. (1996) Impaired tumor growth in colony-stimulating factor 1 (CSF-1)-deficient, macrophage-deficient op/op mouse: evidence for a role of CSF-1-dependent macrophages in formation of tumor stroma. *Int J Cancer*, 65, 122–129.
50. Pei, X.H., Nakanishi, Y., Takayama, K., Bai, F., and Hara, N. (1999) Granulocyte, granulocyte-macrophage, and macrophage colony-stimulating factors can stimulate the invasive capacity of human lung cancer cells. *Br J Cancer*, 79, 40–46.
51. Stanley, E. (1992) Colony-Stimulating Factor-1. In Aggarwal, B. and Gutterman, J. (eds.), *Human Cytokines*. Blackwell, Boston, pp. 196–220.
52. Aharinejad, S., Paulus, P., Sioud, M., Hofmann, M., Zins, K., Schafer, R., Stanley, E.R., and Abraham, D. (2004) Colony-stimulating factor-1 blockade by antisense oligonucleotides and small interfering RNAs suppresses growth of human mammary tumor xenografts in mice. *Cancer Res*, 64, 5378–5384.
53. Russo, J., and Russo, I.H. (2001) The pathway of neoplastic transformation of human breast epithelial cells. *Radiat Res*, 155, 151–154.
54. Aharinejad, S., Abraham, D., Paulus, P., Abri, H., Hofmann, M., Grossschmidt, K., Schafer, R., Stanley, E.R., and Hofbauer, R. (2002) Colony-stimulating factor-1 antisense treatment suppresses growth of human tumor xenografts in mice. *Cancer Res*, 62, 5317–5324.
55. Cox, G.W., Melillo, G., Chattopadhyay, U., Mullet, D., Fertel, R.H., and Varesio, L. (1992) Tumor necrosis factor-alpha-dependent production of reactive nitrogen intermediates mediates IFN-gamma plus IL-2-induced murine macrophage tumoricidal activity. *J Immunol*, 149, 3290–3296.
56. Lejeune, F.J., Ruegg, C., and Lienard, D. (1998) Clinical applications of TNF-alpha in cancer. *Curr Opin Immunol*, 10, 573–580.
57. Balkwill, F. (2002) Tumor necrosis factor or tumor promoting factor? *Cytokine Growth Factor Rev*, 13, 135–141.
58. Saren, P., Welgus, H.G., and Kovanen, P.T. (1996) TNF-alpha and IL-1beta selectively induce expression of 92-kDa gelatinase by human macrophages. *J Immunol*, 157, 4159–4165.
59. Oster, W., Lindemann, A., Horn, S., Mertelsmann, R., and Herrmann, F. (1987) Tumor necrosis factor (TNF)-alpha but not TNF-beta induces secretion of colony stimulating factor for macrophages (CSF-1) by human monocytes. *Blood*, 70, 1700–1703.
60. Zins, K., Abraham, D., Sioud, M., and Aharinejad, S. (2007) Colon cancer cell-derived tumor necrosis factor-alpha mediates the tumor growth-promoting response in macrophages by up-regulating the colony-stimulating factor-1 pathway. *Cancer Res*, 67, 1038–1045.
61. Mroczko, B., Groblewska, M., Wereszczynska-Siemiatkowska, U., Okulczyk, B., Kedra, B., Laszewicz, W., Dabrowski, A., and Szmitkowski, M. (2007) Serum macrophage-colony stimulating factor levels in colorectal cancer patients correlate with lymph node metastasis and poor prognosis. *Clin Chim Acta*, 380, 208–212.
62. Kaminska, J., Nowacki, M.P., Kowalska, M., Rysinska, A., Chwalinski, M., Fuksiewicz, M., Michalski, W., and Chechlinska, M. (2005) Clinical significance of serum cytokine measurements in untreated colorectal cancer patients: soluble tumor necrosis factor receptor type I--an independent prognostic factor. *Tumour Biol*, 26, 186–194.
63. Hagemann, T., Robinson, S.C., Schulz, M., Trumper, L., Balkwill, F.R., and Binder, C. (2004) Enhanced invasiveness of breast cancer cell lines upon co-cultivation with macrophages is due to TNF-alpha dependent up-regulation of matrix metalloproteases. *Carcinogenesis*, 25, 1543–1549.
64. Brown, P.D., Levy, A.T., Margulies, I.M., Liotta, L.A., and Stetler-Stevenson, W.G. (1990) Independent expression and cellular processing of Mr 72,000 type IV collagenase and interstitial collagenase in human tumorigenic cell lines. *Cancer Res*, 50, 6184–6191.

65. Overall, C.M., Wrana, J.L., and Sodek, J. (1991) Transcriptional and post-transcriptional regulation of 72-kDa gelatinase/type IV collagenase by transforming growth factor-beta 1 in human fibroblasts. Comparisons with collagenase and tissue inhibitor of matrix metalloproteinase gene expression. *J Biol Chem*, 266, 14064–14071.
66. Biswas, C., Zhang, Y., DeCastro, R., Guo, H., Nakamura, T., Kataoka, H., and Nabeshima, K. (1995) The human tumor cell-derived collagenase stimulatory factor (renamed EMMPRIN) is a member of the immunoglobulin superfamily. *Cancer Res*, 55, 434–439.
67. Kanekura, T., Chen, X., and Kanzaki, T. (2002) Basigin (CD147) is expressed on melanoma cells and induces tumor cell invasion by stimulating production of matrix metalloproteinases by fibroblasts. *Int J Cancer*, 99, 520–528.
68. Kataoka, H., DeCastro, R., Zucker, S., and Biswas, C. (1993) Tumor cell-derived collagenase-stimulatory factor increases expression of interstitial collagenase, stromelysin, and 72-kDa gelatinase. *Cancer Res*, 53, 3154–3158.
69. Abraham, D., Zins, K., Sioud, M., Lucas, T., and Aharinejad, S. (2008) Host CD147 blockade by small interfering RNAs suppresses growth of human colon cancer xenografts. *Front Biosci*, 13, 5571–5579.
70. Bergers, G., and Benjamin, L.E. (2003) Tumorigenesis and the angiogenic switch. *Nat Rev Cancer*, 3, 401–410.
71. Coussens, L.M., and Werb, Z. (2002) Inflammation and cancer. *Nature*, 420, 860–867.
72. Kurose, K., Gilley, K., Matsumoto, S., Watson, P.H., Zhou, X.P., and Eng, C. (2002) Frequent somatic mutations in PTEN and TP53 are mutually exclusive in the stroma of breast carcinomas. *Nat Genet*, 32, 355–357.
73. Moinfar, F., Man, Y.G., Arnould, L., Bratthauer, G.L., Ratschek, M., and Tavassoli, F.A. (2000) Concurrent and independent genetic alterations in the stromal and epithelial cells of mammary carcinoma: implications for tumorigenesis. *Cancer Res*, 60, 2562–2566.
74. Orimo, A., and Weinberg, R.A. (2006) Stromal fibroblasts in cancer: a novel tumor-promoting cell type. *Cell Cycle*, 5, 1597–1601.
75. Joyce, J.A. (2005) Therapeutic targeting of the tumor microenvironment. *Cancer Cell*, 7, 513–520.
76. Martinez, F.O., Sica, A., Mantovani, A., and Locati, M. (2008) Macrophage activation and polarization. *Front Biosci*, 13, 453–461.
77. McDermott, R.S., Deneux, L., Mosseri, V., Vedrenne, J., Clough, K., Fourquet, A., Rodriguez, J., Cosset, J.M., Sastre, X., Beuzeboc, P., Pouillart, P., and Scholl, S.M. (2002) Circulating macrophage colony stimulating factor as a marker of tumour progression. *Eur Cytokine Netw*, 13, 121–127.
78. Goswami, S., Sahai, E., Wyckoff, J.B., Cammer, M., Cox, D., Pixley, F.J., Stanley, E.R., Segall, J.E., and Condeelis, J.S. (2005) Macrophages promote the invasion of breast carcinoma cells via a colony-stimulating factor-1/epidermal growth factor paracrine loop. *Cancer Res*, 65, 5278–5283.
79. Pollard, J.W. (2004) Tumour-educated macrophages promote tumour progression and metastasis. *Nat Rev Cancer*, 4, 71–78.
80. Allinen, M., Beroukhim, R., Cai, L., Brennan, C., Lahti-Domenici, J., Huang, H., Porter, D., Hu, M., Chin, L., Richardson, A., Schnitt, S., Sellers, W.R., and Polyak, K. (2004) Molecular characterization of the tumor microenvironment in breast cancer. *Cancer Cell*, 6, 17–32.
81. Park, J.E., Lenter, M.C., Zimmermann, R.N., Garin-Chesa, P., Old, L.J., and Rettig, W.J. (1999) Fibroblast activation protein, a dual specificity serine protease expressed in reactive human tumor stromal fibroblasts. *J Biol Chem*, 274, 36505–36512.
82. Sato, T., Sakai, T., Noguchi, Y., Takita, M., Hirakawa, S., and Ito, A. (2004) Tumor-stromal cell contact promotes invasion of human uterine cervical carcinoma cells by augmenting the expression and activation of stromal matrix metalloproteinases. *Gynecol Oncol*, 92, 47–56.
83. De Wever, O., Nguyen, Q.D., Van Hoorde, L., Bracke, M., Bruyneel, E., Gespach, C., and Mareel, M. (2004) Tenascin-C and SF/HGF produced by myofibroblasts in vitro provide convergent pro-invasive signals to human colon cancer cells through RhoA and Rac. *FASEB J*, 18, 1016–1018.
84. Li, G., Satyamoorthy, K., Meier, F., Berking, C., Bogenrieder, T., and Herlyn, M. (2003) Function and regulation of melanoma-stromal fibroblast interactions: when seeds meet soil. *Oncogene*, 22, 3162–3171.
85. Lewis, M.P., Lygoe, K.A., Nystrom, M.L., Anderson, W.P., Speight, P.M., Marshall, J.F., and Thomas, G.J. (2004) Tumour-derived TGF-beta1 modulates myofibroblast differentiation and promotes HGF/SF-dependent invasion of squamous carcinoma cells. *Br J Cancer*, 90, 822–832.
86. Orimo, A., Tomioka, Y., Shimizu, Y., Sato, M., Oigawa, S., Kamata, K., Nogi, Y., Inoue,

S., Takahashi, M., Hata, T., and Muramatsu, M. (2001) Cancer-associated myofibroblasts possess various factors to promote endometrial tumor progression. *Clin Cancer Res*, 7, 3097–3105.

87. Micke, P., and Ostman, A. (2005) Exploring the tumour environment: cancer-associated fibroblasts as targets in cancer therapy. *Expert Opin Ther Targets*, 9, 1217–1233.
88. Carmeliet, P. (2003) Angiogenesis in health and disease. *Nat Med*, 9, 653–660.
89. Rafii, S., Lyden, D., Benezra, R., Hattori, K., and Heissig, B. (2002) Vascular and haematopoietic stem cells: novel targets for anti-angiogenesis therapy? *Nat Rev Cancer*, 2, 826–835.
90. Kowanetz, M., and Ferrara, N. (2006) Vascular endothelial growth factor signaling pathways: therapeutic perspective. *Clin Cancer Res*, 12, 5018–5022.
91. Shibuya, M., and Claesson-Welsh, L. (2006) Signal transduction by VEGF receptors in regulation of angiogenesis and lymphangiogenesis. *Exp Cell Res*, 312, 549–560.
92. Cao, R., Bjorndahl, M.A., Religa, P., Clasper, S., Garvin, S., Galter, D., Meister, B., Ikomi, F., Tritsaris, K., Dissing, S., Ohhashi, T., Jackson, D.G., and Cao, Y. (2004) PDGF-BB induces intratumoral lymphangiogenesis and promotes lymphatic metastasis. *Cancer Cell*, 6, 333–345.
93. Fidler, I.J., and Ellis, L.M. (1994) The implications of angiogenesis for the biology and therapy of cancer metastasis. *Cell*, 79, 185–188.
94. Bergom, C., Gao, C., and Newman, P.J. (2005) Mechanisms of PECAM-1-mediated cytoprotection and implications for cancer cell survival. *Leuk Lymphoma*, 46, 1409–1421.
95. Balkwill, F. (2004) Cancer and the chemokine network. *Nat Rev Cancer*, 4, 540–550.
96. Staller, P., Sulitkova, J., Lisztwan, J., Moch, H., Oakeley, E.J., and Krek, W. (2003) Chemokine receptor CXCR4 downregulated by von Hippel-Lindau tumour suppressor pVHL. *Nature*, 425, 307–312.
97. Kalluri, R. (2003) Basement membranes: structure, assembly and role in tumour angiogenesis. *Nat Rev Cancer*, 3, 422–433.
98. Noel, A., Jost, M., and Maquoi, E. (2007) Matrix metalloproteinases at cancer tumor-host interface. *Semin Cell Dev Biol.* 79, 52–60.
99. Toole, B.P. (2004) Hyaluronan: from extracellular glue to pericellular cue. *Nat Rev Cancer*, 4, 528–539.
100. Bissell, M.J., and Radisky, D. (2001) Putting tumours in context. *Nat Rev Cancer*, 1, 46–54.
101. Andreasen, P.A., Kjoller, L., Christensen, L., and Duffy, M.J. (1997) The urokinase-type plasminogen activator system in cancer metastasis: a review. *Int J Cancer*, 72, 1–22.
102. Deryugina, E.I., and Quigley, J.P. (2006) Matrix metalloproteinases and tumor metastasis. *Cancer Metastasis Rev*, 25, 9–34.
103. Creemers, L.B., Hoeben, K.A., Jansen, D.C., Buttle, D.J., Beertsen, W., and Everts, V. (1998) Participation of intracellular cysteine proteinases, in particular cathepsin B, in degradation of collagen in periosteal tissue explants. *Matrix Biol*, 16, 575–584.
104. Aigner, A. (2006) Gene silencing through RNA interference (RNAi) in vivo: strategies based on the direct application of siRNAs. *J Biotechnol*, 124, 12–25.
105. de Fougerolles, A., Manoharan, M., Meyers, R., and Vornlocher, H.P. (2005) RNA interference in vivo: toward synthetic small inhibitory RNA-based therapeutics. *Methods Enzymol*, 392, 278–296.
106. Urban-Klein, B., Werth, S., Abuharbeid, S., Czubayko, F., and Aigner, A. (2005) RNAi-mediated gene-targeting through systemic application of polyethyleneimine (PEI)-complexed siRNA in vivo. *Gene Ther*, 12, 461–466.
107. Werth, S., Urban-Klein, B., Dai, L., Hobel, S., Grzelinski, M., Bakowsky, U., Czubayko, F., and Aigner, A. (2006) A low molecular weight fraction of polyethyleneimine (PEI) displays increased transfection efficiency of DNA and siRNA in fresh or lyophilized complexes. *J Control Release*, 122, 257–270.
108. Minakuchi, Y., Takeshita, F., Kosaka, N., Sasaki, H., Yamamoto, Y., Kouno, M., Honma, K., Nagahara, S., Hanai, K., Sano, A., Kato, T., Terada, M., and Ochiya, T. (2004) Atelocollagen-mediated synthetic small interfering RNA delivery for effective gene silencing in vitro and in vivo. *Nucleic Acids Res*, 32, e109.
109. Soutschek, J., Akinc, A., Bramlage, B., Charisse, K., Constien, R., Donoghue, M., Elbashir, S., Geick, A., Hadwiger, P., Harborth, J., John, M., Kesavan, V., Lavine, G., Pandey, R.K., Racie, T., Rajeev, K.G., Rohl, I., Toudjarska, I., Wang, G., Wuschko, S., Bumcrot, D., Koteliansky, V., Limmer, S., Manoharan, M., and Vornlocher, H.P. (2004) Therapeutic silencing of an endogenous gene by systemic administration of modified siRNAs. *Nature*, 432, 173–178.
110. Dillon, C.P., Sandy, P., Nencioni, A., Kissler, S., Rubinson, D.A., and Van Parijs, L. (2005) RNAi as an experimental and thera-

peutic tool to study and regulate physiological and disease processes. *Annu Rev Physiol*, 67, 147–173.

111. Leung, R.K., and Whittaker, P.A. (2005) RNA interference: from gene silencing to gene-specific therapeutics. *Pharmacol Ther*, 107, 222–239.

112. Kim, B., Tang, Q., Biswas, P.S., Xu, J., Schiffelers, R.M., Xie, F.Y., Ansari, A.M., Scaria, P.V., Woodle, M.C., Lu, P., and Rouse, B.T. (2004) Inhibition of ocular angiogenesis by siRNA targeting vascular endothelial growth factor pathway genes: therapeutic strategy for herpetic stromal keratitis. *Am J Pathol*, 165, 2177–2185.

113. Sioud, M., and Sørensen, D.R. (2004) Systemic delivery of synthetic siRNAs. *Methods Mol Biol*, 252, 515–522.

114. Schiffelers, R.M., Ansari, A., Xu, J., Zhou, Q., Tang, Q., Storm, G., Molema, G., Lu, P.Y., Scaria, P.V., and Woodle, M.C. (2004) Cancer siRNA therapy by tumor selective delivery with ligand-targeted sterically stabilized nanoparticle. *Nucleic Acids Res*, 32, e149.

115. Arts, G.J., Langemeijer, E., Tissingh, R., Ma, L., Pavliska, H., Dokic, K., Dooijes, R., Mesic, E., Clasen, R., Michiels, F., van der Schueren, J., Lambrecht, M., Herman, S., Brys, R., Thys, K., Hoffmann, M., Tomme, P., and van Es, H. (2003) Adenoviral vectors expressing siRNAs for discovery and validation of gene function. *Genome Res*, 13, 2325–2332.

116. Song, E., Zhu, P., Lee, S.K., Chowdhury, D., Kussman, S., Dykxhoorn, D.M., Feng, Y., Palliser, D., Weiner, D.B., Shankar, P., Marasco, W.A., and Lieberman, J. (2005) Antibody mediated in vivo delivery of small interfering RNAs via cell-surface receptors. *Nat Biotechnol*, 23, 709–717.

117. Santel, A., Aleku, M., Keil, O., Endruschat, J., Esche, V., Durieux, B., Loffler, K., Fechtner, M., Rohl, T., Fisch, G., Dames, S., Arnold, W., Giese, K., Klippel, A., and Kaufmann, J. (2006) RNA interference in the mouse vascular endothelium by systemic administration of siRNA-lipoplexes for cancer therapy. *Gene Ther*, 13, 1360–1370.

118. Tzircotis, G., Thorne, R.F., and Isacke, C.M. (2005) Chemotaxis towards hyaluronan is dependent on CD44 expression and modulated by cell type variation in CD44-hyaluronan binding. *J Cell Sci*, 128, 5129–5128.

119. Simpson, M.A., Wilson, C.M., and McCarthy, J.B. (2002) Inhibition of prostate tumor cell hyaluronan synthesis impairs subcutaneous growth and vascularization in immunocompromised mice. *Am J Pathol*, 161, 849–857.

120. Zukiel, R., Nowak, S., Wyszko, E., Rolle, K., Gawronska, I., Barciszewska, M.Z., and Barciszewski, J. (2006) Suppression of human brain tumor with interference RNA specific for tenascin-C. *Cancer Biol Ther*, 5, 1002–1007.

121. Kunigal, S., Lakka, S.S., Gondi, C.S., Estes, N., and Rao, J.S. (2007) RNAi-mediated downregulation of urokinase plasminogen activator receptor and matrix metalloprotease-9 in human breast cancer cells results in decreased tumor invasion, angiogenesis and growth. *Int J Cancer*, 121, 2307–2316.

122. Guleng, B., Tateishi, K., Ohta, M., Kanai, F., Jazag, A., Ijichi, H., Tanaka, Y., Washida, M., Morikane, K., Fukushima, Y., Yamori, T., Tsuruo, T., Kawabe, T., Miyagishi, M., Taira, K., Sata, M., and Omata, M. (2005) Blockade of the stromal cell-derived factor-1/CXCR4 axis attenuates in vivo tumor growth by inhibiting angiogenesis in a vascular endothelial growth factor-independent manner. *Cancer Res*, 65, 5864–5871.

123. Cheng, J.D., Dunbrack, R.L., Jr., Valianou, M., Rogatko, A., Alpaugh, R.K., and Weiner, L.M. (2002) Promotion of tumor growth by murine fibroblast activation protein, a serine protease, in an animal model. *Cancer Res*, 62, 4767–4772.

124. Kalembeyi, I., Inada, H., Nishiura, R., Imanaka-Yoshida, K., Sakakura, T., and Yoshida, T. (2003) Tenascin-C upregulates matrix metalloproteinase-9 in breast cancer cells: direct and synergistic effects with transforming growth factor beta1. *Int J Cancer*, 105, 53–60.

125. Tanaka, K., Hiraiwa, N., Hashimoto, H., Yamazaki, Y., and Kusakabe, M. (2004) Tenascin-C regulates angiogenesis in tumor through the regulation of vascular endothelial growth factor expression. *Int J Cancer*, 108, 31–40.

126. Shuster, S., Frost, G.I., Csoka, A.B., Formby, B., and Stern, R. (2002) Hyaluronidase reduces human breast cancer xenografts in SCID mice. *Int J Cancer*, 102, 192–197.

127. Zuker, S., Hymowitz, M., Rollo, E.E., Mann, R., Conner, C.E., Cao, J., Foda, H.D., Tompkins, D.C., and Toole, B.P. (2001) Tumorigenic potential of extracellular matrix metalloproteinase inducer. *Am J Pathol*, 158, 1921–1928.

128. Gondi, C.S., Lakka, S.S., Dinh, D.H., Olivero, W.C., Gujrati, M., and Rao, J.S. (2004) RNAi-mediated inhibition of cathepsin

B and uPAR leads to decreased cell invasion, angiogenesis and tumor growth in gliomas. *Oncogene*, 23, 8486–8496.

129. Pillay, V., Dass, C.R., and Choong, P.F. (2007) The urokinase plasminogen activator receptor as a gene therapy target for cancer. *Trends Biotechnol*, 25, 33–39.
130. Yan, S., Sameni, M., and Sloane, B.F. (1998) Cathepsin B and human tumor progression. *Biol Chem*, 379, 123–123.
131. Meryet-Figuieres, M., Resina, S., Lavigne, C., Barlovatz-Meimon, G., Lebleu, B., and Thierry, A.R. (2007) Inhibition of PAI-1 expression in breast cancer carcinoma cells by siRNA at nanomolar range. *Biochimie*, 89, 1228–1233.
132. Bajou, K., Noel, A., Gerard, R.D., Masson, V., Brunner, N., Holst-Hansen, C., Skobe, M., Fusenig, N.E., Carmeliet, P., Collen, D., and Foidart, J.M. (1998) Absence of host plasminogen activator inhibitor 1 prevents cancer invasion and vascularization. *Nat Med*, 4, 923–928.
133. Andreasen, P.A. (2007) PAI-1 - a potential therapeutic target in cancer. *Curr Drug Targets*, 8, 1030–1041.
134. Filleur, S., Courtin, A., Ait-Si-Ali, S., Guglielmi, J., Merle, C., Harel-Bellan, A., Clezardin, P., and Cabon, F. (2003) SiRNA-mediated inhibition of vascular endothelial growth factor severely limits tumor resistance to antiangiogenic thrombospondin-1 and slows tumor vascularization and growth. *Cancer Res*, 63, 3919–3922.
135. Takei, Y., Kadomatsu, K., Yuzawa, Y., Matsuo, S., and Muramatsu, T. (2004) A small interfering RNA targeting vascular endothelial growth factor as cancer therapeutics. *Cancer Res*, 64, 3365–3370.
136. Srinivasan, D.M., Kapoor, M., Kojima, F., and Crofford, L.J. (2005) Growth factor receptors: implications in tumor biology. *Curr Opin Investig Drugs*, 6, 1246–1249.
137. Gillespie, D.L., Whang, K., Ragel, B.T., Flynn, J.R., Kelly, D.A., and Jensen, R.L. (2007) Silencing of hypoxia inducible factor-1alpha by RNA interference attenuates human glioma cell growth in vivo. *Clin Cancer Res*, 13, 2441–2448.
138. Rapisarda, A., Zalek, J., Hollingshead, M., Braunschweig, T., Uranchimeg, B., Bonomi, C.A., Borgel, S.D., Carter, J.P., Hewitt, S.M., Shoemaker, R.H., and Melillo, G. (2004) Schedule-dependent inhibition of hypoxia-inducible factor-1alpha protein accumulation, angiogenesis, and tumor growth by topotecan in U251-HRE glioblastoma xenografts. *Cancer Res*, 64, 6845–6848.
139. Liang, Z., Yoon, Y., Votaw, J., Goodman, M.M., Williams, L., and Shim, H. (2005) Silencing of CXCR4 blocks breast cancer metastasis. *Cancer Res*, 65, 967–971.

Chapter 13

Therapeutic Potential of siRNA-Mediated Targeting of Urokinase Plasminogen Activator, Its Receptor, and Matrix Metalloproteinases

Christopher S. Gondi and Jasti S. Rao

Abstract

Targeting proteases and their activators would retard the invasive ability of cancer cells, and has been shown to induce apoptosis in certain instances. Various methods have been developed to specifically target protease molecules in an attempt to retard invasion and migration. Of these methods, RNA interference (RNAi) holds great therapeutic potential. RNAi technology is now being used to target specific molecules for use as potential anti-cancer agents. RNAi-mediated silencing is almost catalytic when compared to antisense silencing. Of these targets, the uPAR-uPA system and MMPs holds great promise. Targeting uPA/uPAR may provide additive or synergistic treatment benefits if used in combination with conventional therapeutics such as chemotherapy or radiation. Studies point to the fact that specifically targeting MMP-9 or MMP-2 singly or in combination with other proteases could have specific therapeutic implications in the treatment of cancer. In this chapter we discuss the therapeutic potential of siRNA-mediated targeting of the uPAR-uPA system and MMPs as therapeutic agents for the treatment of cancer.

Key words: uPAR, uPA, MMPs, RNAi, siRNAs, gliomas, brain tumors.

1. Introduction

By definition, the term "cancer" encompasses a group of diseases in which cells are aggressive, invasive and/or metastatic. These three malignant properties of cancers differentiate them from benign tumors, which are self-limited in their growth and do not metastasize. Cancers are caused by abnormalities in the genetic material of the transformed cells; cancer-promoting genetic abnormalities may be inherited or randomly acquired through

M. Sioud (ed.), *Methods in Molecular Biology, siRNA and miRNA Gene Silencing, vol. 487*

DOI: 10.1007/978-1-60327-547-7_13

errors in DNA replication. Interactions between carcinogens and the host genome can explain why only some develop cancer after exposure to known carcinogens.

New aspects of the genetics of cancer pathogenesis, such as DNA methylation and microRNAs, are increasingly being recognized as important aspects in cancer development. Genetic abnormalities found in cancer typically affect two general classes of genes: oncogenes and tumor suppressor genes. Oncogenes are often activated in cancer cells and subsequently give those cells new characteristics, such as uncontrolled cell division, apoptotic inhibition, migration and invasion. Tumor suppressor genes are often downregulated or non-functional in cancer cells, thereby resulting in the loss of normal functions in those cells including accurate DNA replication, cell cycle regulation, orientation, tissue adhesion and interaction with protective cells of the immune system.

The ability of cancer cells to migrate and invade surrounding tissues is mediated by molecular interactions of receptors with ligands and various proteases. The most common of these proteases are metalloproteases and serine proteases. Urokinase plasminogen activator (uPA), which activates plasminogen to plasmin, is the most common of these proteases. Targeting of these proteases and their activators would retard the invasive ability of cancer cells and, in some cases, has been shown to induce apoptosis. Various methods have been developed to specifically target these molecules in an attempt to retard invasion and migration. Of these methods, RNA interference (RNAi) holds great therapeutic potential.

Targeted molecular therapies have evolved to concentrate on either post-transcription or post-translation. Post-translational therapies involve the use of specific chemical inhibitors or antibodies, which block or inhibit the activity of the target molecule. This method interferes only at the final step where the involvement of the target molecule is observed. In contrast, post-transcriptional therapies involve targeting of the molecule before the molecule is formed—in other words, the mRNA of the target molecule. This strategy has evolved from antisense techniques to RNAi and includes the possibility of simultaneously targeting multiple molecules (**Fig. 13.1**).

2. What is RNAi?

RNA interference (RNAi; also called "RNA-mediated interference") is a mechanism for RNA-guided regulation of gene expression in which double-stranded ribonucleic acid inhibits the expression of genes with complementary nucleotide sequences. Conserved in most eukaryotic organisms, the RNAi pathway is

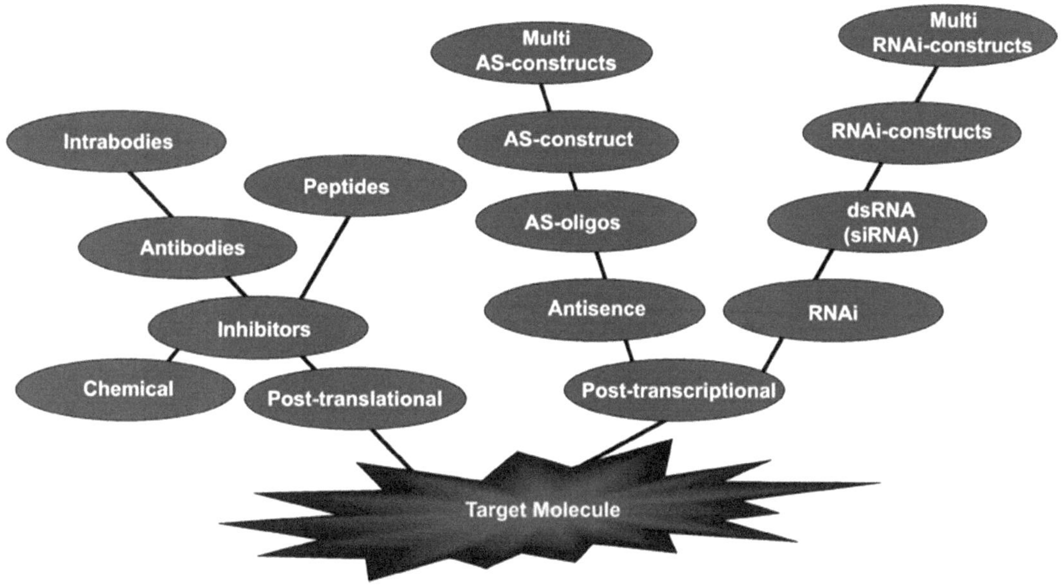

Fig. 13.1. Evolution of targeted molecular therapies. Two approaches have been used to date: post-translation and post-transcription. The post-translational method deals with direct protein or target molecule inhibitors such as chemical drugs, antibodies and inhibitory peptides. The post-transcriptional method deals with targeting the target molecule precursor mRNA; this method includes antisense and RNAi therapies.

thought to have evolved as a form of innate immunity against viruses. This pathway also plays a major role in regulating development and genome maintenance. The RNAi pathway is initiated by the enzyme dicer, which cleaves double-stranded RNA (dsRNA) to short double-stranded fragments of 20–25 base pairs. One of the two strands of each fragment, known as the guide strand, is then incorporated into the RNA-induced silencing complex (RISC) and base pairs with complementary sequences.

The best-studied outcome of this recognition event is a form of post-transcriptional gene silencing. This occurs when the guide strand base pairs with a messenger RNA (mRNA) molecule and induces degradation of the mRNA by argonaute, the catalytic component of the RISC complex. The short RNA fragments are known as small interfering RNA (siRNA), which are perfectly complementary to the gene to which they are suppressing as they are derived from long dsRNA of that same gene or microRNA (miRNA), which are derived from the intragenic regions or an intron and are thus only partially complementary.

The RNAi pathway has been particularly well studied in certain model organisms such as the nematode worm *Caenorhabditis elegans*, the fruit fly *Drosophila melanogaster* and the flowering plant *Arabidopsis thaliana*. RNAi technology is only now being used to target specific molecules for use as potential anti-cancer agents. Of these targets, the uPAR system holds great promise.

3. Advantages of RNAi Over Antisense Therapy

Antisense therapy involves the introduction of single-stranded DNA/RNA complementary to the target native mRNA in question or introduction of a construct (plasmid or viral) that expresses appropriate RNA sequence complementary to the target mRNA driven by an appropriate promoter (e.g., CMV). These complementary DNA/RNA molecules hybridize to the expressed mRNA, thereby blocking the translation of mRNA to proteins. In this case, every molecule of mRNA requires an antisense molecule. Hence, effective silencing is achieved only when equimolar quantities of antisense molecules are present to the corresponding mRNA molecule. In contrast, with RNAi, double-stranded RNA molecules are utilized instead of a single-stranded DNA or RNA molecule. These dsRNA molecules are recycled, thereby effectively silencing multiple mRNA molecules at the same time. RNAi-mediated silencing is almost catalytic when compared to antisense silencing (**Fig. 13.2**).

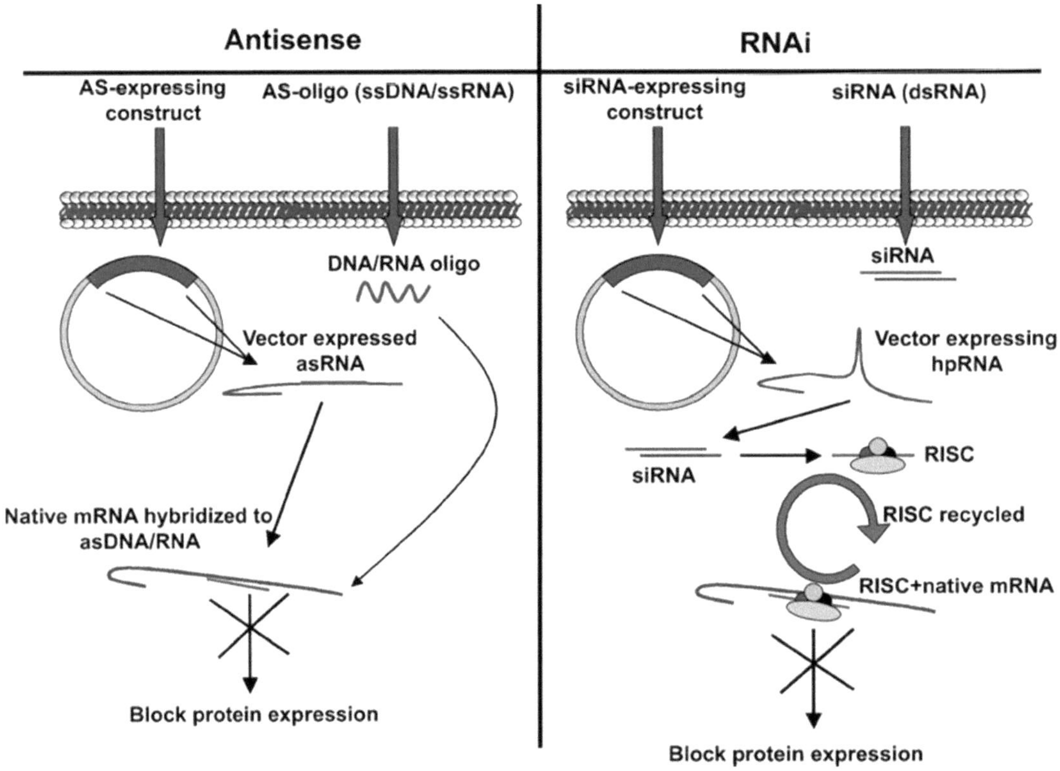

Fig. 13.2. Antisense versus RNAi. Antisense therapy involves the addition of antisense DNA or RNA molecules to block the target mRNA molecule and involves a 1:1 molar ratio of antisense to mRNA molecules. RNAi involves the use of specific siRNA molecules that complex with proteins to form the RNA-induced silencing complex (RISC). In RNAi, one siRNA molecule is capable of targeting multiple mRNA molecules.

4. What Is the uPA-uPAR System?

The uPA-uPAR system is involved in a variety of cell functions including extracellular proteolysis, adhesion, proliferation, chemotaxis, neutrophil priming for oxidant production and cytokine release. All of these processes contribute to tumor development, implantation, angiogenesis, inflammation and metastasis *(1)*. The uPA-uPAR system consists of a serine protease uPA, its cell membrane-associated receptor (uPAR), a substrate plasminogen and plasminogen activator inhibitors (PAI-1 and PAI-2) *(2, 3)*. uPA is produced and secreted as an inactive single-chain polypeptide, called pro-uPA, which lacks plasminogen-activating activity. The binding of pro-uPA to uPAR induces its activation *(3)* which in turn converts plasminogen to the active serine protease plasmin. Plasmin is involved in the degradation of the extracellular matrix (ECM) and basement membranes through direct proteolytic digestion or the activation of other proteases (**Fig. 13.3**) including metalloproteases and collagenases. These processes promote tumor invasion and migration *(4)*.

The binding of uPA to uPAR provides an inducible (**Fig. 13.3**), transient and localized cell surface proteolytic activity *(5)* hereby enabling focused proteolysis of the ECM. Several studies have suggested that uPAR may play a more significant role in the metastatic process *(5)*. Most of the activities of uPA, including its activation by plasmin, are dependent on its binding to uPAR *(2, 3)*. uPAR protein is heavily glycosylated and is covalently linked to the outer layer of the cell membrane via a glycosyl

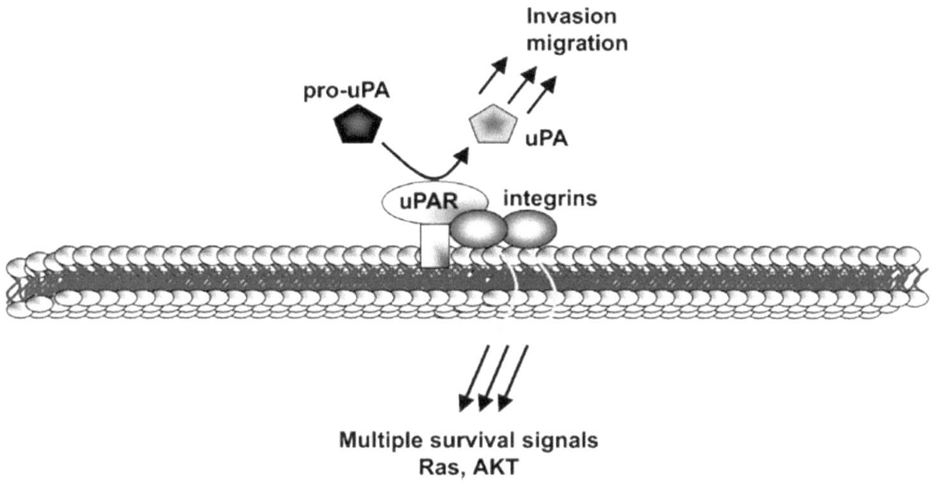

Fig. 13.3. The uPAR system. The uPAR-uPA system deals with uPA and its receptor uPAR. Pro-uPA is activated to active uPA, which in turn, activates plasminogen to plasmin, thereby activating multiple proteases. uPAR is known to be associated with integrins, which initiate multiple survival intracellular signals in cancer cells.

phosphatidylinositol (GPI) anchor with no transmembrane domain *(6, 7)*. uPAR is a >60 kDa glycoprotein, which consists of three cysteine-rich domains (D1, D2 and D3) connected by short linker regions *(8)*. The amino-terminus of the uPAR domain 1 is the primary site for the binding of uPA. However, domains 2 and 3 may be important for high affinity binding of uPA as purified domain 1 has a 1,500-fold lower ligand affinity than the complete tri-domain uPAR *(7)*.

uPAR is a multifunctional protein and is believed to play a role in the regulation of several physiological and pathological conditions that exploit cell adhesion and migration including wound healing, neutrophil recruitment during inflammation as well as tumor invasion and metastasis *(8, 9)*. Several recent studies have shown that the various functions of uPAR are invoked not only by proteolysis but also by intracellular signaling *(3, 10, 11, 12)*. uPAR levels have been strongly correlated with metastatic potential and advanced disease as has been demonstrated in tumor samples obtained from patients with colon and breast cancers *(13–15)*. For example, uPAR is overexpressed in invasive breast cancer tissues, but not in normal and benign breast tumors *(16)*. The importance of the uPAR system makes it a potential target for cancer therapy.

5. Therapeutic RNAi-Mediated Strategies for Targeting the uPAR-uPA System

Several approaches have been employed to target uPAR as a means of cancer therapy. These approaches include small molecule and peptide antagonists of the uPA-uPAR interactions as well as the uPAR interactions that are downstream of uPA binding; these include antibiotics, monoclonal antibodies and antisense technology. To date, antisense technologies used against uPAR include either the classic antisense oligodeoxynucleotides technology, which consists of the injection of antisense DNA strands complementary to uPAR mRNA, or the antisense RNA technology based on cell transfection with a vector capable of expressing the antisense transcript complementary to uPAR mRNA. Research groups investigating antisense RNA technology for downregulation of uPAR *in vivo* have employed both plasmid and adenovirus constructs for this purpose *(17–26)*.

RNAi has provided new avenues for the treatment of cancer. Small interfering RNAs (siRNAs) are believed to be more potent inhibitors of gene expression with less toxicity *(27)*. Our laboratory had already employed shRNA-based RNAi plasmid system for the downregulation of uPAR in prostate cancer *(28)*, glioma *(11–33)* and meningioma *(34, 35)*. We have utilized a plasmid construct expressing the same small hairpin RNA (shRNA) to target uPAR. For the above-referenced studies, human glioblastoma cells were intracranially injected into athymic nude mice. Eight to ten days after tumor growth, mice were implanted with mini-osmotic pumps with a sustained release of 0.25 μl/h of 150 μg of the shRNA-expressing plasmid construct in a subcutaneous sac with a catheter to the intracranial tumor site. The mice were sacrificed and analyzed at the end of the 5-week follow up period or when the control mice started showing symptoms. We reported a 65% regression of pre-established intracranial tumor growth *(23)*. These findings were further confirmed in our laboratory where we reported a 70% inhibition of pre-established intracranial tumor growth *(26)*. Pulukuri et al. *(28)* reported that intratumoral injection with a plasmid construct expressing shRNA for uPAR resulted in partial reduction of pre-established orthotopic prostate tumor in athymic male nude mice with no observable secondary tumor.

Downregulation of more than one component involved in tumor invasion and metastasis may possibly have a synergistic or additive effect in impeding tumor dissemination. We have reported that intracranial injection of human glioma cells infected with a bicistronic adenoviral construct capable of simultaneously expressing antisense uPAR and antisense matrix metalloproteinase-9 (MMP-9) resulted in decreased invasiveness and tumorigenicity in mice *(26)*. Further, subcutaneous injections of the bicistronic construct into established tumors caused tumor regression. MMP-9 is involved with metastasis of various types of cancers, though its inhibition has not led to significant improvements in clinical trials as yet. We therefore hypothesized that a dual targeted approach combining MMP-9 downregulation with that of uPAR has the potential for efficient tumor targeting. Indeed, we found that a bicistronic plasmid construct expressing shRNA simultaneously targeting both uPAR and MMP-9 resulted in total regression of pre-established intracerebral tumor growth in mice.

We have also shown that RNAi-mediated downregulation of uPAR and cathepsin B reduced glioma cell invasion and angiogenesis in *in vivo* models *(29)*. In addition, intratumoral injections of these plasmid vectors expressing shRNA for uPAR and cathepsin B resulted in the regression of pre-established intracranial tumor. Similarly, we have also demonstrated that intraperitoneal injection of a bicistronic plasmid construct expressing shRNA for uPA and uPAR caused the regression of pre-established, intracranial tumors in mice *(33)*. Despite our success thus far, it should be noted that the delivery of uPAR downregulation constructs, whether plasmid vectors, adeno-viral vectors or synthetic strands, still needs to be assessed appropriately in human systems.

Homozygous uPAR-deficient mice display normal growth and fertility, do not show histological abnormalities in tissues and do not differ from wild-type mice for spontaneous lysis of experimental pulmonary plasma clot. This is very similar to what is also noted in uPA-deficient mice. Thus, the apparent lack of toxicity from inhibiting this proteolytic system makes it an ideal candidate for targeting as a cancer therapeutic agent. The uPA-uPAR system plays a very important role in cancer metastasis and may function via a number of signaling pathways. Binding of uPA with its receptor uPAR can activate downstream signaling molecules, including the mitogen-activated protein kinase, signal transducer and activator of transcription (Stat) and the Ras/extracellular signal-regulated kinase pathway, which in turn, lead to cell proliferation, migration and invasion *(3)*. uPA–uPAR-mediated signaling can upregulate the production of MMPs, which induce ECM degradation and, in turn, tumor invasion and metastasis *(3)*. Since uPA-uPAR and their downstream signaling pathways are implicated in many cancers, including essential cellular functions that contribute to the malignancy of tumor cells, targeting uPA–uPAR-mediated signaling pathways may be promising for the treatment of metastatic disease. In addition, targeting uPA/uPAR may provide additive or synergistic treatment benefits if used in combination with conventional therapeutics such as chemotherapy or radiation.

6. Potential for Targeting MMPs

Matrix metalloproteinases (MMPs) are a family of structurally related and highly conserved zinc-dependent endopeptidases collectively capable of degrading most components of the basement membrane and ECM. MMP substrates also include a wide variety of proteins, such as chemotactic molecules, adhesion molecules, proteinase inhibitors, cell-surface receptors, blood clotting factors, latent growth factors and growth factor-binding proteins. Most human MMPs can be divided according to their sequence homology, substrate specificity and cellular location into the following subclasses: collagenases, gelatinases, stromelysins, matrilysins, membrane-type MMPs and others *(36)*. The basic multi-domain structure of MMPs comprises the following: (1) an amino-terminal domain, (2) a catalytic domain and (3) a carboxy-terminal domain. To date, we know of a minimum of 25 secreted or membrane-bound human MMPs. The expression, secretion and activity of MMPs in normal tissues are subject to tight control.

Data generated from intensive studies on MMP activities in different cells and tissues, as well as studies from knock-out animals, illustrate the importance of these enzymes in many normal physiologic processes (e.g., embryonic development, bone resorption, angiogenesis and wound healing) and pathologic processes (rheumatoid arthritis, multiple sclerosis, periodontal disease and tumor growth and metastasis) *(37)*. MMPs are secreted as pro-MMPs and then activated by sequential cleavage steps *(38, 39)*, which involve the removal of signal peptide and pro-peptide domains or a change in configuration, which activates the enzymes. MMP expression and proteolytic activity are tightly regulated at three stages: gene transcription, pro-enzyme activation and activity of natural inhibitors (tissue inhibitors of metalloproteinase: TIMPs). The balance between production, activation and inhibition prevents excessive proteolysis or inhibition. Several factors like cytokines, growth factors, phorbol esters, cell-cell and cell-matrix interactions are thought to control MMP expression *(40)*. Most MMPs are secreted as inactive zymogens, which may be proteolytically activated by different proteinases such as other MMPs, plasmin, trypsin, chymotrypsin and cathepsins. Several cell types produce MMPs including monocytes, macrophages, neutrophils *(41, 42)*, T-lymphocytes *(43)*, endothelial cells *(44)*, fibroblasts *(45)* as well as microglia, astrocytes, oligodendrocytes and neurons in the CNS *(46–48)*. In particular, MMP-2 and MMP-9 are secreted by microglia and astrocytes as active forms *(49)*.

MMPs have been shown to regulate tumor cell invasion through their interactions with extracellular matrix components including cell matrix embedded growth factors and cell adhesion molecules *(15, 50)*. The most important of these metalloproteases are MMP-9 and MMP-2, which have shown to be involved in glioma invasion and angiogenesis *(51, 52)*. MMPs are controlled by enzyme activation to produce a functional form and at the level of gene expression *(53)*. There are also other underlying mechanisms that affect mRNA stability, protein secretion and specific degradation and clearance *(53)*. Growth factors, such as endothelial growth factor (EGF), basic fibroblast growth factor (b-FGF), transforming growth factor (TGF-β1 and β2) and vascular endothelial growth factor (VEGF) have been shown to upregulate MMP-2 and MMP-9 *(54)*. MMP-9 and stromelysin (MMP-3) have been shown to be chiefly transcribed under the influence of various transcription factors commonly found to be involved with cellular stress responses and tissue morphogenesis, including NF-κβ, ETS family members and AP-1 *(55)*. Epidermal growth factor variant subtype III promoted activation of MMP-9, possibly through the activation of MAPK/ERK in glioblastoma *(56)*. We have previously shown that MMP-9 production is induced by cytoskeletal changes involving protein kinase C activation mediated by NF-κβ *(57)*. The mitogen-activated kinase/extracellular signal-regulated kinase (MEK/ERK) signaling pathway is essential for MMP-9 upregulation in astrocytes after PKC induction and TNF-α (cytokine) stimulation. It has been reported that SNB19 glioma cells transfected with dominant negative JNK, MEKK and ERK1 expression vectors decreased MMP-9 expression as well as promoter activity *(58)*. The mt-ERK stable SNB19 cells showed decreased levels of MMP-9 and less invasiveness as compared to parental and vector-transfected stable clones *(25)*. All these studies point to the fact that specifically targeting MMP-9 or MMP-2 singly or in combination with other proteases could have specific therapeutic implications in the treatment of cancer.

7. RNAi-Mediated Strategies for Targeting MMPs

We have previously demonstrated that inhibition of cathepsin B and MMP-9 gene expression via RNA interference reduced tumor cell invasion, tumor growth and angiogenesis in a glioblastoma cell line *(59)*. We have also demonstrated that specific interference of uPAR and MMP-9 gene expression induced by double-stranded RNA resulted in decreased invasion, tumor growth and angiogenesis in gliomas *(52)*. Other researchers have silenced MMP-1 to elucidate the mechanism involved in signaling *(60)*. Wyatt et al. concluded that MMP-1 expression is essential for the ability of MDA-231 cells to invade and destroy a collagen matrix. *In vivo* experiments suggest an important role for MMP-1 in breast tumor growth and have demonstrated the potential of RNAi-mediated targeting of MMP-1 *(61)*. Researchers have also demonstrated that siRNA-mediated blocking of either membrane type-1 MMP (MT1-MMP) or MMP-2 were effective in reducing the hypoxia-induced invasion in MDA-MB-231 and MDA-MB-435 breast carcinoma cell lines *(62)*. Knocking down of MMP-7 by small interfering RNA was shown to suppress lysophosphatidic acid (LPA)-induced invasion in two EOC cell lines (DOV13 and R182). These results show that MMP-7 expression is correlated with EOC invasiveness and LPA-induced MMP-7 secretion/activation and may represent new mechanisms that facilitate ovarian cancer invasion besides the well-known induction of MT1-MMP-mediated pro-MMP-2 activation by LPA *(63)*.

Our studies have reported that the simultaneous targeting of two or more components involved in invasion or migration is significantly more relevant therapeutically than concentrating on one component alone. For example, RNAi-mediated targeting of uPAR and MMP-9 gene expression in the IOMM-LEE malignant meningioma cell line inhibited tumor growth, tumor cell invasion and angiogenesis both *in vitro* and *in vivo*. Our results show that downregulation of uPAR and MMP-9 leads to a decrease in the activation of some of the important enzymes participating in the MAPK and PI3 kinase pathways, which in turn, might decrease cell survival and proliferation. In addition, we have demonstrated the efficiency of RNAi-mediated targeting of uPAR and MMP-9 in pre-established tumor growth *in vivo*. We observed a significant regression of pre-established orthotopic tumors upon RNAi-mediated targeting of uPAR and MMP-9 and have also demonstrated that targeting both the proteins simultaneously augmented the therapeutic treatment of human meningiomas *(35)*.

In another study, we introduced small interfering RNA to downregulate the expression of uPAR and MMP-9 in breast cancer cell lines (MDA MB 231 and ZR 75 1). *In vitro* angiogenesis studies indicated a decrease in the angiogenic and invasive potential of the treated cells. These results suggest a synergistic effect from the simultaneous downregulation of uPAR and MMP-9. We also assessed the levels of phosphorylated forms of MAPK, ERK and AKT signaling pathway molecules and found reduced levels of these molecules in cells treated with the bicistronic construct as compared to the control cells. Furthermore, targeting both uPAR and MMP-9 using RNAi totally regressed orthotopic breast tumors in nude mice, thereby providing evidence that the simultaneous downregulation of uPAR and MMP-9 using RNAi technology may provide an effective tool for breast cancer therapy *(64)*.

In another study, we have demonstrated that the simultaneous targeting of more than two components is significantly superior to targeting two alone. We have showed that direct intratumoral injections of plasmid DNA expressing hpRNA for uPA, uPAR and MMP-9 significantly regressed pre-established intracranial tumors in nude mice as compared to the controls. In addition, cells treated with RNAi for uPAR, uPA and MMP-9 showed reduced pERK levels when compared to parental and EV/SV-treated SNB19 cells. Our results support the therapeutic potential of RNAi as a method for gene therapy in treating gliomas *(30)*. A brief schematic representation of the possible mechanisms involved in MMP-9 and uPAR targeted RNAi therapy is given in **Fig. 13.4**.

The main objective of cancer therapy is to arrest tumor invasion and convert it to a controlled, localized disease. Accumulated lines of evidence indicate that MMPs and the uPAR system play an essential role in tumor invasion and metastasis. Therapeutic strategies that can inhibit a broad spectrum of MMPs and the uPAR system may be beneficial for retarding and preventing tumor progression. To achieve this, further investigation and understanding of proteases at the molecular

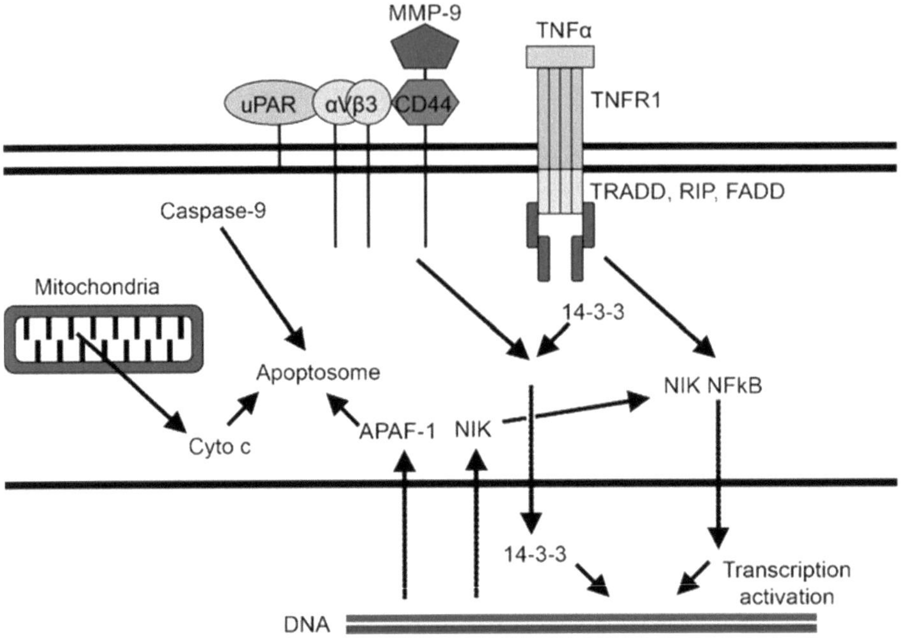

Fig. 13.4. The uPAR system and its association with MMPs. The uPAR system and MMPs are involved in multiple intracellular survival and proliferative signaling events.

level should play an important role in the future development of new, target-selective treatments. It is sufficiently clear that the simultaneous targeting of multiple systems is more synergistic than additive.

References

1. Ge, Y. and Elghetany, M. T. (2003) Urokinase plasminogen activator receptor (CD87): something old, something new. *Lab Hematol.* 9, 67–71.
2. Mazar, A. P., Henkin, J. and Goldfarb, R. H. (1999) The urokinase plasminogen activator system in cancer: Implications for tumor angiogenesis and metastasis. *Angiogenesis* 3, 15–32.
3. Rao, J. S. (2003) Molecular mechanisms of glioma invasiveness: the role of proteases. *Nat. Rev. Cancer* 3, 489–501.
4. Mazzieri, R., Masiero, L., Zanetta, L., Monea, S., Onisto, M., Garbisa, S. and Mignatti, P. (1997) Control of type IV collagenase activity by components of the urokinase-plasmin system: a regulatory mechanism with cell-bound reactants. *EMBO J.* 16, 2319–2332.
5. Wang, Y. (2001) The role and regulation of urokinase-type plasminogen activator receptor gene expression in cancer invasion and metastasis. *Med. Res. Rev.* 21, 146–170.
6. Alfano, D., Franco, P., Vocca, I., Gambi, N., Pisa, V., Mancini, A., Caputi, M., Carriero, M. V., Iaccarino, I. and Stoppelli, M. P. (2005) The urokinase plasminogen activator and its receptor: role in cell growth and apoptosis. *Thromb. Haemost.* 93, 205–211.
7. Wang, Y. (2001) The role and regulation of urokinase-type plasminogen activator receptor gene expression in cancer invasion and metastasis. *Med. Res. Rev.* 21, 146–170.
8. Ploug, M. (2003) Structure-function relationships in the interaction between the urokinase-type plasminogen activator and its receptor. *Curr. Pharm. Des.* 9, 1499–1528.
9. Ploug, M. (1998) Identification of specific sites involved in ligand binding by photoaffinity labeling of the receptor for the urokinase-type plasminogen activator. Residues located at equivalent positions in

uPAR domains I and III participate in the assembly of a composite ligand-binding site. *Biochemistry* 37, 16494–16505.

10. Gondi, C. S., Kondraganti, S., Dinh, D., Olivero, W., Gujrati, M., and Rao, J. S. (2005) RNAi-mediated simultaneous down-regulation of uPAR and cathepsin B induces the accumulation of FasL on the cell surface and induces caspase 8-mediated apoptosis in SNB19 human glioma cells. *Proceedings 96th American Association for Cancer Research* 46, 1269.
11. Gondi, C. S., Kandhukuri, N., Kondraganti, S., Gujrati, M., Olivero, W. C., Dinh, D. H. and Rao, J. S. (2006) Down-regulation of uPAR and cathepsin B retards cofilin dephosphorylation. *Int. J. Oncol.* 28, 633–639.
12. Adachi, Y., Lakka, S. S., Chandrasekar, N., Yanamandra, N., Gondi, C. S., Mohanam, S., Dinh, D. H., Olivero, W. C., Gujrati, M., Tamiya, T., Ohmoto, T., Kouraklis, G., Aggarwal, B. and Rao, J. S. (2001) Down-regulation of integrin alpha(v)beta(3) expression and integrin-mediated signaling in glioma cells by adenovirus-mediated transfer of antisense urokinase-type plasminogen activator receptor (uPAR) and sense p16 genes. *J. Biol. Chem.* 276, 47171–47177.
13. Aguirre-Ghiso, J. A., Alonso, D. F., Farias, E. F., Gomez, D. E. and Kier Joffe, E. B. (1999) Deregulation of the signaling pathways controlling urokinase production. Its relationship with the invasive phenotype. *Eur. J. Biochem.* 263, 295–304.
14. Cantero, D., Friess, H., Deflorin, J., Zimmermann, A., Brundler, M. A., Riesle, E., Korc, M. and Buchler, M. W. (1997) Enhanced expression of urokinase plasminogen activator and its receptor in pancreatic carcinoma. *Br. J. Cancer* 75, 388–395.
15. Kruger, A., Soeltl, R., Lutz, V., Wilhelm, O. G., Magdolen, V., Rojo, E. E., Hantzopoulos, P. A., Graeff, H., Gansbacher, B. and Schmitt, M. (2000) Reduction of breast carcinoma tumor growth and lung colonization by overexpression of the soluble urokinase-type plasminogen activator receptor (CD87). *Cancer Gene Ther.* 7, 292–299.
16. Tan, X., Egami, H., Nozawa, F., Abe, M. and Baba, H. (2006) Analysis of the invasion-metastasis mechanism in pancreatic cancer: involvement of plasmin(ogen) cascade proteins in the invasion of pancreatic cancer cells. *Int. J. Oncol.* 28, 369–374.
17. Dass, C. R., Nadesapillai, A. P., Robin, D., Howard, M. L., Fisher, J. L., Zhou, H. and Choong, P. F. (2005) Downregulation of uPAR confirms link in growth and metastasis of osteosarcoma. *Clin. Exp. Metastasis* 22, 643–652.
18. Dass, C. R. and Choong, P. F. (2006) Biophysical delivery of peptides: Applicability for cancer therapy. *Peptides* 27, 3479–3488.
19. Wang, Y. (2001) The role and regulation of urokinase-type plasminogen activator receptor gene expression in cancer invasion and metastasis. *Med. Res. Rev.* 21, 146–170.
20. Go, Y., Chintala, S. K., Mohanam, S., Gokaslan, Z., Venkaiah, B., Bjerkvig, R., Oka, K., Nicolson, G. L., Sawaya, R. and Rao, J. S. (1997) Inhibition of in vivo tumorigenicity and invasiveness of a human glioblastoma cell line transfected with antisense uPAR vectors. *Clin. Exp. Metastasis* 15, 440–446.
21. Kook, Y. H., Adamski, J., Zelent, A. and Ossowski, L. (1994) The effect of antisense inhibition of urokinase receptor in human squamous cell carcinoma on malignancy. *EMBO J.* 13, 3983–3991.
22. Gondi, C. S., Lakka, S. S., Yanamandra, N., Siddique, K., Dinh, D. H., Olivero, W. C., Gujrati, M. and Rao, J. S. (2003) Expression of antisense uPAR and antisense uPA from a bicistronic adenoviral construct inhibits glioma cell invasion, tumor growth, and angiogenesis. *Oncogene* 22, 5967–5975.
23. Gondi, C. S., Lakka, S. S., Yanamandra, N., Olivero, W. C., Dinh, D. H., Gujrati, M., Tung, C. H., Weissleder, R. and Rao, J. S. (2004) Adenovirus-mediated expression of antisense urokinase plasminogen activator receptor and antisense cathepsin B inhibits tumor growth, invasion, and angiogenesis in gliomas. *Cancer Res.* 64, 4069–4077.
24. Lakka, S. S., Rajagopal, R., Rajan, M. K., Mohan, P. M., Adachi, Y., Dinh, D. H., Olivero, W. C., Gujrati, M., Ali-Osman, F., Roth, J. A., Yung, W. K., Kyritsis, A. P. and Rao, J. S. (2001) Adenovirus-mediated antisense urokinase-type plasminogen activator receptor gene transfer reduces tumor cell invasion and metastasis in non-small cell lung cancer cell lines. *Clin. Cancer Res.* 7, 1087–1093.
25. Lakka, S. S., Rajan, M., Gondi, C. S., Yanamandra, N., Chandrasekar, N., Jasti, S. L., Adachi, Y., Siddique, K., Gujrati, M., Olivero, W., Dinh, D. H., Kouraklis, G., Kyritsis, A. P. and Rao, J. S. (2002) Adenovirus-mediated expression of antisense MMP-9 in glioma cells inhibits tumor growth and invasion. *Oncogene* 21, 8011–8019.
26. Lakka, S. S., Gondi, C. S., Yanamandra, N., Dinh, D. H., Olivero, W. C., Gujrati, M. and Rao, J. S. (2003) Synergistic down-regulation of urokinase plasminogen activator

receptor and matrix metalloproteinase-9 in SNB19 glioblastoma cells efficiently inhibits glioma cell invasion, angiogenesis, and tumor growth. *Cancer Res.* 63, 2454–2461.

27. McCaffrey, A. P., Meuse, L., Pham, T. T., Conklin, D. S., Hannon, G. J. and Kay, M. A. (2002) RNA interference in adult mice. *Nature* 418, 38–39.
28. Pulukuri, S. M., Gondi, C. S., Lakka, S. S., Jutla, A., Estes, N., Gujrati, M. and Rao, J. S. (2005) RNA Interference-directed Knockdown of Urokinase Plasminogen Activator and Urokinase Plasminogen Activator Receptor Inhibits Prostate Cancer Cell Invasion, Survival, and Tumorigenicity in Vivo. *J. Biol. Chem.* 280, 36529–36540.
29. Gondi, C. S., Lakka, S. S., Dinh, D. H., Olivero, W. C., Gujrati, M. and Rao, J. S. (2004) RNAi-mediated inhibition of cathepsin B and uPAR leads to decreased cell invasion, angiogenesis and tumor growth in gliomas. *Oncogene* 23, 8486–8496.
30. Gondi, C. S., Lakka, S. S., Dinh, D., Olivero, W., Gujrati, M. and Rao, J. S. (2004) Downregulation of uPA, uPAR and MMP-9 using small, interfering, hairpin RNA (siRNA) inhibits glioma cell invasion, angiogenesis and tumor growth. *Neuron Glia Biology* 1, 165–176.
31. Gondi, C. S., Kandhukuri, N., Kondraganti, S., Gujrati, M., Olivero, W. C., Dinh, D. H. and Rao, J. S. (2006) RNA interference-mediated simultaneous down-regulation of urokinase-type plasminogen activator receptor and cathepsin B induces caspase-8-mediated apoptosis in SNB19 human glioma cells. *Mol. Cancer Ther.* 5, 3197–3208.
32. Gondi, C. S., Kandhukuri, N., Dinh, D. H., Gujrati, M. and Rao, J. S. (2007) Down-regulation of uPAR and uPA activates caspase-mediated apoptosis and inhibits the PI3K/AKT pathway. *Int. J. Oncol.* 31, 19–27.
33. Gondi, C. S., Lakka, S. S., Dinh, D. H., Olivero, W. C., Gujrati, M., and Rao, J. S. (2007) Intraperitoneal injection of an hpRNA-expressing plasmid targeting uPAR and uPA retards angiogenesis and inhibits intracranial tumor growth in nude mice. *Clin. Cancer Res.* 13, 4051–4060.
34. Kondraganti, S., Gondi, C. S., McCutcheon, I., Dinh, D. H., Gujrati, M., Rao, J. S. and Olivero, W. (2006) RNAi-mediated downregulation of urokinase plasminogen activator and its receptor in human meningioma cells inhibits tumor invasion and growth. *Int. J. Oncol.* 28, 1353–1360.
35. Tummalapalli, P., Gondi, C. S., Dinh, D. H., Gujrati, M. and Rao, J. S. (2007) RNA interference-mediated targeting of urokinase plasminogen activator receptor and matrix metalloproteinase-9 gene expression in the IOMM-Lee malignant meningioma cell line inhibits tumor growth, tumor cell invasion and angiogenesis. *Int. J. Oncol.* 31, 5–17.
36. Nagase, H. and Woessner, J. F., Jr. (1999) Matrix metalloproteinases. *J. Biol. Chem.* 274, 21491–21494.
37. Stefanidakis, M. and Koivunen, E. (2006) Cell-surface association between matrix metalloproteinases and integrins: role of the complexes in leukocyte migration and cancer progression. *Blood* 108, 1441–1450.
38. Murphy, G., Knauper, V., Cowell, S., Hembry, R., Stanton, H., Butler, G., Freije, J., Pendas, A. M., and Lopez-Otin, C. (1999) Evaluation of some newer matrix metalloproteinases. *Ann. NY Acad. Sci.* 878, 25–39.
39. Shapiro, S. D., Fliszar, C. J., Broekelmann, T. J., Mecham, R. P., Senior, R. M. and Welgus, H. G. (1995) Activation of the 92-kDa gelatinase by stromelysin and 4-aminophenylmercuric acetate. Differential processing and stabilization of the carboxyl-terminal domain by tissue inhibitor of metalloproteinases (TIMP). *J. Biol. Chem.* 270, 6351–6356.
40. Westermarck, J. and Kahari, V. M. (1999) Regulation of matrix metalloproteinase expression in tumor invasion. *FASEB J.* 13, 781–792.
41. Nielsen, B. S., Timshel, S., Kjeldsen, L., Sehested, M., Pyke, C., Borregaard, N. and Dano, K. (1996) 92 kDa type IV collagenase (MMP-9) is expressed in neutrophils and macrophages but not in malignant epithelial cells in human colon cancer. *Int. J. Cancer* 65, 57–62.
42. Welgus, H. G., Campbell, E. J., Cury, J. D., Eisen, A. Z., Senior, R. M., Wilhelm, S. M. and Goldberg, G. I. (1990) Neutral metalloproteinases produced by human mononuclear phagocytes. Enzyme profile, regulation, and expression during cellular development. *J Clin. Invest.* 86, 1496–1502.
43. Leppert, D., Waubant, E., Galardy, R., Bunnett, N. W. and Hauser, S. L. (1995) T cell gelatinases mediate basement membrane transmigration in vitro. *J. Immunol.* 154, 4379–4389.
44. Haas, T. L. and Madri, J. A. (1999) Extracellular matrix-driven matrix metalloproteinase production in endothelial cells: implications for angiogenesis. *Trends Cardiovasc. Med.* 9, 70–77.
45. Unemori, E. N., Bair, M. J., Bauer, E. A. and Amento, E. P. (1991) Stromelysin expression regulates collagenase activation in human

fibroblasts. Dissociable control of two metalloproteinases by interferon-gamma. *J. Biol. Chem.* 266, 23477–23482.

46. Gottschall, P. E. and Deb, S. (1996) Regulation of matrix metalloproteinase expressions in astrocytes, microglia and neurons. *Neuroimmunomodulation* 3, 69–75.
47. Maeda, A. and Sobel, R. A. (1996) Matrix metalloproteinases in the normal human central nervous system, microglial nodules, and multiple sclerosis lesions. *J. Neuropathol. Exp. Neurol.* 55, 300–309.
48. Uhm, J. H., Dooley, N. P., Oh, L. Y. and Yong, V. W. (1998) Oligodendrocytes utilize a matrix metalloproteinase, MMP-9, to extend processes along an astrocyte extracellular matrix. *Glia* 22, 53–63.
49. Muir, E. M., Adcock, K. H., Morgenstern, D. A., Clayton, R., von Stillfried, N., Rhodes, K., Ellis, C., Fawcett, J. W., and Rogers, J. H. (2002) Matrix metalloproteases and their inhibitors are produced by overlapping populations of activated astrocytes. *Brain Res. Mol. Brain Res.* 100, 103–117.
50. Itoh, T., Tanioka, M., Yoshida, H., Yoshioka, T., Nishimoto, H. and Itohara, S. (1998) Reduced angiogenesis and tumor progression in gelatinase A-deficient mice. *Cancer Res.* 58, 1048–1051.
51. Lakka, S. S. and Rao, J. S. (2005) Role and regulation of proteases in human glioma. in *Proteases in the Brain*, Lendeckel, U. and Hooper, N. M., eds., Springer, New York, pp. 151–177.
52. Lakka, S. S., Gondi, C. S., Dinh, D. H., Olivero, W. C., Gujrati, M., Rao, V. H., Sioka, C. and Rao, J. S. (2005) Specific interference of uPAR and MMP-9 gene expression induced by double-stranded RNA results in decreased invasion, tumor growth and angiogenesis in gliomas. *J. Biol. Chem.* 280, 21882–21892.
53. Sternlicht, M. D. and Werb, Z. (2001) How matrix metalloproteinases regulate wcell behavior. *Annu. Rev. Cell Dev. Biol.* 17, 463–516.
54. Rooprai, H. K., Rucklidge, G. J., Panou, C. and Pilkington, G. J. (2000) The effects of exogenous growth factors on matrix metalloproteinase secretion by human brain tumour cells. *Br. J. Cancer* 82, 52–55.
55. VanMeter, T. E., Rooprai, H. K., Kibble, M. M., Fillmore, H. L., Broaddus, W. C. and Pilkington, G. J. (2001) The role of matrix metalloproteinase genes in glioma invasion: co-dependent and interactive proteolysis. *J. Neurooncol.* 53, 213–235.
56. Choe, G., Park, J. K., Jouben-Steele, L., Kremen, T. J., Liau, L. M., Vinters, H. V., Cloughesy, T. F. and Mischel, P. S. (2002) Active matrix metalloproteinase 9 expression is associated with primary glioblastoma subtype. *Clin. Cancer Res.* 8, 2894–2901.
57. Chintala, S. K., Sawaya, R., Aggarwal, B. B., Majumder, S., Giri, D. K., Kyritsis, A. P., Gokaslan, Z. L. and Rao, J. S. (1998) Induction of matrix metalloproteinase-9 requires a polymerized actin cytoskeleton in human malignant glioma cells. *J. Biol. Chem.* 273, 13545–13551.
58. Lakka, S. S., Jasti, S. L., Kyritsis, A. P., Yung, W. K., Ali-Osman, F., Nicolson, G. L. and Rao, J. S. (2000) Regulation of MMP-9 (type IV collagenase) production and invasiveness in gliomas by the extracellular signal-regulated kinase and jun amino-terminal kinase signaling cascades. *Clin. Exp. Metastasis* 18, 245–252.
59. Lakka, S. S., Gondi, C. S., Yanamandra, N., Olivero, W. C., Dinh, D. H., Gujrati, M. and Rao, J. S. (2004) Inhibition of cathepsin B and MMP-9 gene expression in glioblastoma cell line via RNA interference reduces tumor cell invasion, tumor growth and angiogenesis. *Oncogene* 23, 4681–4689.
60. Yuan, J., Dutton, C. M., and Scully, S. P. (2005) RNAi mediated MMP-1 silencing inhibits human chondrosarcoma invasion. *J. Orthop. Res.* 23, 1467–1474.
61. Wyatt, C. A., Geoghegan, J. C., and Brinckerhoff, C. E. (2005) Short hairpin RNA-mediated inhibition of matrix metalloproteinase-1 in MDA-231 cells: effects on matrix destruction and tumor growth. *Cancer Res.* 65, 11101–11108.
62. Munoz-Najar, U. M., Neurath, K. M., Vumbaca, F. and Claffey, K. P. (2006) Hypoxia stimulates breast carcinoma cell invasion through MT1-MMP and MMP-2 activation. *Oncogene* 25, 2379–2392.
63. Wang, F. Q., Smicun, Y., Calluzzo, N. and Fishman, D. A. (2006) Inhibition of matrilysin expression by antisense or RNA interference decreases lysophosphatidic acid-induced epithelial ovarian cancer invasion. *Mol. Cancer Res.* 4, 831–841.
64. Kunigal, S., Lakka, S. S., Gondi, C. S., Estes, N. and Rao, J. S. (2007) RNAi-mediated downregulation of urokinase plasminogen activator receptor and matrix metalloprotease-9 in human breast cancer cells results in decreased tumor invasion, angiogenesis and growth. *Int. J. Cancer* 121, 2307–2316.

Chapter 14

Silencing of HIF-1α by RNA Interference in Human Glioma Cells In Vitro and In Vivo

David L. Gillespie, Jeannette R. Flynn, Brian T. Ragel, Maria Arce-Larreta, David A. Kelly, Sheryl R. Tripp, and Randy L. Jensen

Abstract

Higher-grade gliomas are distinguished by increased vascular endothelial cell proliferation and peritumoral edema. These are thought to be instigated by vascular endothelial growth factor, which in turn is regulated by cellular oxygen tension. Hypoxia inducible factor-1α (HIF-1α) is a main responder to intracellular hypoxia and is overexpressed in many human cancers, including gliomas. Here we present methods for investigating the role of HIF-1α in glioma growth in vivo and in vitro using RNA interference in U251, U87, and U373 glioma cells.

Key words: HIF-1, glioma, VEGF, RNA interference, siRNA, hypoxia, gene therapy.

1. Introduction

Overexpression of hypoxia-inducible factor-1α (HIF-1α) has been described in many common human cancers and their metastases *(1, 2, 3)*. The role of HIF-1 in solid tumor growth is still not entirely clear, but previous work suggests that this transcription factor is necessary for growth and angiogenesis of these tumors *(4, 5)*. A direct correlation between tumor grade and HIF-1 expression in glioblastoma multiforme (GBM) has been demonstrated *(6)*. We and others have published data supporting the proposition that HIF-1α expression represents an "angiogenic switch" that facilitates the progression of a low-grade astrocytoma to a GBM and promotes cell survival in hypoxic conditions by elevating glycolysis and angiogenesis *(2, 7)*. It has also been

M. Sioud (ed.), *Methods in Molecular Biology, siRNA and miRNA Gene Silencing, vol. 487*

DOI: 10.1007/978-1-60327-547-7_14

demonstrated that reducing HIF-1α expression using RNAi and other methods leads to a reduction of glioma growth in vivo and in vitro *(5, 8–10)*.

The transcription factor HIF-1 is composed of two heterodimeric subunits, HIF-1α and HIF-1β (known as aryl hydrocarbon receptor nuclear translocator (ARNT)). At the mRNA levels, HIF-1α and HIF-1β are both constitutively expressed and do not seem to be significantly modified by hypoxia *(11)*. Whereas HIF-1β protein is found in normoxic cells, HIF-1α is rapidly degraded by proteasomal degradation. During low oxygen tension conditions (1–2% O_2), this degradation is inhibited, leading to increased HIF-1 *(12)*. HIF-1 binds to DNA hypoxia response elements (HREs) and induces the transcription of a number of well-characterized genes that help cells survive low oxygen conditions *(13)*. These genes include vascular endothelial growth factor (VEGF), erythropoietin, transferrin, GLUT-1, and almost every gene in the glycolytic pathway *(14)*. Stabilization of HIF-1α promotes cell survival via adaptive modifications of cellular metabolism that increase these glycolytic enzymes and hence the glycolysis rate. This adaptation of cancer cells through increased glycolysis was first proposed by Warburg *(15)* as a necessary step toward an aggressive phenotype. The results of recent studies of HIF-1 have indicated a possible link between it and the Warburg effect in various cell types *(16, 17)* and prompted the proposal that aerobic glycolysis could be controlled by dysregulation of HIF-1α *(18)*. Here we describe methods for promoting HIF-1α stabilization using hypoxia to assist in studying the effects of treatment and subsequent detection and analysis of HIF-1α and other downstream targets in response to RNAi targeting HIF-1α.

2. Materials

2.1. Cell Culture and Hypoxia Chamber

1. Dulbecco's Modified Eagle's Medium (DMEM) supplemented with 10% fetal bovine serum.
2. BD BBL GasPak Jar system.
3. GasPak Plus Hydrogen-CO_2 generator envelopes.
4. Oxygen indicator strips.

2.2. Cell Lysis and Protein Isolation

1. Phosphatase inhibitor buffer (PIB), 40x: 250 mM NaF, 500 mM β-glycerophosphate, 50 mM Na_3VO_4. Filter and store in aliquots at –20°C indefinitely or at 4°C for up to 1 month.
2. Hypotonic buffer (HB): 0.5 M HEPES, pH 7.5, 0.5 M NaF, 0.5 M EDTA, 0.1 M Na_2MoO_4. Filter and store at 4°C for up to 6 months.

3. 10% NP-40.
4. Lysis buffer: 400 mM NaCl, 20 mM HEPES (pH 7.5), 10 mM NaF, 10 mM PNPP, 1 mM Na_2VO_4, 0.1 mM EDTA, 10 μM Na_2MoO_4, 10 mM β-glycerophosphate, 20% glycerol. Store at 4°C for up to 6 months. Directly before use, add 0.1 M dithiothreitol (DTT), 100x protease inhibitor cocktail (Sigma-Aldrich, St Louis, MO) for final concentrations of 1 mM DTT and 1x protease inhibitor. Lysis buffer that has had DTT and protease inhibitor added is referred to as "complete lysis buffer" in this protocol. Complete lysis buffer should be discarded after 24 h (*see* **Note 1**).
5. Polyethylene cell lifters (Corning, Corning, NY).

2.3. Protein Isolation from Tumor Tissues

1. Mihir's homogenation buffer: 10 mM HEPES (pH 7.6), 1 mM Na_2VO_4, 100 mM NaF, 0.4 mM PMSF, 0.1 mM EGTA, 10 mM $Na_4P_2O_7$. Store at 4°C for 6 months. Just prior to using add 100x protease inhibitor (Sigma-Aldrich) for final concentration of 1x and 0.1 M DTT for a final concentration of 2 mM DTT. Buffer should be discarded after 24 h once the protease inhibitors are added.
2. Lysis buffer: *see* **Sect. 2.2**.
3. Scalpel blades, size #10.
4. Polytron PT-1200 homogenizer (or similar rotor-stator homogenizer).

2.4. SDS-Polyacrylamide Gel Electrophoresis (SDS-PAGE)

1. Running buffer: NuPAGE MOPS 20x SDS running buffer diluted to 1x.
2. Loading buffer: 4xNuPAGE LDS buffer, 1% β-mercapto-ethanol (BME).
3. Precast NuPAGE Bis-Tris 4–12% gradient gel, 12 well.
4. Prestained molecular weight markers: Kaleidoscope markers (Bio-Rad, Hercules, CA).

2.5. Western Blot

1. Transfer buffer: NuPAGE 20x Transfer buffer.
2. Hybond-P PVDF membrane from Amersham Biosciences, Piscataway, NJ, and Gel Blot paper from ISC Bioexpress (Kaysville, UT).
3. TBS-T (Tris-buffered saline with Tween-20: Prepare 10x stock with 1.37 M NaCl, 27 mM KCl, 250 mM Tris-HCl (pH 7.4), 1% Tween-20. Dilute 100 ml with 900 ml of water for use.
4. Blocking buffer: Make up 5% non-fat powdered milk in 1x TBS-T. Need 100 ml/membrane (small gel). Keep at 4°C for no more than 48 h.

5. Primary antibody: Anti-human HIF-1 mouse monoclonal IgG from Novus Biologicals (Littleton, CO).
6. Secondary antibody: Anti-mouse IgG conjugated to horse radish peroxidase (Amersham Biosciences).
7. Enhanced chemiluminescent plus (ECL+) reagents from Amersham Biosciences (Piscataway, NJ) and Bio-Max light film (Kodak, Rochester, NY).
8. Straight-edge platform paper cutter.

2.6. Enzyme-Linked Immunosorbent Assay for VEGF

1. R&D systems Quantikine VEGF Enzyme-linked immunosorbent assay (ELISA) kit (Minneapolis, MN).
2. Benchmark microplate reader (Bio-Rad Laboratories, Hercules, CA).
3. Bio-Rad 1575 ImmunoWash plate washer.

2.7. RNAi Design

1. Silencer siRNA construction kit (Ambion, Austin, TX).
2. http://www.rnaiweb.com/
3. http://www.ncbi.nlm.nih.gov/projects/genome/RNAi/
4. http://rnai.cs.unm.edu/projects/
5. http://www.ambion.com/techlib/resources/RNAi/index.html

2.8. Immunohistochemistry for MIB-1 and GLUT-1

1. Citrate buffer: 0.01 M citric acid (Sigma, St. Louis, MO), 0.025 M NaOH (Sigma), dH_2O (pH 6.0).
2. 4% potassium iodine (Fisher), 2% iodine, dH_2O.
3. 5% sodium thiosulfate, dH_2O.
4. Antibodies for Ki-67 (MIB-1) and GLUT-1 (Dako Cytomation, Carpinteria, CA).
5. IView DAB detection kit (Ventana Medical Systems, Tucson, AZ).
6. Electric pressure cooker DC2000 (BioCare Medical, Concord, CA).
7. Dawn dish-washing detergent, 0.2%.
8. Automated BenchMark XT immunostainer from Ventana Medical Systems (Tucson, AZ).

2.9. Cell Transfection and Imaging

1. Lipofectamine 2000 (Invitrogen, Carlsbad, CA).
2. Dead Cell Reagent (Invitrogen).
3. Silencer siRNA Labeling kit (Ambion).
4. Optimem serum-free media (Invitrogen).

2.10. Real Time PCR

1. RNeasy Mini spin column kit (Qiagen, Valencia, CA).
2. Superscript III first-strand synthesis for RT-PCR kit (Invitrogen).

3. IQ SYBR Green supermix kit (Bio-Rad, Hercules, CA).
4. Bio-Rad MyIQ iCycler.

2.11. Cell Inoculation

1. Matrigel (BD Biosciences, Franklin Lakes, NJ).
2. Lo-dose 0.5 ml insulin syringe (BD Biosciences).
3. Trypsin-EDTA, 0.25% (Invitrogen).

2.12. siRNA Treatment

1. Annealed siRNAs at 150 μM in sterile annealing buffer (50 mM Tris, 150 mM NaCl).
2. 4x JetPEI (Q-Biogene, Irvine CA).
3. Sterile 150 mM NaCl.
4. Lo-dose 0.5 ml insulin syringe (BD Biosciences).

3. Methods

Because HIF-1α is a nuclear protein that is almost completely regulated post-translationally *(11)*, it is important to assay nuclear protein levels. Assays involving transcription activity or mRNA levels (such as Real Time PCR) are only informative when they are used to assess RNAi knockdown effects or treatments that affect general transcription activity. Because HIF-1α is degraded quickly in the presence of oxygen, it is extremely important to do all steps quickly with protease inhibitors and on wet ice to slow down the degradation process. To increase HIF-1α levels and ease detection, it is important to expose cells to hypoxia during experimentation. This also creates a broader range of HIF-1α protein levels to observe response to treatments. The GasPak jar and pouch systems have been used for many years to produce an inexpensive, reliable hypoxic atmosphere for short-term cell culture and HIF-1α stabilization *(8, 19, 20)*. Many more intricate and expensive oxygen-controlling environments are also commercially available. Because inappropriate sample handling can have significant effects on the results, all experiments should be repeated several times. It is also helpful to verify the results by assessing downstream HIF-1 transcription targets such as VEGF, Ki-67 (MIB-1), and GLUT-1, which are much more stable and easily measured (*see* **Note 2**). There are several methods for creating a positive HIF-1α control. We have used 18 h hypoxic U251 glioma cells or treatment with cobalt chloride.

3.1. Hypoxia Chamber Cell Growth Studies

1. Cells are plated on 60 or 100 mm tissue culture dishes until they reach 70–80% confluence. It is important that the cells be actively growing to get maximum HIF-1α production. We use

glioma cell lines U251 and U87 for most hypoxia experiments because they produce high levels of HIF-1α after a short 4 h exposure. Maximum levels for these cells are observed between 12 and 18 h exposures, which are best done overnight.

2. The culture medium is changed immediately before the dishes are placed in a GasPak Plus anaerobic culture chamber. Two GasPak catalyst envelopes are taped to the inside of the chamber to reduce the oxygen concentration below 1%. One oxygen indicator strip is taped to the inside to verify the oxygen concentration. Cells can be treated for 4–48 h.

3.2. Protein Isolation from Cell Lines

This procedure can be used for an 80% confluent cell layer of 21 cm^2 (plate 5–8 × 10^6 cells/60 mm dish). Volumes can be scaled up for use with 100 mm plates. It is absolutely essential to proceed quickly and keep everything on ice. The following steps should be done before starting:

(a) Locate/reserve 4°C swing-bucket centrifuge and 4°C fixed-rotor centrifuge, and make up phosphate-buffered saline (PBS)/PIB using cold PBS. Put in ice bucket. Add 250 μl of PIB to 10 ml of PBS, needs 8 ml/plate.

(b) Make complete lysis buffer, 30 μl/sample. Add 1 μl of 0.1 M DTT and 1 μl of protease inhibitor to 100 μl of lysis buffer). Put HB buffer on ice.

(c) Make up ice/water bath in flat ice tray (*see* **Note 3**). Label one 1.7 ml tube/sample, and one 15 ml tube/sample, and put on ice (*see* **Note 4**).

(d) Assemble one 25 ml pipette (for adding PBS/PIB), four 1 ml pipettes (for aspirating media/wash), one 5 ml pipette, and one cell lifter/plate, 10% NP-40.

1. Put cell plates in ice tray and aspirate media.
2. Wash cells with 4 ml of ice-cold PBS/PIB. Swirl and aspirate (*see* **Note 5**).
3. Add 4 ml of ice-cold PBS/PIB.
4. Scrape the cells off the dish with a cell lifter. Tip plate and use a 5 ml pipette to transfer cells into a pre-chilled 15 ml tube (*see* **Note 6**).
5. Centrifuge at 386 g for 5 min at 4°C, then aspirate the supernatant.
6. Resuspend the pellet in 0.6 ml of ice-cold HB buffer by gentle pipetting.
7. Allow the cells to swell on ice for 15 min.
8. Add 30 μl of 10% NP-40 (0.5% final), and vortex for 10 s.
9. Centrifuge the homogenate at 500 g for 5 min at 4°C (*see* **Note 7**). For a cytoplasmic VEGF assay, transfer 600 μl of

supernatant to 1.5 ml tube, keep on ice for protein assay (Step 12). Place tubes upside down in ice to drain all residual liquid, aspirate supernatant from side of tube and lid.

10. Resuspend the nuclear pellet in 30 μl of complete lysis buffer and shake vigorously on ice (or in cold room) for 30 min (use a vortexer with tube holder, medium setting). During the incubation time, label one 1.7 ml tube/sample for protein, and keep in ice.
11. Centrifuge at 30,000 g for 20 min at 4°C and save the supernatant (nuclear cell extract). Transfer into prechilled tube.
12. Assay protein concentration using the standard Bradford method.
13. Aliquot if desired and store at −80°C. Avoid freeze/thaw cycles.

3.3. Protein Isolation from Tumors

Tumors should be placed in liquid nitrogen immediately after resection. If the tumors are harvested from mice or other small animals, the animals should not be sacrificed using CO_2 because this method can lead to elevated levels of HIF-1 in tumor samples. All steps are to be done on ice.

1. Place frozen tissue (300–500 mg) in 3 ml of ice cold Mihir's homogenation buffer in a 15 ml centrifuge tube on ice. Slice into small pieces with a #10 scalpel (*see* **Note 8**).
2. Homogenize for 2 min. If using a variable speed homogenizer, start at the lowest speed and gradually increase to the maximum.
3. Centrifuge at 850 g for 10 min (*see* **Note 9**).
4. Remove supernatant to clean 15 ml tube, save pellet.
5. Repeat spin with the supernatant. During this step, make up complete nuclear lysis buffer, add 300 μl to pellet and resuspend.
6. Remove supernatant to clean 15 ml tube containing 600 μl 100% glycerol (final glycerol concentration is 20%). Vortex for 30 s, then hold on ice for 5 min.
7. Combine pellet with resuspended pellet from first spin, and hold on ice.
8. Centrifuge supernatant and glycerol at 15,344 g for 15 min. Save the pellet (this is the nuclei) and supernatant if needed for VEGF assay (*see* **Note 10**).
9. Resuspend pelleted nuclei in three volumes complete nuclear lysis buffer (~200 μl).
10. Mix vigorously at 4°C for 30 min.

11. Spin at 30,000 g for 20 min. Aliquot supernatant into 1.7 ml tubes, determine protein concentration of nuclear and whole cell extract (from Step 7), and store at –80°C.

3.4. SDS-PAGE

These instructions assume the use of a Novex X-cell SureLock electrophoresis system (Invitrogen, Carlsbad, CA) using the NuPAGE pre-cast 12-well gel and buffers.

1. Calculate protein volume for 25–50 μg of total protein. Use lysis buffer to adjust for a final volume of 16 μl per well.
2. Make up loading dye—50 μl 4x LDS dye + 5 μl BME per gel.
3. Label one 0.7 μl tube/sample. Add 4 μl loading dye to each tube.
4. Add complete lysis buffer, then add protein to tubes. Heat at 75°C for 10 min.
5. While heating, make up running buffer—40 ml 20x MOPS NuPage buffer + 760 ml dH_2O. Remove gel from wrapper, rinse in dH_2O, remove well comb and bottom adhesive strip. Assemble gel, fill inner chamber with buffer, and rinse out wells with a transfer pipette.
6. Spin down samples to collect condensation and load gel.
7. Fill outer gel box chamber with remaining buffer.
8. Run at 200 V until blue (20 kD) standard runs off gel—about 2 h.

3.5. Western Blot

All rinses and washes are done at room temperature using TBS-T, unless otherwise noted. HIF-1α appears as a doublet at ~120 kD (**Fig. 14.1**). Do not mistake the two bands at ~100 kD for HIF-1α.

1. Cut one corner of the membrane for ease in orientation.
2. Activate membrane in 100% MEOH for 20 s, then rinse in dH_2O for 2 min. Make up 250 ml of NuPage transfer buffer: 25 ml MeOH and 12.5 ml 20x buffer.
3. Equilibrate membrane for 2 min in 10 ml of transfer buffer. While equilibrating, assemble pads and filter paper in the cathode side of the transfer module and soak in transfer buffer.
4. Place gel large plate down, break open, and remove the small plate, being careful to leave gel on large plate. Cut off and remove wells.
5. Place wet filter paper on gel, remove bubbles, and flip over. Push the "foot" of the gel out of slot and carefully remove gel from plate. Cut off the gel "foot."
6. Assemble module in the following order, from the bottom (cathode) side—2 pads, paper, gel, membrane, paper, 2 pads, top cover (anode).

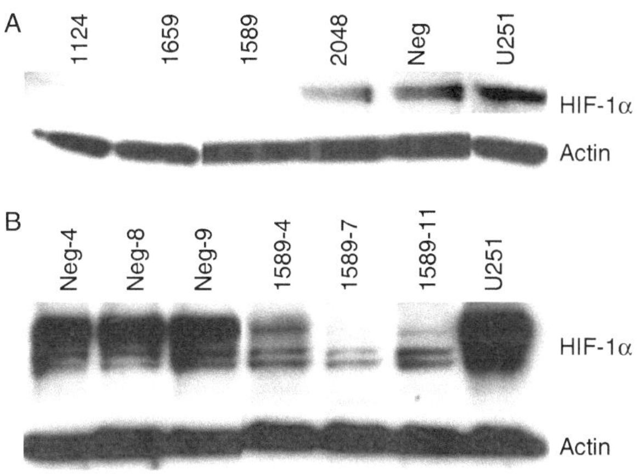

Fig. 14.1. (**a**) Western blot of protein isolated from transfected cells. The cells were allowed to recover for 30 h post-transfection and then exposed to hypoxia (<2% O_2) for 18 h. Protein was isolated 48 h post-transfection. Cells were transfected with siRNA-1124 (*lane 1*); siRNA-1659 (*lane 2*); siRNA-1589 (*lane 3*); siRNA-2048 (*lane 4*); and siRNA-Neg (*lane 5*). (**b**) Western blot of nuclear extract protein from subcloned U251 glioma cells with stably integrated shRNA plasmids pSil2.1_h-Neg or pSil2.1_h-1589 following 18-h hypoxia exposure. Lanes 1–3 show three different clones expressing shRNA-Neg. Lanes 4–6 show clones expressing shRNA-1589. Lane 7 is the wild-type U251 control. This Western blot shows some of the nonspecific bands that appear just below HIF-1α that can sometimes be confused with it. Running the gel out as long as possible will help separate these bands, as in (a)

7. Place in NuPage electrophoresis module, fill inner chamber with transfer buffer to just over the pads, and check for leaks.
8. Cover the bottom of an ice bucket (large enough to hold the entire gel apparatus) with ice and place the gel box inside. Fill the outer gel chamber with ice and ddH_2O. Attach lid and pack ice around sides and top, check for level. Transfer for 1 h at 40 V.
9. Remove the membrane from the gel, fold over a corner so that the protein side of the membrane can be identified and block overnight in 50 ml of 5% milk TBS-T, covered, at room temperature with gentle mixing.
10. The next day, remove membrane from blocking buffer and place it on plastic wrap. Then wrap membrane and smooth out bubbles and wrinkles.
11. Locate the 75 and 50 kD protein bands. Carefully cut the membrane in half just below the 75 kD marker. The top of the membrane (>75 kD) is probed for HIF-1, the bottom (<75 kD) is probed for actin simultaneously (*see* **Note 11**).
12. Dilute each primary antibody in 15 ml 5% milk/TBS-T (for HIF-1, 1:1500 dilution; for actin, 1:10,000) (*see* **Note 12**).
13. Place membrane in a new container with the primary antibody dilution and mix for 30 min.

14. Remove primary antibody and drain membrane container well. Rinse two times in 30 ml for 15 s each.
15. Rinse one time in 50 ml for 15 min, and then three times in 30 ml for 5 min each. During the final rinse, dilute secondary antibody 1:10,000 in 5% milk/TBS-T and bring ECL+ reagents to room temperature.
16. Add secondary antibody solution and mix for 20 min.
17. Rinse two times in 30 ml for 15 s each.
18. Rinse one time in 50 ml for 15 min.
19. Rinse three times in 30 ml for 5 min each. During final rinse, get plastic wrap cut to size and spread out on bench top and secure to bench with tape. Also, make up ECL+ detection solution—2 ml solution A + 50 μl solution B per small blot.
20. Drain membrane well and place on plastic wrap.
21. Add ECL+ drop-wise to membrane and activate for 5 min.
22. Drain membrane well and cover in new plastic wrap. Remove all bubbles and place in film cassette and expose to film for 30 s to 20 min.

3.6. Enzyme-Linked Immunosorbent Assay for VEGF

There are several different ELISA kits for VEGF, including kits from R&D Systems (Minneapolis, MN), GE Healthcare/Amersham (Piscataway, NJ), and Biosource (Camarillo, CA). While all of these work adequately, the R&D systems kit gave the most consistent results, though it is also the most expensive. The following protocol is based on its use.

VEGF is a main target of HIF-1 and, because it has a much longer half life, it can be measured in samples in which HIF-1 has been degraded or it can be used as supporting data for HIF-1 activity. We have found VEGF levels in glioma cell lines begin to be increased by 15 h of hypoxia in culture but do not reach their highest levels until 48 h *(7)*. The source of protein for this assay can be extracellular (cell culture medium) or cytoplasmic, such as from step 9 in the cell protein extract or steps 7 and 8 in the tumor extraction protocol. Care should be taken when using cell culture medium, because VEGF will accumulate throughout the hypoxia exposure and is not directly correlated with HIF-1 activity. The cytoplasmic VEGF levels tend to have a more linear relation to HIF-1 activation. For this reason we have generally assayed cell culture media only when VEGF levels are particularly low.

1. The calibrator diluent RD5K should be used when assaying cytoplasmic samples. Some tumors have high hemoglobin content in the cytoplasmic fraction, in which case the RD6U diluent may work better.

2. Load 100 μl of cytoplasmic protein, and the recommended 50 μl for cell culture media.
3. Continue with the protocol as outlined in the kit instructions. A plate washer (*see* **Sect. 2.6**) will speed up the process and help with reproducibility.
4. Absorbance at 450 nm is measured and corrected using the 540 nm reading on a Benchmark microplate reader (Bio-Rad Laboratories, Hercules, CA). Data analysis is performed using Microplate Manager III software (Bio-Rad Laboratories).
5. For comparison between experiments, calculate [VEGF]/[protein].

3.7. RNAi Design

For the latest siRNA design rules see Chap. 1. In addition, there are many Internet resources dedicated to siRNA design. For example, see the URLs listed in **Sect. 2.8**. In addition, a good overview of the considerations needed for designing RNAi experiments has been reviewed recently *(21)*. Many companies sell pre-tested siRNAs for most common known genes, or will design and synthesize siRNAs for any DNA sequence. For our experiments, siRNAs were designed by searching the coding sequence of HIF-1α for two adenines followed by 19 nucleotides with a GC content below 45% that did not contain more than 3 thymines or adenines in a row. These sequences were tested for possible homology to other human and mouse genes with BLAST (http://www.ncbi.nlm.nih.gov/BLAST). Four potential siRNA sequences were selected and prepared using Ambion's Silencer siRNA construction kit. After screening, siRNA 1589 (AAUUCAAGUUGGAAUUGGUAG) was selected for further use (**Fig. 14.1**). Two previously published siRNAs (1124 and 1659) *(22)* were used for validation. A negative control was designed by randomizing the sequence of siRNA1589 (AAUUAGCGUAGAUGUAAUGUG) and checking for non homology to any human or mouse gene by BLAST. siRNAs are named based on the nucleotide position in the HIF-1α mRNA sequence. Because of the potential for off-target effects, it is essential that all data are verified using at least three different siRNAs targeting the same gene. An alternate method is to use siRNAs designed in the 3′ untranslated region of a gene and rescue the knock-down with a cDNA expression vector *(23)*.

3.8. Immunohistochemistry for MIB-1 and GLUT-1

Several studies have shown that GLUT-1 and MIB-1 protein levels are regulated in part by HIF-1 *(8, 24–26)*. These molecules are good markers for HIF-1 levels when studying fixed or sectioned tissues where HIF-1 stability during sample handling is a factor (**Fig. 14.2**). While we do not present the application here, VEGF is also a good immunohistochemical marker. We use the automated BenchMark XT immunostainer from Ventana

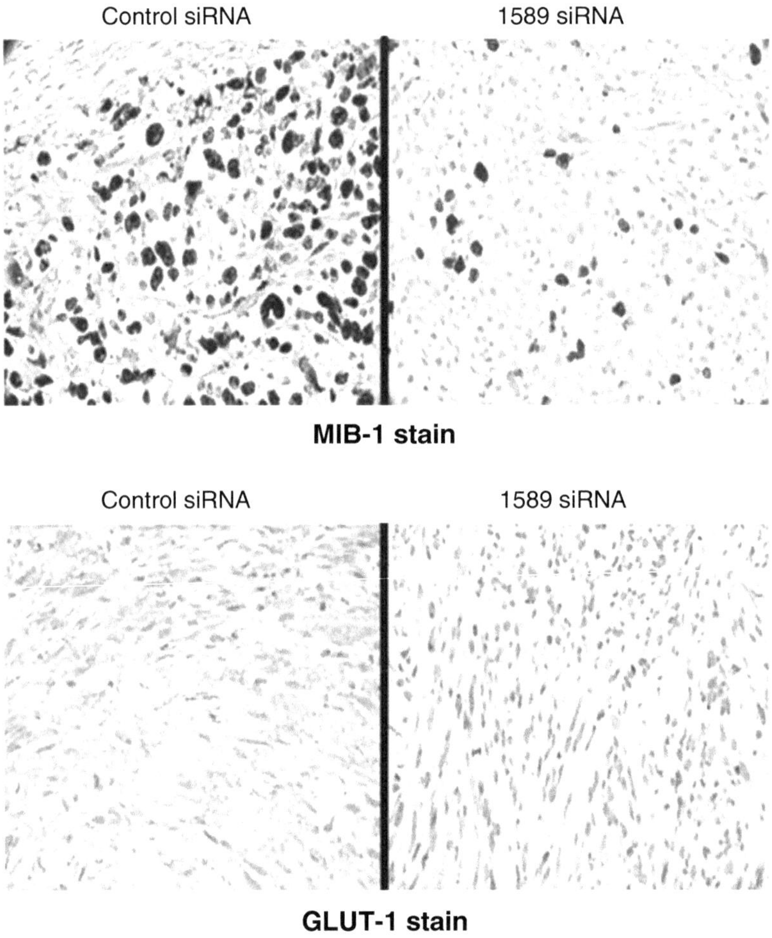

Fig. 14.2. Representative immunohistochemistry results for MIB-1 (*top*) and GLUT-1 (*bottom*) positive (control siRNA) and negative (1589 siRNA) stains of U251 tumors, grown in mouse flanks. The siRNA was complexed with 4x JetPEI and injected three times per week for 60 days. Pictures were taken at 400x. Brown color indicates MIB-1 or GLUT-1, blue color is the counter stain.

Medical Systems, but these protocols can be easily adapted to manual staining methods that are readily available from Ventana, Dako, and others.

1. Four-micron thick sections are cut from each sample and placed on glass slides.
2. The slides are allowed to air dry at room temperature.
3. The slides are melted in a 60°C oven for 30 min.
4. The sections are de-paraffinized in 3 changes of xylene for 5 min each and hydrated in graded alcohols for 1 min each (100% × 2, 95% × 2 and 70%), then placed in dH_2O (*see* **Note 13**).
5. The sections undergo HIER (heat-induced epitope retrieval) in citrate buffer (pH 6.0).

a. GLUT-1: Slides are microwaved for 15 min at half power. Slides are cooled in hot buffer for an additional 15 min.

b. MIB-1: Slides are placed in the DC2000 electric pressure cooker for 4 min. The total time in the pressure cooker is approximately 35–40 min.

6. The slides are taken from the hot buffer/pressure cooker and placed in dH_2O.
7. All of the staining steps are performed at 37°C on the automated BenchMark XT immunostainer.
8. The primary antibody dilution is applied for 32 min: GLUT-1: 1: 850; MIB-1: 1:200 (mAb Ki-67).
9. GLUT-1 only: Apply the general amplification reagent (Ventana Medical Systems).
10. The sections are detected using the IView DAB detection kit.
11. The slides are counterstained for 4 min with hematoxylin (Ventana Medical Systems).
12. The slides are removed from the autostainer and placed in a 0.2% Dawn/dH_2O mixture.
13. The slides are gently washed with the mixture to remove any coverslip oil applied by the automated instrument.
14. The slides are gently rinsed in dH_2O until all the Dawn/dH_2O mixture is removed.
15. The slides are placed in iodine for 30 s to remove any metal precipitates (from fixation).
16. The slides are placed in 5% sodium thiosulfate for 30 s to clear any iodine.
17. The slides are dehydrated in graded alcohols for 1 min each [70%, 95% (twice), and 100% (twice)] and dipped 10 times each in four changes of xylene.
18. The slides are coverslipped and allowed to air dry.

3.9. Cell Transfection and Imaging

There are many transfection reagents available for in vitro delivery of siRNAs. Because it is common for different cell types to have varying responses to different transfection reagents, the best one is often determined by the specific cells to be transfected. We have found the glioma cell lines U251 and U87 are transfected effectively with siRNA using Lipofectamine 2000 (Invitrogen) following the manufacturer's protocol (*see* **Note 14**). The following protocol has been used effectively with glioma cells in 6-well plates. All liquid handling is done in the tissue culture hood with a sterile technique.

1. Plate the cell at 30 50% confluence and allow them to recover for 24 h.

2. Mix 50–100 pmol siRNA with Optimem serum-free media to a final volume of 250 µl per well. It is helpful to combine replicate wells in one master mix.
3. In a separate tube, combine 5 µl of Lipofectamine 2000 per well with Optimem for a final volume of 250 µl per well. Again, a master mix will simplify this protocol, i.e., 30 µl of Lipofectamine 2000 in 1470 µl of Optimem for one 6-well plate.
4. Mix each siRNA/Optimem solution with equal volume of Lipofectamine/Optimem and hold for 20 min at room temperature.
5. Remove media from cell culture plates and add 500 µl of siRNA/Lipofectamine mix per well.
6. Hold plate in cell culture incubator for 18–24 h.
7. Add 3 ml of regular cell culture media (DMEM) and let recover for 18–24 h.
8. Change the media. Cells can now be exposed to hypoxia or other procedures.

3.10. Real-Time PCR

Assessing the relative levels of *HIF-1α* mRNA is the fastest way to screen candidate siRNAs for function. Because *HIF-1α* transcription levels are relatively unchanged between hypoxia and normoxia, cells do not need to be exposed to hypoxic conditions, although it is suggested all real-time results be checked in hypoxia. Because the reagents and methods used for real-time PCR are specific for the detection instrument, we only present a brief outline of our procedure, using the Bio-Rad MyIQ iCycler.

1. Cytoplasmic RNA is isolated from 60–80% confluent cells using the Qiagen RNeasy Mini spin column kit.
2. The Invitrogen Superscript III first strand synthesis for RT-PCR kit is then used with oligo-dT primers for cDNA synthesis. The cDNA is quantified by spectrophotometer and diluted to 500 ng/µl (*see* **Note 15**).
3. Real-time PCR is done using the IQ SYBR Green supermix kit (Bio-Rad), using the Bio-Rad MyIQ iCycler. The primer sequences are 10-11f- CCTGAGCCTAATAGTCCc; 11-12r-TACTCAGGACACAGATTTAGAC. Cycle times are 95°C, 5 min; 40 cycles: 95°C, 30 s, 58°C, 30 s, 72°C, 30 s; melt 95°C, 1 min. Actin is used for positive control.

3.11. Xenograft Flank Tumor Model

The xenograft flank tumor model is a relatively simple system for testing treatments in vivo *(27)*. The glioma cell lines U251, U87, and U373 have all been grown successfully by us in flanks of CD-1 (*nu/nu* genotype) athymic nude mice (Charles River, Wilmington, MA). It is important that these next two in vivo procedures receive Institutional Animal Care and Use Committee (IACUC) board approval before proceeding.

1. Cells are grown to 60–70% confluence (*see* **Note 16**).
2. Thaw Matrigel on ice. Place all tubes and syringes in ice to pre-cool (*see* **Note 17**).
3. Trypsinize 1 × 10^6 cells per flank injection. Thoroughly de-clump cells with pipetting, centrifuge, and remove supernatant and place on ice (*see* **Note 18**).
4. Cells are resuspended in 100 μl of 75% Matrigel/media solution per injection. To avoid bubbles, add the 25% volume of media first to resuspend the pellet, then add Matrigel (*see* **Note 19**).
5. Draw up 100 μl into 0.5 ml syringes, one syringe per flank (*see* **Note 20**).
6. Cells are injected subcutaneously on the mouse flank (*see* **Note 21**).
7. Tumors are allowed to establish for up to two weeks, depending on the cell growth, and then their size measured with calipers twice weekly.
8. Tumor volume is calculated as length × width × height and expressed as cubic millimeters.
9. Tumor volume measurements should be made by a single blinded observer to prevent observer bias and to avoid interpersonal differences in caliper tumor measurement.

3.12. siRNA Mouse Treatment

It is extremely important to assure all solutions and instruments used are sterile. All liquid handling is done in a sterile hood. It is also important to remove RNases from the work area, and follow basic RNA handling techniques. We have used both 4x JetPEI (Q-Biogene, Irvine CA) *(8)* and a noncommercial multifunctional carrier (MFC) developed by Z.R. Lu in the Department of Pharmaceutics and Pharmaceutical Chemistry, University of Utah *(28)*. Because MFC is not commercially available at this time, only the protocol for JetPEI is presented, following the manufacturer's instructions, using the nitrogen-to-phosphate (N/P) ratio of 8. It is assumed that the siRNAs have been annealed previously.

1. For each siRNA injection, add 4 μl of 150 μM siRNA (600 pmol) to 46 μl of 150 mM NaCl and hold for 5 min.
2. Prepare a master mix of 4x JetPEI by adding 8–42 μl of 150 mM NaCl per injection.
3. Add equal volumes of the PEI master mix to each siRNA; adjust for a final volume of 100 μl per injection. Hold at room temperature for 30 min.
4. Using 0.5 ml syringes, inject 100 μl per mouse intratumorally. (*see* **Note 22**).

5. Injections can be repeated daily, though a schedule of every 2 days has been adequate in our experience.

4. Notes

1. The protease inhibitor freezes at 4°C, so it must be thawed at room temperature and not kept on ice.
2. An ELISA for HIF-1 in 96-well format is available from Active Motif (Carlsbad, CA) that is useful for quick screens, verifying results, and determining the amount of transcriptionally active HIF-1. Care should be taken when interpreting the results, because the assay only detects HIF-1 that is bound to a short segment of DNA. Because HIF-1α requires HIF-1β to bind DNA, the results can be affected by overall HIF-1β protein levels or its ability to bind HIF-1α in solution. It is much faster than a Western blot, especially when screening many samples, but it is also fairly expensive and requires a 96-well plate reader.
3. We use a standard 9-L insulated lab tray that is available from Fisher Scientific. It is approximately 16" by 14". To maximize the ability to cool the cell culture plates quickly, it is important to assemble the ice/water bath correctly. Fill the tray approximately half full with crushed ice, then divide the ice such that one long side of the tray has twice the thickness of ice as the rest. This will accommodate holding tubes in place. Add water to the ice tray until the water level is slightly below that of the ice. Use an empty 60 mm plate (or whatever size you are using) to "pre-mark" where the cell plates will be placed. A well-formed ring of ice, approximately the height of the plate bottom, around each plate will make tipping and suctioning easier. Be careful that the water level is not above the lowest level of ice, or it can spill into the plates during washing and suctioning. If the water level is correct, the entire bottom surface of each plate will be in contact with it, but not the sides. The 9-L tray can fit eight 60 mm plates or six 100 mm plates, in a slightly off-set pattern of two rows.
4. To decrease the chance for error, place the tubes in the same order and location that the plates will be in.
5. To wash cells quickly, aliquot 4 ml into each plate from one pipette filled to 32 ml (assuming 8 plates). Discard the lid so as to not waste time removing and replacing it later. Slide the entire ice tray to swirl all plates simultaneously. To aspirate quickly, tilt plates using ice for support instead of picking them up.

6. To facilitate a quick and clean transfer of cells, unwrap and place the 5 ml pipette in the pipettor prior to scraping the cells. Then the plate can be continuously tilted while scraping and pipetting, so all the cells collect in the plate bottom.
7. If the pellet is large and white instead of small and clear, the addition of NP-40 was skipped.
8. Frozen tissue is often difficult to cut. Sectioning the tissue into approximate 500 mg pieces and freezing in separate tubes makes this step easier. If a portion still needs to be removed, hemostats or small scissors work well for breaking off sections of the frozen tissue.
9. For some tumors it is necessary to centrifuge longer. If the supernatant is still very cloudy with suspended pieces after the second spin, increase this time to 15 min for both spins.
10. The pellet from Step 7 can be used for VEGF ELISA, but some tumors do not yield consistent results. In this case, it is possible to use this supernatant for the ELISA, but only if the VEGF concentrations are high.
11. Most HIF-1 antibodies we have tried detect a very strong nonspecific band at ~66 kD. On occasion this band can make detection of the actual HIF-1 band (at ~120 kD) difficult, especially when HIF-1 concentration is low. Cutting the membrane also halves the time required for blotting HIF-1 and actin, because the blots can be done simultaneously.
12. The volume may need to be adjusted for the particular blotting container used. The volume should be such that the membrane is completely covered and the solution can mix freely around it.
13. The slides may sit in dH_2O until the next step is performed, up to 1 week, although proceeding to the next step within 60 min is suggested.
14. For determining the best ratio of Lipofectamine 2000 to siRNA, it is helpful to label some siRNAs with FAM fluorescent dye using the Silencer siRNA Labeling kit (Ambion) and visualize the transfection efficiency. Dead cells can be readily labeled using Invitrogen's Dead Cell reagent.
15. For some cell types with low quantities of mRNA, spiking the oligo-dT primers with a Hif-1α-specific primer can improve the real-time results. We have used RT-ex12- 5′-TCTGTTT-GGTGAGGCTGTCCG.
16. It is important that the cells be in logarithmic growth before being implanted so the tumors begin to grow quickly.
17. Matrigel is a liquid at 4°C but solidifies at room temperature. It is very important to keep all Matrigel containers cold for ease in handling.

18. Always calculate for one extra injection per five injections to avoid running out of cells.
19. A larger volume may be used with slower-growing cells, up to 250 μl. Care should be taken when using any more than this because it can cause undue stress on the mouse during injection.
20. It is tempting to fill up the 0.5 ml syringe for five injections; however, this leads to an unequal cell number injected over multiple mice and thus large disparities in tumor growth.
21. To ease later measuring, we have found injection just above the rib cage, which sticks out prominently on the side of the mouse when it is held by standard tail and nape immobilization, is best.
22. When injecting highly vascular tumors, excessive bleeding can occur. To reduce this, the injection can be done subcutaneously on the surface of the tumor, or slightly to the side. Applying direct pressure to the injection site with a paper towel may be necessary.

Acknowledgments

We thank Kristin Kraus for her excellent editorial guidance.

References

1. Zhong, H., De Marzo, A. M., Laughner, E., Lim, M., Hilton, D. A., Zagzag, D., Buechler, P., Isaacs, W. B., Semenza, G. L., and Simons, J. W. (1999) Overexpression of hypoxia-inducible factor 1alpha in common human cancers and their metastases. *Cancer Res* 59, 5830–5835.
2. Kaur, B., Khwaja, F. W., Severson, E. A., Matheny, S. L., Brat, D. J., and Van Meir, E. G. (2005) Hypoxia and the hypoxia-inducible-factor pathway in glioma growth and angiogenesis. *Neurooncol* 7, 134–153.
3. Vaupel, P., Thews, O., and Hoeckel, M. (2001) Treatment resistance of solid tumors: role of hypoxia and anemia. *Med Oncol* 18, 243–259.
4. Maxwell, P. H., Dachs, G. U., Gleadle, J. M., Nicholls, L. G., Harris, A. L., Stratford, I. J., Hankinson, O., Pugh, C. W., and Ratcliffe, P. J. (1997) Hypoxia-inducible factor-1 modulates gene expression in solid tumors and influences both angiogenesis and tumor growth. *Proc Natl Acad Sci U S A* 94, 8104–8109.
5. Ryan, H. E., Poloni, M., McNulty, W., Elson, D., Gassmann, M., Arbeit, J. M., and Johnson, R. S. (2000) Hypoxia-inducible factor-1alpha is a positive factor in solid tumor growth. *Cancer Res* 60, 4010–4015.
6. Zagzag, D., Zhong, H., Scalzitti, J. M., Laughner, E., Simons, J. W., and Semenza, G. L. (2000) Expression of hypoxia-inducible factor 1alpha in brain tumors: association with angiogenesis, invasion, and progression. *Cancer* 88, 2606–2618.
7. Jensen, R. L., Ragel, B. T., Whang, K., and Gillespie, D. (2006) Inhibition of hypoxia inducible factor-1alpha (HIF-1alpha) decreases vascular endothelial growth factor (VEGF) secretion and tumor growth in malignant gliomas. *J Neurooncol* 78, 233–247.

8. Gillespie, D. L., Whang, K., Ragel, B. T., Flynn, J. R., Kelly, D. A., and Jensen, R. L. (2007) Silencing of hypoxia inducible factor-1alpha by RNA interference attenuates human glioma cell growth in vivo. *Clin Cancer Res* 13, 2441–2448.
9. Li, L., Lin, X., Staver, M., Shoemaker, A., Semizarov, D., Fesik, S. W., and Shen, Y. (2005) Evaluating hypoxia-inducible factor-1alpha as a cancer therapeutic target via inducible RNA interference in vivo. *Cancer Res* 65, 7249–7258.
10. Rapisarda, A., Zalek, J., Hollingshead, M., Braunschweig, T., Uranchimeg, B., Bonomi, C. A., Borgel, S. D., Carter, J. P., Hewitt, S. M., Shoemaker, R. H., and Melillo, G. (2004) Schedule-dependent inhibition of hypoxia-inducible factor-1alpha protein accumulation, angiogenesis, and tumor growth by topotecan in U251-HRE glioblastoma xenografts. *Cancer Res* 64, 6845–6848.
11. Semenza, G. L. (1999) Regulation of mammalian O2 homeostasis by hypoxia-inducible factor 1. *Annu Rev Cell Dev Biol* 15, 551–578.
12. Richard, D. E., Berra, E., and Pouyssegur, J. (1999) Angiogenesis: how a tumor adapts to hypoxia. *Biochem Biophys Res Commun* 266, 718–722.
13. Wang, G. L., Jiang, B. H., Rue, E. A., and Semenza, G. L. (1995) Hypoxia-inducible factor 1 is a basic-helix-loop-helix-PAS heterodimer regulated by cellular O2 tension. *Proc Natl Acad Sci U S A* 92, 5510–5514.
14. Forsythe, J. A., Jiang, B. H., Iyer, N. V., Agani, F., Leung, S. W., Koos, R. D., and Semenza, G. L. (1996) Activation of vascular endothelial growth factor gene transcription by hypoxia-inducible factor 1. *Mol Cell Biol* 16, 4604–4613.
15. Warburg, O. (1956) On the origin of cancer cells. *Science* 123, 309–314.
16. Robey, I. F., Lien, A. D., Welsh, S. J., Baggett, B. K., and Gillies, R. J. (2005) Hypoxia-inducible factor-1alpha and the glycolytic phenotype in tumors. *Neoplasia* 7, 324–330.
17. Xu, R. H., Pelicano, H., Zhou, Y., Carew, J. S., Feng, L., Bhalla, K. N., Keating, M. J., and Huang, P. (2005) Inhibition of glycolysis in cancer cells: a novel strategy to overcome drug resistance associated with mitochondrial respiratory defect and hypoxia. *Cancer Res* 65, 613–621.
18. Lu, H., Forbes, R. A., and Verma, A. (2002) Hypoxia-inducible factor 1 activation by aerobic glycolysis implicates the Warburg effect in carcinogenesis. *J Biol Chem* 277, 23111–23115.
19. Kendler, A., and Dawson, G. (1990) Progressive hypoxia inhibits the de novo synthesis of galactosylceramide in cultured oligodendrocytes. *J Biol Chem* 265, 12259–12266.
20. Steinbach, J. P., Wolburg, H., Klumpp, A., Probst, H., and Weller, M. (2003) Hypoxia-induced cell death in human malignant glioma cells: energy deprivation promotes decoupling of mitochondrial cytochrome c release from caspase processing and necrotic cell death. *Cell Death Differ* 10, 823–832.
21. Cullen, B. R. (2006) Enhancing and confirming the specificity of RNAi experiments. *Nat Methods* 3, 677–681.
22. Yeung, H. Y., Lai, K. P., Chan, H. Y., Mak, N. K., Wagner, G. F., and Wong, C. K. (2005) Hypoxia-inducible factor-1-mediated activation of stanniocalcin-1 in human cancer cells. *Endocrinology* 146, 4951–4960.
23. Lassus, P., Rodriguez, J., and Lazebnik, Y. (2002) Confirming specificity of RNAi in mammalian cells. Sci STKE 2002, PL13.
24. Tan, C., de Noronha, R. G., Roecker, A. J., Pyrzynska, B., Khwaja, F., Zhang, Z., Zhang, H., Teng, Q., Nicholson, A. C., Giannakakou, P., Zhou, W., Olson, J. J., Pereira, M. M., Nicolaou, K. C., and Van Meir, E. G. (2005) Identification of a novel small-molecule inhibitor of the hypoxia-inducible factor 1 pathway. *Cancer Res* 65, 605–612.
25. Chau, N. M., Rogers, P., Aherne, W., Carroll, V., Collins, I., McDonald, E., Workman, P., and Ashcroft, M. (2005) Identification of novel small molecule inhibitors of hypoxia-inducible factor-1 that differentially block hypoxia-inducible factor-1 activity and hypoxia-inducible factor-1alpha induction in response to hypoxic stress and growth factors. *Cancer Res* 65, 4918–4928.
26. Pansare, V., Munkarah, A. R., Schimp, V., Haitham Arabi, M., Saed, G. M., Morris, R. T., and Ali-Fehmi, R. (2007) Increased expression of hypoxia-inducible factor 1alpha in type I and type II endometrial carcinomas. *Mod Pathol* 20, 35–43.
27. Jensen, R.L., Leppla, D., Rokosz, N., and Wurster, R.D. (1998) Matrigel augments xenograft transplantation of meningioma cells into athymic mice. *Neurosurgery* 42, 130–135; discussion 35–36.
28. Wang, X. L., Nguyen, T., Gillespie, D., Jensen, R., and Lu, Z. R. (2008) A multifunctional and reversibly polymerizable carrier for efficient siRNA delivery. *Biomaterials* 29, 15–22.

Chapter 15

RNA Interference-Mediated Validation of Genes Involved in Telomere Maintenance and Evasion of Apoptosis as Cancer Therapeutic Targets

Marco Folini, Marzia Pennati, and Nadia Zaffaroni

Abstract

The discovery of new cancer-related therapeutic targets is mainly based on the identification of genes involved in pathways selectively exploited in cancer cells, including those leading to unlimited replicative potential, evasion of apoptosis, angiogenesis, tissue invasion and metastatic spread. Potentially, a gene – or a gene product – is recognized as a cancer target whether its modulation in experimental models can specifically modify or revert the cancer phenotype. As soon as RNA interference (RNAi) – a natural gene silencing mechanism – was demonstrated in mammalian cells, it rapidly became an essential means for gene knockdown in preclinical models, making it possible to define the role of several human genes and to identify those specifically involved in the onset and progression of cancer. Owing to its powerful gene-silencing properties, RNAi has been proposed as a useful tool to validate new therapeutic targets and to develop innovative anticancer therapies. This chapter summarizes the findings from recent studies relying on the use of RNAi-based approaches to functionally validate therapeutic targets related to two tumor hallmarks: the unlimited replicative potential (i.e., activation of telomere maintenance mechanisms) and evasion of apoptosis (i.e., up-regulation of anti-apoptotic factors).

Key words: Human cancer, RNA interference, telomeres, telomerase, telomere-related proteins, inhibitors of apoptosis proteins, apoptosis.

1. Introduction

1.1. Telomere Maintenance Mechanisms

Telomeres are essential genetic elements consisting of repeated G-rich DNA arrays and associated proteins located at the ends of eukaryotic chromosomes (**Fig. 15.1**) *(1)*. Mammalian telomeres contain from <1 to >50 kb of telomeric DNA, depending on the species and cell type. Functional telomeres are essential for

M. Sioud (ed.), *Methods in Molecular Biology, siRNA and miRNA Gene Silencing, vol. 487*

DOI: 10.1007/978-1-60327-547-7_15

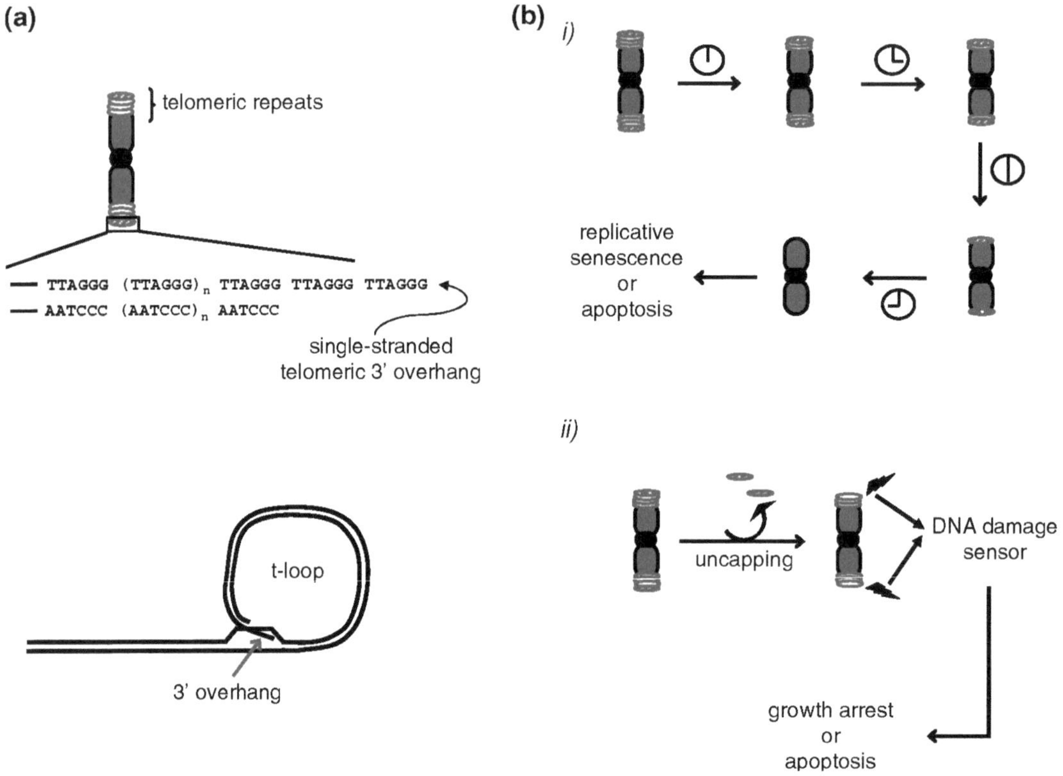

Fig. 15.1. (**a**) Schematic representation of a metaphase chromosome. The sequence of human telomeric repeats and a draft of the telomeric t-loop are indicated. (**b**) Representation of the pathways leading to cell growth arrest or death following interference with telomere structure and function: (i) progressive telomere shortening within each round of cell division as a result of the lack of telomere maintenance mechanisms; (ii) telomere uncapping arising from the loss of telomere protection as a consequence of the interference with telomere-associated proteins or destabilization of telomeric t-loops.

cell survival and continued proliferation (**Fig. 15.1**): they act as a "mitotic clock" that counts the number of divisions a cell can undergo (the telomere hypothesis) *(2)*. Telomeres progressively shorten and 50–200 bp of telomeric DNA are lost with every round of replication in all somatic cells because of the end-replication problem (i.e., the incomplete replication of lagging-strand DNA synthesis and chromosome-end processing). When telomeres become critically short (Hayflick's limit), they trigger a cellular response that leads to cell growth arrest or death *(3)*. Besides their ability to limit the number of cell divisions, telomeres play an essential role to preserve the genome stability *(4,5)*. Human telomeres are composed of a double-stranded repetitive sequence $(TTAGGG)_n$ and a short 3′ single-stranded overhang *(4,5)*. Chromosome integrity is guaranteed by the 3′ overhang that invades the double-stranded telomeric DNA, displaces the G-strand and anneals with the C-strand giving rise to a large duplex lariat structure (**Fig. 15.1**) *(6)*. This telomeric structure, known as t-loop, physically plugs up the end of chromosomes and protects them from being recognized as double-strand breaks and from the action of endonucleases. The loss of telomere protection (i.e., uncapped chromosome ends resulting from the disruption of telomere structure or as a consequence of telomere attrition) is usually referred to as telomere dysfunction *(7)*. Dysfunctional telomeres are at great risk of degradation, recombination or fusion by cellular DNA repair systems, and therefore can lead to genomic instability, a common cause and hallmark of cancer.

Telomere length maintenance is likely to be important for cellular immortalization *(8)*, and hence for cancer biology *(8,9)*. In most (85–90%) human tumors, telomere shortening is counterbalanced by de novo synthesis of telomeric DNA catalyzed by telomerase *(9,10)*. Some tumors (up to 15%), however, do not have telomerase activity and maintain their telomeres by one or more mechanisms referred to as alternative lengthening of telomeres (ALT) *(11,12)*. Telomere dynamics in ALT cells are consistent with a recombination-based mechanism, and characteristics of ALT cells include unusually long and heterogeneous telomeres and nuclear structures called ALT-associated promyelocytic leukemia bodies (APBs) that contain telomeric DNA, telomere-specific binding proteins and proteins involved in DNA recombination and replication *(13)*. Since the presence of telomere maintenance mechanisms has been identified as one of the six hallmarks of cancer *(9)*, approaches aimed at interfering with such mechanisms could represent novel and promising anticancer therapies *(14)*.

Human telomerase is a ribonucleoprotein complex and its main core consists of a catalytic subunit, the human telomerase reverse transcriptase (hTERT) *(15)*, and the human telomerase RNA component (hTR) *(16)*. The *hTR* gene is located on chromosome 3 and encodes for the telomerase RNA component, which consists of a 451-nucleotide long RNA. hTR contains a sequence located at its 5′ end which acts as a template for the addition of telomeric repeats at the 3′ terminus of the linear chromosomes during the enzyme's catalytic cycle *(16)*. The human telomerase RNA component is consistently expressed in almost all human tissues and, for this reason, does not represent a limiting factor for telomerase activity but is essential for the enzyme's catalytic activity through its association with the catalytic subunit *(17)*. hTERT is a 127 kDa protein comprising a specific telomerase domain (T-motif) and shares structural and functional properties with reverse transcriptases. It is encoded by a 37-kb long gene located on chromosome 5 and composed of 16 exons and 15 introns. The catalytic component of telomerase is typically expressed in telomerase-positive tumor tissues and during embryonic development, while its expression is progressively lost during differentiation of somatic cells *(17)*. The protein is almost undetectable in most normal human somatic cells, with the exception of the proliferative cells of renewal tissues *(17)*.

Such evidence has suggested that hTERT is the limiting factor for the restoration of telomerase activity and its expression is strictly regulated at multiple levels *(17)*. The *hTERT* gene undergoes transcriptional regulation that is mediated by different transcription factors *(17)*. The hTERT pre-mRNA is post-transcriptionally modified by alternative splicing, a process which generates different hTERT transcripts with opposite functions. In particular, the α-variant (which lacks conserved residues from the catalytic core of the protein) acts as a dominant negative *(18)*. Additional post-translational mechanisms such as the assembly of telomerase in a large complex holoenzyme mediated by hsp90 and p23 *(17)* and the phosphorylation/dephosphorylation by protein kinases (PKC, Akt and c-Abl) and protein phosphatase 2A *(17)* are involved in the regulation of hTERT activity. The hTERT protein undergoes cellular relocalization from the cytoplasm to the nucleus (a process presumably mediated by the 14-3-3 protein) and can be sequestered in a form of enzymatically inactive complex into the nucleolus through its interaction with PinX1 *(19)*. Recent findings have contributed to uncover unexpected functions of telomerase. Specifically, it has been demonstrated that hTERT preserves the integrity of the ends of chromosomes through its capping function by protecting telomeres in unclosed loops and preventing them from resembling DNA double-strand breaks *(19)*. In addition, novel functions of telomerase that might have a potentially important role in tumor cells are related to the ability of hTERT to cross-link telomeres and enhance genomic stability and DNA repair as well as to maintain tumor cell survival and proliferation via enzymatic activity-independent intermolecular interactions involving p53 and poly-(ADP-ribose) polymerase *(20,21)*.

Other proteins are involved in the regulation of telomerase expression and activity. Specifically, telomerase-associated protein 1 (TEP1) was found to interact with hTR *(22)*, even though its specific role in the telomerase complex is still unclear. The "Ever Shorter Telomeres 1" (EST1) protein is an essential regulatory component of the telomerase holoenzyme *(23)*. It has been shown that a primary function of this subunit is to recruit telomerase to chromosome termini *(24)*. In addition EST1 may act as a positive regulator of telomerase once the enzyme has been brought to the telomere *(24)*. A nucleolar protein altered in dyskeratosis congenita, dyskerin, carries out two separate functions, both fundamental for proliferating cells: the pseudourydilation of ribosomal RNA and the stabilization of telomerase enzymatic complex through the binding of a specific sequence within hTR *(23)*. It has been demonstrated that dyskerin alterations result in the reduction of ribosome rRNA processing rate, in the degradation of hTR and subsequent impairment of telomerase enzymatic activity *(25)*. Such alterations can explain the reduced proliferative capacity and increased susceptibility to cancer observed in patients affected by dyskeratosis congenita *(26)*, thus suggesting a specific role of the protein in tumorigenesis.

The telomerase complex is not the unique factor required to maintain telomere length and structure. There are three further proteins directly associated to telomeric DNA: telomeric repeat binding factor 1 and 2 (TRF1 and TRF2) and human protection of telomeres 1 (POT1). These proteins, interconnected to additional factors (i.e., TIN2, Rap1, TPP1), form a protein complex referred to as shelterin *(27)*. Shelterin exerts a protective function at the telomere level by avoiding chromosome ends from being recognized as sites of DNA damage and inappropriately processed by DNA repair systems *(27)*. TRF1, TRF2 and POT1 are essential and crucial regulators of telomere structure, capping and length control, respectively. TRF1 and TRF2 specifically bind to double-stranded telomeric repeats through a Myb-like helix/turn/helix motif whereas POT1 interacts with the 3′ overhang by its oligonucleotide binding domain *(28,29)*. TRF1 is a ubiquitously expressed protein that negatively regulates telomere length by physically preventing telomerase from acting on telomere ends *(30)*. Its function is regulated by TIN2, which interacts directly with TRF1, thus enhancing TRF1-dependent pairing of telomeric repeats *(31)*. The number of TRF1 proteins present on the end of a chromosome correlates with the length of the telomere. TRF1 also bind proteins that are thought to have non-telomeric functions, such as the poly-ADP ribosylase (PARP), tankyrase 1 (TRF1-interacting ankyrin-related ADP-ribose polymerase 1), a component of the human telomeric complex *(32)*. It has been demonstrated that telomere elongation by tankyrase 1 requires the catalytic activity of its PARP domain and does not occur in telomerase-negative primary human cells *(33)*. Tankyrase 1 lacks a nuclear localization signal, shows a predominantly perinuclear and cytoplasmic distribution and it has been found to localize at the telomeres of metaphase chromosomes in cells that over-express TRF1 *(34)*. The precise functions of tankyrase 1 are not completely understood. However, it has been proven that tankyrase 1 can ADP-ribosylate TRF1, thus interfering with its negative regulation of telomere length *(35)*. Moreover, the localization of tankyrase 1 at the nuclear periphery and its PARP activity – commonly associated with DNA repair – raises the possibility that tankyrase 1 participates in repairing or signaling the occurrence of telomere dysfunctions *(35)*. Whereas TRF1 is directly involved in telomerase regulation, TRF2 has unique functions at telomere. It stabilizes the 3′ overhang and helps to form the t-loop structure at the end of the telomere, thus playing a key role in the protection of chromosome ends *(36)*. TRF2 brings to telomeres a variety of proteins involved in DNA damage response, including MRN complex (MRE11/Rad50/Nbs1, key players of the homologous recombination and non-homologous end joining), Ku heterodimer and ATM kinase *(5)*, and it is considered an early component of the DNA repair damage response system *(7)*. POT1 is a single-stranded telomeric DNA binding protein, which specifically binds to the G-rich 3′ telomeric overhang *(29)*. Even though the function of POT1 remains uncertain, it has been demonstrated that, in association with TRF2, it participates in the protection of telomeres by contributing to the t-loop formation as well as by regulating the nucleolytic processing responsible for the 3′ overhang formation *(37)*. In addition, POT1 acts as a negative regulator of telomere length, presumably by blocking telomerase from gaining access to 3′ telomere ends *(27)*.

1.2. The Inhibitors of Apoptosis Protein (IAP) Family

Apoptosis, an active form of cell death, plays a central role in the development and homeostasis of multicellular organisms. This process is tightly regulated at various levels and its deregulation is involved in a wide spectrum of disease such as proliferative and degenerative disorders. Cancer cells are often characterized by increased resistance to apoptosis *(38)* which enables their survival under abnormal growth stimulation and mediates their increased resistance to various forms of cellular stress, such as DNA damage, hypoxia or nutrient deprivation *(38)*. Dysfunction of the apoptotic pathway is considered to be a major cause of tumor resistance to treatment, since many chemotherapeutic agents and radiation act by inducing apoptosis. Therefore, specific targeting of factors involved in the apoptotic pathways could represent a promising approach to enhance cancer cell sensitivity to chemotherapy and radiotherapy.

Apoptotic cell death occurs primarily through two well-characterized pathways: the extrinsic or death receptor pathway and the intrinsic or mitochondrial pathway (**Fig. 15.2**). In the death receptor pathway, ligands such as tumor-necrosis factor (TNF), FAS ligand or TNF-related apoptosis-inducing ligand (TRAIL) interact with their respective death receptors. These interactions ultimately lead to the activation of the proteases caspase-8 and caspase-10 *(39)*. The intrinsic pathway involves the release from the intermembrane space of the mitochondria of cytochrome *c* which interacts with apoptotic protease-activating factor-1 (Apaf-1) in the cytoplasm. This interaction causes the assembly of a multiprotein caspase activating complex (apoptosome) and leads to activation of caspase-9 *(40)*. The intrinsic and extrinsic pathways for apoptosis converge on downstream effector caspases (i.e., caspases-3, -6 and -7), resulting in a proteolytic cascade that leads to the cell death phenotype (i.e., DNA fragmentation, chromatin condensation, cell shrinkage and membrane blebbing) *(39,40)*. Caspase-3 and caspase-7 are targets of suppression by an endogenous family of antiapoptotic proteins called inhibitor of apoptosis proteins (IAPs), which also interfere with caspase-9 processing (**Fig. 15.2**). To date, eight IAP-family members have been identified by the presence of one to three tandem baculoviral IAP repeats (BIRs): NAIP, c-IAP1 (MIHB, HIAP-2), c-IAP2 (HIAP-1, MIHC, API2), XIAP (hILP, MIHA, ILP-1), survivin, Apollon/BRUCE, livin (ML-IAP, KIAP) and ILP-2 *(41)*. The IAP family members have widespread anti-apoptotic functions and some of these proteins are highly expressed in human tumor cells *(42– 44)*. Structure-function studies have demonstrated the requirement for at least one BIR domain for suppression of apoptosis *(41)*.

The first mammalian IAP to be identified was the neuronal apoptosis inhibitory protein (NAIP), which was isolated during a positional cloning effort to identify the causative gene for spinal muscular atrophy *(45)*. NAIP contains three BIR domains and a very large and unique C-terminus including a nucleotide-binding oligomerization domain (NOD) *(41)*. The anti-apoptotic activity of NAIP is probably achieved through either the interaction of its BIR domains with cellular caspases, including caspase-3 and caspase-7 *(46)*, or the ATP-dependent inhibition of effector caspase-9 *(47)*. NAIP also appears to protect cells from apoptosis through selective activation of the mitogen-activated protein (MAP) kinase, JNK1 *(48)*. It has been demonstrated that NAIP

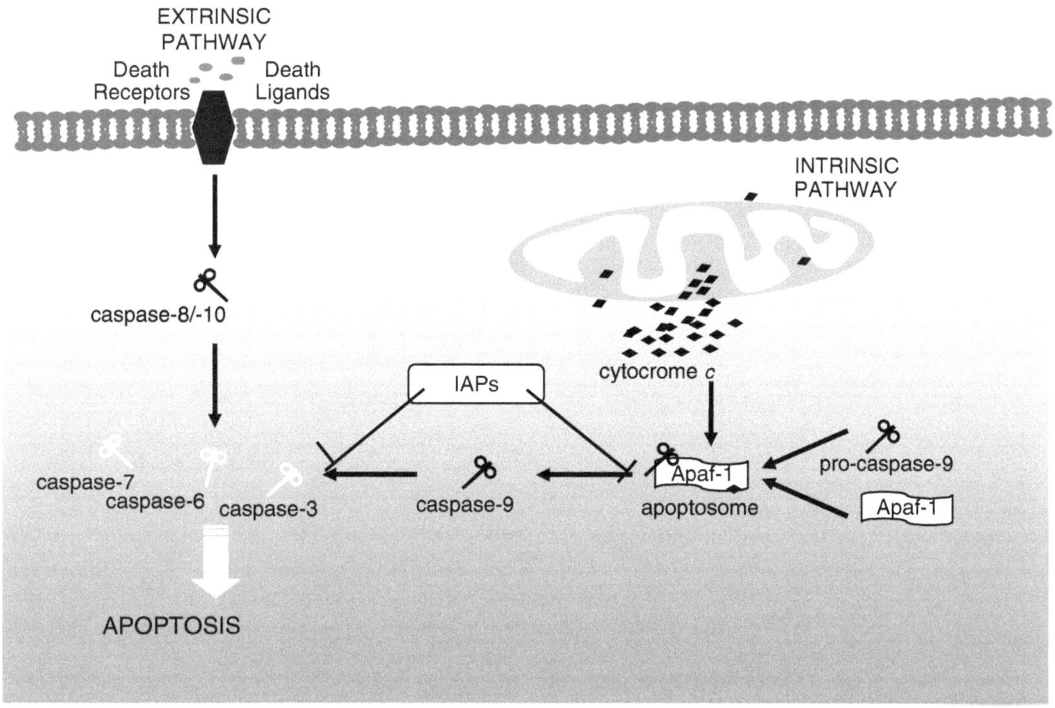

Fig. 15.2. Schematic representation of the extrinsic (death receptor-mediated) and the intrinsic (mitochondrial) pathways leading to apoptosis in human cells.

is up-regulated in some human malignancies such as acute myelogenous leukemia and esophageal cancer *(49–51)*.

c-IAP1 and c-IAP2 proteins possess three BIR domains and a RING domain required for the regulation of the self-ubiquitination and protein degradation *(41)*. Unique to c-IAP1 and c-IAP2 is the presence of a caspase recruitment domain (CARD), which typically mediates oligomerization with other CARD-containing proteins. The function of the CARD domain in the IAPs is not yet demonstrated but probably it allows IAP to form protein-protein interactions with Apaf-1 and some death domain-containing proteins *(41)*. Both c-IAP1 and c-IAP2 are not potent inhibitors of caspases-3, -7 and -9. In fact, despite their protein scaffold being suitable for direct caspase inhibition, they do not possess specific caspase inhibitory interaction sites *(52)*. Overexpression of both c-IAP1 and c-IAP2 has been observed in a variety of malignancies including medulloblastomas, glioblastomas and renal-cell, gastric, non-small cell lung and esophageal squamous cell carcinomas *(41)*.

XIAP contains three BIR domains (BIR 1–3), followed by a RING-finger domain with E3 ubiquitin ligase activity *(41)*. Structural and functional studies have demonstrated that the anti-apoptotic activity of XIAP is related to its ability to directly bind and inhibit caspases-3, -7 and -9, but not caspases-1, -6, -8 or -10 *(53)*. Specifically, the BIR3 domain of XIAP binds to the N-terminus of active caspase-9, preventing its dimerization and inhibiting its protease activity *(54)*, whereas the BIR2, which acts in tandem with its N-terminal linker, binds to and blocks the active site of caspase-3 and -7 *(55)*. The function of the BIR1 domain is unknown, but it has been suggested that it could increase the XIAP efficiency to bind caspases *(56)*. In addition, the RING domain can promote the degradation of several proteins by marking them with ubiquitin molecules and reduce the anti-apoptotic effects of XIAP when mutated *(57)*. XIAP has also been implicated in a variety of intracellular signaling events including NF-κB *(58)*, TGF-β *(59)* and c-Jun-N-terminal kinase pathways *(48)*. Furthermore, evidence indicates XIAP as a regulator of the cell cycle through its binding to MAGE-D1 and NRAGE cell cycle regulators. It has been also recently identified a role for XIAP in copper homeostasis. The E3 ligase in the RING finger promotes the ubiquitination and degradation of COMMD1, a protein involved in the efflux of copper from the cell *(60)*. The direct interaction between copper and XIAP induces a substantial conformational change so that copper-bound XIAP is unable to inhibit caspases, and cells that express this form of the protein exhibit increased rates of cell death in response to apoptotic stimuli. Aberrant expression of XIAP has been observed in several types of human malignancies and seems to influence patient's outcome *(61–65)*. Several studies have demonstrated that upregulation of XIAP markedly reduces the apoptotic response of tumor cells to chemotherapy and radiotherapy *(66,67)*. Such evidence indicates XIAP as an attractive target for cancer therapy. In this context, AEG35156, a second-generation antisense oligonucleotide targeting the coding region of XIAP, is currently being evaluated in phase I/II clinical trials *(68)*.

The smallest IAP, survivin, is composed of a single BIR domain and an extended C-terminal α-helical coiled-coil domain *(41)*. In addition to its anti-apoptotic function, survivin plays a major role in the control of mitosis. To exert its function in cell division, survivin localizes to the mitotic apparatus and its physical association with INCENP, Aurora B *(69,70)* and Borealin/Dasra B *(71)* proteins is required to target the chromosomal passenger complex to the kinetochore, properly form the bipolar spindle, and complete cytokinesis *(69–71)*. Such a function of preservation of genome fidelity and regulation of microtubule dynamics requires a sharp cell-cycle-dependent transcription of the survivin gene during the mitotic phase as well as post-translational modifications of the protein including phosphorylation by the $p34^{cdc2}$ and Aurora B kinases and monoubiquitination *(72)*. This pathway seems to be dominant in normal cells and constitutes the primary function of survivin in adult tissues *(72)*. Other non-cell-cycle-dependent mechanisms driving *survivin* gene transcription independent of mitosis have been described, which involve tissue patterning circuits, cytokine activation, costimulatory messages and pleiotropic signaling mechanisms that are operative during development and generally upregulated in cancer cells *(73)*. Such non-cell-cycle-dependent pathways are thought to be dominant in tumors *(73)*. Furthermore, subcellular compartmentalization of survivin seems to play a role in the anti-apoptotic function of the protein. Specifically, the existence of a mitochondrial pool of survivin was reported to be exclusively associated with tumor transformation *(73)*. It has been recently found that survivin has a nuclear export signal and that in cancer cells the anti-apoptotic and mitotic roles of survivin can be separated through mutation of its nuclear export signal, which abrogates the cytoprotective activity of the protein but still allows mitosis to proceed *(74,75)*. Alternative splicing of the human survivin gene gives rise to five different isoforms *(76)*. Little is known about the functions of alternative spliced forms of survivin which are generally expressed at lower levels than the wild-type survivin. As survivin forms homodimers, it has been suggested that these isoforms could interfere with wild-type survivin function by forming heterodimers *(76)*. It has beeen shown that survivin-ΔEx3, survivin-2β and survivin-3β do not play a role in mitosis, but preliminary data suggest that heterodimerization of wild-type survivin with survivin-ΔEx3 is essential for the inhibition of mitochondrial-dependent apoptosis *(76)* and that the survivin-

2β splice variant attenuates the antiapoptotic activity of wild-type survivin *(72,76)*. Survivin expression in normal tissues is developmentally regulated and the protein was found to be absent or low in most terminally differentiated tissues. However, recent studies attribute a role to survivin in certain physiological processes including hematopoietic cell survival/proliferation, T-cell development/maturation and vascular endothelial cell homeostasis *(72)*. Several studies have demonstrated that survivin is overexpressed in most human solid tumor types and hematologic malignancies, and that such an overexpression is associated with clinicopathologic variables of aggressiveness and may represent a prognostic marker for patient's outcome *(72)*. Moreover, there is evidence that survivin plays an important role in the chemo- and radio-resistant phenotype of human cancer cells *(77–79)*. In recent years considerable efforts have been made to validate survivin as a new target in cancer therapy *(72)* and the applicability of survivin-targeted strategies for the clinical treatment of human tumors is currently under investigation as the first survivin inhibitors recently entered phase I and II clinical trials.

Apollon is the largest member of the IAP family. It contains only one BIR domain at its N-terminal region and it also includes a C-terminal E2 motif, which can form thioester bonds with ubiquitin *(41)*. Apollon has also been proposed to function as an E3 ligase and itself is regulated by ubiquitin-dependent degradation, mediated by E2 UbcH5 and E3 Nrdp1 *(80)*. It exerts its anti-apoptotic activity by promoting ubiquitination and degradation of the mature Smac/DIABLO *(81)*. In addition, Apollon binds to pro-caspase-9 and inhibits its cleavage *(80)*. It has been reported that Apollon is up-regulated in some brain tumor cell lines that are resistant to certain DNA-damaging agents, and that antisense oligonucleotide-mediated down-regulation of Apollon enhances the extent of apoptosis induced by these agents *(82)*.

Livin is composed of a single BIR domain and a C-terminal RING finger domain *(41)*. The protein has little direct effect on caspase activity and it is assumed that its anti-apoptotic effect may be due to its antagonist function on the interaction between XIAP and the pro-apoptotic protein Smac/DIABLO *(41)*. There are two different isoforms of the protein, livin-α and livin-β, which strongly differ in their anti-apoptotic properties in tumor cells *(83)*. Initially, livin had been linked to malignant melanoma *(83)*. However, there is accumulating evidence that it is also expressed in other tumor types, including breast, cervix, prostate and lung cancers as well as lymphomas *(83)*. Interestingly, livin exhibits a restricted expression pattern, since it has been found to be expressed in tumor cells but not, or to substantially lesser amounts, in most normal adult tissues *(83)*. Therefore, livin may contribute to tumorigenesis and inhibition of its expression may represent an interesting therapeutic strategy. In this context, different methods to counteract livin in tumor cells have been proposed to inhibit tumor growth and enhance tumor cell response to apoptosis-inducing agents *(83)*.

The IAP-like protein 2 (ILP-2) is the most recently identified member of the human IAP family *(41)*. ILP-2 is the product of a human testis-specific mRNA and its coding sequence is very similar to the C-terminal half of XIAP, the region containing the BIR3 and RING domains, and lacks the first two N-terminal BIR domains present in XIAP *(84)*. ILP-2 is an unstable molecule and its mechanisms of action are still largely unknown *(84)*.

2. RNA Interference

Drug discovery requires the identification and preclinical validation of biological targets. This process usually implies the use of molecular techniques that allow to analyze the expression and function of genes and proteins. In this context, RNA interference (RNAi) has proven to be a powerful tool in elucidating gene functions, as well as in the identification of unknown genes with putative functions relevant to disease *(85)*. RNAi is a regulatory mechanism of most eukaryotic cells that relies on the use of small double-stranded RNAs (dsRNA) to control gene expression *(85)*. The effectors of this pathway, known as small interfering RNAs (siRNA), are 21–22 bp long dsRNAs harboring a 2-nucleotide long 3′ overhang. SiRNAs are recognized by the RNAi machinery, which in turn leads to degradation of target mRNAs. Briefly, in mammalian cells siRNAs are produced from long dsRNA precursors after the cleavage by RNase III endonuclease Dicer, which hands off the siRNAs to the RNA-induced silencing complex (RISC). The main component of RISC is Argonaute (Ago-2), which possesses an inherent cleavage activity *(85)*. Once loaded onto RISC, double-stranded siRNAs are unwound, leaving the antisense strand to guide RISC to its homologous mRNA. Target mRNAs, with perfect or nearly perfect complementary to the antisense strand of siRNA, are recognized, cleaved by Ago-2 and hence degraded. In some instances, partial complementary between siRNA and target mRNA may also induce the repression of translation *(86)*. Because of its potency and selectivity, RNAi has been harnessed for silencing specific genes in several experimental models, becoming a useful tool for the identification and validation of genes involved in human diseases, including cancer, autoimmune and dominant genetic disorders and viral infections *(87–89)*. Such a broad spectrum of action makes RNAi an attractive choice for the development of future therapeutic interventions. In this context, it is possible to exploit this naturally occurring gene silencing mechanism to modulate the expression of a given gene. RNAi can be artificially triggered by two different approaches: (1) effector siRNAs are delivered to target cells as preformed duplexes and (2) effector siRNAs (i.e., short hairpin RNA, shRNA) are produced after intracellular processing of longer dsRNAs transcribed from plasmid or viral vectors, under the control of RNA polymerase III promoters as well as tissue- and tumor-specific promoters *(86)*.

2.1. Targeting Telomerase by RNAi

The first study dealing with the use of RNAi to interfere with telomerase function in cancer cells was reported by Kosciolek et al. *(90)* in 2003. Two types of RNA molecules targeted to hTR and hTERT components – chemically synthesized siRNAs and a long dsRNA expressed in target cells as a hairpin construct – were used. The ability of chemically synthesized siRNAs to inhibit telomerase activity was assessed in a panel of human cancer cell lines. Specifically, siRNA targeting the hTR component was more effective in inducing inhibition of the enzyme's catalytic activity than that designed to target the catalytic component hTERT. The anti-telomerase effect was concentration-dependent and relied on the transfection schedule. Furthermore, transfectant clones expressing the siRNA construct directed against the hTR subunit were characterized by a decreased hTR expression, a marked inhibition of the enzyme's catalytic activity and a reduced telomeric DNA content *(90)*. More recently, it has been shown that suppression of hTERT by a shRNA-expressing retroviral vector affected the proliferative potential of HT29 colorectal adenocarcinoma *(91)* and HeLa cells *(92, 93)* as a consequence of telomerase activity inhibition, telomere shortening and loss of telomeric 3′ overhang. In addition, hTERT-specific shRNA was able to reduce the tumorigenicity of HeLa cells in nude mice and to significantly increase the sensitivity of cervical cancer cells to conventional chemotherapeutic agents and radiation *(92)*. An enhancement of the radiosensitivity – via down-regulation of hTERT mRNA and protein expression – has been also recently demonstrated in HeLa cells by using a survivin promoter-driven siRNA expression vector targeting hTERT, suggesting that targeted tumor gene silencing systems could be a useful therapeutic approach for radiosensitization of human cervical carcinomas *(94)*.

Thus far, an impressive number of studies have clearly demonstrated that the interference with telomerase activity could be a promising approach to limit the growth of cancer cells. In addition, an ever-increasing body of evidence has helped to better dissect the functions and identify novel roles of telomerase in human cancers. In fact, data generated from RNAi-based experiments, together with those obtained in different studies dealing, for example, with the use of antisense-mediated approaches to target hTERT *(14)*, point to a pro-survival and anti-apoptotic role of telomerase, which would be independent of its ability to guarantee cell proliferation via telomere-elongating activity. In our lab, we recently demonstrated that targeting hTERT by different chemically synthesized siRNAs resulted in a variable degree of telomerase activity inhibition in PC-3 and DU145 prostate cancer cells *(95)*. Such an effect was paralleled by a marked reduction of hTERT mRNA and protein expression levels and inhibition of cell proliferation, which occurred without a concomitant impairment of telomere length and 3′ overhang. Moreover, significant reduction of the tumorigenic potential was observed when PC-3 cells were xenotransplanted into nude mice *(95)*. It has also been recently reported that transient transfection of T24 bladder cancer cells, and HepG2 and SMMC-7721 hepatocarcinoma cells with a DNA-based vector carrying shRNAs against hTERT resulted in the suppression of cell growth in vitro and in vivo as the consequence of hTERT down-regulation and concomitant reduction of c-myc expression levels *(96, 97)*. Very recently, Wang et al. *(98)* showed that shRNA-mediated hTERT knockdown induced significant inhibition of telomerase activity, down-regulation of c-myc and proliferating cell nuclear antigen and up-regulation of caspase-3 in nasopharyngeal carcinoma cells. In addition, hTERT knockdown in TB10 and U87MG human glioblastoma cell lines by an RNAi-based approach significantly inhibited the development of tumors in subcutaneously and intracranially grafted nude mice *(99)*. In accordance with these data, in a recent study it has been demonstrated that injection of lentiviral vectors encoding anti-hTERT siRNA significantly inhibited the growth of pre-established macroscopic xenograft tumors arising from U87MG glioblastoma cells *(100)*. The inhibitory effect on in vivo tumor growth was somewhat in contrast with the lack of effects on cell growth, cell cycle progression and telomere length observed in anti-hTERT siRNA expressing U87MG cells during short-term in vitro cultures *(100)*. Since in both studies *(99,100)* the in vivo effect occurred within a relatively small number of population doublings, hTERT down-regulation could have impaired in vivo cancer cell growth prior to significantly affecting telomere length and, presumably, as a consequence of the inhibition of its extra-telomeric functions. In this context, evidence that hTERT could exert functions other than its inherent telomere elongating activity has been further provided. RNAi-mediated down-regulation of endogenous hTERT in MCF-7 cells markedly increased apoptosis induced by both the 4625 Bcl-2/Bcl-X_L bi-specific antisense oligonucleotide and the HA14-1 Bcl-2 inhibitor *(101)*. Conversely, the ectopic expression of hTERT blocked Bcl-2-dependent apoptosis, suggesting that hTERT is involved in the mitochondrial pathway of programmed cell death. The role of hTERT as an anti-apoptotic factor has been recently highlighted by Massard et al. *(102)*. Specifically, they showed that hTERT suppression through RNAi sensitizes cancer cells to mitochondria-dependent apoptosis induced by DNA-damaging agents and reactive oxygen species via an increased activation of Bax protein *(102)*. Interestingly, no relationship was found between hTERT depletion and expression of Bax itself or Bax-regulatory proteins, suggesting that post-translational modifications in the Bax interactome accounted for the enhanced activation of Bax in hTERT-depleted cells *(102)*. Furthermore, using transformed human embryonic kidney (HEK 293) cells, p53- and p16^{INK4a}-null human ovarian cancer SKOV-3 cells and p53-null MDA-MB-157 human breast cancer cells, the role of hTERT in the regulation of cell cycle has been investigated *(103)*. In HEK 293 cells, hTERT knockdown resulted in increased p53 and p21 transcription and decreased cell proliferation. In p53-null MDA-MB-157 cells transfected with shRNA-expressing vectors targeting hTERT, down-regulation of telomerase catalytic component was paralleled by marked impairment of cell growth and increase in p21 expression, suggesting the ability of hTERT to regulate p21 expression in a p53-independent manner. Conversely, hTERT knockdown in p53- and p16^{INK4a}-null SKOV-3 cells, resulted in a decreased expression of p21 and did not affect cell growth *(103)*. Even though data from this study did not elucidate the exact mechanisms by which hTERT regulates p21, they point to a possible involvement of telomerase in a complex signaling network that can sustain cancer cell growth *(103)*.

Until a few years ago, it was believed that the impairment of cancer cell growth induced by targeting telomerase was ascribable to telomere shortening – as a result of inhibition of telomere lengthening activity of telomerase accomplished through the interference with hTR expression/function – or, alternatively, to the loss of telomere lengthening-independent functions of telomerase as a results of hTERT expression impairment. This widely accepted scenario has been partially modified on the basis of recent data obtained in Dr. Blackburn's lab. Specifically, it has been showed that a shRNA targeting hTR, expressed from a lentiviral vector, quickly inhibited the growth of HCT116 colon cancer cells and LOX melanoma cells, independently of p53 status or telomere length, and without bulk telomere shortening *(104)*. By contrast, no effect was detected in the immortalized, telomerase-negative VA13 cell line. Moreover, hTR down-regulation did not cause telomere dysfunctions in these experimental models, but induced a modulation of the global gene expression profile, including suppression of specific genes implicated in angiogenesis and metastasis *(105)*. Such evidence could be indicative of a novel response pathway distinct from the expression profile changes previously reported by the same authors and induced by a mutant-template telomerase RNA causing telomere dysfunctions *(104,105)*.

Taken together, these data – along with results obtained from studies based on others telomerase inhibitory strategies *(14)* – provide compelling evidence about telomerase as a promising anticancer target for the development of innovative clinical interventions. In this context, the first telomerase inhibitor, GRN163L, an oligonucleotide N3′-P5′ thiophosphoramidate targeting hTR, recently entered clinical trials *(106)*.

2.2. Targeting Telomere-Related Proteins by RNAi

Due to their essential role in regulation of telomere length and in protection of chromosome ends *(3–5)*, telomere-related proteins have been attracting increasing interest as potential cancer therapeutic targets, although their specific role in tumor cells with respect to normal cells is still to be disclosed. In fact, available information about the specific expression/function of telomere-related proteins in tumor tissue specimens is still scanty. It has been recently reported that genes encoding for these proteins display different patterns of expression in human breast cancer specimens and in normal breast tissues, suggesting different and sometimes opposing roles in mammary carcinogenesis *(107)*. For example, hTERT, hTR, tankyrase 1, EST1 and TEP1 were found to be up-regulated in tumor specimens, with hTERT and TEP1 correlating with clinical outcome. Conversely, POT1 transcription levels demonstrated a compelling trend to be lower in malignant than in normal tissues and much lower in those patients who develop recurrent disease. *(107)*. However, the role of these proteins in cancer development and progression is far from being understood and need to be deeply investigated in tumors of different histologic origin.

In this paragraph we will focus on recent data obtained by RNAi-mediated approaches to target telomere-related proteins. It has been showed that siRNA-mediated knockdown of TRF2 in combination with hTR down-regulation (accomplished through the use of a lentiviral vector bearing a mutant hTR template and a shRNA targeting hTR) resulted in an addictive inhibitory effect on cell proliferation of LOX melanoma cells *(108)*. In addition, it has been reported that the inhibition of TRF2 expression by RNAi partially reversed the resistance phenotype of multidrug-resistant variant SGC7901 gastric cancer cells suggesting a possible role of TRF2 in drug resistance of gastric cancer *(109)*.

As previously reported, TRF2 cooperates with POT1 to maintain telomere integrity *(110)*. In fact, the over-expression of exogenous POT1 can block the erosive effect on telomeric overhang and the occurrence of chromosome abnormalities detectable in different cells expressing a dominant negative form of TRF2 *(110)*. The role of POT1 in protecting telomeres has been evaluated in MCF-7 breast cancer cells exposed to anti-POT1 siRNAs *(110)*. Specifically, the down-regulation of POT1 mRNA resulted in the induction of apoptosis as a consequence of telomere dysfunction, increased expression of p53 and Bax and concomitant decrease of Bcl-2 levels *(110)*. The effects of RNAi-mediated POT1 down-regulation were also determined by Hockemeyer et al. *(37)* in a panel of human cell lines. These authors showed a significant increase in the levels of telomere-induced foci (TIFs) – assessed by the co-localization of 53BP1 and γ-H2AX with TRF1 – in the G_1 phase of the cell cycle in Hela cells and in primary hTERT-expressing BJ fibroblasts. Despite the occurrence of TIFs, cancer cells did not show significant growth alterations following POT1 knockdown. In contrast, fibroblasts responded to POT1 depletion with strongly reduced proliferation and induction of senescence, independently of the presence of telomerase *(37)*.

To address the role of tankyrase 1 as a positive regulator of telomere length, tankyrase 1-specific shRNAs were expressed in BJ-hTERT fibroblasts, characterized by an inherent and gradual telomere shortening rate *(111)*. Such shRNAs were able to induce a partial depletion of tankyrase 1 protein and a significant increase of BJ-hTERT telomere shortening rate but did not affect cell proliferation. By contrast, it has been demonstrated that an almost complete siRNA-mediated depletion of tankyrase 1 in HeLaI.2.11 cells induced a dramatic mitotic arrest and a predominance of abnormal mitotic figures, with chromosomes aligned on the metaphase plate, unable to proceed into anaphase and characterized by persistent association of sister chromatids at the telomere level *(112)*. Such a phenotype was efficiently rescued in siRNA-treated cells over-expressing wild-type tankyrase 1 *(112)* and seemed to be attributable to the reduced tankyrase-mediated PARsylation of nuclear mitotic apparatus protein *(113)*.

Dyskerin is involved in the stabilization of the telomerase enzymatic complex *(25,26)*. Dyskerin mRNA expression levels were found to be directly correlated to those of hTR and high dyskerin levels being associated with a poor clinical outcome in patients with breast carcinomas *(25)*, suggesting that dyskerin expression levels might influence tumor progression by affecting telomerase activity. To get an insight into the role of dyskerin in breast cancer cells, dyskerin mRNA levels were knocked-down by RNA interference in the MCF-7 breast cancer cells *(25)*. Forty-eight hours after transfection of siRNAs, the levels of dyskerin mRNA were significantly reduced compared with MCF-7 cells transfected with control sequences, and dyskerin down-regulation was paralleled by a reduction of hTR levels, telomerase activity as well as rRNA pseudo-uridylation *(25)*.

In an attempt to identify candidate ALT genes, Jiang et al. *(114)* demonstrated that the formation of APBs was significantly reduced in ALT-positive IIICF/c cells depleted by RNAi for telomere associated proteins (TRF1, TRF2, Tin2, RAP1, MRE11 RAD50 and NBS1). Because of the close linkage between ALT-mediated telomere maintenance and the ability to form APBs, these data suggest these telomere-associated proteins are also required for the maintenance of the ALT mechanism *(114)*. In accordance with this hypothesis, Zhong et al. *(115)* recently reported that shRNA-mediated depletion of NBS1 – with or without concomitant depletion of other members of the MRN complex (MRE11 and RAD50) – resulted in inhibition of ALT-mediated telomere maintenance, as evidenced by decreased telomere length, in ALT-positive IIICF/c cells. Specifically, in some clones there was an initial period of rapid shortening followed by stabilization of telomere length, whereas in other clones a continuous shortening (at a rate within the reported range for normal human somatic cells lacking a telomere maintenance mechanism) was observed. Conversely, depletion of NBS1 in telomerase-positive cells did not result in telomere shortening, suggesting the MRE11/RAD50/NBS1 complex as a possible specific target for ALT-positive tumors *(115)*.

2.3. Targeting IAPs by RNAi

Lima et al. *(116)* first demonstrated that down-regulation of XIAP by RNAi increased cellular apoptosis in MCF-7 breast cancer cells and sensitized tumor cells to etoposide and doxorubicin. Successively, several studies showed that inhibition of XIAP remarkably suppressed tumor cell growth and induced apoptosis in tumor models of different histologic origin, both in vitro and in vivo *(117–122)*. Moreover, XIAP down-regulation was able to overcome the resistance of tumor cells to chemotherapeutic agents *(120–122)* and APO2L/TRAIL *(117,118,123,124)*, and to enhance x-ray sensitivity of nonsmall cell lung cancer cells independently of *p53* gene status *(125)*. The importance of XIAP and c-IAP2 for the survival of pancreatic cancer cells following treatment with anti-cancer drugs was also recently confirmed by Lopes et al. *(126)*. Although down-regulation of these IAPs did not seem to have a cytotoxic effect per se, inhibition of XIAP and c-IAP2 was very efficient in increasing tumor cell sensitivity to cisplatin, doxorubicin and paclitaxel. Finally, Pan et al. *(127)* used an oncolytic adenoviral vector (ZD55) to express a shRNA targeting XIAP, and showed that infection of hepatocellular carcinoma cells with the combination of ZD55-XIAP-shRNA and ZD55-expressing TRAIL resulted in significant reduction of XIAP expression and potent antitumor activity both in vitro and in vivo.

Carvalho et al. *(128)* first showed that survivin was no longer detectable in HeLa cells 60 h after transfection with specific siRNAs and that survivin-depleted cells were delayed in mitosis and accumulated in prometaphase with misaligned chromosomes. In this model, loss of survivin activated the mitotic checkpoint, which was mediated by induction of p53 and was associated with the increase of its downstream target, p21. Several studies dealing with the use of chemically synthesized siRNAs or plasmid/viral vectors encoding shRNAs showed that RNAi-mediated survivin knockdown was able to reduce tumor cell proliferative potential and induce caspase-dependent apoptosis in a variety of human tumor cell models *(72,129–131)* as well as to decrease the formation of new tumors and the growth of already established lesions in nude mice *(72)*. Recently, Sarty et al. *(132)* showed that siRNAs targeting survivin significantly reduced the survival of activated K-Ras-transformed colon cancer cells compared with its normal isogenic counterpart in which the mutant K-Ras gene had been disrupted, suggesting that tumors expressing the activated K-Ras oncogene may be particularly sensitive to inhibitors of the protein. Survivin down-regulation also induced an enhanced apoptotic response of tumor cells of different histologic origin treated with diverse anticancer agents, TNFα and APO2L/TRAIL, and caused radiosensitization of a sarcoma cell line expressing wild-type p53 *(72,118,133)*. Moreover, transfection of endothelial cells with survivin-specific siRNAs induced a marked increase in the rate of apoptosis, a dose-dependent inhibition of their migration on vitronectin, and a decrease in capillary formation *(72)*. To verify whether alternative splicing variants of survivin are suitable targets in cancer treatment, the effects induced by specific inhibition of survivin-ΔEx3 and survivin-2β have been recently investigated in bladder cancer cells *(134)*. siRNA-mediated inhibition of survivin-ΔEx3 or survivin-2β caused a less pronounced antiproliferative effect (in term of induction of apoptosis, cell cycle arrest, and reduced colony formation) than that observed after simultaneous down-regulation of multiple survivin transcripts, including wild-type survivin. Moreover, a marked reduction of cell viability was observed in cells depleted for all the survivin transcripts after exposure to cisplatin, mitomycin C or gemcitabine. Conversely, the inhibition of survivin-ΔEx3 and survivin-2β did not result in the enhancement of tumor cell chemosensitivity, suggesting that these two survivin transcripts do not represent useful targets for anticancer treatment *(134)*. It has been also demonstrated that only the siRNA construct targeting wild-type survivin was able to reduce colony formation under hypoxic or normoxic conditions, whereas a less pronounced reduction of tumor cell plating efficiency was observed after specific knockdown of survivin-ΔEx3 and survivin-2β independently of oxygen concentration *(135)*. In addition, survivin-ΔEx3 and survivin-2β knockdown did not sensitize sarcoma cells to ionizing radiation neither under hypoxic nor under normoxic conditions, whereas down-regulation of wild-type survivin significantly enhanced the radiosensitivity of these cells *(135)*. To prevent non-physiological responses associated with persistent suppression of a gene that is essential for cell survival and cell cycle progression, systems allowing an inducible regulation of RNAi have been developed. In this context, Coumoul et al. *(136)* demonstrated that inducible suppression of survivin was efficiently achieved in mouse embryonic stem (ES) cells using a Cre-LoxP approach, as indicated by the decreased gene expression levels and the reduced proliferative potential.

Crnkovic-Mertens et al. *(137)* reported that RNAi-mediated silencing of livin led to caspase-3 activation, and increased apoptotic rate in response to different proapoptotic stimuli, including exposure to doxorubicin, etoposide, UV-irradiation and TNFα. A short-term interference with livin following transient transfection of cancer cells with chemically synthesized siRNAs only induced a modest increase of the apoptotic rate. By contrast, stable transfection with shRNA-expressing vector inhibited the growth of livin-expressing tumor cells in colony formation assays with high efficiency *(137)*. Recently, the same research group used an isoform-specific RNAi-mediated strategy to assess the individual role of the two endogenous splicing variants of livin *(138)*. They found that selective inhibition of livin-β, but not of livin-α, reduced tumor cell growth and sensitized HeLa cells to different proapoptotic stimuli, including UV irradiation, TNF-α and etoposide. Moreover, such a proapoptotic sensitization was specifically reverted by ectopic expression of livin-β but not of livin-α, suggesting that livin-β isoform plays a key role for the protection of HeLa cells from apoptosis *(138)*.

It has been reported that RNAi-mediated down-regulation of Apollon promotes apoptotic cell death as a consequence of a reduced degradation of Smac/DIABLO and caspase-9 *(80)* and sensitizes p53-mediated apoptosis in p53 wild-type tumors by maintaining high levels of p53 protein *(139)*. Qiu et al. *(140)* demonstrated that siRNAs targeting Apollon markedly reduced the levels of the protein in HeLa cells and increased the rate of spontaneous and TNFα-induced apoptosis, in terms of caspase-3 activation. Furthermore, in 293T cells, which are known to be resistant to apoptotic stimuli, the exposure to Apollon-specific siRNAs had no effect on the levels of active caspase-3 by itself but caused a marked increase in the apoptotic response in the presence of TNFα *(140)*.

Overall, results obtained in the different studies indicate IAPs as potential cellular determinants of the chemo- and radio-resistant phenotype of human tumor cells, and suggest that approaches aimed at down-regulating their expression may be relevant for the treatment of human tumors with the dual aim to inhibit tumor cell growth and enhance tumor cell response to apoptosis-inducing agents *(72,141,142)*.

3. Concluding Remarks

In the post-genomic era an increasing body of evidence about the molecular causes of human diseases has widened, at least in theory, the number of available targets to treat a complex malignancy such as cancer. As reported herein, siRNA-based approaches have been demonstrated to be efficient tools for the validation of tumor-related targets and have contributed to disclose the role of many genes in the onset of cancer *(86)*. It is now well established that telomerase represents a suitable target to impair cancer cell growth *(143,144)*. The enzyme is expressed in a wide variety of human cancers and its down-regulation can lead to cell growth arrest and programmed cell death as a consequence of the interference with telomerase enzymatic activity, which is responsible for the maintenance of telomere length and integrity, or with its pro-survival and anti-apoptotic functions (**Fig. 15.1**) *(14,144)*. However, prolonged treatments with telomerase inhibitors could lead to the occurrence of specific mechanisms of resistance, such as the expression of ALT phenotype *(13)*. Such a possibility together with the evidence that ALT mechanisms are typically present in a small but defined fraction of tumors (mainly of mesenchymal origin) and may coexist with telomerase within a tumor and in the same cancer cell as well *(11, 12)* emphasizes the notion that all tumor cells require a solution to the “end replication problem” *(2)*. Hereupon, it is reasonable to raise questions about the effects that could result from the interference with telomere structure, rather than telomere maintenance mechanisms, in cancer cells with respect to normal cells, and whether telomere-related proteins could represent potential targets for new anticancer interventions. In our opinion, the validation of telomere-related proteins as cancer-specific targets still requires a comprehensive survey of their expression and correlation with clinical parameters in tumors of different histologic origin as well as the establishment of sensible experimental models to properly study the biological consequences of their modulation and to get new insights into their role – if any – in cancer development and progression.

Increasing evidence indicates that the presence of mechanisms of cytoprotection and resistance to apoptosis confers high survival capability and low chemo- and radiosensitivity to many cancer cell types. Therefore, modulation of apoptosis by targeting anti-apoptotic proteins may be an important approach for treating cancer. Apoptosis inducers have been already introduced in the clinical armamentarium to treat cancer. However, such compounds are not devoid of collateral toxic effects because of the similarity of pathways/genes responsible for apoptosis control and execution in normal and tumor cells *(145)*. Therefore, the development of new apoptosis inducers with a strong selectivity for tumors is needed to minimize side effects and maximize the therapeutic efficacy. The overexpression of some IAPs, such as XIAP, survivin and livin, in tumors, and their very low or undetectable expression levels in corresponding normal tissues, suggests these proteins as promising and specific molecular targets for cancer therapy *(41)*. A better understanding of the effects exerted by IAP-family members on normal versus malignant cells should be important in identifying strategies that maximally disrupt IAP functions in cancer cells while minimally affecting normal tissues. In this context, in accord with the more recent view suggesting that the two main functions of survivin (i.e., spindle monitoring at mitosis and the ability to counteract apoptosis) are differentially exploited in normal and tumor cells, the identification of points at which these functions can be separated could open new possibilities for strategies aimed at eliminating cancer cells while preserving normal cell viability. Thus far, results obtained by different experimental studies aimed at targeting specific IAPs by means of RNAi-based approaches clearly demonstrated that the interference with the expression of these proteins results in marked anti-proliferative and pro-apoptotic activity, even though the therapeutic utility of their down-regulation remain to be further elucidated.

In conclusion, turning-off a pathogenic gene by RNAi-based technologies represents an appealing approach for the validation of cancer targets and for the development of innovative therapies (*see* Chaps. 11–14 and 22). There is a widely diffused opinion that RNAi provides a powerful tool for targeted inhibition of gene expression with respect to conventional antisense strategies, presumably because it relies on a natural process. Despite the unique assumed potential of RNAi, limitations in the use of such an approach have been described *(86)*. The design of siRNA has to be conducted carefully in order to achieve specific mRNA knockdown while minimizing undesired effects. To provide effector molecules with both selectivity and specificity, nontarget sequence homologies of both strands of an siRNA must be avoided *(87)*. Moreover, thermodynamic parameters that can influence the efficiency with which siRNA is unwound and assembled into the RISC complex need to be carefully determined *(146)*. The presence of a suitable target site in a given RNA represents a major determinant for the biological activity of siRNAs: intramolecular folding of target sequence within an mRNA or its association to regulatory proteins may hamper the binding of RISC *(147,148)*. Moreover, the choice of an adequate control siRNA is of outstanding importance in obtaining meaningful data from siRNA knockdown experiments *(148)*. Even the most carefully designed siRNA sequence might show a considerable amount of off-target effects or cause deregulation of micro-RNA pathway – another natural mechanism of gene regulation – as a result of competitive binding to proteins shared by both gene-silencing mechanisms *(149)*. In addition, cellular uptake represents a main hurdle that has to be overcome for an efficient therapeutic inhibition of gene expression and the exploitation of siRNAs in clinical trials *(88,148)*. Different strategies have been developed for efficient delivery of siRNA in vitro and in vivo, mainly dealing with the use of cationic-lipid or liposome formulations, electroporation, conjugation to peptides or aptamers *(87)*, and hydrodynamic delivery systems *(88)*. The stability in biological fluids also represents a difficult issue to be addressed. Naked siRNA half-life varies from seconds to minutes, due to degradation by endogenous nucleases and rapid elimination by kidney clearance. The introduction of a phosphorothioate backbone linkage at the 3′-end protects against exonuclease degradation and 2′-sugar modification (e.g., 2′-*O*-methyl or 2′-fluoro) provides endonucleases resistance while maintaining silencing activity (87,148, *see* Chap. 2). It should be also taken into account that the impact of a chemical modification on silencing properties of siRNAs may depend on its position in the sequence, whether it occurs on the sense or antisense strand, and the particular residue that is altered (88, *see* Chap. 9). Alternatively, the siRNA half-life can be increased by conjugating them with other molecules (e.g., cholesterol) or by incorporating siRNA into different types of particles (e.g., polyethyleneimine and liposomes) *(87,88)*. As soon as these challenges are addressed, the next step will be to harness this powerful technology for therapeutic purposes especially for those malignancies that hitherto remain untreatable or poorly responsive to conventional therapies.

References

1. Blasco, M.A. (2005) Telomeres and human disease: ageing, cancer and beyond. *Nat. Rev. Genet.* 6, 611–622.
2. Autexier, C. and Greider, C.W. (1996) Telomerase and cancer: revisiting the telomere hypothesis. *Trends Biochem. Sci.* 21, 387–391.
3. Kim, S.H., Kaminker, P., and Campisi, J. (2002) Telomeres, aging and cancer: in search of a happy ending. *Oncogene* 21, 503–511.
4. de Lange, T. (2005) Telomere-related genome instability in cancer. *Cold Spring Harb. Symp. Quant. Biol.* 70, 197–204.
5. Rodier, F., Kim, S.H., Nijjar, T.et-al., . (2005) Cancer and aging: The importance of telomeres in genome maintenance. *Int. J. Biochem. Cell. Biol.* 37, 977–990.
6. de Lange, T. (2004) T-loops and the origin of telomeres. *Nat. Rev. Mol. Cell. Biol.* 5, 323–329.
7. de Lange, T. (2002) Protection of mammalian telomeres. *Oncogene* 21, 532–540.
8. Hanahan, D., and Weinberg, R.A. (2000) The hallmarks of cancer. *Cell* 100, 57–70.
9. Hahn, W.C. and Meyerson, M. (2001) Telomerase activation, cellular immortalization and cancer. *Ann. Med.* 2, 123–129.
10. Shay, J.W., and Bacchetti, S. (1997) A survey of telomerase activity in human cancer. *Eur. J. Cancer* 33, 787–791.
11. Henson, J.D., Neumann, A.A., Yeager, T.R. and Reddel, R.R. (2002) Alternative lengthening of telomeres in mammalian cells. *Oncogene* 21, 598–610.
12. Reddel, R.R., and Bryan, T.M. (2003) Alternative lengthening of telomeres: dangerous road less travelled. *Lancet* 361, 1840.
13. Muntoni, A. and Reddel, R.R. (2005) The first molecular details of ALT in human tumor cells. *Hum. Mol. Genet.* 14(Spec No. 2), R191–196.
14. Folini, M., and Zaffaroni, N. (2005) Targeting telomerase by antisense-based approaches: perspectives for new anti-cancer therapies. *Curr. Pharm. Des.* 11, 1105–1117.
15. Harrington, L., Zhou, W., McPhail, T.et al., . (1997) Human telomerase contains evolutionarily conserved catalytic and structural subunits. *Genes Dev.* 11, 3109–3115.
16. Feng, J., Funk, W.D., Wang, S.S.et al., . (1995) The RNA component of human telomerase. *Science* 269, 1236–1241.
17. Cong, Y.S., Wright, W.E., and Shay, J.W. (2002) Human telomerase and its regulation. *Microbiol. Mol. Biol. Rev.* 66, 407–425.
18. Yi, X., White, D.M., Aisner, D.L.et-al., . (2000) An alternate splicing variant of the human telomerase catalytic subunit inhibits telomerase activity. *Neoplasia* 2, 433–440.

19. Blackburn, E.H. (2005) Telomeres and telomerase: their mechanisms of action and the effects of altering their functions. *FEBS Lett.* 579, 859–862.
20. Cao, Y., Li, H., Deb, S., and Liu, J.P. (2002) TERT regulates cell survival independent of telomerase enzymatic activity. *Oncogene* 21, 3130–3138.
21. Sharma, G.G., Gupta, A., Wang, H.et al., . (2003) hTERT associates with human telomeres and enhances genomic stability and DNA repair. *Oncogene* 22, 131–146.
22. Harrington, L., McPhail, T., Mar, V.et-al., . (1997) A mammalian telomerae-associated protein. *Science* 275, 973–977.
23. Smogorzewska, A. and de Lange, T. (2004) Regulation of telomerase by telomeric proteins. *Annu. Rev. Biochem.* 73, 177–208.
24. Evans, S.K., and Lundblad, V. (1999) Est1 and Cdc13 as comediators of telomerase access. *Science* 286, 117–120.
25. Montanaro, L., Brigotti, M., Clohessy, J. et-al., . (2006) Dyskerin expression influences the level of ribosomal RNA pseudouridylation and telomerase RNA component in human breast cancer. *J. Pathol.* 210, 10–18.
26. Marrone, A., and Mason, P.J. (2003) Dyskeratosis congenita. *Cell. Mol. Life. Sci.* 60, 507–517.
27. de Lange, T. (2005) Shelterin: the protein complex that shapes and safeguards human telomeres. *Genes Dev.* 19, 2100–2110.
28. Broccoli, D., Smogorzewska, A., Chong, L., and de Lange, T. (1997) Human telomeres contain two distinct Myb-related proteins, TRF1 and TRF2. *Nat. Genet.* 17, 231–235.
29. Baumann, P., Podell, E., and Cech, T.R. (2002) Human Pot1 (protection of telomeres) protein: cytolocalization, gene structure, and alternative splicing. *Mol. Cell. Biol.* 22, 8079–8087.
30. van Steensel, B. and de Lange, T. (1997) Control of telomere length by the human telomeric protein TRF1. *Nature* 385, 740–743.
31. Kim, S.H., Kaminker, P., and Campisi, J. (2003) TIN2, a new regulator of telomere length in human cells. *Nat. Genet.* 23, 405–412.
32. Seimiya, H. (2006) The telomeric PARP, tankyrases, as targets for cancer therapy. *Br. J. Cancer* 94, 341–345.
33. Cook, B.D., Dynek, J.N., Chang, W.et al., . (2002) Role for the related poly(ADP-Ribose) polymerases tankyrase 1 and 2 at human telomeres. *Mol. Cell. Biol.* 22, 332–342.
34. Smith, S., Giriat, I., Schitt, A., and de Lange, T. (1998) Tankyrase, a poly (ADP-ribose) polymerase at human telomeres. *Science* 282, 1484–1487.
35. Donigian, J.R. and de Lange, T. (2007) The role of the poly(ADP-ribose) polymerase tankyrase1 in telomere length control by the TRF1 component of the shelterin complex. *J. Biol. Chem.* 282, 22662–22667.
36. Stansel, R.M., de Lange, T., and Griffith, J.D. (2001) T-loop assembly in vitro involves binding of TRF2 near the 3′ telomeric overhang. *EMBO J.* 20, 5532–5540.
37. Hockemeyer, D., Sfeir, A.J., Shay, J.W.et al., . (2005) POT1 protects telomeres from a transient DNA damage response and determines how human chromosomes ends. *EMBO J.* 24, 2667–2678.
38. Green, D.R., and Evan, G.I. (2002) A matter of life and death. *Cancer Cell* 1, 19–30.
39. Nagata, S. (1997) Apoptosis by death factor. *Cell* 88, 355–365.
40. Cryns, V., and Yuan, J. (1999) Proteases to die for. *Genes Dev.* 12, 1551–1570.
41. Hunter, A., LaCasse, E.C., and Korneluk, R.G. (2007) The inhibitors of apoptosis (IAPs) as cancer targets. *Apoptosis* 12, 1543–1568.
42. Tanaka, K., Iwamoto, S., Gon, G.et al., . (2000) Expression of survivin and its relationship to loss of apoptosis in breast carcinomas. *Clin. Cancer Res.* 6, 127–134.
43. Yang, L., Cao, Z., Yan, H., and Wood, W.C. (2003) Coexistence of high levels of apoptotic signaling and inhibitor of apoptosis proteins in human tumor cells: implication for cancer specific therapy. *Cancer Res.* 63, 6815–6824.
44. Hong, X., Lei, L., and Glas, R. (2003) Tumors acquire inhibitor of apoptosis protein (IAP)-mediated apoptosis resistance through altered specificity of cytosolic proteolysis. *J. Exp. Med.* 197, 1731–1743.
45. Roy, N., Mahadevan, M.S., McLean, M. et al., . (1995) The gene for neuronal apoptosis inhibitory protein is partially deleted in individuals with spinal muscular atrophy. *Cell* 80, 167–178.
46. Maier,J.K., Lahoua, Z., Gendron, N.H. et al., . (2002) The neuronal apoptosis inhibitory protein is a direct inhibitor of caspases 3 and 7. *J. Neurosci.* 22, 2035–2043.
47. Davoodi, J., Lin, L., Kelly, J.et al., . (2004) Neuronal apoptosis-inhibitory protein does not interact with Smac and requires ATP to bind caspase-9. *J. Biol. Chem.* 279, 40622–40628.

48. Sanna, M.G., de Silva Correia, J., Ducrey, O. et al., . (2002) IAP supression of apoptosis involves distinct mechanisms: the TAK1/JNK1 signaling cascade and caspase inhibition. *Mol. Cell Biol.* 22, 1754–1766.
49. Yamamoto, K., Abe, S., Nakagawa, Y.et al., . (2004) Expression of IAP family proteins in myelodysplastic syndromes transforming to overt leukemia. *Leukemia Res.* 28, 1203–1211.
50. Nakagawa, Y., Hasegawa, M., Kurata, M. et al., . (2005) Expression of IAP-Family Proteins in Adult Acute Mixed Lineage Leukemia (AMLL). *Am. J. Hematol.* 78, 173–180.
51. Nemoto, T., Kitagawa, M., Hasegawa, M. et al., . (2004) Expression of IAP family proteins in esophageal cancer. *Exp. Mol. Pathol.* 76, 253–259.
52. Eckelman, B.P. and Salvesen, G.S. (2006) The human anti apoptotic proteins cIAP1 and cIAP2 bind but do not inhibit caspases. *J. Biol. Chem.* 281, 3254–3260.
53. Deveraux, Q.L., Roy, N., Stennicke, H.R. et al., . (1998) IAPs block apoptotic events induced by caspase-8 and cytochrome c by direct inhibition of distinct caspases. *EMBO J.* 17, 2215–2223.
54. Srinivasula, S.M., Hegde, R., Saleh, A. et al., . (2001) A conserved XIAP-interaction motif in caspase-9 and Smac/DIABLO regulates caspase activity and apoptosis. *Nature* 410, 112–116.
55. Stennicke, H.R., Ryan, C.A., and Salvesen, G.S. (2002) Reprieval from execution: the molecular basis of caspase inhibition. *Trends Biochem. Sci.* 27, 94–101.
56. Suzuki, Y., Nakabayashi, Y., Nakata, K. et al., . (2001) X-linked inhibitor of apoptosis protein (XIAP) inhibits caspase-3 and -7 in distinct modes. *J. Biol. Chem.* 276, 27058–27063.
57. Suzuki, Y., Nakabayashi, Y., and Takahashi, R. (2001) Ubiquitin-protein ligase activity of X-linked inhibitor of apoptosis protein promotes proteasomal degradation of caspase-3 and enhances its anti-apoptotic effect in Fas-induced cell death. *Proc. Natl. Acad. Sci. USA* 98, 8662–8667.
58. Levkau,B., Garton, K.J., Ferri, N.et al., . (2001) XIAP induces cell-cycle arrest and activates nuclear factorkappaB: new survival pathways disabled by caspase-mediated cleavage during apoptosis of human endothelial cells. *Circ. Res.* 88, 282–290.
59. Kaur, S., Wang, F., Venkatraman, M., and Arsura, M. (2005) X-linked inhibitor of apoptosis (XIAP) inhibits c-Jun N-terminal kinase 1 (JNK1) activation by transforming growth factor beta1 (TGF-beta1) through ubiquitin-mediated proteosomal degradation of the TGF-beta1-activated kinase 1 (TAK1). *J. Biol. Chem.* 280, 38599–38608.
60. Mufti, A.R., Burstein, E., and Duckett, C.S. (2007) XIAP: cell death regulation meets copper homeostasis. *Arch. Biochem. Biophys.* 463, 168–174.
61. Tamm, I., Richter, S., Oltersdorf, D.et al., . (2004) High expression levels of x-linked inhibitor of apoptosis protein and survivin correlate with poor overall survival in childhood de novo acute myeloid leukemia. *Clin. Cancer Res.* 10, 3737–3744.
62. Mizutani, Y., Nakanishi, H., Li, Y.N.et al., . (2007) Overexpression of XIAP expression in renal cell carcinoma predicts a worse prognosis. *Int. J. Oncol.* 30, 919–925.
63. Seligson, D.B., Hongo, F., Huerta-Yepez, S. et al., . (2007) Expression of X-Linked Inhibitor of Apoptosis Protein Is a Strong Predictor of Human Prostate Cancer Recurrence. *Clin. Cancer Res.* 13, 6056–6063.
64. Ferreira, C.G., van der Valk, P., Span, S.W. et al., . (2001) Expression of x-linked inhibitor of apoptosis as a novel prognostic marker in radically resected non-small cell lung cancer patients. *Clin. Cancer Res.* 7, 2468–2474.
65. Ferreira, C.G., van der Valk, P., Span, S.W. et al., . (2001) Assessment of IAP (inhibitor of apoptosis) proteins as predictors of response to chemotherapy in advanced non–small-cell lung cancer patients. *Ann. Oncol.* 12, 799–805.
66. Schimmer, A.D., Dalili, S., Batey, R.A., and Riedl, S.J. (2006) Targeting XIAP for the treatment of malignancy. *Cell Death Differ.* 13, 179–188.
67. Holcik, M., Yeh, C., Korneluk, R.G., and Chow, T. (2000) Translational upregulation of X-linked inhibitor of apoptosis (XIAP) increases resistance to radiation induced cell death. *Oncogene* 19, 4174–4177.
68. Cheung, H.H., LaCasse, E.C., and Korneluk, R.G. (2006) X-Linked Inhibitor of Apoptosis Antagonism: Strategies in Cancer Treatment. *Clin. Cancer Res.* 12, 3238–3242.
69. Honda, R., Korner, R., and Nigg, E.A. (2004) Exploring the functional interactions between aurora B, INCENP, and survivin in mitosis. *Mol. Biol. Cell* 14, 3325–3341.
70. Wheatley, S.P., Carvalho, A., Vagnarelli, P., and Earnshaw, W.C. (2001) INCENP is required for proper targeting of survivin to

the centromeres and the anaphase spindle during mitosis. *Curr. Biol.* 11, 886–890.

71. Gassmann, R., Carvalho, A., Henzing, A.J. et al.. (2004) Borealin: a novel chromosomal passenger requied for stability of the bipolar mitotic spindle. *J. Cell. Biol.* 166, 179–191.
72. Pennati, M., Folini, M., and Zaffaroni, N. (2007) Targeting survivin in cancer therapy: fulfilled promises and open questions. *Carcinogenesis* 28, 1133–1139.
73. Altieri, D.C. (2006) The case of survivin as a regulator of microtubule dynamics and cell-death decisions. *Curr. Opin. Cell Biol.* 18, 609–615.
74. Colnaghi, R., Connell, C.M., Barrett, R.M.A., and Wheatley, S.P. (2006) Separating the anti-apoptotic and mitotic roles of survivin. *J. Biol. Chem.* 281, 33450–33456.
75. Stauber, R.H., Mann, W., and Knauer, S.K. (2007) Nuclear and cytoplasmic survivin: molecular mechanisms, prognostic, and theraputic potential. *Cancer Res.* 67, 5999–6002.
76. Caldas, H., Jiang, Y., Holloway, M.P.et al., . (2005) Survivin splice variants regulate the balance between proliferation and cell death. *Oncogene* 24, 1994–2007.
77. Zaffaroni, N., Pennati, M., Colella, G.et al., . (2002) Expression of the anti-apoptotic gene survivin correlates with taxol resistance in human ovarian cancer. *Cell Mol. Life Sci.* 59, 1406–1412.
78. Zhang, M., Mukherjee, N., Bermudez, R.S.et al., . (2005) Adenovirus-mediated inhibition of survivin expression sensitizes human prostate cancer cells to paclitaxel in vitro and in vivo. *Prostate* 64, 293–302.
79. Asanuma, K., Moriai, R., and Yajima, T. (2000) Survivin as a radio-resistance factor in pancreatic cancer. *Jap. J. Cancer Res.* 91, 1204–1209.
80. Qiu, X.B. and Goldberg, A.L. (2005) The membrane-associated inhibitor of apoptosis protein, BRUCE/Apollon, antagonizes both the precursor and mature forms of Smac and caspase-9. *J. Biol. Chem.* 280, 174–182.
81. Hao, Y., Sekine, S., Kawabata, A.et al., . (2004) Apollon ubiquitinates SMAC and caspase-9, and has an essential cytoprotection function. *Nat. Cell Biol.* 6, 849–860.
82. Chen, Z., Naito, M., Hori, S.et al., . (1999) A human IAP-family gene, apollon, expressed in human brain cancer cells. *Biochem. Biophys. Res. Commun.* 264, 847–854.
83. Liu, B., Han, M., Wen, J.K., and Wang, L. (2007) Livin/ML-IAP as a new target for cancer treatment. *Cancer Lett.* 250, 168–176.
84. Shin, H., Renatus, M., Eckelman, B.P.et al., . (2005) The BIR domain of IAP-like protein 2 is conformationally unstable: implications for caspase inhibition. *Biochem. J.* 385, 1–10.
85. Chatterjee-Kishore, M., and Miller, C.P. (2005) Exploring the sounds of silence: RNAi-mediated gene silencing for target identification and validation. *Drug Discov. Today* 10, 1559–1565.
86. Aagaard, L. and Rossi, J.J. (2007) RNAi therapeutics: principles, prospects and challenges. *Adv. Drug Deliv. Rev.* 59, 75–86.
87. de Fougerolles, A., Vornlocher, H.P., Maraganore, J., and Lieberman, J. (2007) Interfering with disease: a progress report on siRNA-based therapeutics. *Nat. Rev. Drug Discov.* 6, 443–453.
88. Dykxhoorn, D.M., Palliser, D., and Lieberman, J. (2006) The silent treatment: siRNAs as small molecule drugs. *Gene Ther.* 13, 541–552.
89. Dykxhoorn, D.M. and Lieberman, J. (2006) Knocking down disease with siRNAs. *Cell* 126, 231–235.
90. Kosciolek, B.A., Kalantidis, K., Tabler, M., and Rowley, P.T. (2003) Inhibition of telomerase activity in human cancer cells by RNA interference. *Mol. Cancer Ther.* 2, 209–216.
91. de Souza Nascimento, P., Alves, G., and Fiedler, W. (2006) Telomerase inhibition by an siRNA directed against hTERT leads to telomere attrition in HT29 cells. *Oncol. Rep.* 16, 423–428.
92. Nakamura, M., Masutomi, K., Kyo, S.et al., . (2005) Efficient inhibition of human telomerase reverse transcriptase expression by RNA interference sensitizes cancer cells to ionizing radiation and chemotherapy. *Hum. Gene Ther.* 16, 859–868.
93. Kurvinen, K., Syrjanen, S., and Johansson, B. (2006) Long-term suppression of telomerase expression in HeLa cell clones, transfected with an expression vector carrying siRNA targeting hTERT mRNA. *Int. J. Oncol.* 29, 279–288.
94. Wang, R., Lin, F., Wang, X.et al., . (2007) The therapeutic potential of survivin promoter-driven siRNA on suppressing tumor growth and enhancing radiosensitivity of human cervical carcinoma cells via downregulating hTERT gene expression. *Cancer Biol. Ther.* 6(8), 2582.
95. Gandellini, P., Folini, M., Bandiera, R.et al., . (2007) Down-regulation of human telomerase reverse transcriptase through specific activation of RNAi pathway quickly results

in cancer cell growth impairment. *Biochem Pharmacol.* 73, 1703–1714.

96. Zou, L., Zhang, P., Luo, C., and Tu, Z. (2006) shRNA-targeted hTERT suppress cell proliferation of bladder cancer by inhibiting telomerase activity. *Cancer Chemother. Pharmacol.* 57, 328–334.
97. Zhang, P.H., Zou, L., and Tu, Z.G. (2006) RNAi-hTERT inhibition hepatocellular carcinoma cell proliferation via decreasing telomerase activity. *J. Surg. Res.* 131, 143–149.
98. Wang, Y., Duan, H.G., Chen, S.M.et al., . (2007) Effect of RNA interference targeting human telomerase reverse transcriptase on telomerase and its related protein expression in nasopharyngeal carcinoma cells. *J. Laryngol. Otol.* 121, 476–482.
99. Pallini, R., Sorrentino, A., Pierconti, F. et al., . (2006) Telomerase inhibition by stable RNA interference impairs tumor growth and angiogenesis in glioblastoma xenografts. *Int. J. Cancer* 118, 2158–2167.
100. Zhao, P., Wang, C., Fu, Z.et-al., . (2007) Lentiviral vector mediated siRNA knock-down of hTERT results in diminished capacity in invasiveness and in vivo growth of human glioma cells in a telomere length-independent manner. *Int. J. Oncol.* 31, 361–368.
101. Del Bufalo, D., Rizzo, A., Trisciuoglio, D.et-al., . (2005) Involvement of hTERT in apoptosis induced by interference with bcl-2 expression and function. *Cell Death Differ.* 12, 1429–1438.
102. Massard, C., Zermati, Y., Pauleau, A.L. et al., . (2006) hTERT: A novel endogenous inhibitor of the mitochondrial cell death pathway. *Oncogene* 25, 4505–4514.
103. Lai, S.R., Cunningham, A.P., Huynh, V.Q.et-al., . (2007) Evidence of extra-telomeric effects of hTERT and its regulation involving a feedback loop. *Exp Cell Res.* 313, 322–330.
104. Li, S., Rosenberg, J.E., Donjacour, A.A. et al., . (2004) Rapid inhibition of cancer cell growth induced by lentiviral delivery and expression of mutant-template telomerase RNA and anti-telomerase short-interfering RNA. *Cancer Res.* 64, 4833–4840.
105. Li, S., Crothers, J., Haqq, C.M., and Blackburn, E.H. (2005) Cellular and gene expression responses involved in the rapid growth inhibition of human cancer cells by RNA interference-mediated depletion of telomerase RNA. *J. Biol. Chem.* 280, 23709–23717.
106. Kelland, L. (2007) Targeting the limitless replicative potential of cancer: the telomerase/telomere pathway. *Clin. Cancer Res.* 13, 4960–4963.
107. Salhab, M., Jiang, W.G., Newbold, R.F., and Mokbel, K. (2007) The expression of gene transcripts of telomere-associated genes in human breast cancer: Correlation with clinico-pathological parameters and clinical outcome. Breast Cancer Res. Treat. doi:10.1007/s10549-007-9622-8.
108. Goldkorn, A., and Blackburn, E.H. (2006) Assembly of mutant-template telomerase RNA into catalytically active telomerase ribonucleoprotein that can act on telomeres is required for apoptosis and cell cycle arrest in human cancer cells. *Cancer Res.* 66, 5763–5771.
109. Ning, H., Li, T., Zhao, L.et al., . (2006) TRF2 promotes multidrug resistance in gastric cancer cells. *Cancer Biol. Ther.* 5, 950–956.
110. Yang, Q., Zheng, Y.L., and Harris, C.C. (2005) POT1 and TRF2 cooperate to maintain telomeric integrity. *Mol. Cell. Biol.* 25, 1070–1080.
111. Donigian, J.R., and de Lange, T. (2007) The role of the poly(ADP-ribose) polymerase Tankyrase 1 in telomere length control by TRF1 componenet of the shelterin complex. *J. Biol. Chem.* 282, 22662–22667.
112. Dynek,J.N. and Smith, S. (2004) Resolution of sister telomere association is required for progression through mitosis. *Science* 304, 60–62.
113. Chang, W., Dynek, J.N., and Smith, S. (2005) NuMA is a major acceptor of poly(ADP-ribosyl)ation by tankyrase 1 in mitosis. *Biochem. J.* 391(Pt 2), e5–6.
114. Jiang, W.Q., Zhong, Z.H., Henson, J.D., and Reddel, R.R. (2007) Identification of by methionine restriction and RNA interference. *Oncogene* 26, 4635–4647.
115. Zhong, Z.H., Jiang, W.Q., Cesare, A.J. et al., . (2007) Disruption of telomere maintenance by depletion of the MRE11/RAD50/NBS1 complex in cells that use alternative lengthening of telomeres. *J. Biol. Chem.* 282, 29314–29322.
116. Lima, R.T., Martins, L.M., Guimaraes, J.E.et al., . (2004) Specific downregulation of bcl-2 and xIAP by RNAi enhances the effects of chemotherapeutic agents in MCF-7 human breast cancer cells. *Cancer Gene Ther.* 11, 309–316.
117. Chawla-Sarkar, M., Bae, S.I., Reu, F.J., Jacobs, B.S., Lindner, D.J., and Borden, E.C. (2004) Downregulation of Bcl-2, FLIP or IAPs (XIAP and survivin) by siRNAs sensitizes resistant melanoma cells to Apo2L/

TRAIL-induced apoptosis. *Cell Death Differ.* 11, 915–923.
118. Yamaguchi, Y., Shiraki, K., Fuke, H.et al., . (2005) Targeting of X-linked inhibitor of apoptosis protein or survivin by short interfering RNAs sensitize hepatoma cells to TNF-related apoptosis-inducing ligand- and chemotherapeutic agent-induced cell death. *Oncol. Rep.* 14, 1311–1316.
119. Zhang, Y., Wang, Y., Gao, W.et al., . (2006) Transfer of siRNA against XIAP induces apoptosis and reduces tumor cells growth potential in human breast cancer in vitro and in vivo. *Breast Cancer Res. Treat.* 96, 267–277.
120. Shrikhande, S.V., Kleeff, J., Kayed, H. et al., . (2006) Silencing of X-linked inhibitor of apoptosis (XIAP) decreases gemcitabine resistance of pancreatic cancer cells. *Anticancer Res.* 26, 3265–3273.
121. Chen, J., Xiao, X.Q., Deng, C.M.et al., . (2006) Downregulation of xIAP expression by small interfering RNA inhibits cellular viability and increases chemosensitivity to methotrexate in human hepatoma cell line HepG2. *J. Chemother.* 18, 525–531.
122. Zhang, S., Ding, F., Luo, A.et al., . (2007) XIAP is Highly Expressed in Esophageal Cancer and its Downregulation by RNAi Sensitizes Esophageal Carcinoma Cell Lines to Chemotherapeutics. *Cancer Biol Ther.* 6, 973–980.
123. Lee, T.J., Jung, E.M., Lee, J.T.et al., . (2006) Mithramycin A sensitizes cancer cells to TRAIL-mediated apoptosis by downregulation of XIAP gene promoter through Sp1 sites. *Mol. Cancer Ther.* 5, 2737–2746.
124. Shrader, M., Pino, M.S., Lashinger, L.et al., . (2007) Gefitinib reverses TRAIL resistance in human bladder cancer cell lines via inhibition of AKT-mediated X-linked inhibitor of apoptosis protein expression. *Cancer Res.* 67, 1430–1435.
125. Ohnishi, K., Scuric, Z., Schiestl, R.H.et al., . (2006) siRNA targeting NBS1 or XIAP increases radiation sensitivity of human cancer cells independent of TP53 status. *Radiat. Res.* 166, 454–462.
126. Lopes, R.B., Gangeswaran, R., McNeish, I.A.et al., . (2007) Expression of the IAP protein family is dysregulated in pancreatic cancer cells and is important for resistance to chemotherapy. *Int. J. Cancer.* 120, 2344–2352.
127. Pan, Q., Liu, B., Liu, J.et al., . (2007) Synergistic antitumor activity of XIAP-shRNA and TRAIL expressed by oncolytic adenoviruses in experimental HCC. *Acta Oncol.* 11, 1–10.
128. Carvalho, A., Carmena, M., Sambade, C. et al., . (2003) Survivin is required for stable checkpoint activation in taxol-treated HeLa cells. *J. Cell Sci.* 116, 2987–2998.
129. Tsuji, N., Asanuma, K., Kobayashi, D. et al., . (2005) Introduction of a survivin gene-specific small inhibitory RNA inhibits growth of pancreatic cancer cells. *Anticancer Res.* 25, 3967–3972.
130. Ai, Z., Yin, L., Zhou, X.et al., . (2006) Inhibition of survivin reduces cell proliferation and induces apoptosis in human endometrial cancer. *Cancer* 107, 746–756.
131. Congmin, G., Mu, Z., Yihui, M., and Hanliang, L. (2006) Survivin attractive target for RNAi in non-Hodgkin's lymphoma, Daudi cell line as a model. *Leuk. Lymphoma* 47, 1941–1948.
132. Sarthy, A.V., Morgan-Lappe, S.E., Zakula, D. et al., . (2007) Survivin depletion preferentially reduces the survival of activated K-Ras-transformed cells. *Mol Cancer Ther.* 6, 269–276.
133. Sato, A., Ito, K., Asano, T.et al., . (2007) Synergistic effect of survivin-specific small interfering RNA and topotecan in renal cancer cells: topotecan enhances liposome-mediated transfection by increasing cellular uptake. *Int J Oncol.* 30, 695–700.
134. Wuttig, D., Kunze, D., Fuessel, S.et al., . (2007) Are overexpressed alternative survivin transcripts in human bladder cancer suitable targets for siRNA-mediated in vitro inhibition. *Int. J. Oncol.* 30, 1317–1324.
135. Kappler, M., Rot, S., Taubert, H., et al. (2007) The effects of knockdown of wild-type survivin, survivin-2B or survivin-Delta3 on the radiosensitization in a soft tissue sarcoma cells in vitro under different oxygen conditions. Cancer Gene Ther. doi: 10.1038.
136. Coumoul,X.,Li,W.,Wang,R.H.,andDeng,C. (2004) Inducible suppression of Fgfr2 and Survivin in ES cells using a combination of the RNA interference (RNAi) and the Cre-LoxP system. Nucleic Acids Res. 32, e85.
137. Crnkovic-Mertens, I., Muley, T., Meister, M.et al., . (2006) The anti-apoptotic livin gene is an important determinant for the apoptotic resistance of non-small cell lung cancer cells. *Lung Cancer* 54, 135–142.
138. Crnkovic-Mertens, I., Semzow, J., Hoppe-Seyler, F., and Butz, K. (2006) Isoform-specific silencing of the Livin gene by RNA interference defines Livin beta as key mediator of apoptosis inhibition in HeLa cells. *J. Mol. Med.* 84, 232–240.

139. Ren, J., Shi, M., Liu, R.et al., . (2005) The Birc6 (Bruce) gene regulates p53 and the mitochondrial pathway of apoptosis and is essential for mouse embryonic development. *Proc. Natl. Acad. Sci. USA* 102, 565–570.
140. Qiu, X.B., Markant, S.L., Yuan, J., and Goldberg, A.L. (2004) Nrdp1-mediated degradation of the gigantic IAP, BRUCE, is a novel pathway for triggering apoptosis. *EMBO J.* 23, 800–810.
141. Fischer, U. and Schulze-Osthoff, K. (2005) New approaches and therapeutics targeting apoptosis in disease. *Pharmacol. Rev.* 57, 187–215.
142. Chang, H. and Schimmer, A.D. (2007) Livin/melanoma inhibitor of apoptosis protein as a potential therapeutic target for the treatment of malignancy. *Mol. Cancer Ther.* 6, 24–30.
143. Boukamp, P. and Mirancea, N. (2007) Telomeres rather than telomerase a key target for anti cancer therapy. *Exp. Dermatol.* 16, 71–79.
144. Kelland,L.R. (2005) Overcoming the immortality of tumour cells by telomere and telomerase based cancer therapeutics – current status and future prospcts. *Eur. J. Cancer.* 41, 971–979.
145. Bremer, E., van Dam, G., Kroesen, B.J. et al., . (2006) Targeted induction of apoptosis for cancer therapy: current progress and prospects. *Trends Mol. Med.* 12, 382–393.
146. Birmingham, A., Anderson, E., Sullivan, K. et al., . (2007) A protocol for designing siRNAs with high functionality and specificity. *Nat. Protoc.* 2, 2068–2078.
147. Schubert, S., Grünweller, A., Erdmann, V.A., and Kurreck, J. (2005) Local RNA target structure influences siRNA efficacy: Systematic analysis of intentionally designed binding regions. *J. Mol. Biol.* 348, 883–893.
148. Cejka, D., Losert, D., and Wacheck, V. (2006) Short interfering RNA (siRNA): Tool or therapeutic. *Clin. Sci.* 110, 47–58.
149. Joshua-Tor, L. (2006) The Argonautes. *Cold Spring Harb. Symp. Quant. Biol.* 71, 67–72.

Chapter 16

Treating Respiratory Viral Diseases with Chemically Modified, Second Generation Intranasal siRNAs

Sailen Barik

Abstract

Chemically synthesized short interfering RNA (siRNA) of pre-determined sequence has ushered a new era in the application of RNA interference (RNAi) against viral genes. We have paid particular attention to respiratory viruses that wreak heavy morbidity and mortality worldwide. The clinically significant ones include respiratory syncytial virus (RSV), parainfluenza virus (PIV) and influenza virus. As the infection by these viruses is clinically restricted to the respiratory tissues, mainly the lungs, the logical route for the application of the siRNA was also the same, i.e., via the nasal route. Following the initial success of intranasal siRNA against RSV, second-generation siRNAs were made against the viral polymerase large subunit (L) that were chemically modified and screened for improved stability, activity and pharmacokinetics. 2′-*O*-methyl (2′-*O*-Me) and 2′-deoxy-2′-fluoro (2′-F) substitutions in the ribose ring were incorporated in different positions of the sense and antisense strands and the resultant siRNAs were tested with various transfection reagents intranasally against RSV. Based on these results, we propose the following consensus for designing intranasal antiviral siRNAs: (i) modified 19–27 nt long double-stranded siRNAs are functional in the lung, (ii) excessive 2′-OMe and 2′-F modifications in either or both strands of these siRNAs reduce efficacy, and (iii) limited modifications in the sense strand are beneficial, although their precise efficacy may be position-dependent.

Key words: siRNA, intranasal, RNAi, antiviral, RSV, parainfluenza, influenza.

1. Introduction

Double-stranded short interfering RNAs (siRNAs) induce post-transcriptional gene silencing in a variety of metazoan cells and tissues *(1)*. Successful use of synthetic siRNA in 2001 *(1, 2)*, targeting cellular and viral genes in cell culture, opened the door to the siRNA as prospective antivirals and drugs for gene therapy.

M. Sioud (ed.), *Methods in Molecular Biology, siRNA and miRNA Gene Silencing, vol. 487*

DOI: 10.1007/978-1-60327-547-7_16

The "first-generation" siRNAs were designed to mimic the products of the Dicer endonuclease cleavage; they were 19-nt long duplexes with 3′-terminal 2-nt overhangs and contained natural, unmodified ribose and bases. As reported by us *(3, 4)* and numerous other laboratories *(5 – 7)*, these siRNAs proved effective both in cell culture and in animals.

Of particular mention is the antiviral activity of siRNAs (*see* this Chapter), first documented by us against respiratory syncytial virus (RSV) as proof-of-concept. In this study, synthetic unmodified siRNAs, designed against essential viral genes, prevented RSV growth in cultured lung epithelial cells *(3)*. To translate this success to the animal model *(4)*, we reasoned that the most logical route for delivery of these siRNAs should be the nose (intranasal). First, it is noninvasive. Second, it should involve relatively simple application procedures, such as nasal drops or aerosol inhalation. Third, the siRNA drug will follow the same route as the virus and reach the same tissues, including the lungs. Lastly, our subsequent results showed that the intranasally administered siRNA remains in the lung and does not leach into blood or other tissues, minimizing the risk of systemic side effects (data not shown). In this chapter, we have, therefore, focused on the intranasal delivery of anti-respiratory viral siRNAs.

We will first summarize the progress made in four major respiratory viruses: RSV, parainfluenza virus (PIV), influenza virus (Flu), and severe acute respiratory syndrome (SARS) coronavirus (SARS-CoV) *(4, 5, 8–10)*. The first three viruses are clinically the most significant, claiming a large number of human lives each year throughout the world. Annual flu epidemics alone affect 10–20% of the U.S. population, averaging about 114,000 hospitalizations. Partly due to the high mutation rate of RNA genomes, there is no definitive vaccine or reliable antiviral treatment for these viruses. Ribavirin and IFN, although sometimes used, are both relatively non-specific and toxic *(11)*. Regarding the molecular feature of their genomes, RSV and PIV are nonsegmented negative-strand RNA viruses, belonging to different genera of the *Paramyxoviridae* family *(5)*. Flu is an orthomyxovirus and contains segmented negative-strand RNA genome *(8, 9)*. Coronoviruses such as SARS-CoV contain positive-strand, enveloped RNA genomes *(10)*. A commonality among these RNA viruses is that they encode genes for RNA-dependent polymerase (RdRP) to transcribe and replicate their genomes, because the host animal cells are devoid of such activity. Many of the potent antiviral siRNAs have, therefore, targeted viral genes coding for RdRP subunits or related to RdRP function *(3, 4, 5, 8–10)*. They include, e.g., P (Phosphoprotein) *(3, 4)*, L (Large) (*see* **Table 16.1**), and N (Nucleocapsid) genes of RSV; P and L genes of PIV; PA and N genes of Flu; and the replicase gene of SARS-CoV.

Table 16.1
Examples of RSV L gene siRNA sequences and modification formats

siRNA sequence and modification

Target sequences:	
Example #1:	5′ AAUGGCAGAUGGAUAAUUCUAUU 3′ (-4.2 kcal)
Example #2:	5′ AACCCUAAUCAUGUGGUAUCUUU 3′ (-3.5 kcal)
siRNA:	
Sense:	5′ CCCUAAUCAUGUGGUAUCUdTdT 3′
Antisense:	3′ TdTdGGGAUUAGUACACCAUAGA 5′
Modification format #1	
Sense:	5′ **CCCUAAUCAUGUGGUAUCU**dTdT 3′
Antisense:	3′ TdTdGGGAUUAGUACACCAUAGA 5′
Modification format #2	
Sense:	5′ CC**C**UA**A**UC**A**U**G**U**G**GU**A**U**C**UdTdT 3′
Antisense:	3′ TdTdG**G**G**A**UU**AG**U**A**C**A**CC**A**U**A**G**A** 5′

Two target sequences in the RSV L gene mRNA (Long strain) with the $AAN_{19}TT$ motif (the N_{19} regions underlined) are shown on top. The thermodynamic difference (ΔG) of stability of the two termini is shown in parenthesis. We tend to prefer #2 because of the CCC stretch at the 5′-end of the N_{19} sequence despite the slightly more favorable ΔG value of #1. All subsequent siRNAs were based on sequence #2. Bold = 2′-*O*-Me modifications; underlined = 2′-F modifications. We have confirmed the antiviral efficacy of both of these modified sequences in cell culture as well as in mice (data not shown). The dTdT part adds stability to the siRNA.

In all cases, siRNAs were designed against essential viral gene(s), optimized in cell culture and then used intranasally in the appropriate animal model. Antiviral activity was shown against RSV and parainfluenza virus (PIV) using the BALB/c mouse model as well *(4)*. In these studies, the siRNAs were delivered in complex with either Oligofectamine (Invitrogen, Carlsbad, CA) or Mirus Transit TKO reagent (*see* **Sect. 2**), although uncomplexed intranasal siRNA also showed significant efficacy when compared to untreated mice *(4)*. Intranasal siRNA, complexed with oligofectamine or polyethyleneimine (PEI), was also protective against highly pathogenic influenza A viruses of the H5 and H7 subtypes in mice *(8, 9)*. Notwithstanding their success, the activity of these siRNAs was transient, lasting only a few days. Therefore, enhancement of the intracellular and extracellular stability of synthetic siRNAs while increasing (or without compromising) their RNAi activity is a continuing goal for therapeutic translation of RNAi.

A variety of chemical modifications, including terminal and internal ones, have been added to the first-generation siRNA sequences to improve stability and delivery, leading to what we call "second-generation" siRNAs. Advantage has been taken of the free 2′-OH group of the ribose moiety of RNA (in contrast to DNA that lacks this OH group), to which various substituents were added. We have pursued two promising ones, namely 2′-*O*-methyl (2′-*O*-Me) and 2′-flouride (2′-F). The latter modification is placed on pyrimidine nucleosides (C, U), leading to 2′-FC and 2′-FU residues. In a number of previous studies *(12–15)*, these substitutions were introduced to various extents in the antisense strand ("guide" strand) or both strands of the siRNA and were shown to enhance stability and potency, although intranasal application was not tested. Additionally, they tend to reduce siRNA-driven innate immune response *(16, 17)*. In systematic studies in cell culture, the biochemical and functional activity of the siRNA was vindicated but found to be affected by the position of the modifications in the sequence *(12–15)*. Generally speaking, those with the modified ribonucleotides at the 5′-end of the antisense strand were less active relative to the 3′-modified ones. Internally, while 2′-F was generally well-tolerated on the antisense strand, 2′-*O*-Me showed significant shift in activity depending on the position. In contrast, incorporation of 2′-*O*-Me in the sense strand of siRNA did not show a strong positional preference. In a comprehensive study, however, internal 2′-*O*-Me modifications in either or both strands actually made the siRNA less active *(18)*. In an animal experiment *(19)*, all the 2′-OH residues in siRNAs against hepatitis B virus were substituted with 2′-F and 2′-*O*-Me. When administered intravenously (i.v.) as lipid complexes, the 2′-*O*-Me, 2′-F siRNAs showed improved efficacy and longer half-life in plasma and liver. When these siRNAs were additionally Cy3-labeled, it revealed their accumulation in the liver and spleen, but not in the lung, explaining the success of the i.v. administration against hepatitis while suggesting that it is an ineffective route against lung infections.

Based on the absence of a uniform pattern in these studies, we reasoned that substituted second-generation siRNAs against respiratory viruses should be individually optimized through the following steps: (i) Design the first-generation siRNA following the generally accepted sequence rules, (ii) select the ones with lowest IC_{50} (preferably below ~20 nM) in a cell culture assay for virus growth, (iii) add OMe and F substitutions in various "format" (i.e., number and placement), (iv) ensure that the substitutions either improved or did not reduce knockdown efficiency by screening in cell culture, (v) confirm efficacy and lack of toxicity in animal model, and (vi) test improved stability in serum and blood in vitro. Finally, if deemed necessary and resources permit, a low- or high-throughput assay for off-target effects can be performed to further ensure target specificity. Here, we present a consensus procedure tested in our laboratory that follows the above steps. Essentially the same protocol can be used to test other modifications as long as the synthetic chemistry is available.

2. Materials

The reagents described below have been used successfully by us but various equivalents are available commercially that can be optimized.

2.1. siRNAs and Their Use

1. Synthetic siRNA of the chosen sequence, lyophilized, purchased without any modification, 2 nmol (Dharmacon, Lafayette, CO).
2. Selected, most effective siRNA(s) from above, purchased with 2′-*O*-Me, 2′-FC, 2′-FU modifications in various formats (Dharmacon), in both sense and antisense strands.
3. RNase-free ART aerosol resistant pipet tips (Molecular Bio-Products, San Diego, CA).
4. RNase-free microfuge tubes (Ambion, Austin, TX).
5. siRNA buffer: 20 mM KCl, 6 mM HEPES-KOH pH 7.5, 0.2 mM $MgCl_2$. This is usually supplied by the siRNA manufacturer (Dharmacon) as 6x stock; dilute as and when needed.

2.2. Cell Culture and Virus Growth

We assume that the reader has access to an appropriate cell culture facility, consisting of incubators and culture hoods, and there is available expertise on virus growth and assay. In this chapter, we will only cover specific issues related to the testing of antiviral siRNA against RSV. Common respiratory viruses, such as RSV, PIV and human influenza are biosafety level 2 (BSL2) pathogens. Obey all institutional regulations. The following materials are needed:

1. Suitable cell line, e.g., A549 or HEp-2 cells for RSV.
2. Standard cell culture media. RSV and the host cell lines grow virtually in any cell culture media. Use D-MEM with glucose, supplemented with glutamine and 10% fetal bovine serum (heat-inactivated) with or without penicillin/streptomycin.
3. Standard disposable sterile plastic ware for cell culture.

2.3. Animals and Related

1. BALB/c mice, 8–10 weeks old, weighing 16–20 g (Charles River Laboratories, Wilmington, MA)
2. Five mg/ml sodium pentobarbital (Nembutal)
3. 25-gauge single-use hypodermic needles (VWR, West Chester, PA).
4. One cc single-use syringes with BD Luer-Lok tip (VWR).

2.4. siRNA Transfection

1. TransIT-TKO siRNA transfection reagent (Mirus Bio Corporation, Madison, WI).
2. Opti-MEM I Reduced Serum Medium (Gibco, Invitrogen Corporation, Carlsbad, CA).

3. RNase-free gel-loading microcapillary tips (VWR).
4. RNase-free microfuge tubes (Ambion).

3. Methods

3.1. siRNA Sequence Design

1. Use a free online siRNA design program for your target mRNA sequence (*see* **Note 1**). We generally use the one available on the Whitehead Institute (MIT) server at the following URL (http://jura.wi.mit.edu/bioc/siRNAext/); registration is required but free. Copy and paste the target sequence in the box and initially choose $AAN_{19}TT$ in the "recommended patterns." Usually, there is no need to change the "Filter criteria" below, i.e., leave them as in the default. Click on the Search button. With RSV L gene, we found many prospective ones (**Table 16.1**), but if no siRNA is obtained with your gene sequence, repeat the procedure, this time choosing NAN_{21}.
2. This will lead to a table of prospective siRNA sequences that are ordered according to "Thermodynamic values" by the default setting. The goal is to select sequences with high negative values, i.e., sequences closer to the top of the table. In addition, visually examine the central N_{19} part (i.e., ignore the first two and the last two nucleotides) and pick the ones that are AT-rich at the 3′-end (right hand side) and GC-rich at the 5′-end (left hand side). The number of sequences one can test really depends on available resources, but order at least 3 for a given target. Order the sense strand as $N_{19}dTdT$ and the antisense strand as $(N')_{19}dTdT$, such that the N′ part is complementary to the N_{19} sequence. Here, we provide an example of tested functional siRNA sequences based on the L polymerase gene of RSV (**Table 16.1**). Assume that you will receive the siRNA roughly a week after ordering. Purchase the smallest amount initially (to save money)—generally 2 nmol of each.
3. After receiving the lyophilized RNA (2 nmol), briefly spin the tubes to ensure that the RNA is at the bottom. Add 42 μl 1x siRNA buffer to each RNA. Pipette the solution up and down 8–10 times, avoiding the introduction of bubbles.
4. Place the solution on an orbital mixer/shaker for 30 min at room temperature for complete mixing. Briefly centrifuge tubes containing siRNA to ensure that the solution is collected at the bottom of the tube.
5. Combine the volumes of the complementary strands of RNA (42 + 42 = 84 μl), vortex for 10 s and centrifuge for 30 s. Add 16 μl of 6x siRNA buffer to make 100 μl, mix.

6. Heat the mixture at 60°C for 45 min.
7. Remove from heat and centrifuge briefly, 5–10 s.
8. Allow solution to cool to room temperature over 30 min to allow formation of the double-stranded siRNA, which is now ready to use. The siRNA concentration of this stock solution is 20 μM (*see* **Notes 2, 3**).
9. Aliquot the siRNA into small volumes and store at –20 to –80°C. For best results, do not freeze-thaw a tube more than four times.

3.2. Reiterated Testing of Modified and Unmodified siRNA

1. Trypsinize cells and seed ~5 × 10^4 cells in 500 μl complete growth medium per well of a 24-well plate. Incubate for 24 h to achieve a confluency of 60–70%.
2. Immediately before transfection, add 50 μl of Opti-MEM I to a sterile Eppendorf tube. To this, add 2.5 μl TransIT-TKO reagent, mix thoroughly by vortexing, and incubate at room temperature for 10–20 min.
3. Dilute the siRNA 20-fold by mixing 3.5 μl of the stock with 66.5 μl siRNA buffer. Test three concentrations of each siRNA, and each concentration at least in duplicate, and if possible, triplicate. The final concentrations and the corresponding volumes are: 5 nM (1.5 μl), 10 nM (3 μl), and 50 nM (15 μl). Add these volumes to the diluted transfection reagent made in Step 2 above and mix by gentle pipetting. Incubate at room temperature for 10–20 min.
4. Adjust the medium volume in each well of the 24-well cell monolayer (Step 1 above) to 250 μl by removing half of the original volume (250 μl).
5. Add the transfection-ready siRNA mixture (from Step 3) dropwise to the cells. Gently tilt and rock the plate to evenly distribute the complexes.
6. Incubate for 18 h in a cell culture incubator.
7. Add challenge virus to the cells. For RSV, add about 10^4 virions (1 μl of stock 10^7 pfu/ml) in each well. For each siRNA concentration, keep an uninfected well as control (to check if the siRNA itself will cause cellular death by off-target effects).
8. Incubate another 24 h if virus growth is monitored by Western blot or another 72 h if virus growth is to be monitored by extracellular titer. For the latter, replace old media with equal volume of fresh prewarmed media (for better virus growth). For Western, remove media, wash the monolayer with 0.5 ml PBS twice, and add 20 μL of 2x SDS sample buffer. Scrap the monolayer, mix well by pipetting, boil, analyze in SDS-PAGE and Western with anti-RSV antibody. For titer, collect media from the wells. Expect ~10^6 pfu/ml from uninhibited cells

and 1–3 $\log_{10}$ lower if the siRNA works well. Do serial dilutions of each sample accordingly (in fresh complete medium), plate on either A549 or HEp-2 monolayers, incubate for 48 h, count plaques under microscope (20x magnification). Take the average of triplicate plaque assays for each sample.

9. Select the most effective siRNA sequences (strongest antiviral effect) as candidates for 2′-*O*-Me and 2′-F modifications. For reasons mentioned in the Introduction, there is no universal consensus regarding the format. Thus, the guidelines offered here should be considered suggestions only, although the sequences shown have been tested to work (**Table 16.1**). The reader is encouraged to conduct more exhaustive modifications and test other sequences if resources permit.
   (a) Introduce 2′-*O*-Me modifications in all sites in the sense strand initially, but then try less extensive modifications (at various positions), which are sometimes more effective.
   (b) Introduce alternate modifications in sense and antisense strands. Start with 2′-F (in available C's and U's) followed by 2′-*O*-Me in the sense strand, and reciprocate in the antisense strand (**Table 16.1**). Order the modified siRNA from commercial sources (e.g., Dharmacon) and then process them as in Steps 3–9 (**Sect. 3.2**).

3.3. siRNA Testing in Mice

Prior to the intranasal administration of siRNA, the mouse (*see* **Note 2**) must be anesthetized by using any standard procedure available in the laboratory, e.g., by administering Nembutal via intraperitoneal (IP) injection. The recommended drug dosage for mice is 50 mg/kg (*see* **Note 3**).

1. Gently lift the mouse by the tail and place it on a cage lid.
2. Grip the loose skin of the neck to immobilize the head of the mouse. Extend the tail to draw the skin tight over the abdomen by gripping the tail with your little finger.
3. Hold mouse in a head-down position and disinfect injection site (the lower right or left quadrant of the abdomen); use a hypodermic needle to administer the anesthetic. The mouse is ready for siRNA administration when no voluntary movement is observed.
4. Place the anesthetized mouse on a lab towel facing up; with its head immobilized, insert microcapillary tip containing siRNA/transfection reagent complexes (*see* **Note 4**) into the nostril. Use a range of 2–20 nmol siRNA per mouse (e.g., 2, 6, 10 nmol). Instill solution slowly over a 2–3 min period, allowing the animal to breathe the liquid in (*see* **Note 5**).
5. Return the mouse to the cage and monitor for at least 45 min to avoid depression of cardiac and/or respiratory functions.

6. Test for the desired RNAi effect at appropriate intervals. For antiviral studies, instill virus through the nostril as well. For human RSV, which does not infect mice well, use 10^7–10^8 virus particles per animal, and measure standard lung titer assay and/or clinical symptoms (such as body weight, respiration rate etc) (*see* **Note 6**).

4. Notes

1. Recently, relatively large (26–28 nt) long double-stranded RNAs that act as Dicer substrates (D-siRNA) have been shown to be more potent than the regular 19-nt siRNAs used here *(20, 21)*. We have also found them to be at least as potent as the first and second generation 19-mer siRNA in terms of intranasal anti-RSV activity without increased immune reactions (data not shown). In preliminary experiments, they also lent well to 2′-*O*-Me and 2′-F modifications for intranasal antiviral activity (data not shown), but it is recommended that the exact format be optimized for each sequence. Follow the published D-siRNA design guidelines *(20, 22)*.
2. Although we have described the laboratory mouse model here, intranasal dosage and delivery can be easily scaled up or down for other laboratory animals.
3. siRNA concentration: The pharmaceutical industry prefers expressing drug concentrations in wt/vol or wt/body weight (e.g., mg/kg). For the researcher, however, it is easier and more useful to express siRNA concentrations in molar units (e.g., μM or nM), since this allows direct comparison between the potency of different siRNAs even when they differ in base composition or modifications (and hence formula weight).
4. Naked siRNA (without transfection reagent), with or without chemical modifications, seems to be about 80% as potent as reagent-complexed siRNA in intranasal antiviral tests, although the mechanism for their entry is unknown. Since naked siRNAs would be free of reagent toxicity, it is worth testing each modified second-generation siRNA in naked application as well. We do not recommend using PEI because of its apparent toxicity in intranasal application *(4)*.
5. Avoid using excessive liquid because it may suffocate the animal and cause death. As a rule, keep the total instilled volume under 45 μl for BALB/c mice, but higher volumes may be tolerated by larger species.

6. The protocol described here may be modified for aerosolized siRNA using an enclosure to house the anesthetized animal and a handheld nebulizer (the common type used as an inhaler by asthmatics). A larger amount of siRNA is needed because only a fraction of the aerosol is actually inhaled by the animal. If used routinely, consider optimizing a commercial motorized nebulizer. Check with the local pediatricians for the exact model, vendor and usage. Modify the system by removing the facial mask at the delivery end and inserting the tube directly into the enclosure. A snug fit of the mask should reduce siRNA waste.

References

1. Fire, A., Xu, S., Montgomery, M. K., Kostas, S. A., Driver, S. E., and Mello, C. C. (1998) Potent and specific genetic interference by double-stranded RNA in *Caenorhabditis elegans*. *Nature* 391, 806–811.
2. Elbashir, S. M., Martinez, J., Patkaniowska, A., Lendeckel, W., and Tuschl, T. (2001) Functional anatomy of siRNAs for mediating efficient RNAi in *Drosophila melanogaster* embryo lysate. *EMBO J.* 20, 6877–6888.
3. Bitko, V. and Barik, S. (2001) Phenotypic silencing of cytoplasmic genes using sequence-specific double-stranded short interfering RNA and its application in the reverse genetics of wild type negative-strand RNA viruses. *BMC Microbiol.* 1, 34.
4. Bitko, V., Musiyenko, A., Shulyayeva, O., and Barik, S. (2004) Inhibition of respiratory viruses by nasally administered siRNA. *Nat. Med.* 11, 50–55.
5. Barik, S. (2004) Control of nonsegmented negative-strand RNA virus replication by siRNA. *Virus Res.* 102, 27–35.
6. Barik, S. (2005) Silence of the transcripts: RNA interference in medicine. *J. Mol. Med.* 83, 764–773.
7. Bitko, V. and Barik, S. (2007) Intranasal antisense therapy: preclinical models with a clinical future? *Curr. Opin. Mol. Ther.* 9, 119–125.
8. Tompkins, S. M., Lo, C. Y., Tumpey, T. M., and Epstein, S. L. (2004) Protection against lethal influenza virus challenge by RNA interference in vivo. *Proc. Natl. Acad. Sci. USA* 101, 8682–8686.
9. Ge, Q., Filip, L., Bai, A., Nguyen, T., Eisen, H. N., and Chen, J. (2004) Inhibition of influenza virus production in virus-infected mice by RNA interference. *Proc. Natl. Acad. Sci. USA* 101, 8676–8681.
10. Li, B. J., Tang, Q., Cheng, D., Qin, C., Xie, F. Y., Wei, Q., Xu, J., Liu, Y., Zheng, B. J., Woodle, M. C., Zhong, N., and Lu, P. Y. (2005) Using siRNA in prophylactic and therapeutic regimens against SARS coronavirus in Rhesus macaque. *Nat. Med.* 11, 944–951.
11. Maggon, K., Barik, S. (2004) New drugs and treatment for respiratory syncytial virus. *Rev. Med. Virol.* 14, 149–168.
12. De Paula, D., Bentley, M. V., and Mahato, R. I. (2007) Hydrophobization and bioconjugation for enhanced siRNA delivery and targeting. *RNA* 13, 431–456.
13. Koller, E., Propp, S., Murray, H., Lima, W., Bhat, B., Prakash, T. P., Allerson, C. R., Swayze, E. E., Marcusson, E. G., and Dean, N. M. (2006) Competition for RISC binding predicts in vitro potency of siRNA. *Nucleic Acids Res.* 34, 4467–4476.
14. Kraynack, B. A., Baker, B. F. (2006) Small interfering RNAs containing full 2′-O-methylribonucleotide-modified sense strands display Argonaute2/eIF2C2-dependent activity. *RNA* 12, 163–176.
15. Czauderna, F., Fechtner, M., Dames, S., Aygün, H., Klippel, A., Pronk, G. J., Giese, K., and Kaufmann, J. (2003) Structural variations and stabilising modifications of synthetic siRNAs in mammalian cells. *Nucleic Acids Res.* 31, 2705–2716.
16. Sioud, M., Furset, G., and Cekaite, L. (2007) Suppression of immunostimulatory siRNA-driven innate immune activation by 2′-modified RNAs. *Biochem. Biophys. Res. Commun.* 361, 122–126.

17. Sioud, M. (2007) RNA interference and innate immunity. *Adv. Drug Deliv. Rev.* 59, 153–163.
18. Chiu, Y. L., Rana, T. M. (2003) siRNA function in RNAi: a chemical modification analysis. *RNA* 9, 1034–1048.
19. Morrissey, D. V., Lockridge, J. A., Shaw, L., Blanchard, K., Jensen, K., Breen, W., Hartsough, K., Machemer, L., Radka, S., Jadhav, V., Vaish, N., Zinnen, S., Vargeese, C., Bowman, K., Shaffer, C. S., Jeffs, L. B., Judge, A., MacLachlan, I., Polisky, B. (2005) Potent and persistent in vivo anti-HBV activity of chemically modified siRNAs. *Nat. Biotechnol.* 23, 1002–1007.
20. Kim, D. H., Behlke, M. A., Rose, S. D., Chang, M. S., Choi, S., and Rossi, J. J. (2005) Synthetic dsRNA Dicer substrates enhance RNAi potency and efficacy. *Nat. Biotechnol.* 23, 222–226.
21. Siolas, D., Lerner, C., Burchard, J., Ge, W., Linsley, P. S., Paddison, P. J., Hannon, G. J., and Cleary, M. A. (2005) Synthetic shRNAs as potent RNAi triggers. *Nat. Biotechnol.* 23, 227–231.
22. Amarzguioui, M. and Rossi, J. J. (2008) Principles of Dicer substrate (D-siRNA) design and function. *Methods Mol. Biol.* 442, 3–10.

Chapter 17

Progress in the Therapeutic Applications of siRNAs Against HIV-1

Miguel Angel Martínez

Abstract

Therapeutic options against the human immunodeficiency virus type 1 (HIV-1) continue to expand with the development of new drugs and new therapeutic strategies. Nevertheless, management of HIV-1 infected individuals has become increasingly complex. The emergence of drug-resistant variants, the growing recognition of the long term toxicity of antiretroviral therapies and the persistence of viral reservoirs justify the continued efforts to develop new anti-HIV-1 strategies. Recent advances regarding the utility of RNA-mediated interference (RNAi) to specifically inhibit HIV-1 replication have opened new possibilities for the development of gene-based therapies against HIV-1 infection. Here, the recent advances in siRNA-based therapies are reviewed.

Key words: HIV-1, RNAi, gene therapy.

1. Introduction: Alternatives to Conventional Antiretroviral Therapy

Since the first report on HIV-1-infected individuals in the *New England Journal of Medicine* in 1981 *(1)*, nearly 40 million individuals have been infected. In 2006 alone four million people were infected, and over three million died of the acquired immunodeficiency syndrome (AIDS) (www.unaids.org). A unique feature of HIV-1 is the establishment of a pool of latently infected cells very early during primary infection, resulting in the indefinite establishment of HIV-1 infection in all infected individuals. This feature of HIV-1 infection places it in sharp contrast with almost all other viral infections, in which the initial rounds of viral replication do not establish a permanent reservoir of infection.

M. Sioud (ed.), *Methods in Molecular Biology, siRNA and miRNA Gene Silencing, vol. 487*

DOI: 10.1007/978-1-60327-547-7_17

Although a few HIV-1 infected individuals have remained healthy for nearly 18 years *(2, 3)* to date not a single spontaneous cure of HIV-1 infection has been confirmed, and infected individuals have either died or remained infected.

Therefore, HIV-1 poses a greater challenge to the development of classic vaccination or antiviral strategies geared to ultimately lead to the eradication of the virus. A safe and effective HIV-1 vaccine to stop the spread of HIV-1 has not yet been developed and the possibility of developing a successful vaccine in the near future remains to be seen *(4)*.

The introduction of potent antiretroviral therapies has been an important achievement towards the control of HIV-1 infection and AIDS *(5, see* **Table 17.1**). Therapies to combat HIV-1 infection have resulted in a dramatic decrease in AIDS-associated morbidity and mortality in developed countries *(6)*. Despite the development of successful therapeutic strategies employing treatment combinations, or highly active antiretroviral therapy (HAART), it has not been possible to eradicate HIV-1 in infected individuals, mainly due to the persistence of viral reservoirs *(7, 8, 9, 10)*. The HIV-1 reservoir is not eradicated even after extended antiretroviral therapy that can reduce viremia to undetectable levels (<50 viral RNA copies per milliliter of plasma). Therefore, individuals infected with HIV-1 need to receive antiretroviral therapy for many years, if not for life. Current treatments target viral enzymes, such as the reverse transcriptase (RT) and protease, as well as the envelope glycoprotein gp41 (**Table 17.1**). The most recent antiretroviral approved for clinical use blocks the binding of HIV-1 to its co-receptor CCR5 in host cells *(11)*. Despite the success of potent combination regimens, the development of HIV-1 drug resistance constitutes a major hurdle towards long-term efficacy of current antiretroviral therapies *(12)*. Moreover, just several years after the implementation of HAART, the increase of HIV-1 drug resistance has required the initiation of a continuous effort in the development of new drugs, therapy strategies and viral targets. So far, 22 antiretroviral drugs have been approved and several others are in clinical or preclinical development (**Table 17.1**). Current limitations of HAART are not restricted to drug resistance. Toxicity and side effects, such as hyperlipidemia, hyperglycemia, hypersensitivity, pancreatitis, lipoatrophy, anemia and neutropenia, rash, diarrhoea, gastrointestinal distress, insulin resistance, immune reconstitution syndrome, renal dysfunction, hepatotoxicity and an increased risk of liver cirrhosis and myocardial infarction have all been described in response to treatment with antiretroviral drugs *(13)*. Moreover, the likelihood of side effects increases due to the improved lifespan as a result of the success of HAART. Finally, coinfection with hepatitis C, hepatitis B, tuberculosis or malaria complicates the treatment regimens for HIV-1 infected individuals.

Table 17.1
Approved drugs for the treatment of HIV-1 infected individuals

Nucleoside and nucleotide reverse transcriptase inhibitors (nRTIs)
Abacavir
Didanosine
Emtricitabine
Lamivudine
Stavudine
Tenofovir
Zidovudine
Nonnucleoside and nucleotide reverse transcriptase inhibitors (NNRTIs)
Efavirenz
Etravirine
Nevirapine
Atazanavir
Fosamprenavir
Darunavir
Indinavir
Lopinavir
Nelfinavir
Saquinavir
Tipranavir
Ritonavir
Entry inhibitors
Enfuvirtide
Maraviroc
Integrase inhibitors
Raltegravir

Based on *(12)*.

The limitations and challenges of current antiretroviral therapy justify the continued effort to develop new anti-HIV-1 strategies. Fire et al. first discovered that introducing long double-stranded RNA (dsRNA) into the nematode *Caenorhabditis elegans* led to the targeted degradation of homologous mRNA, revealing the existence of a fundamental mechanism now known as RNAi through which gene expression can be regulated *(14)*. Later findings by Elbashir et al. showed that RNAi also occurs in mammalians cells *(15)*. Importantly, this study made the extraordinary demonstration that cell transfection of synthetic 21 base pairs (bp) short interfering RNA (siRNA) duplexes can mediate RNAi in a sequence-specific manner; this finding enabled the specific regulation of gene expression in a variety of biological systems, including diseased cells.

In this review, I will focus on the progress in research employing RNAi to disrupt the disease process caused by HIV-1. Given that an exogenous RNA can be targeted without affecting cellular functions, one of the most promising applications of RNAi is in the treatment of infectious diseases. Nevertheless, there are some obstacles to the development of an anti-HIV-1 RNAi-based therapeutic agent. One major problem of all antiretroviral therapies is the emergence of resistant variants, and the enormous genomic heterogeneity of HIV-1 may hinder the efficacy of single defined siRNAs. Coexpression of multiple siRNAs could reduce the emergence of single-siRNA-resistant viruses, with an effect comparable to that achieved by HAART, which combines three- or four-anti-HIV-1 drugs in treatment. An alternative strategy to counter the possibility of siRNA-resistant viruses would be to target transcripts of cellular cofactors essential for HIV-1 infectivity or replication (e.g., specific cell surface receptors). The side effects of downregulating cellular targets over the long term, however, are currently unknown. Delivery remains a major hurdle for RNAi therapy, as siRNAs are unable to cross the mammalian cell membrane without aid. Gene-therapy vectors are capable of stably expressing siRNA precursors, although well-documented hazards have been observed after the introduction of foreign vector sequences into chromosomal DNA. Finally, off-target effects, such as the targeting of genes sharing partial homology to the siRNA and the stimulation of the innate immune system, need to be considered and avoided. Although the innate immune system is efficiently triggered only by dsRNAs more than 30 bp, high concentrations of smaller siRNAs may be able to activate this pathway. Nevertheless, there is no evidence that the activation of the innate immune system influences the degree or specificity of siRNAs.

2. Targeting HIV-1 Replication by RNAi

The HIV-1 life cycle begins by viral binding via the viral envelope to the cellular receptor CD4 in conjunction with a coreceptor, either CXCR4 in the case of T-cell-tropic virus or CCR5 for macrophage-tropic virus (**Fig. 17.1**). HIV-1 infects cells of the immune system, specifically CD4+ cells, which include T lymphocytes, monocytes and macrophages. As a result, in the absence of effective vaccination or therapy, a slow and continued depletion

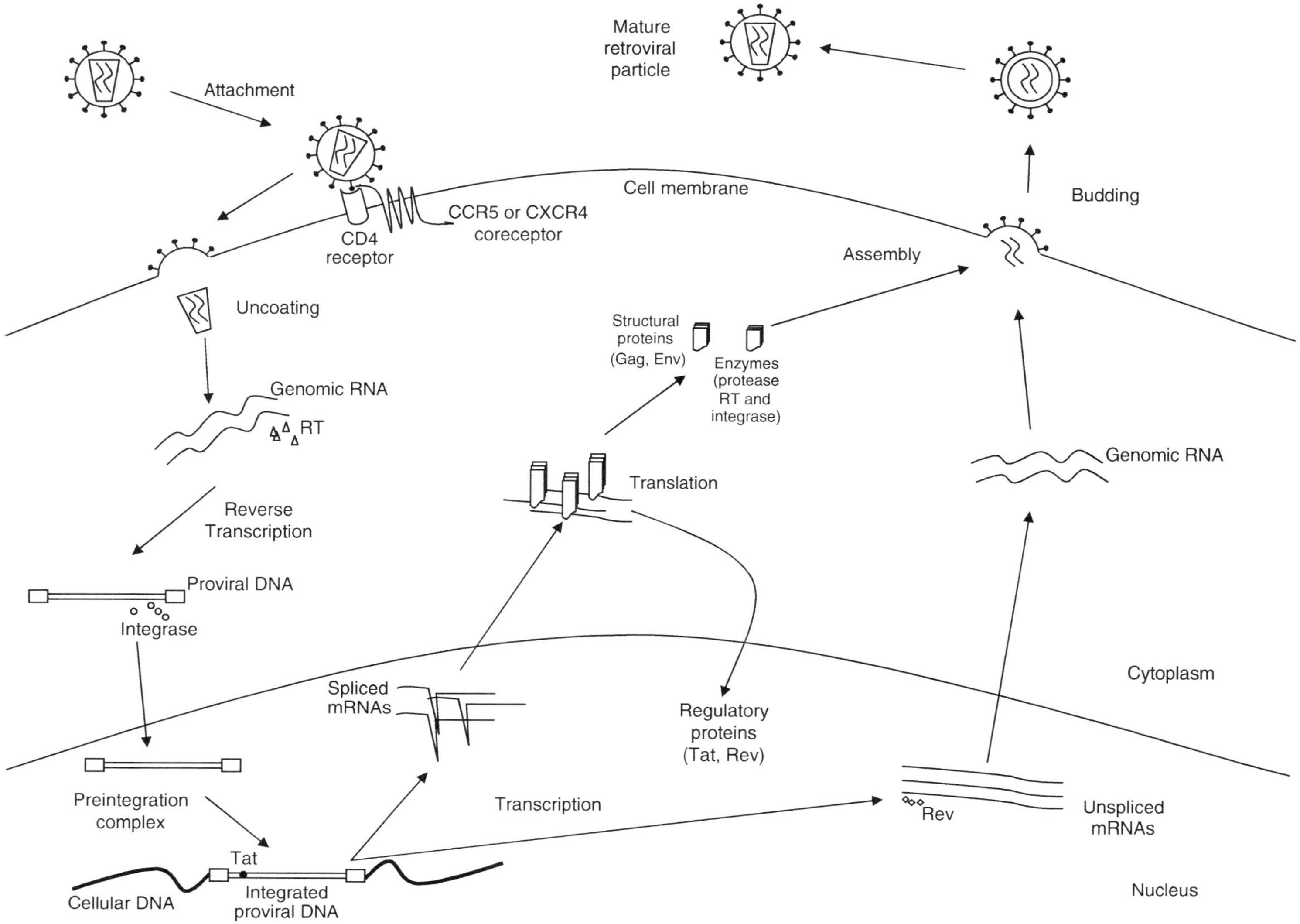

Fig. 17.1. Schematic representation of the HIV-1 replication cycle. Virus attachment, reverse transcription, integration into the cellular genome, transcription, translation, virus assembly, virus budding and virus maturation are consecutively shown by arrows. siRNAs or shRNAs that target HIV-1 RNA might induce the cleavage of preintegrated genomic RNA or interfere with HIV-1 RNA transcripts postintegration and block progeny virus production.

of CD4+ T cells ensues and there is a progression towards AIDS. Fusion of the viral and cellular membranes provides the HIV-1 genome, two RNA molecules of positive polarity, access to the interior of the cell. The genomic viral RNA is then converted into double-stranded DNA by the HIV-1 RT (**Fig. 17.1**). The viral DNA forms a pre-integration complex along with the viral integrase, which is then transported to the nucleus where the DNA is integrated into the host genome as a provirus. HIV-1 replication can be divided into early and late phases. During the early phase, spliced transcripts encoding two essential proteins, Tat and Rev, are synthesized. Tat further activates viral transcription, whereas Rev interacts with Rev-responsive elements (RRE) to facilitate nuclear export of unspliced and singly spliced mRNAs. These late mRNAs encode the remaining viral proteins, including Gag, Pol and Env, that assemble at the cell surface together with the viral genome to form virions that are released from the cell by budding. The viral protease continues to process the viral polyproteins into their mature form, condensing the viral genomic RNA core and yielding infectious particles (**Fig. 17.1**).

Soon after the demonstration that synthetic siRNAs were able to induce the RNAi mechanism in mammalian cells *(15)*, several studies reported that HIV-1 gene expression and replication ex vivo could be inhibited by virus-specific synthetic siRNAs *(16–22)* or expressed siRNAs *(16, 18)* that were targeted to early or late phases of virus replication. siRNAs against several HIV-1 components (Gag, Env, Pol, the HIV-1 long terminal repeat, Vif, Nef, Tat and Rev) were confirmed to effectively target these genes. Moreover, CD4+ T cells *(18, 23)* and macrophages *(24, 25)*, which are the natural targets for HIV-1 infection, were also found to be functional for RNAi. However, an important question arose from these pioneer studies: is the virion-associated incoming genomic viral RNA targeted by RNAi? Since HIV-1 is capable of being integrated in the host genome as well as infecting resting CD4+ T cells to create a pool of latently infected cells *(4)*, the capacity to target incoming viral RNA by siRNAs has important therapeutic implications (**Fig. 17.1**) *(26)*. If it is not possible to target incoming viral genomic RNA, it will be virtually impossible to prevent the formation of integrated provirus and, as a result, to sterilize cells from infection. Targeting the incoming viral RNA may have several advantages, as the RNAi machinery would only have to deal with two or a few (in the case of superinfection) genomic viral RNAs. However, after a provirus is established, several thousands of viral transcripts are generated de novo in the infected cell and the targeted degradation of these may be more difficult for the RNAi machinery. Data has been conflicting as to whether RNAi can target the incoming genomic viral RNA of infecting HIV-1 particles. Several studies have reported degradation of the incoming RNA genome in cells transfected with artificial siRNAs or stably expressing siRNAs *(18–20, 27)*. However, other studies have reported an absence of RNAi-mediated degradation of the genomic viral RNA in siRNAs-transfected or siRNA-producing cells *(22, 28, 29)*. Moreover, a study with the non-segmented, negative-strand RNA virus respiratory syncytial (RSV) showed that genomic and antigenomic viral RNAs encapsidated in the virus nucleocapsid protein were not susceptible to siRNA-mediated silencing, but viral mRNAs were highly susceptible *(30)*.

The mechanisms underscoring the efficiency of specific siRNAs in targeting genes is not completely known. RNAi efficiency is known to be influenced by local RNA structure of the target sequence *(31)*. Randomly selected siRNAs against a target sequence showed a large variation in their efficiency *(32)*. It has been argued that some regions of genomic viral RNA might be less accessible to siRNAs when the RNA is contained in the virus reverse transcription complex; therefore, some siRNAs might be more effective than others in inhibiting this early step in virus replication *(26)*. Recently, a more exhaustive study was performed to address this issue *(33)*. In this study, the transduction efficiency of a lentiviral vector, as measured by the number of successful integration events, was determined in a cell line stably expressing an siRNA targeting the HIV-1 Nef sequence. The results indicated a similar transduction efficiency for vectors regardless of the presence or absence of the Nef target sequence in their genome. Moreover, no reduced transduction efficiencies were observed in the presence of multiple other stably expressed siRNAs targeting the vector genome, or when synthetic siRNA targeting Nef was transiently transfected prior to transduction. These results highlight the difficulties of designing therapeutic RNAi strategies aimed at preventing HIV-1 proviral integration. Silencing Tat, a factor that must be expressed before efficient provirus transcription can occur, may make it easier to target other viral transcripts by virtue of their lower abundance when Tat function is compromised *(26)*.

Synthetic siRNAs mediate strong and specific but transient suppression of gene expression *(18–20, 28)*. The transient reduction in gene expression by siRNAs severely restricts its applications in a therapeutic setting of a persistent infection, such as HIV-1 infection. However, this limitation was quickly overcome with the use of plasmids or viral vectors that expressed siRNAs from polymerase-III transcription units. The sense and antisense strands of an siRNA were expressed from two different promoters and RNAi was triggered upon annealing of the two strands *(16)*. A more potent gene silencing was achieved when the sense and antisense strands were expressed as a single transcript with the ability to form a duplex hairpin structure *(34)*. Indeed, HIV-1 replication could be efficiently inhibited with such short hairpin RNAs (shRNAs) in stably transduced cell lines *(34–37)*. An additional benefit in using shRNAs is that these can be processed by Dicer like the endogenous microRNAs (miRNAs). Tat siRNA delivered as a pre-miRNA precursor was 80% more effective in reducing HIV-1 p24 antigen production than Tat siRNA expressed as a conventional shRNA. Modeling shRNAs after endogenous miRNAs can increase the antiviral potency of RNAi *(38)*.

As discussed in this section, an exogenously induced RNAi response in mammalian cells mediated by siRNAs or shRNAs can inhibit HIV-1 replication. These results raise the question of whether HIV-1 interacts with the cellular RNAi machinery, and understanding the potential physiological interaction between HIV-1 and the RNAi machinery could significantly contribute to the design of an effective RNAi-based therapy. RNAi constitutes a key component of the innate immune response to viral infection in both plants and invertebrate animals *(39)*, and has been postulated to have a similar protective function in mammals *(40)*. In plants and invertebrate animals, long dsRNA serves as the initial trigger for the RNAi mechanism; the initial dsRNA is then processed by Dicer into siRNAs. However, in mammalian cells, long dsRNA sequences (more than 30 bp in length) are potent inducers of the interferon (IFN) response *(41)*. It remains unclear whether the introduction of long dsRNA into mammalian somatic cells is capable of resulting in the production of siRNAs *(39)*. Nevertheless, RNAi can be induced in mammalian cells by the nuclear endogenous expression of RNAs that are intermediates in the miRNA biogenesis pathway (i.e., siRNAs and shRNAs). The human genome encodes several hundred different miRNA molecules that are believed to be key players in the posttranscriptional regulation of many aspects of cellular differentiation, and RNAi has been postulated to exist in mammals only as a mechanism of post-transcriptional regulation "programmed" by endogenously encoded miRNA, rather than being involved in intrinsic antiviral immunity *(39)*.

All RNA viruses, except retroviruses, produce long dsRNA molecules in infected cells that represent essential intermediates in genomic RNA production. Many DNA viruses also generate long dsRNA, as their small packed genomes have convergent transcription promoters. Some studies have attempted to identify siRNAs in virus-infected human cells. Pfeffer et al. *(42)* failed to identify any viral siRNA in cells infected by DNA viruses, such as human cytomegalovirus, Kaposi sarcoma-associated herpes virus, Epstein-Barr virus and mouse herpes virus, as well as HIV-1 and the RNA viruses for yellow fever and hepatitis C. This report, however, did identify several virally encoded miRNA molecules in the DNA virus. In contrast, Bennasser et al. *(43)* reported that HIV-1 gives rise to an siRNA that can inhibit HIV-1 replication, and additionally proposed that the HIV-1 Tat protein relieves this inhibition by blocking the function of the cellular Dicer, which plays a key role in both siRNA and miRNA biogenesis. Moreover, the Ebola virus VP35 protein has been shown to be a suppressor of RNAi in mammalian cells and its suppressor activity is functionally equivalent to that of the HIV-1 Tat protein *(44)*. VP35 can replace HIV-1 Tat and thereby support the replication of an HIV-1 variant deficient for Tat. These results support the hypothesis that RNAi is indeed part of the innate antiviral response in mammalian cells. However, in a recent report, the stable expression of physiological levels of Tat did not globally inhibit miRNA production or expression in HIV-1 infected human cells *(45)*. Strong evidence for RNAi as part of the antiviral innate immune in plants comes from the demonstration that almost all plant viruses encode one or more suppressor of RNA silencing (SRS) proteins, which target several key steps in the RNAi response *(40)*. Similarly, several viruses that target invertebrates (specifically nematodes and insects) also encode SRS proteins. Flock house virus, a member of the nodavirus family that infect both insects and vertebrate cells, encodes a viral SRS that inhibits Dicer function *(46)*. Further evidence of the controversial relationship between HIV-1 and cellular RNAi machinery comes from the recent report by Triboulet et al. *(47)*,which provides evidence for a physiological role of the miRNA-silencing machinery in controlling HIV-1 replication. This study showed that HIV-1 infection and replication were more efficient in peripheral blood mononuclear cells from HIV-1-infected donors and latently infected cells knocked-down for Dicer or Drosha, a miRNA processing factor. Moreover, HIV-1 actively suppressed the expression of the polycistronic miRNA cluster miR-17/92, and this suppression was found to be required for efficient viral replication. As suggested in this study, these findings may help address the current challenge of how to activate latent viral reservoirs in HIV-1 therapy. Virally encoded miRNAs have been discovered in herpes viruses *(39)*, and human cytomegalovirus has been shown to evade the host immune system by targeting host genes with virally encoded miRNAs *(48)*. Whether RNA viruses, including HIV-1, encode miRNAs still remains controversial *(48)*. The therapeutic implications of this molecular feature are intriguing, as targeting these viral miRNAs might constitute an antiviral therapy while mimicking their role could provide a means of immunosuppressive therapy *(48)*. Latent infection is one of the most important characteristics required for the in vivo survival of all HIV-1 strains and the major obstacle in eradicating virus infection with HAART *(7–10)*. A recent study showed that cellular miRNAs potently inhibit HIV-1 production in resting primary CD4+ T cells *(49)*; the 3′ ends of HIV-1 mRNAs were found to be targeted by a cluster of cellular miRNAs including miR-28, miR-125b, miR-150, miR-223 and miR-382, which are enriched in resting

CD4+ T cells compared to activated CD4+ T cells (see Chap. 20). Because specific inhibitors of these miRNAs substantially counteracted their effects on the target mRNAs *(49)*, a combined miRNA inhibitor panel could be used to activate latent HIV-1 for therapeutic purposes. In agreement with the former report, the cellular miRNAs miR-196, miR-296, miR-351, miR-431 and miR-448 have been shown to be capable of inhibiting hepatitis C virus replication and infection *(50)*. These miRNAs are upregulated by IFNα/β, suggesting that cellular miRNAs may be components of the mammalian innate immune response *(50)*.

3. Obstacles to Therapy

3.1. HIV-1 Escaping Inhibition by RNAi

This review highlights the evidence showing that RNAi provides a robust method for specifically inhibiting the expression of targeted HIV-1 genes, and its promise as a novel and broadly applicable approach to antiviral therapy. However, clinical application of RNAi faces several challenges, specifically the potential for viral escape. One of the main advantages of the RNAi mechanism is that it is highly sequence specific. Pioneer studies confirmed the specificity of RNAi by evaluating the activity of an siRNA with one or more mismatches relative to the target RNA sequence *(18)*; these studies showed that a mismatch may be sufficient to reduce the silencing effect, suggesting that an unspecific antiviral response induced by siRNA was not involved in the silencing effect, but also showing that HIV-1 may easily escape inhibition by siRNAs. Indeed, HIV-1 has been observed to promptly escape suppression by effective siRNAs *(36, 51, 52)*.

Unlike eukaryotic DNA polymerases, the HIV-1 RT lacks proofreading activity, and its error rate has been estimated at 10^{-4} to 10^{-5} mutations per nucleotide and cycle of replication (**Table 17.2**) *(53)*. If one assumes that 10^{9}–10^{10} viral particles are produced each day in an infected person *(54, 55)*, these must be the product of at least 10^{7}–10^{8} replication cycles. Given the length of the HIV-1 genome (approximately 10,000 nucleotides) and its high recombination rate *(56)*, it is likely that every single possible point mutation (and likely many double mutations) will occur at least once each day in an infected individual *(57, 58)*. Although specific combinations of multiple mutations may be rare, it is clear that the degree of potential genetic change drives the diversification of HIV-1 in response to the selective pressure of host immune responses or antiretroviral therapy (**Table 17.2**).

HIV-1 strains have diversified extensively through mutation and recombination since their initial transmission to humans many decades ago in central Africa. Phylogenetic analysis of numerous

Table 17.2
Important parameters that influence variability and adaptability of HIV-1 populations

1. Average number of mutations per genome within the mutant spectrum of an infected individual viral population
Generally this averages at 1–100 (if not longer in some cases) mutations per genome.
2. Mutation rate
Error rate has been estimated at between 10^{-4} to 10^{-5} mutations per nucleotide and cycle of replication.
4. Genome length
10 kb
3. Virus population size and fecundity
Variable but 10^{9}–10^{10} viral particles can be produced each day in an infected individual. Every single possible point mutation (and probably many double mutations) will occur at least once each day in an infected individual.
5. Mutations needed for a phenotypic change
Many recorded adaptive changes depend on one or a few mutations.

Based on *(57, 58)*

isolates obtained from diverse geographic origins has allowed the division of HIV into types, groups, subtypes, sub-subtypes, circulating recombinant forms (CRFs) and unique recombinant forms (URFs). HIV-1 is divided into three groups, M, O and N *(59)*. Most sequences within HIV-1 group M, which accounts for the majority of infections worldwide, fall into a limited number of discrete clades, allowing the classification into subtypes and sub-subtypes. Near 30 circulating genetic forms of the HIV-1 group M are presently recognized, including 11 subtypes and sub-subtypes, and nearly 20 CRF. The diverse HIV-1 subtypes, CRFs and the inter-subtype mosaic genomes further exacerbate the problem of designing effective siRNAs.

Boden et al. *(51)* characterized the potency and durability of virus-specific RNAi in cell lines that stably expressed shRNA targeting the HIV-1 transactivator protein gene Tat. They found that the antiviral activity of Tat shRNA was abolished due to the emergence of viral species harboring a point mutation in the shRNA target region. During the first three weeks, HIV-1 replication was reduced by 95% in cells expressing Tat shRNA compared to control cells. By day 25, however, viral titers increased, indicating loss of Tat shRNA-mediated antiviral activity, and sequencing of this virus stock revealed that the emerged viral species contained a nonsynonymous mutation at nucleotide position 9 of the targeted sequence. Das et al. *(36)* used retroviral transduction to stably introduce vectors expressing siRNAs directed against the HIV-1 Nef gene into human T cells; HIV-1 escape variants that were resistant to siRNA-Nef appeared after several weeks of culture, and these RNAi-resistant viruses contained point mutations, double point mutations or partial or complete deletion of the Nef gene target sequence. The complete inactivation of the accessory Nef gene has a relatively minor impact on the replication capacity of HIV-1 ex vivo. Interestingly, Westerhout et al. *(52)* used human T cells that stably express an siRNA-Nef target sequence to show that HIV-1 can also escape siRNA-mediated Nef inhibition by a point mutation outside the target sequence. This mutation changed local RNA folding, such that the target sequence becomes inaccessible to the RNAi machinery. Recently, Sabariegos et al. *(60)* showed that optimal HIV-1 gene silencing by siRNA requires precise complementarity with most of the target sequence and that only a few substitutions at the 5′ and 3′ ends are partially tolerated. Viral escape was simulated by systematically introducing single-nucleotide substitutions in all 19 HIV-1 residues targeted by an effective siRNA directed against the RT coding region. All mutant viruses that were tested replicated better in the presence of the siRNA-RT than in the presence of the wild type virus. The antiviral activity of the siRNA-RT was completely abolished by single substitutions in 10 (positions 4 to 11, 14, and 15) out of 16 positions tested (substitutions at 3 of the 19 positions rendered nonviable viruses). With the exception of one substitution, substitutions at either the 5′ or 3′ end were better tolerated by the RNA interference machinery and only partially affected siRNA-RT inhibition.

Since the emergence of resistant virus variants poses a serious problem for using RNAi in a therapeutic setting, different strategies have been developed to counteract viral escape. One obvious strategy that will successfully both inhibit a diversity of HIV-1 isolates and protect against the emergence of viral escape would be to identify viral conserved sequences for targeting. Targeting highly conserved sequences may hamper the development of escape mutants, as these variants may be highly compromised for viral fitness. Indeed, Lee et al. *(37)* targeted highly conserved HIV-1 Vif sequences and successfully suppressed a variety of primary viral isolates from five different viral clades. This study also showed that tolerance to target sequence mismatches may depend on the sequence of the siRNA tested *(37)*. Nevertheless, our knowledge of HIV-1 variability and experience from over one decade of antiretroviral therapy have taught us that is unrealistic to try to target HIV-1 with a unique siRNA because the rapid development of resistance will occur. The high error rate of HIV-1 RT and the rapid viral turnover means that for each of the 19 nucleotides targeted by an si- or shRNA, several potential escape variants will be already present before the start of therapy *(57)*. Moreover, HIV-1 may escape RNAi through silent mutations, which, in most cases, will not have any cost on the viral fitness. To counteract this strategic weakness, co-expression of multiple siRNAs or shRNAs that target conserved RNA sequences could reduce the emergence of single siRNA-resistant virus with a comparable effect to that achieved by the multiple anti-HIV drug combination approach employed by HAART. Delivery of multiple siRNAs could induce highly active antiretroviral gene silencing (HAAGS) *(61)*.

The combination of multiple shRNAs against conserved HIV-1 genomic regions effectively inhibits HIV-1 replication. Expression of three different shRNAs from a single lentiviral vector resulted in similar levels of inhibition per shRNA compared to single shRNA vectors *(62)*. In this study, the three shRNAs targeted Gag and Pol HIV-1 coding regions; moreover, when cells transduced with a double shRNA viral vector were infected, virus escape was delayed *(62)*. There are several methods currently in use that are capable of expressing multiple effective siRNAs. One possibility is by inserting multiple shRNA-expression cassettes into a viral vector. However, repeats of the same regulatory sequences (e.g., the U6 or H1 polymerase III promoter) may cause genetic instability and reduced titer of the vector system *(63)*. Moreover, some studies have shown that it is possible to saturate the RNAi pathway by high levels of expression of shRNAs *(64, 65)* resulting in cellular toxicity, particularly with the U6 promoter [64]. Polymerase III promoters other than U6 and H1 has been used *(66)*. Another approach has been the use of long-hairpin RNAs (lhRNAs), from which multiple siRNAs can be produced. lhRNAs greater than 50 bp in length can be expressed in cells and create multiple siRNAs via Dicer-mediated processing without inducing the IFN pathway *(67)*. Several reports described efficient RNAi induction by lhRNAs against HIV-1 *(68, 69)*. Nishitsuji et al. *(69)* showed that a 50 bp lhRNA against a conserved HIV-1 integrase region suppressed HIV-1 replication in a variant resistant to a shorter shRNA for the same target. Similarly, lentiviral vectors expressing 50, 53, or 80 bp lhRNAs targeting contiguous sequences within the Tat and Rev genes inhibited viral replication against both non-mutant and mutant variants of HIV-1 *(70)*.

An alternative to target multiple conserved viral genomic regions is to use a second generation of siRNAs that recognize the mutated target sites *(60, 69, 71)* by using siRNAs or shRNAs that target the most likely escape variants. Sabariegos et al. tested a second generation siRNA that compensates for a fully resistant mutation at position 9 of the target sequence *(60)*. Nishitsuji et al. isolated single point mutation resistant viruses that emerged in lentiviral vector transduced cells expressing shRNAs against the HIV-1 U3 region, integrase and Tat genes *(69)* and restored viral inhibition by using a second generation shRNA that matched escape mutants *(69)*. The use of cell culture infections to determine which target site mutations result in replication competent escape variants will allow the design of siRNAs capable of inhibiting these escape variants. To reduce the number of second generation siRNAs to be designed, this strategy should also target a conserved region of the HIV-1 coding sequence and the new siRNAs should also be capable of inducing RNAi. Recently, *Escherichia coli* endoribonuclease III (RNase III) or mammalian Dicer was used to cleave dsRNA into endoribonuclease-prepared siRNA (esiRNA) *(72)*, which generate a variety of siRNAs that efficiently and specifically target multiple sites in the cognate RNA *(73, 74)*. In this study, esiRNAs targeting the HIV-1 RT coding region reduced viral replication by 90% in a dose dependent and sequence specific manner. Importantly, esiRNAs obtained from the prototypic RT sequence of the HXB2 strain and from highly mutated RT sequences showed similar degrees of viral inhibition, suggesting that the heterogeneous population of esiRNAs could overcome individual mismatches in the RT sequence. These results demonstrated the ability of esiRNAs to function as potent HIV-1 inhibitors. Moreover, sequence targets do not need to be highly conserved to reach a high level of viral replication inhibition. Nevertheless, this work was performed in cell culture and its translation to an in vivo setting may be difficult.

Another recently explored strategy has been to combine an shRNA with an anti-HIV-1 ribozyme and an RNA decoy. A triple combination lentiviral construct comprised of a U6-driven TAR RNA decoy with a U16 snoRNA for nucleolar localization, a U6 promoted shRNA targeted to both Tat and Rev, and a VA1 promoted chimeric anti-CCR5 trans-cleaving hammerhead ribozyme efficiently transduced human progenitor CD34+ cells and improved suppression of HIV-1 over 42 days compared to a single anti-Tat/Rev shRNA or double combinations of shRNA/ribozyme or decoy *(75)*. This triple combination is about to enter human clinical trials for AIDS/lymphoma patients using autologous hematopoietic progenitor cells as the targets for vector insertion. A second trial in which the same construct will be inserted in autologous T lymphocytes will most likely initiate in late 2007 or early 2008 *(66)*.

In order to prevent provirus establishment and to augment the genetic barrier for viral escape, cellular receptors or co-factors involved in the initial phase of infection may be targeted. The combination of siRNAs or shRNAs directed against both HIV-1 and cellular transcripts may result in a more potent and robust inhibition of viral replication. Nevertheless, targeting a cellular cofactor required by the virus should be performed with caution as this approach has the potential to harm host cells. Several cellular factors involved in the HIV-1 life-cycle have been explored as possible RNAi targets: the CD4 receptor and coreceptors CCR5 and

CXCR4 *(17, 76, 77)*, integration factors like BAF-1, Emerin and LEDGF/75 *(78, 79, 80)*, transcriptional factors such as NF-κB, PAK-1 and cyclin *(22, 81, 82)*, and Furin which is involved in Env maturation *(81)*. Since HIV-1 actively suppressed the expression of the polycistronic miRNA cluster miR-17/92 and this suppression was found to be required for efficient viral replication *(47)*, nuclear expression of some miRNAs from this cluster may be another alternative to inhibit HIV-1 replication *(66)*. In vivo suppression of genes such as CD4 could be limited due to roles in normal immune function, making HIV-1 coreceptors more attractive alternatives to target host proteins. CCR5 may be a potential coreceptor target, as a homozygous mutation in CCR5 effectively conferred protection from HIV-1 without any serious deleterious effects in human immune function, and heterozygous individuals with 50% decrease in CCR5 surface expression have lower plasma viral load and a substantially prolonged course of disease *(83)*. A potent and noncytotoxic shRNA directed to CCR5 stably down-regulates CCR5 when introduced via CD34+ hematopoietic stem cell transplant in non-human primates *(84)* and these cells were less susceptible to simian immunodeficiency virus (SIV) infection ex vivo. Recently, several CCR5 antagonists examined in clinical trials were able to reduce plasma viral loads by more than one order of magnitude during treatment *(11)*. Although, CCR5 antagonists hold great promise, the long-term toxicity associated with the impairment of CCR5 function remains to be addressed. Targeting cellular factors will require extensive toxicity studies.

3.2. In Vivo Delivery

Delivery remains a major hurdle for RNAi therapy, as siRNAs are unable to cross the mammalian cell membrane without aid. Most of the ex vivo transfection methods used for delivering siRNAs can not be used in vivo. There are two strategies for delivering siRNAs in vivo. As discussed in the previous section, one is to stably express siRNA precursors, such as shRNAs, from viral vectors using gene therapy technology; the other is to deliver synthetic siRNAs naked or by complexing or covalently linking the siRNA with lipids, aptamers, peptides or proteins *(85)*. Intranasal administration of naked siRNAs either in saline or with excipients such as 5% dextrose or lung surfactants reduced the viral load of respiratory syncytial virus (RSV) and parainfluenza virus in pediatric or immuno-compromised individuals by more than three orders of magnitude *(86*, see Chap. 15). Similarly, siRNA intranasally administered in a non-human primate model of severe acute respiratory syndrome (SARS) corona virus infection significant inhibited viral replication in the lung *(87)*. These studies clearly demonstrated the utility of RNAi to treat viral respiratory infections. Liposomes, vesicles with an aqueous compartment enclosed in a phospholipid bilayer that can fuse with cell membranes and enhance drug delivery into cells, have been extensively used to deliver siRNA in vitro and in vivo. In vitro transfection of siRNA using lipid-based delivery agents is a routine laboratory procedure. Lipid-based formulations effectively delivered siRNA to the liver in animal models of hepatitis B and Ebola virus infection *(88, 89)*. Locally injected liposomes have also delivered siRNA effectively to target cells in the eye and nervous system, and to tumors *(85)*. Topical delivery of siRNAs using liposomes may be especially effective. Intravaginal application of siRNAs protected mice from lethal herpes simple virus type 2 (HSV-2) sexually transmitted infection *(90)*. The cervical/vaginal mucosa is the main port of HIV-1 entry in women, thus an effective topical microbicide against HIV-1 may be useful to prevent sexual transmission. Since the cellular coreceptor CCR5 is required for infection by the majority of primary HIV-1 isolates, a lipid-formulated siRNA against the CCR5 transcripts will be an excellent candidate for anti-HIV-1 microbicide.

siRNAs can be complexed with cationic peptides and polymers to form stable nanoparticles via ionic interactions with their negatively charged phosphate backbones *(91)*. Similarly, one study showed that a protamine antibody fusion protein delivered non-covalently bound siRNA to HIV-1-infected primary CD4+ T cells in vitro with high efficiency *(92)*. In this approach, the Fab fragment of an HIV-1 envelope antibody mediates receptor-specific binding to cells expressing the HIV-1 envelope protein, illustrating the potential for antibodies to also direct siRNA selectively into cells in vivo.

Given the HIV-1 life cycle, in which the persistence of a latent reservoir of resting infected cells makes eradication of the virus from infected individuals extremely difficult, the most promising strategy for in vivo delivery of siRNA is the use of gene therapy approaches. Because HIV-1 predominately infects T lymphocytes and macrophages, blood CD34+ hematopoietic stem cell transplant could be the best strategy, in which progeny cells would stably express siRNAs that down-regulate viral or cellular genes that are targets for HIV-1 infection. As noted before, HIV-1 can infect resting non-dividing CD4+ T cells, thus lentiviral vectors should be employed over murine retroviral vectors, as HIV-1-based lentiviral vectors have been shown to be particularly suited for the transduction of non-dividing cells, such as hematopoietic progenitor cells *(93)*. These HIV-1-based vectors seem to be preferable candidates for gene therapy development in the treatment of HIV-1-infected individuals *(94)*. However, shRNAs targeting highly conserved HIV-1 sequences may be an obstacle in vector production, as it has been reported that expression of shRNAs that target also the delivering vector can result in a reduction of the transduction titer *(23)*, suggesting that these shRNAs could cross-react with critical sequences of the vector. Nevertheless, new lentiviral vectors have been developed to overcome this issue, and the vector backbone may be modified by point mutations to resist RNAi-mediated degradation during vector production *(95)*. However, introducing RNAi-resistant mutations within the lentiviral vector backbone adds the potential risk of transfer of resistance to the wild type HIV-1 genome. One solution that has been suggested would be to incorporate stop codons within these vector sequences that should not affect the RNA/DNA function, yet would confer full resistance to RNAi *(96)*. Recombination will result in an RNAi-resistant yet replication-defective HIV-1 variant due to the stop codons. Alternatively, it has also been shown that the incorporation of target sequences for endogenous miRNAs within the lentiviral genome can severely reduce potential recombination or mobilization *(97)*. The HIV-1 genome is highly prone to recombination, and recombinants arise very frequently *(56)*. Lentiviruses, unlike murine retroviruses, are more prone to integrate distally from promoters within intron sequences, potentially limiting their overall oncogenicity *(98)*. One of the main concerns with current gene therapy approaches is insertional mutagenesis from preferential integration of retroviral vectors into actively transcribed genes, including proto-oncogenes. This has led to hematological malignancies in young individuals with severe immunodeficiency (SCID) disease treated with retroviral-based gene therapy *(99)*. This problem could be addressed by designing vectors that integrate into specific, well-defined regions of the genome. The partially random insertion of transgenes into chromosomal DNA of hematopoietic cells may induce clonal competition, which could potentially trigger leukemia or sarcoma *(100, 101)*.

Lentiviral vectors derived from HIV-1, HIV-2/SIV, or feline immunodeficiency virus (FIV) have been shown to be capable of stably transducing many cell types, including hematopoietic stem cells *(102, 103)*. Interestingly, HIV-1 vectors have been shown to cross-package by FIV and are capable of stably transducing and protecting human primary blood mononuclear cells from HIV-1 infection *(104)*. FIV packaged HIV-1 vectors reduce the likelihood of immune recognition, or seroconversion, due to exposure to HIV-1 structural proteins. A continuous effort is focused on producing effective, safe and high-titer lentiviral vectors to deliver RNAi.

3.3. Immunostimulation and Off-Target Toxicity by RNAi

Several important issues related to in vivo safety, toxicity or side effects warrant close attention before RNAi may be considered a valid alternative for treating HIV-1 infection or other diseases. In mammalian cells, dsRNA is recognized by dsRNA sensors such as Toll-like receptors (TLRs), the dsRNA-dependent protein kinase R (PKR) and retinoic-acid-inducible gene-I (RIG-I), which are components of the innate immune system. This recognition leads to the activation of the IFN-regulatory transcription factors and NF-kB, which in turn results in the expression of the IFNs *(41)* that subsequently activate the transcription of hundreds of IFN-stimulated genes (ISGs) through the JAK-STAT pathway. Many ISGs encode proteins with antiviral activities, including PKR. The IFN response plays a crucial role in antiviral immunity in vertebrates, particularly against RNA viruses that generate dsRNA molecules during their life cycle. The IFN response could significantly influence the in vivo application of siRNA owing to the off-target effects and toxicities associated with immune stimulation. siRNA molecules less than 30 bp in length are generally considered incapable of inducing IFN pathways *(15)*. However, synthetic siRNAs formulated in non-viral delivery vehicles can be potent inducers of IFNs and inflammatory cytokines both in vivo in mouse and ex vivo in human blood cells *(105)*. The immunostimulatory activity of formulated siRNAs and the associated toxicities may be dependent on the nucleotide sequence *(106)*. Although some of the off-target effects can be reduced using lower concentration of siRNAs, the IFN response was observed even at low concentrations *(107)*. These studies emphasize the relevance of examining the immunostimulatory effects of any potential therapeutic siRNA in human immune cells prior to clinical applications. Replacement of the 2′-hydroxyl uridines with either 2′-fluoro, or 2′-deoxy or 2′-*O* methyl uridines can abrogate immune recognition of siRNAs by TLRs without compromising siRNA silencing potency *(108, 109, 110)*. In addition, modified nucleotides may protect siRNAs from nuclease degradation and ameliorate their pharmacokinetic parameters in vivo *(111)*.

Similar to synthetic siRNAs, endogenously expressed shRNAs also activated the IFN pathway *(112)*. Of particular concern is the development of a multiple shRNA or lhRNAs approach against HIV-1 because these molecules are longer than 30 bp. Endogenously (nuclear) expressed shRNAs were recently shown to evade detection by TLRs, RIG-1 and PKR when integrated in CD34+ progenitor hematopoietic stem cells *(113)*. Endogenously expressed lhRNA encoding an effective Nef-specific shRNA was capable of inhibiting HIV-1 production without inducing the type I IFN genes *(68)*. Endogenously produced dsRNA is suggested to be less active than exogenous dsRNA in inducing the IFN response *(114)*. Extended shRNAs (e-shRNAs) of 40–44 bp that encode two effective siRNAs against conserved HIV-1 sequences, Pol and Nef, did not induce the IFN response in e-shRNA transfected 293T cells *(63)*. A simple strategy for avoiding activation of the IFN response by dsRNA has been described *(67)*, where modified hairpin-RNAs (mhRNAs) of more than 100 bp with multiple specific point-mutations within the sense strand and transcribed from the U6 or tRNA(Val) promoters can produce RNAi without inducing the IFN pathway genes *(67, 115)*. The expression of lhRNAs also appears to be well tolerated and does not induce IFN gene activation in vivo when delivered to mice via by hydrodynamic tail-vein injection *(116)*.

Another potential side effect of siRNAs and shRNAs is lethality associated with overdosing. A previous study in which mice were treated with high doses of a viral vector expressing shRNAs that resulted in high copy numbers per cell led to fatality due to oversaturation of the mi/siRNA pathways *(65)*. This potential risk should be assessed prior to an eventual clinical RNAi application. Cells should be transduced with few or a single vector copy to avoid high expression levels that may induce unwanted side effects. Overdosing could also induce off-target effects in which siRNAs or shRNAs silence partially complementary transcripts through a miRNA-like mechanism. Cytotoxic effects of shRNAs in human T lymphocytes as a result of overexpression have also been reported *(64)*. The risk of adverse effects of siRNAs would increase in situations where shRNA expression is to be maintained in a living organism for a long period, such as for intracellular immunization against HIV-1. In addition, transduced T cell lines and differentiated primary human T cells are relatively different from the hematopoietic CD34+ stem cells that will be transduced in an ultimate gene therapy therapeutic setting. Stem cells will develop into different lineages, and expression of shRNAs could influence their development. As suggested before *(96)*, saturation of the miRNA pathway may disturb hematopoiesis, and indeed, miRNAs are involved in the regulation of genes that control hematopoiesis in the mouse *(117)*. Whether in vivo gene silencing by siRNA or shRNAs can disrupt the endogenous miRNA pathway remains to be addressed. Recently, target genes were shown to be effectively silenced in the mouse and hamster liver by systemic administration of synthetic siRNA without any demonstrable effect on miRNA levels or activity *(118)*. siRNAs targeting two hepatocyte-specific genes (apolipoprotein B and factor VII) were administered to mice and achieved efficient (80%) silencing of mRNA transcripts without significant changes in the levels of three hepatocyte-expressed miRNAs (miR-122, miR-16 and let-7a) *(118)*. Moreover, multiple administrations of an siRNA targeting the hepatocyte-expressed gene Scap in hamsters achieved long-term mRNA silencing without significant changes in miR-122 levels *(118)*. This study may advance the use of siRNAs as a safe therapeutic alternative.

Another potential side effect for siRNAs and shRNAs is the silencing of cellular mRNAs that share partial homology to the siRNA or shRNA sequences through an miRNA-like mechanism. This off-target effect requires complementarity between the siRNA seed region and the 3′untranslated region of a target mRNA *(119, 121, 122)*. Any effective siRNA or shRNA may in theory have numerous potential off-target mRNAs, thus off-target silencing may not be easily eliminated by siRNA sequence selection. Moreover, the combining of multiple shRNAs would increase the number of off-target mRNAs potentially affected. Off-target transcript silencing may limit the specificity of siRNAs for therapeutic applications. One study showed that 2′-uridine modifications of siRNAs can reduce siRNAs off-target effects *(123)*, and similarly, chemical modification also reduced off-target phenotypes in growth inhibition studies. Key to the modification was a 2′-*O*-methyl ribosyl substitution at position 2 in the guide strand, which reduced the silencing of most off-target transcripts exhibiting complementarity to the seed region of the siRNA guide strand *(124)*.

Ex vivo selection of siRNAs to verify the absence of unwanted off-targets should help to define strategies to either enhance or avoid the non-specific effects of siRNAs in order to develop safe therapeutics. Studies with an appropriate animal model should also help preclinical assessment of safety and efficacy. HIV-1 research is largely restricted to in vitro, ex vivo or clinical studies, all limited in their ability to rapidly assess new strategies to treat virus infection. The humanized Rag2(−/−) gammac(−/−) SCID mouse model sustains long-term multi-lineage hematopoiesis and is capable of mounting immune responses, achieved by the intraperitoneal injection of CD34+ precursor cells into a newborn Rag2(−/−) gammac(−/−) mouse *(125, 126)*. In this model, injecting p53 shRNA-transduced CD34+ cells resulted in stable expression and down-modulation of p53 in the mature T-cell offspring. Infection of HIV-1 in this mouse model resulted in high viremia and CD4+ T cell depletion *(127)*. Therefore, this in vivo animal model and/or others developed in the near future should be valuable to evaluate RNAi-based strategies aiming to prevent or treat HIV-1 infection.

4. Conclusions

The discovery of RNAi has provided us with powerful new tools for biological research and drug discovery, and RNAi is currently advancing from basic research to clinical trials. It has also expanded the direction of a new field of therapeutic strategies with the potential to treat a wide range of different diseases, including HIV-1. Several decades ago, adaptive immunity allowed the possibility of preventing the spread of many pathogenic human viruses. Nevertheless, the discovery of an effective preventive or therapeutic vaccine against HIV-1 remains elusive. RNAi therapeutics is rapidly arriving into clinical studies for many diseases, including several which are currently untreatable or difficult to treat with conventional drugs. Although promising studies and results have emerged over the recent years, the challenge remains as to whether RNAi will be suitable for treating or preventing HIV-1 infection.

Acknowledgements

MAM is supported by the Spanish *Minsiterio de Educación y Ciencia* (MEC) project BFU2006-01066/BMC, *Fondo de Investigación Sanitaria* (FIS) project PI070098, and Fundación para la investigación y la prevención del SIDA en España (FIPSE 36549/06).

References

1. Gottlieb, M.S., Schroff, R., Schanker, H.M. et al. (1981) Pneumocystis carinii pneumonia and mucosal candidiasis in previously healthy homosexual men: evidence of a new acquired cellular immunodeficiency. *N Engl J Med* 305, 1425–1431.
2. Kirchhoff, F., Greenough, T.C., Brettler, D.B., Sullivan, J.L., and Desrosiers, R.C. (1995) Brief report: absence of intact nef sequences in a long-term survivor with non-progressive HIV-1 infection. *N Engl J Med* 332, 228–232.
3. Learmont, J.C., Geczy, A.F., Mills, J. et-al. (1999) Immunologic and virologic status after 14 to 18 years of infection with an attenuated strain of HIV-1. A report from the Sydney Blood Bank Cohort. *N Engl J Med* 340, 1715–1722.
4. Johnston, M.I. and Fauci, A.S. (2007) An HIV vaccine-evolving concepts. *N Engl J Med* 356, 2073–2081.
5. Ho, D.D. (1995) Time to hit HIV, early and hard. *N Engl J Med* 333, 450–451.
6. Walensky, R.P., Paltiel, A.D., Losina, E. et-al. (2006) The survival benefits of AIDS treatment in the United States. *J Infect Dis* 194, 11–19.
7. Chun, T.W., Stuyver, L., Mizell, S.B. et-al. (1997) Presence of an inducible HIV-1 latent reservoir during highly active antiretroviral therapy. *Proc Natl Acad Sci U S A* 94, 13193–13197.
8. Finzi, D., Hermankova, M., Pierson, T. et-al. (1997) Identification of a reservoir for HIV-1 in patients on highly active antiretroviral therapy. *Science* 278, 1295–1300.
9. Wong, J.K., Hezareh, M., Gunthard, H.F. et-al. (1997) Recovery of replication-competent HIV despite prolonged suppression of plasma viremia. *Science* 278, 1291–1295.
10. Ibanez, A., Puig, T., Elias, J., Clotet, B., Ruiz, L., and Martinez, M.A. (1999) Quantification of integrated and total HIV-1 DNA after long-term highly active antiretroviral therapy in HIV-1-infected patients. *AIDS* 13, 1045–1049.
11. Este, J.A. and Telenti, A. (2007) HIV entry inhibitors. *Lancet* 370, 81–88.
12. Johnson, V.A., Brun-Vezinet, F., Clotet, B. et al. (2007) Update of the Drug Resistance Mutations in HIV-1: 2007. *Top HIV Med* 15, 119–125.
13. Deeks, S.G. (2006) Antiretroviral treatment of HIV infected adults. *BMJ* 332, 1489.
14. Fire, A., Xu, S., Montgomery, M.K., Kostas, S.A., Driver, S.E., and Mello, C.C. (1998) Potent and specific genetic interference by double-stranded RNA in Caenorhabditis elegans. *Nature* 391, 806–811.
15. Elbashir, S.M., Harborth, J., Lendeckel, W., Yalcin, A., Weber, K., and Tuschl, T. (2001) Duplexes of 21-nucleotide RNAs mediate RNA interference in cultured mammalian cells. *Nature* 411, 494–498.
16. Lee, N.S., Dohjima, T., Bauer, G. et al. (2002) Expression of small interfering RNAs targeted against HIV-1 rev transcripts in human cells. *Nat Biotechnol* 20, 500–505.

17. Novina, C.D., Murray, M.F., Dykxhoorn, D.M. et al. (2002) siRNA-directed inhibition of HIV-1 infection. *Nat Med* 8, 681–686.
18. Jacque, J.M., Triques, K., and Stevenson, M. (2002) Modulation of HIV-1 replication by RNA interference. *Nature* 418, 435–438.
19. Coburn, G.A. and Cullen, B.R. (2002) Potent and specific inhibition of human immunodeficiency virus type 1 replication by RNA interference. *J Virol* 76, 9225–9231.
20. Capodici, J., Kariko, K., and Weissman, D. (2002) Inhibition of HIV-1 infection by small interfering RNA-mediated RNA interference. *J Immunol* 169, 5196–5201.
21. Park, W.S., Miyano-Kurosaki, N., Hayafune, M. et-al. (2002) Prevention of HIV-1 infection in human peripheral blood mononuclear cells by specific RNA interference. *Nucleic Acids Res* 30, 4830–4835.
22. Surabhi, R.M. and Gaynor, R.B. (2002) RNA interference directed against viral and cellular targets inhibits human immunodeficiency virus Type 1 replication. *J Virol* 76, 12963–12973.
23. Banerjea, A., Li, M.J., Bauer, G. et al. (2003) Inhibition of HIV-1 by lentiviral vector-transduced siRNAs in T lymphocytes differentiated in SCID-hu mice and CD34+ progenitor cell-derived macrophages. *Mol Ther* 8, 62–71.
24. Song, E., Lee, S.K., Dykxhoorn, D.M. et al. (2003) Sustained small interfering RNA-mediated human immunodeficiency virus type 1 inhibition in primary macrophages. *J Virol* 77, 7174–7181.
25. Lee, M.T., Coburn, G.A., McClure, M.O., and Cullen, B.R. (2003) Inhibition of human immunodeficiency virus type 1 replication in primary macrophages by using Tat- or CCR5-specific small interfering RNAs expressed from a lentivirus vector. *J Virol* 77, 11964–11972.
26. Stevenson, M. (2003) Dissecting HIV-1 through RNA interference. *Nat Rev Immunol* 3, 851–858.
27. Joshi, P.J., North, T.W., and Prasad, V.R. (2005) Aptamers directed to HIV-1 reverse transcriptase display greater efficacy over small hairpin RNAs targeted to viral RNA in blocking HIV-1 replication. *Mol Ther* 11, 677–686.
28. Hu, W.Y., Myers, C.P., Kilzer, J.M., Pfaff, S.L., and Bushman, F.D. (2002) Inhibition of retroviral pathogenesis by RNA interference. *Curr Biol* 12, 1301–1311.
29. Nishitsuji, H., Ikeda, T., Miyoshi, H., Ohashi, T., Kannagi, M., and Masuda, T. (2004) Expression of small hairpin RNA by lentivirus-based vector confers efficient and stable gene-suppression of HIV-1 on human cells including primary non-dividing cells. *Microbes Infect* 6, 76–85.
30. Bitko, V. and Barik, S. (2001) Phenotypic silencing of cytoplasmic genes using sequence-specific double-stranded short interfering RNA and its application in the reverse genetics of wild type negative-strand RNA viruses. *BMC Microbiol* 1, 34–40
31. Westerhout, E.M. and Berkhout, B. (2007) A systematic analysis of the effect of target RNA structure on RNA interference. *Nucleic Acids Res* 35, 4322–4330.
32. Holen, T., Amarzguioui, M., Wiiger, M.T., Babaie, E., and Prydz, H. (2002) Positional effects of short interfering RNAs targeting the human coagulation trigger Tissue Factor. *Nucleic Acids Res* 30, 1757–1766.
33. Westerhout, E.M., ter Brake, O., and Berkhout, B. (2006) The virion-associated incoming HIV-1 RNA genome is not targeted by RNA interference. *Retrovirology* 3, 57.
34. Brummelkamp, T.R., Bernards, R., and Agami, R. (2002) A system for stable expression of short interfering RNAs in mammalian cells. *Science* 296, 550–553.
35. Boden, D., Pusch, O., Lee, F., Tucker, L., and Ramratnam, B. (2004) Efficient gene transfer of HIV-1-specific short hairpin RNA into human lymphocytic cells using recombinant adeno-associated virus vectors. *Mol Ther* 9, 396–402.
36. Das, A.T., Brummelkamp, T.R., Westerhout, E.M. et-al. (2004) Human immunodeficiency virus type 1 escapes from RNA interference-mediated inhibition. *J Virol* 78, 2601–2605.
37. Lee, S.K., Dykxhoorn, D.M., Kumar, P. et-al. (2005) Lentiviral delivery of short hairpin RNAs protects CD4 T cells from multiple clades and primary isolates of HIV. *Blood* 106, 818–826.
38. Silva, J.M., Li, M.Z., Chang, K. et al. (2005) Second-generation shRNA libraries covering the mouse and human genomes. *Nat Genet* 37, 281–1288.
39. Cullen, B.R. (2006) Is RNA interference involved in intrinsic antiviral immunity in mammals. *Nat Immunol* 7, 563–567.
40. Voinnet, O. (2005) Induction and suppression of RNA silencing: insights from viral infections. *Nat Rev Genet.* 6, 206–220.
41. Garcia-Sastre, A. and Biron, C.A. (2006) Type 1 interferons and the virus-host relationship: a lesson in detente. *Science* 312, 879–882.

42. Pfeffer, S., Sewer, A., Lagos-Quintana, M. et al. (2005) Identification of microRNAs of the herpesvirus family. *Nat Methods* 2, 269–276.
43. Bennasser, Y., Le, S.Y., Benkirane, M., and Jeang, K.T. (2005) Evidence that HIV-1 encodes an siRNA and a suppressor of RNA silencing. *Immunity* 22, 607–619.
44. Haasnoot, J., de Vries, W., Geutjes, E.J., Prins, M., de Haan, P., and Berkhout, B. (2007) The Ebola virus VP35 protein is a suppressor of RNA silencing. *PLoS Pathog* 3, e86.
45. Lin, J. and Cullen, B.R. (2007) Analysis of the Interaction of Primate Retroviruses with the Human RNA Interference Machinery. J Virol 81, 12218–12226
46. Li, H., Li, W.X., and Ding, S.W. (2002) Induction and suppression of RNA silencing by an animal virus. *Science* 296, 1319–1321.
47. Triboulet, R., Mari, B., Lin, Y.L. et-al. (2007) Suppression of microRNA-silencing pathway by HIV-1 during virus replication. *Science* 315, 1579–1582.
48. Stern-Ginossar, N., Elefant, N., Zimmermann, A. et al. (2007) Host immune system gene targeting by a viral miRNA. *Science* 317, 376–381.
49. Huang, J., Wang, F., Argyris, E. et al. (2007) Cellular microRNAs contribute to HIV-1 latency in resting primary CD4(+) T lymphocytes. *Nat Med* 13, 1241–1247.
50. Pedersen, I.M., Cheng, G., Wieland, S., et al. (2007) Interferon modulation of cellular microRNAs as an antiviral mechanism. *Nature* 449, 919–922.
51. Boden, D., Pusch, O., Lee, F., Tucker, L., and Ramratnam, B. (2003) Human immunodeficiency virus type 1 escape from RNA interference. *J Virol* 77, 11531–11535.
52. Westerhout, E.M., Ooms, M., Vink, M., Das, A.T., and Berkhout, B. (2005) HIV-1 can escape from RNA interference by evolving an alternative structure in its RNA genome. *Nucleic Acids Res* 33, 796–804.
53. Mansky, L.M. and Temin, H.M. (1995) Lower in vivo mutation rate of human immunodeficiency virus type 1 than that predicted from the fidelity of purified reverse transcriptase. *J Virol* 69, 5087–5094.
54. Ho, D.D., Neumann, A.U., Perelson, A.S., Chen, W., Leonard, J.M., and Markowitz, M. (1995) Rapid turnover of plasma virions and CD4 lymphocytes in HIV-1 infection. *Nature* 373, 123–126.
55. Wei, X., Ghosh, S.K., Taylor, M.E. et al. (1995) Viral dynamics in human immunodeficiency virus type 1 infection. *Nature* 373, 117–122.
56. Jung, A., Maier, R., Vartanian, J.P. et al. (2002) Multiply infected spleen cells in HIV patients. *Nature* 418, 144.
57. Coffin, J.M. (1995) HIV population dynamics in vivo: implications for genetic variation, pathogenesis, and therapy. *Science* 267, 483–489.
58. Domingo, E., Martin, V., Perales, C., Grande-Perez, A., Garcia-Arriaza, J., and Arias, A. (2006) Viruses as quasispecies: biological implications. *Curr Top Microbiol Immunol* 299, 51–82.
59. Leitner, T., Foley, B., Hahn, B. et al. (2007) HIV Sequence Compendium 2006/2007. Published by Theoretical Biology and Biophysics Group, Los Alamos National Laboratory, LA-UR number 07-4826. 2007.
60. Sabariegos, R., Gimenez-Barcons, M., Tapia, N., Clotet, B., and Martinez, M.A. (2006) Sequence homology required by human immunodeficiency virus type 1 to escape from short interfering RNAs. *J Virol* 80, 571–577.
61. Martinez, M.A., Clotet, B., and Este, J.A. (2002) RNA interference of HIV replication. *Trends Immunol* 23, 559–561.
62. ter Brake, O., Konstantinova, P., Ceylan, M., and Berkhout, B. (2006) Silencing of HIV-1 with RNA interference: a multiple shRNA approach. *Mol Ther* 14, 883–892.
63. Liu, Y.P., Haasnoot, J., and Berkhout, B. (2007) Design of extended short hairpin RNAs for HIV-1 inhibition. *Nucleic Acids Res* 35, 5683–5693.
64. An, D.S., Qin, F.X., Auyeung, V.C. et al. (2006) Optimization and functional effects of stable short hairpin RNA expression in primary human lymphocytes via lentiviral vectors. *Mol Ther* 14, 494–504.
65. Grimm, D., Streetz, K.L., Jopling, C.L. et al. (2006) Fatality in mice due to oversaturation of cellular microRNA/short hairpin RNA pathways. *Nature* 441, 537–541.
66. Scherer, L.J., Frank, R., and Rossi, J.J. (2007) Optimization and characterization of tRNA-shRNA expression constructs. *Nucleic Acids Res* 35, 2620–2628.
67. Akashi, H., Miyagishi, M., Yokota, T. et al. (2005) Escape from the interferon response associated with RNA interference using vectors that encode long modified hairpin-RNA. *Mol Biosyst* 1, 382–390.

68. Konstantinova, P., de Vries, W., Haasnoot, J., ter Brake, O., and de Haan, P., Berkhout, B. (2006) Inhibition of human immunodeficiency virus type 1 by RNA interference using long-hairpin RNA. *Gene Ther* 13, 1403–1413.
69. Nishitsuji, H., Kohara, M., Kannagi, M., and Masuda, T. (2006) Effective suppression of human immunodeficiency virus type 1 through a combination of short- or long-hairpin RNAs targeting essential sequences for retroviral integration. *J Virol* 80, 7658–7666.
70. Sano, M., Li, H., Nakanishi, M., and Rossi, J.J. (2008) Expression of long anti-HIV-1 hairpin RNAs for the generation of multiple siRNAs: Advantages and limitations. *Mol Ther* 16, 170–177.
71. ter Brake, O. and Berkhout, B. (2005) A novel approach for inhibition of HIV-1 by RNA interference: counteracting viral escape with a second generation of siRNAs. *J. RNAi Gene Silencing* 1, 56–65.
72. Gimenez-Barcons, M., Clotet, B., and Martinez, M.A. (2007) Endoribonuclease-Prepared Short Interfering RNAs Induce Effective and Specific Inhibition of Human Immunodeficiency Virus Type 1 Replication. *J Virol* 81, 10680–10686.
73. Kawasaki, H., Suyama, E., Iyo, M., and Taira, K. (2003) siRNAs generated by recombinant human Dicer induce specific and significant but target site-independent gene silencing in human cells. *Nucleic Acids Res* 31, 981–987.
74. Myers, J.W., Jones, J.T., Meyer, T., and Ferrell, J.E. Jr. (2003) Recombinant Dicer efficiently converts large dsRNAs into siRNAs suitable for gene silencing. *Nat Biotechnol* 21, 324–328.
75. Li, M. and Rossi, J.J. (2005) Lentiviral vector delivery of siRNA and shRNA encoding genes into cultured and primary hematopoietic cells. *Methods Mol Biol* 309, 261–272.
76. Martinez, M.A., Gutierrez, A., Armand-Ugon, M. et-al. (2002) Suppression of chemokine receptor expression by RNA interference allows for inhibition of HIV-1 replication. *AIDS* 16, 2385–2390.
77. Qin, X.F., An, D.S., Shen, I.S., and Baltimore, D. (2003) Inhibiting HIV-1 infection in human T cells by lentiviral-mediated delivery of small interfering RNA against CCR5. *Proc Natl Acad Sci U S A* 100, 183–188.
78. Llano, M., Saenz, D.T., Meehan, A. et-al. (2006) An essential role for LEDGF/p75 in HIV integration. *Science* 314, 461–464.
79. Maertens, G., Cherepanov, P., Pluymers, W. et-al. (2003) LEDGF/p75 is essential for nuclear and chromosomal targeting of HIV-1 integrase in human cells. *J Biol Chem* 278, 33528–33539.
80. Jacque, J.M. and Stevenson, M. (2006) The inner-nuclear-envelope protein emerin regulates HIV-1 infectivity. *Nature* 441, 641–645.
81. Nguyen, D.G., Wolff, K.C., Yin, H., Caldwell, J.S., and Kuhen, K.L. (2006) "UnPAKing" human immunodeficiency virus (HIV) replication: using small interfering RNA screening to identify novel cofactors and elucidate the role of group I PAKs in HIV infection. *J Virol* 80, 130–137.
82. Chiu, Y.L., Cao, H., Jacque, J.M., Stevenson, M., and Rana, T.M. (2004) Inhibition of human immunodeficiency virus type 1 replication by RNA interference directed against human transcription elongation factor P-TEFb (CDK9/CyclinT1). *J Virol* 78, 2517–2529.
83. Berger, E.A., Murphy, P.M., and Farber, J.M. (1999) Chemokine receptors as HIV-1 coreceptors: roles in viral entry, tropism, and disease. *Annu Rev Immunol* 17, 657–700.
84. An, D.S., Donahue, R.E., Kamata, M. et al. (2007) Stable reduction of CCR5 by RNAi through hematopoietic stem cell transplant in non-human primates. *Proc Natl Acad Sci U S A* 104, 13110–13115.
85. de Fougerolles, A., Vornlocher, H.P., Maraganore, J., and Lieberman, J. (2007) Interfering with disease: a progress report on siRNA-based therapeutics. *Nat Rev Drug Discov* 6, 443–453.
86. Bitko, V., Musiyenko, A., Shulyayeva, O., and Barik, S. (2005) Inhibition of respiratory viruses by nasally administered siRNA. *Nat Med* 11, 50–55.
87. Li, B.J., Tang, Q., Cheng, D. et al. (2005) Using siRNA in prophylactic and therapeutic regimens against SARS coronavirus in Rhesus macaque. *Nat Med* 11, 944–951.
88. Morrissey, D.V., Lockridge, J.A., Shaw, L. et al. (2005) Potent and persistent in vivo anti-HBV activity of chemically modified siRNAs. *Nat Biotechnol* 23, 1002–1007.
89. Geisbert, T.W., Hensley, L.E., Kagan, E. et-al. (2006) Postexposure protection of guinea pigs against a lethal Ebola virus challenge is conferred by RNA interference. *J Infect Dis* 193, 1650–1657.
90. Palliser, D., Chowdhury, D., Wang, Q.Y. et al. (2006) An siRNA-based microbicide protects mice from lethal herpes simplex virus 2 infection. *Nature* 439, 89–94.

91. Juliano, R.L. (2005) Peptide-oligonucleotide conjugates for the delivery of antisense and siRNA. *Curr Opin Mol Ther* 7, 132–136.
92. Song, E., Zhu, P., Lee, S.K. et al. (2005) Antibody mediated in vivo delivery of small interfering RNAs via cell-surface receptors. *Nat Biotechnol* 23, 709–717.
93. Naldini, L., Blomer, U., Gallay, P. et al. (1996) In vivo gene delivery and stable transduction of nondividing cells by a lentiviral vector. *Science* 272, 263–267.
94. Morris, K.V. and Rossi, J.J. (2006) Lentivirus-mediated RNA interference therapy for human immunodeficiency virus type 1 infection. *Hum Gene Ther* 17, 479–486.
95. Ter Brake, O. and Berkhout, B. (2007) Lentiviral vectors that carry anti-HIV shRNAs: problems and solutions. *J Gene Med* 9, 743–750.
96. Ter Brake, O. (2007) Development of an RNAi based gene therapy against HIV-1. Amsterdam: Doctoral Thesis, University of Amsterdam.
97. Brown, B.D., Venneri, M.A., Zingale, A., Sergi Sergi, L., and Naldini, L. (2006) Endogenous microRNA regulation suppresses transgene expression in hematopoietic lineages and enables stable gene transfer. *Nat Med* 12, 585–591.
98. Wu, X., Li, Y., Crise, B., and Burgess, S.M. (2003) Transcription start regions in the human genome are favored targets for MLV integration. *Science* 300, 1749–1751.
99. Hacein-Bey-Abina, S., von Kalle, C., Schmidt, M. et-al. (2003) A serious adverse event after successful gene therapy for X-linked severe combined immunodeficiency. *N Engl J Med* 348, 255–256.
100. Baum, C. (2007) Insertional mutagenesis in gene therapy and stem cell biology. *Curr Opin Hematol* 14, 337–342.
101. Kustikova, O.S., Geiger, H., Li, Z. et-al. (2007) Retroviral vector insertion sites associated with dominant hematopoietic clones mark "stemness" pathways. *Blood* 109, 1897–1907.
102. Gervaix, A., Schwarz, L., Law, P. et al. (1997) Gene therapy targeting peripheral blood CD34+ hematopoietic stem cells of HIV-infected individuals. *Hum Gene Ther* 8, 2229–2238.
103. Yam, P.Y., Li, S., Wu, J., Hu, J., Zaia, J.A., and Yee, J.K. (2002) Design of HIV vectors for efficient gene delivery into human hematopoietic cells. *Mol Ther* 5, 479–484.
104. Morris, K.V., Gilbert, J., Wong-Staal, F., Gasmi, M., and Looney, D.J. (2004) Transduction of cell lines and primary cells by FIV-packaged HIV vectors. *Mol Ther* 10, 181–190.
105. Sioud, M. (2007) RNA interference and innate immunity. *Adv Drug Deliv Rev* 59, 153–163.
106. Judge, A.D., Sood, V., Shaw, J.R., Fang, D., McClintock, K., and MacLachlan, I. (2005) Sequence-dependent stimulation of the mammalian innate immune response by synthetic siRNA. *Nat Biotechnol* 23, 457–462.
107. Sioud, M. and Sorensen, D.R. (2003) Cationic liposome-mediated delivery of siRNAs in adult mice. *Biochem Biophys Res Commun* 312, 1220–1225.
108. Chiu, Y.L. and Rana, T.M. (2003) siRNA function in RNAi: A chemical modification analysis. *RNA* 9, 1034–1048.
109. Layzer, J.M., McCaffrey, A.P., Tanner, A.K., Huang, Z., Kay, M.A. and Sullenger, B.A. (2004) In vivo activity of nuclease-resistant siRNAs. *RNA* 10, 766–771.
110. Sioud, M. (2006) Single-stranded small interfering RNA are more immunostimulatory than their double-stranded counterparts: a central role for 2′-hydroxyl uridines in immune responses. *Eur J Immunol* 36, 1222–1230.
111. Sioud, M. (2005) On the delivery of small interfering RNAs into mammalian cells. *Expert Opin Drug Deliv* 2, 639–651.
112. Kariko, K., Bhuyan, P., Capodici, J., and Weissman, D. (2004) Small interfering RNAs mediate sequence-independent gene suppression and induce immune activation by signaling through toll-like receptor 3. *J Immunol* 172, 6545–6549.
113. Robbins, M.A., Li, M., Leung, I. et-al. (2006) Stable expression of shRNAs in human CD34+ progenitor cells can avoid induction of interferon responses to siRNAs in vitro. *Nat Biotechnol* 24, 566–571.
114. Konstantinova, P., ter Brake, O., Haasnoot, J., de Haan, P., and Berkhout, B. (2007) Trans-inhibition of HIV-1 by a long hairpin RNA expressed within the viral genome. *Retrovirology* 4, 15.
115. Watanabe, T., Sudoh, M., Miyagishi, M. et-al. (2006) Intracellular-diced dsRNA has enhanced efficacy for silencing HCV RNA and overcomes variation in the viral genotype. *Gene Ther* 13, 883–892.
116. Weinberg, M.S., Ely, A., Passman, M., Mufamadi, S.M., and Arbuthnot, P. (2007) Effective anti hepatitis B virus hammerhead ribozymes derived from multimeric precursors. *Oligonucleotides* 17, 104–112.

117. Chen, C.Z., Li, L., Lodish, H.F., and Bartel, D.P. (2004) MicroRNAs modulate hematopoietic lineage differentiation. *Science* 303, 83–86.

118. John, M., Constien, R., Akinc, A. et al. (2007) Effective RNAi-mediated gene silencing without interruption of the endogenous microRNA pathway. *Nature* 449, 745–747.

119. Jackson, A.L., Bartz, S.R., Schelter, J. et al. (2003) Expression profiling reveals off-target gene regulation by RNAi. *Nat Biotechnol* 21, 635–637.

120. Jackson, A.L., Burchard, J., Schelter, J. et al. (2006) Widespread siRNA "off-target" transcript silencing mediated by seed region sequence complementarity. *RNA* 12, 1179–1187.

121. Birmingham, A., Anderson, E.M., Reynolds, A. et al. (2006) 3′ UTR seed matches, but not overall identity, are associated with RNAi off-targets. *Nat Methods* 3, 199–204.

122. Lin, X., Ruan, X., Anderson, M.G. et al. (2005) siRNA-mediated off-target gene silencing triggered by a 7 nt complementation. *Nucleic Acids Res* 33, 4527–4535.

123. Cekaite, L., Furset, G., Hovig, E., and Sioud, M. (2007) Gene expression analysis in blood cells in response to unmodified and 2′-modified siRNAs reveals TLR-dependent and independent effects. *J Mol Biol* 365, 90–108.

124. Jackson, A.L., Burchard, J., Leake, D. et al. (2006) Position-specific chemical modification of siRNAs reduces "off-target" transcript silencing. *RNA* 12, 1197–1205.

125. Gimeno, R., Weijer, K., Voordouw, A. et al. (2004) Monitoring the effect of gene silencing by RNA interference in human CD34+ cells injected into newborn RAG2−/− gammac−/− mice: functional inactivation of p53 in developing T cells. *Blood* 104, 3886–3893.

126. Legrand, N., Weijer, K., and Spits, H. (2006) Experimental models to study development and function of the human immune system in vivo. *J Immunol* 176, 2053–2058.

127. Baenziger, S., Tussiwand, R., and Schlaepfer, E. (2006) Disseminated and sustained HIV infection in CD34+ cord blood cell-transplanted Rag2−/−gamma c−/− mice. *Proc Natl Acad Sci U S A* 103, 15951–15956.

Chapter 18

Protein Components of the microRNA Pathway and Human Diseases

Marjorie P. Perron and Patrick Provost

Abstract

MicroRNAs (miRNAs) are key regulators of messenger RNA (mRNA) translation known to be involved in a wide variety of cellular processes. In fact, their individual importance is reflected in the diseases that may arise upon the loss, mutation or dysfunction of specific miRNAs. It has been appreciated only recently that diseases may also develop when the protein components of the miRNA machinery itself are affected. The core enzymes of the major protein complexes involved in miRNA biogenesis and function, such as the ribonucleases III (RNases III) Drosha and Dicer as well as Argonaute 2 (Ago2), appear to be essential. However, the accessory proteins of the miRNA pathway, such as the DiGeorge syndrome critical region gene 8 (DGCR8) protein, Exportin-5 (Exp-5), TAR RNA binding protein (TRBP) and fragile X mental retardation protein (FMRP), are each related, in various ways, to specific genetic diseases.

Key words: RNA silencing, microRNA, gene expression, diseases.

1. Role of the microRNA-Guided RNA Silencing Pathway

The microRNA (miRNA)-guided RNA silencing pathway is a gene regulatory process present in most eukaryotic cells which is involved in the repressive control of messenger RNA (mRNA) translation. miRNAs are encoded in the genome and the regulation of their expression is crucial for the maintenance of cellular homeostasis. Although estimated to represent ~2% of the genome, miRNA gene products have been proposed to regulate as many as 92% of the genes in human *(1)*! Several characteristics of miRNAs and their recognition of binding sites through which they regulate specific mRNAs are conserved throughout evolution. miRNAs are a well-defined family of non-coding, ~21- to

M. Sioud (ed.), *Methods in Molecular Biology, siRNA and miRNA Gene Silencing, vol. 487*

DOI: 10.1007/978-1-60327-547-7_18

23-nucleotide (nt) RNA species that can bind their target mRNA through imperfect complementarity usually in the 3′ non-translated region (NTR), causing specific translational repression. miRNA-based gene regulation has been shown to play an important role in development, cell growth, apoptosis and other cellular processes (for a review, see Ouellet et al. *(2)*).

A single miRNA can regulate several mRNAs through recognition of a specific binding site that is believed to require the perfect pairing of its miRNA seed region, referred to as nt 2 to 8 from the 5′ end *(3)*. The 3′ region of the miRNA can exhibit imperfect complementarity with the mRNA target, but may compensate for a weaker binding of the miRNA seed region. Conversely, more than one miRNA can recognize the same mRNA target and strengthen its translational repression. Elucidation of the architecture of the well-studied interaction between let-7 and lin-41 in *C. elegans* has allowed a better understanding of the pairing determinants of a miRNA with its mRNA target *(3,4)*. The let-7 miRNA forms an imperfect duplex with six complementary sites present in the lin-41 3′ NTR, and the occupancy of only two of the sites, separated by a 27-nt sequence, are sufficient to induce silencing of lin-41. The 27-nt sequence between the two binding sites for let-7 also seems to be important, as deletional mutagenesis or nucleotide substitutions in the sequence prevent lin-41 silencing *(3)*. However, the characteristics of an experimentally validated miRNA:target pair are not sufficient by themselves to establish the rules governing mRNA recognition by all existing miRNAs, which may require elaborated studies of several different miRNA:mRNA combinations.

In fundamental research, the properties of this unique gene silencing machinery can be exploited by synthetic, miRNA mimicking small interfering RNA (siRNA) to specifically downregulate the expression of our gene of interest to study its function. Having the ability to structurally mimic miRNAs and to hijack the endogenous pathway of the cell in which they are introduced, siRNAs represent a powerful research tool with a promising future in human therapy, with the possibility to downregulate a protein found to be overexpressed in specific diseases, such as cancer.

Recent advances have led to the identification of miRNA targets, some of which are related to specific diseases. Among the different scenarios that may be envisioned, overexpression of a given miRNA may accentuate the translational repression of its target mRNAs, whereas the downregulation of another miRNA may lead to an enhanced translation of its target mRNAs. For instance, deregulated expression of specific miRNAs has been linked to certain cases of cancer. The miR-17-92 cluster, a region that encodes for seven miRNAs, is overexpressed in human lung cancer cell lines *(5)*. On the opposite, frequent deletions and downregulation of miR-15 and miR-16 genes at 13q14 occur in chronic lymphocytic leukemia *(6)*. These two miRNAs have been shown to negatively regulate the antiapoptotic Bcl2 protein at the posttranscriptional level *(7)*. These two examples illustrate the importance of a highly regulated miRNA expression in the maintenance of normal cell function and the development of specific diseases that can result from an imbalanced gene regulation due to miRNA dysfunction.

Involved in the biogenesis and action of miRNAs, the protein components of the miRNA-guided RNA silencing machinery may also be susceptible to defects, the occurrence of which could lead to important genetic disorders and viral infection. Indeed, some protein components of the miRNA pathway have now been linked to specific diseases.

2. Major Protein Components of the microRNA-Guided RNA Silencing Pathway

Involving only a few proteins, miRNA biogenesis and function are well-orchestrated processes relying on numerous types of protein-protein, protein-RNA and RNA-RNA interactions that are acting in coordination for the generation of thousands of different miRNAs and regulation of tens of thousands of different mRNAs. The most notable feature of this machinery remains its ability to recognize and process specific RNA structures independently of their sequences in miRNA biogenesis, allowing any miRNAs to be synthesized and any mRNA to be regulated by miRNAs. For a recent review, see Perron and Provost *(8)*.

miRNA genes are transcribed by RNA polymerase II into a long non-coding RNA known as the primary miRNA (pri-miRNA) *(9,10)*. Harboring a 5′ methylated cap and a 3′ poly(A) tail, this primary transcript folds on itself to form hairpin-loop structures that can be recognized and cleaved by the nuclear Microprocessor complex, composed of Drosha and the DiGeorge syndrome critical region gene 8 (DGCR8) protein, into a miRNA precursor (pre-miRNA) *(11–16)*. Following its export from the nucleus mediated by Exportin-5 (Exp-5) *(17)*, the pre-miRNA is cleaved by the pre-miRNA processing complex, formed by Dicer and TAR RNA binding protein (TRBP), into a miRNA:miRNA* duplex *(18–20)*. This complex then assembles with Argonaute 2 (Ago2) protein to form a miRNA-containing ribonucleoprotein (miRNP) complex *(21)*, after which the mature miRNA strand of ~21- to 23-nt is selected. Depending on whether the complementarity of the miRNA to its target is perfect or not, the miRNP complex can lead either to mRNA cleavage and degradation or to initial translational inhibition *(22)*. In this latter case, the repressed mRNA is translocated to the P-bodies, after which the mRNA can either be destroyed or relocalized to the translational machinery for expression upon a specific cellular signal (*see* **Fig. 18.1**) *(23,24)*.

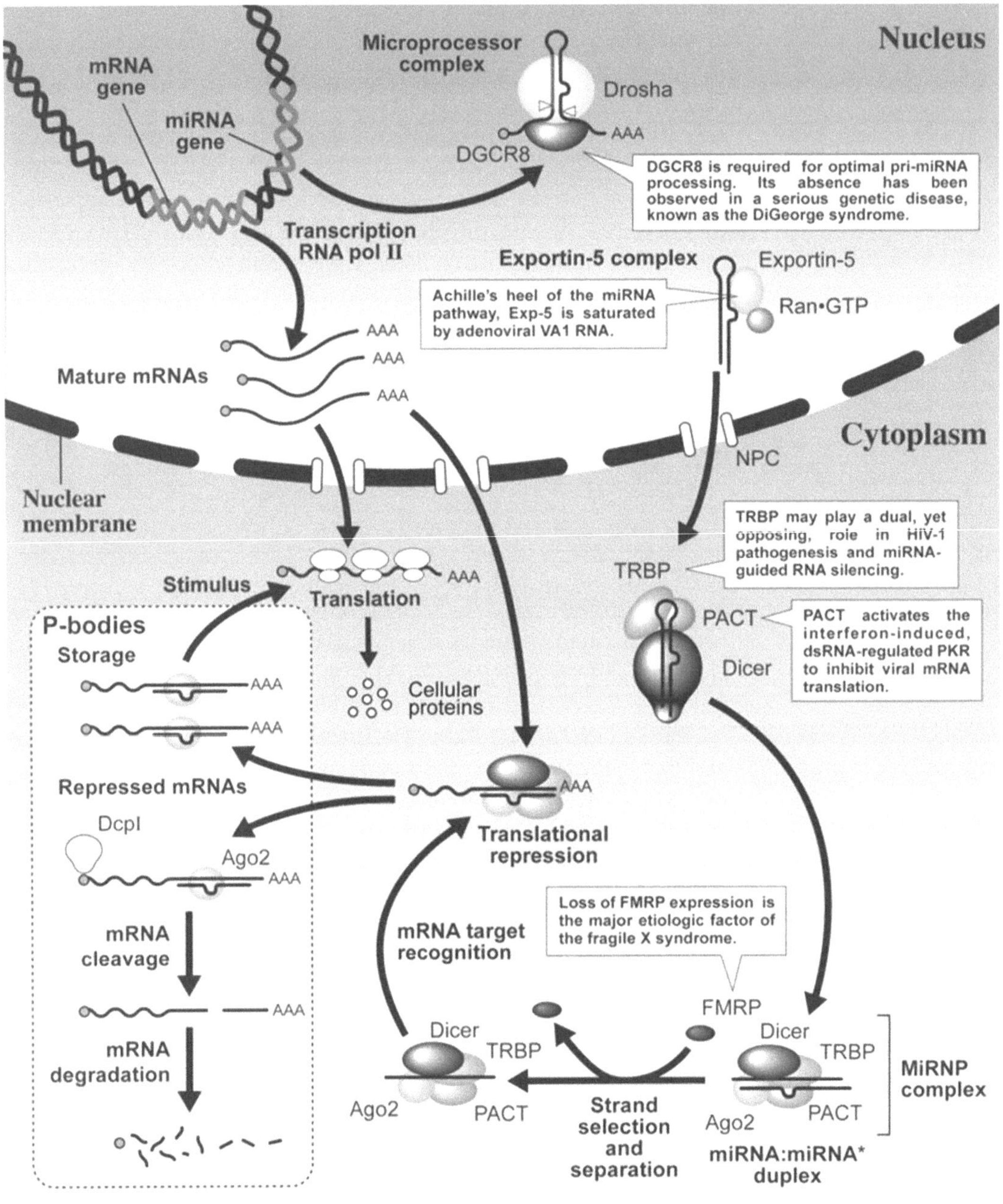

Fig. 18.1. Protein components of the microRNA (miRNA)-guided RNA silencing pathway and human diseases. Encoded by the genome, miRNA and messenger RNA (mRNA) genes are transcribed by RNA polymerase II in the nucleus. The primary miRNA (pri-miRNA) is recognized and cleaved by the Microprocessor complex, which contains Drosha and the DiGeorge syndrome critical region gene 8 (DGCR8) protein, to form the miRNA precursor (pre-miRNA). Following its export into the cytoplasm via Exportin-5 (Exp-5), in a Ran•GTP-dependent manner through the nuclear pore complex (NPC), the pre-miRNA is cleaved by the Dicer•TAR RNA binding protein (TRBP) complex into a miRNA:miRNA* duplex. This complex is then joined by Argonaute 2 (Ago2) to form a miRNA-containing ribonucleoprotein (miRNP) complex, after which the miRNA strand is selected. Depending on the degree of complementarity between the miRNA and its mRNA target, the miRNP complex will either mediate mRNA cleavage if the complementarity is perfect, or initially inhibit mRNA translation if the complementarity is imperfect. In this latter case, the repressed mRNA is translocated to the P-bodies, after which the mRNA can either be destroyed or relocalized to the translational machinery for expression upon a specific cellular signal.

2.1. Drosha

The initial step in miRNA biogenesis is the recognition of the pri-miRNA structure by the Microprocessor complex. Drosha is the core enzyme of this nuclear complex. This ribonuclease III (RNase III) enzyme binds to pri-miRNAs independently of their sequences and excises ~70-nt pre-miRNAs. Drosha was initially reported to have a role in ribosomal RNA processing *(25)*. We anticipate Drosha to be essential for cell viability, since its downreglation by RNA interference (RNAi) in embryonic stem (ES) cells correlated with a proportional decrease in mature miRNA production, thereby indicating its absolute requirement for miRNA biogenesis *(11)*.

2.2. Dicer

The second major protein complex of the miRNA-guided RNA silencing pathway is the pre-miRNA processing complex containing the RNase III Dicer. This enzyme is responsible for the cleavage of the pre-miRNA into a miRNA:miRNA* duplex with extremities harboring 2-nt 3′ overhangs, the signature characteristic of members of the RNases III family of enzymes *(26)*. In mammals, the presence of Dicer is essential, as Dicer-deficient mice die at the embryonic stage, suggesting that Dicer is required for mammalian development *(27,28)*. Dicer-deficient mouse ES cells are defective in differentiation and centromeric silencing. The role of miRNAs in ES cell differentiation has been recently studied by generating an inducible Dicer knockout model *(29)*. Analysis of Dicer-null ES cells revealed an impairment in miRNA biogenesis and a severe defect in differentiation both in vivo and in vitro. Epigenetic silencing of centromeric repeat sequences and concomitant expression of homologous small double-stranded RNAs (dsRNAs) were also markedly reduced. Noticeably, the phenotype was rescued by the re-expression of Dicer in these cells *(29)*. These results suggest that Dicer participates in multiple, fundamental biological processes in mammals, ranging from stem cell differentiation to the maintenance of centromeric heterochromatin structure and centromeric silencing *(29)*.

Deregulation of Dicer expression has also been observed in cases of cancer. A reduced expression of Dicer in non-small cell lung cancer *(30)* or an overexpression of Dicer in prostate adenocarcinoma and in precursor lesions of lung adenocarcinoma *(31,32)* have been reported, indicating that an adequate level of Dicer expression may be required for maintaining normal cell functions.

2.3. Argonaute 2

The Ago2 protein is the major component of the miRNP effector complex. Ago2 is a member of the PAZ and PIWI domain (PPD) protein family *(33)*. A remarkable feature of the miRNP complex remains its versatility in inducing either mRNA cleavage or translational repression on the sole basis of base pairing between the miRNA and its mRNA target. This notable aspect resides in the structure of the Ago2 protein, as determined from the structure of the *D. melanogaster* Ago1 PAZ domain solved by nuclear magnetic resonance *(34,35)*. The PAZ domain specifically recognizes siRNAs via their characteristic 2-nt 3′ overhangs *(34, 35, 36, 37)*. The 3′ end of the guide siRNA strand is anchored in the PAZ domain, whereas the PIWI domain, acting in concert with the PAZ, cleaves the mRNA strand at the phosphodiester bond facing that present between nucleotides 10 and 11 of the siRNA. An active RNA-induced silencing complex (RISC) can then be regenerated and initiate, armed with the same siRNA, a new round of mRNA cleavage *(38)*. When a miRNA perfectly complementary to its mRNA target is loaded into a miRNP, the cleavage occurs at precisely the same site as that seen for siRNA-guided cleavage *(39)*. After cleavage of the mRNA, the miRNA remains intact and can guide the recognition and destruction of additional mRNAs *(39)*.

Translational repression induced by miRNAs implicates that the target mRNA is not cleaved by Ago2. In fact, structural studies have shown that the imperfect complementarity occurring between a miRNA and its mRNA target prevents mRNA cleavage by Ago2 by moving away from the catalytic site portion of the mRNA facing nucleotides 10 and 11 of the miRNA. The mechanism underlying translational inhibition likely involves multiple binding sites for the same or different miRNAs in the mRNA 3′ NTR and cooperation among the different miRNPs attached to these binding sites on the targeted messages. These miRNPs may prevent initiation of translation or block active translation through combinatorial control and steric hindrance of the ribosomal machinery *(40,41)*. This scenario is supported by the relatively high degree of conservation among the different miRNAs recognizing the same target throughout evolution.

Ago2 is essential for mouse development, and cells lacking this protein are unable to mount an experimental response to siRNAs *(42)*. Recently, another strategy of disruption of Ago2 in mice indicates that the absence of this protein leads to embryonic lethality early in development, i.e., after the implantation stage *(43)*. It was also shown that Ago2 deficiency impairs miRNA biogenesis from pre-miRNA followed by a reduction in miRNA expression levels *(44)*. Using hematopoiesis in mice as a model system to study the physiological function of Ago2 in vivo, it was found that Ago2 controls early development of lymphoid and erythroid cells *(44)*. Thus, there is no doubt that Ago2 is essential for the miRNA-guided RNA silencing pathway and there is no homologue in the cell capable of compensating the loss of Ago2. These results suggest that among the Ago family members, only Ago2 seems to have mRNA cleavage activity *(43)*.

2.4. The P-bodies

The P-bodies are specific cytoplasmic foci of aggregated mRNA-containing RNP (mRNP) complexes associated with the translation repression and mRNA decay machineries *(45,46)*. P-bodies are referred to as the GW182-containing bodies, because they contain the GW182 RNA-binding protein as well as other proteins implicated in protein degradation, such as the decapping enzyme Dcp1 *(47,48)*. Recently, a link between these bodies and the miRNA-guided RNA silencing pathway has been established. In fact, Ago proteins have been localized to these protein complexes *(45)* and found to interact directly with GW182 *(23)*. Indeed, silencing of GW182 delocalizes resident P-body proteins and impairs silencing of a miRNA reporter system. The importance of P-bodies in mRNA repression is supported by the findings that mutations that prevent Ago proteins from localizing to P-bodies also prevent translational repression of mRNAs *(23)*.

miRNAs are also present in P-bodies, as the formation of this structure seems to be a consequence, rather than the cause, of miRNA biogenesis *(49)*. The authors reported that endogenous let-7 miRNA co-precipitates with a GW182 protein complex. In addition, knockdown of two proteins, Drosha and DGCR8, which are essential for the generation of mature miRNAs, results in a loss of the P-bodies *(49)*. P-bodies thus represent the cellular site where repressed mRNAs accumulate and are ultimately either degraded or rescued and redirected to the translational machinery.

miRNA-guided mRNA repression appears to be a reversible process. Recently, it was observed that upon a specific stress, a repressed mRNA can be released and translated back into proteins *(24)*. The authors showed that the cationic amino acid transporter 1 (CAT-1) mRNA and reporter genes bearing its 3′ NTR can be relieved from miR-122-induced inhibition in human hepatocarcinoma cells subjected to amino acid deprivation. The derepression of CAT-1 mRNA is accompanied by its release from cytoplasmic P-bodies and its recruitment to polysomes. mRNA derepression seems to require binding of HuR, an AU-rich element (ARE) binding protein, to the 3′ NTR of CAT-1 mRNA. The authors proposed that proteins interacting with the 3′ NTR of mRNAs will generally act as modifiers, altering the potential of miRNAs to repress gene expression *(24)*.

Interestingly, autoantibodies directed against protein components of the miRNA pathway have been detected in human patients with immune diseases. First, anti-GW182 antibodies were found in patients harboring Sjögren's syndrome, which is the most common clinical diagnosis, followed by mixed motor/sensory neuropathy and systemic lupus erythematosus *(50)*. Recently, it was reported that autoantibodies from patients with rheumatic diseases as well as from a mouse model of autoimmunity recognize Ago2, a component of the P-bodies *(51)*. Indirect immunofluorescence studies demonstrated that these autoantibodies target the P bodies. These autoantibodies were also capable of immunoprecipitating additional components of the miRNA pathway, including Dicer *(51)*.

3. The Accessory Proteins of the microRNA Pathway and Human Diseases

Assisting the major enzymes in their function, different accessory proteins play an important role in miRNA biogenesis and action. Of these accessory proteins, some have been shown to be implicated in genetic disorders or viral infection.

3.1. The DiGeorge Syndrome Critical Region Gene 8 Protein and the DiGeorge Syndrome

The DGCR8 gene is present in a common monoallelic deleted genomic region containing ~30 genes located in the q11.2 region of the human chromosome 22 *(52)*. There are two types of deletion, a 3.0-Mb deletion and a 1.5-Mb deletion. Although most patients have in common the 1.5-Mb deletion, some patients harbor chromosomal deletions that have either no overlap with those two types of deletions *(53,54)* or no detectable deletion in the q11 region of chromosome 22 *(55, 56, 57)*. Heterozygous deletion of this locus leads to the most common human genetic deletion and patients presenting this deletion display clinical phenotypes defined as the DiGeorge syndrome, Conotruncal anomaly face syndrome and Velocardiofacial syndrome *(52)*. Patients carrying this deletion demonstrate various conditions, ranging from congenital heart defects and characteristic facial appearance to immunodeficiency and behavioral problems *(58)*. Many efforts have been made to identify the genes located in this region that could be correlated with these phenotypes.

The DGCR8 protein interacts directly with the RNase III Drosha *(14)* within the Microprocessor complex. DGCR8 has been proposed to guide Drosha by functioning as a molecular anchor that determines the position where Drosha cleaves its pri-miRNA substrate, since Drosha lacks processing specificity in the absence of DGCR8 *(14,15)*. Knockdown of DGCR8 by RNAi, like its partner Drosha, decreased the level of mature miRNAs, indicating that both proteins are required for miRNA biogenesis *(14)*. These findings were confirmed in DGCR8 knockout ES mouse cell *(59)*.

The DGCR8 gene is present in the most common 1.5-Mb deletion *(52)*. The deletion encompasses ~30 genes, thus it is difficult to conclude which gene(s) is(are) involved in the disease. It is probably due to a combinational effect between all or some of these genes. Since the deletion occurs in one allele, the cell still possesses one copy of the genes, such as that encoding for DGCR8. Whether this single DGCR8 gene copy is sufficient for an efficient miRNA-guided RNA silencing pathway remains to be elucidated.

3.2. Exportin-5 and Adenovirus Infection

First discovered as a new family member of the nuclear karyopherin β transporter family, Exp-5 was found to mediate nuclear export of dsRNA binding proteins *(60)*. The initial model stipulates that Exp-5 and Ran•GTP associate with a dsRBD-containing protein in the nucleus and that this exporting complex translocates through the nuclear pore complex (NPC) to the cytoplasm *(60)*.

Then, Ran•GTP hydrolysis releases in the cytoplasm the dsRBD-containing protein that will allow it to interact with regulatory elements of its mRNA target. Further release or degradation of the mRNA target would permit import of the dsRBD-containing protein back to the nucleus *(60)*. More recently, Exp-5 was found to interact directly with pre-miRNAs and to mediate their transport from the nucleus to the cytoplasm in a Ran•GTP hydrolysis dependent manner *(17,61)*.

Downregulation of Exp-5 by RNAi showed that this transporter is required for efficient inhibition of gene expression induced by a pre-miRNA or a pre-miRNA mimetic short hairpin RNA (shRNA), but not by a siRNA, in dual luciferase reporter gene assay in 293T cells *(17)*, suggesting that the pre-miRNA or shRNA export precedes their processing. Entering in the RNAi pathway downstream of Exp-5, siRNAs do not required Exp-5 for their activity. Moreover, miRNA biogenesis is dependent on the presence of Exp-5 in HeLa cells *(61)*. Indeed, upon depletion of Exp-5 by RNAi for 48–72 h, the levels of mature miRNAs was reduced by 40–60% *(61)*. On the other hand, episomal overexpression of Exp-5 enhances the capacity for RNAi induced either by miRNAs or shRNAs, but not siRNAs *(62)*. The expression of endogenous pre-miRNAs and mature miRNAs is also enhanced upon Exp-5 overexpression, indicating that Exp-5 is the rate-limiting component of miRNA biogenesis.

The structural VA1 RNA overexpressed by adenoviruses was found to be able to saturate the transporter Exp-5 *(63)*. VA1 RNAs are 160-nt long and can be recognized by Exp-5 via their minihelix RNA motif *(63)*, which are dsRNA molecules similar in structure to pre-miRNAs *(64)*. Exp-5 directly interacts with VA1 RNA in a Ran•GTP-dependent manner *(63)*. VA1, which is expressed at very high levels in adenovirus-infected cells, potently inhibited RNAi induced by shRNAs or pre-miRNAs, without affecting RNAi induced by siRNA duplexes *(65)*. Competition binding for the Exp-5 and inhibition of Dicer function via direct binding of VA1 RNA appear to be the cause of the inhibition of RNAi function in adenovirus-infected cells *(65)*.

Later, it was found that small RNAs derived from VA1 RNAs, similar to miRNAs, can be found in adenovirus-infected cells *(64)*. These small RNAs are efficiently bound by Ago2, and behave as functional siRNAs, in that they inhibit the expression of reporter genes with complementary sequences. Inhibition of small VA1 RNA function negatively affected adenovirus production, indicating that they are required for optimal virus replication *(64)*. Adenoviruses thus appear to utilize the endogenous miRNA-guided RNA silencing pathway to their advantage in viral production.

3.3. TAR RNA Binding Protein and HIV-1

Initially discovered in 1991, TRBP is a cellular factor acting in synergy with the viral TAT protein in the transactivation of the long terminal repeat (LTR) of human immunodeficiency virus type 1 (HIV-1), a process that results in the transcription of viral genes *(66)*. TRBP is also known to interact with the protein Tax encoded by the human T-cell leukemia virus type 1 (HTLV-1) and inhibits its function of activation of the transcription from the LTR via the association with host cellular factors *(67)*. TRBP exerts other functions, including inhibition of the interferon (IFN)-induced dsRNA-regulated protein kinase R (PKR) *(68)*, as well as a growth promoting role with oncogenic potential activity *(69)*. The tumor suppressor Merlin regulates the oncogenic activities of TRBP through direct interaction *(70)*.

A link has been established between HIV-1 and protein components of the miRNA-guided RNA silencing pathway. TRBP was found to interact directly with Dicer in co-immunoprecipitation experiments *(19,21)*. TRBP possesses three dsRBD and is found in two isoforms in the cell, with TRBP2 being 21 amino acids longer than TRBP1 *(68,71,72)*. The structure involved in the interaction with Dicer includes the third C-terminal dsRBD of TRBP, as determined by deletional analysis in the yeast two-hybrid system *(19)*. Furthermore, pre-miRNA processing is affected when TRBP is depleted in cells *(19)*, suggesting that its presence is required for an efficient miRNA-guided RNA silencing pathway. Although TRBP appears to be an accessory protein of Dicer, in vitro analyses demonstrated that Dicer can efficiently cleave its substrate in the absence of TRBP *(21)*, suggesting that TRBP is not as essential to Dicer as DGCR8 is for Drosha. TRBP is also part of a ternary complex comprising Dicer and Ago2 in a sequence of events in which TRBP is required for the recruitment of Ago2 to the siRNA bound by Dicer *(21)*, thereby coupling the initiation step of the miRNA processing and the effector steps of the miRNA pathway. TRBP may thus exert a dual role in HIV-1 pathogenesis and RNA silencing *(73)*, but also in other important physiological processes.

3.4. PKR-Activating Protein and the PKR Signaling Pathway

Recently, a new protein has been reported to interact with a complex containing Dicer, Ago2 and TRBP: the PKR-activating protein (PACT). The direct interaction between PACT and Dicer involves the third dsRBD domain of PACT and the N-terminal region of Dicer containing the putative helicase motif *(74)*. The accumulation of mature miRNAs in vivo and the efficiency of siRNA-induced RNAi are affected upon depletion of PACT *(74)*. These findings were confirmed in shRNA-induced RNAi, but not when using siRNAs, suggesting that TRBP and PACT function primarily in siRNA production *(75)*. Moreover, the presence of both TRBP and PACT increased the ability of Dicer to cleave a 566-bp long dsRNA in vitro, as compared to TRBP and PACT added individually *(75)*.

PACT and TRBP share 44% of sequence identity and both possess three similar dsRBD domains. They bind to PKR, although exerting opposing effects, i.e., PACT activates PKR, whereas TRBP inhibits it *(76, 77, 78)*. The third dsRBD of PACT and TRBP, which is devoid of any detectable dsRNA binding activity *(78)*, mediates their interaction with Dicer as well as their regulation of PKR *(19,21,78)*. The C-terminal domain of PACT can also mediate homomultimerization *(79)*. Interestingly, a recent study demonstrated that PACT directly interacts with TRBP and that this complex associates with Dicer to facilitate production of siRNAs *(75)*.

PACT and TRBP may establish an interesting, but intriguing redundant link between the miRNA-guided RNA silencing pathway and the PKR signaling pathway. PKR is a dsRNA-dependent serine/threonine protein kinase that phosphorylates the translation initiation factor eIF2 to cause a general reduction of protein synthesis *(80)*. PKR is activated in response to dsRNA of cellular, viral or synthetic origin, with a size greater than ~30 nt, but not by siRNA. PKR thus mediates a critical role in response to dsRNA, acting as a sensor of viral infections (for review see García et al. *(81)*). For example, VA1 is known to play a key role in blocking activation of protein PKR during the adenoviral replication cycle, presumably by viral dsRNAs. In the absence of VA1, activation of PKR induces phosphorylation of the translation factor eIF-2α and, hence, inhibition of viral mRNA translation *(81,82)*. It is not surprising that viruses evolved in a way that they can inhibit PKR signaling.

3.5. Fragile X Mental Retardation Protein and the Fragile X Syndrome

In humans, loss of expression of the *FMR1* gene product is the etiologic factor of the fragile X syndrome, the most frequent cause of inherited mental retardation *(83,84)*. The *FMR1* gene, which spans ~38 kb, is located in the q27.3 region located at the tip of the X chromosome long arm *(85)*. The syndrome is transmitted as an X-linked dominant trait and it affects about 1 in 4000 males, who will develop in almost all cases moderate to severe mental retardation (IQ $\leq$ 50), and about 1 in 7000 females, who present in general a milder mental handicap *(85)*. FMRP has been detected in practically every tissue in humans and rodents, with high levels in the brain, testes, esophagus, lung and kidney *(86)*.

FMRP has been identified in a miRNP complex containing Dicer and Ago2 proteins in mammalian cells in vivo *(87)*. FMRP is an RNA-binding protein *(88,89)* known to be involved in the regulation of mRNA translation, mRNA transfer and local modulation of synaptic mRNA translation *(90–94)*. FMRP has been reported to behave as a negative regulator of translation, both in vitro and in vivo *(90– 94)*.

Our group recently showed that human FMRP can act as a miRNA acceptor protein for the RNase III Dicer and facilitate assembly of miRNAs on specific target RNA sequences *(95)*. We have proposed a model in which FMRP could facilitate miRNA assembly on target mRNAs. Functioning within a duplex miRNP, FMRP may also mediate mRNA targeting through a strand exchange mechanism, in which the miRNA* of the duplex is swapped for the mRNA *(96)*. Furthermore, FMRP may contribute to the relief of miRNA-guided mRNA repression through a reverse strand exchange reaction, possibly initiated by a specific cellular signal, that would liberate the mRNA for translation *(96)*. Although the intracellular sites hosting these events remain to be determined, we cannot exclude the possible involvement of the P-bodies.

We hypothesized that the absence of FMRP expression may result in suboptimal miRNA assembly on, and/or disassembly from, their natural mRNA targets, leading to a perturbed protein expression profile *(96)*. This may be expected given the requirement of FMRP for efficient small RNA-guided gene regulation *(95)*. Elucidation of the exact role and function of FMRP in miRNA-guided gene regulation may hold the key to determining the molecular basis of the fragile X syndrome and establishing a causal link between dysfunction of the RNA-silencing machinery and a human genetic disease.

4. Conclusion

The core protein components of the miRNA-guided RNA silencing machinery are essential for cellular homeostasis. However, deregulation of some accessory proteins can be tolerated by the organism, yet resulting in a specific disease. Although attractive, hypotheses and correlations have been established between the role exerted by these components in the miRNA pathway and the phenotype related to their deregulation, these issues remain to be validated. Only then we could examine whether the clinical portrait of the patients harboring a defective protein component of the miRNA pathway could be improved upon a therapy aimed at restoring the functionality of the miRNA silencing machinery.

Acknowledgments

We are grateful to Gilles Chabot for the graphic illustration. M. P. P. was supported by a doctoral studentship from the Natural Sciences and Engineering Research Council of Canada (NSERC). P. P. is a New Investigator of the Canadian Institutes of Health Research and Junior 2 Scholar from the Fonds de la Recherche en Santé du Québec. This work was financially supported by a Discovery grant from NSERC (262938-2008).

References

1. Miranda, K.C., Huynh, T., Tay, Y., Ang, Y.S., Tam, W.L., Thomson, A.M., Lim, B., and Rigoutsos, I. (2006) A pattern-based method for the identification of MicroRNA binding sites and their corresponding heteroduplexes. *Cell*, 126, 1203–1217.
2. Ouellet, D.L., Perron, M.P., Gobeil, L.A., Plante, P., and Provost, P. (2006) MicroRNAs in Gene Regulation: When the Smallest Governs It All. *J Biomed Biotechnol*, 2006, 69616.
3. Vella, M.C., Reinert, K., and Slack, F.J. (2004) Architecture of a validated microRNA:target interaction. *Chem Biol*, 11, 1619–1623.
4. Vella, M.C., Choi, E.Y., Lin, S.Y., Reinert, K., and Slack, F.J. (2004) The C. elegans microRNA let-7 binds to imperfect let-7 complementary sites from the lin-41 3 UTR. *Genes Dev*, 18, 132–137.
5. Hayashita, Y., Osada, H., Tatematsu, Y., Yamada, H., Yanagisawa, K., Tomida, S., Yatabe, Y., Kawahara, K., Sekido, Y., and Takahashi, T. (2005) A polycistronic micro-RNA cluster, miR-17-92, is overexpressed in human lung cancers and enhances cell proliferation. *Cancer Res*, 65, 9628–9632.
6. Calin, G.A., Dumitru, C.D., Shimizu, M., Bichi, R., Zupo, S., Noch, E., Aldler, H., Rattan, S., Keating, M., Rai, K. (2002) Frequent deletions and down-regulation of micro- RNA genes miR15 and miR16 at 13q14 in chronic lymphocytic leukemia. *Proc Natl Acad Sci U S A*, 99, 15524–15529.
7. Cimmino, A., Calin, G.A., Fabbri, M., Iorio, M.V., Ferracin, M., Shimizu, M., Wojcik, S.E., Aqeilan, R.I., Zupo, S., Dono, M. (2005) miR-15 and miR-16 induce apoptosis by targeting BCL2. *Proc Natl Acad Sci U S A*, 102, 13944–13949.
8. Perron, M.P. and Provost, P. (2008) Protein interactions and complexes in human micro-RNA biogenesis and function. *Front Biosci*, 13, 2537–2547.
9. Lee, Y., Jeon, K., Lee, J.T., Kim, S., and Kim, V.N. (2002) MicroRNA maturation: stepwise processing and subcellular localization. *Embo J*, 21, 4663–4670.
10. Lee, Y., Kim, M., Han, J., Yeom, K.H., Lee, S., Baek, S.H., and Kim, V.N. (2004) MicroRNA genes are transcribed by RNA polymerase II. *Embo J*, 23, 4051–4060.
11. Lee, Y., Ahn, C., Han, J., Choi, H., Kim, J., Yim, J., Lee, J., Provost, P., Radmark, O., Kim, S. (2003) The nuclear RNase III Drosha initiates microRNA processing. *Nature*, 425, 415–419.
12. Basyuk, E., Suavet, F., Doglio, A., Bordonne, R., and Bertrand, E. (2003) Human let-7 stem-loop precursors harbor features of RNase III cleavage products. *Nucleic Acids Res*, 31, 6593–6597.
13. Denli, A.M., Tops, B.B., Plasterk, R.H., Ketting, R.F., and Hannon, G.J. (2004) Processing of primary microRNAs by the Microprocessor complex. *Nature*, 432, 231–235.
14. Gregory, R.I., Yan, K.P., Amuthan, G., Chendrimada, T., Doratotaj, B., Cooch, N., and Shiekhattar, R. (2004) The Microprocessor complex mediates the genesis of microRNAs. *Nature*, 432, 235–240.
15. Han, J., Lee, Y., Yeom, K.H., Kim, Y.K., Jin, H., and Kim, V.N. (2004) The Drosha-DGCR8 complex in primary microRNA processing. *Genes Dev*, 18, 3016–3027.
16. Landthaler, M., Yalcin, A., and Tuschl, T. (2004) The human DiGeorge syndrome critical region gene 8 and Its D. melanogaster homolog are required for miRNA biogenesis. *Curr Biol*, 14, 2162–2167.
17. Yi, R., Qin, Y., Macara, I.G., and Cullen, B.R. (2003) Exportin-5 mediates the nuclear export of pre-microRNAs and short hairpin RNAs. *Genes Dev*, 17, 3011–3016.
18. Provost, P., Dishart, D., Doucet, J., Frendewey, D., Samuelsson, B., and Radmark, O. (2002) Ribonuclease activity and RNA binding of recombinant human Dicer. *Embo J*, 21, 5864–5874.
19. Haase, A.D., Jaskiewicz, L., Zhang, H., Laine, S., Sack, R., Gatignol, A., and Filipowicz, W. (2005) TRBP, a regulator of cellular PKR and HIV-1 virus expression, interacts with Dicer and functions in RNA silencing. *EMBO Rep*, 6, 961–967.
20. Zhang, H., Kolb, F.A., Brondani, V., Billy, E., and Filipowicz, W. (2002) Human Dicer preferentially cleaves dsRNAs at their termini without a requirement for ATP. *Embo J*, 21, 5875–5885.
21. Chendrimada, T.P., Gregory, R.I., Kumaraswamy, E., Norman, J., Cooch, N., Nishikura, K., and Shiekhattar, R. (2005) TRBP recruits the Dicer complex to Ago2 for microRNA processing and gene silencing. *Nature*, 436, 740–744.
22. Bartel, D.P. (2004) MicroRNAs: genomics, biogenesis, mechanism, and function. *Cell*, 116, 281–297.
23. Liu, J., Rivas, F.V., Wohlschlegel, J., Yates, J.R., 3rd, Parker, R., and Hannon, G.J. (2005) A role for the P-body component

GW182 in microRNA function. *Nat Cell Biol*, 7, 1261–1266.
24. Bhattacharyya, S.N., Habermacher, R., Martine, U., Closs, E.I., and Filipowicz, W. (2006) Relief of microRNA-mediated translational repression in human cells subjected to stress. *Cell*, 125, 1111–1124.
25. Wu, H., Xu, H., Miraglia, L.J., and Crooke, S.T. (2000) Human RNase III is a 160-kDa protein involved in preribosomal RNA processing. *J Biol Chem*, 275, 36957–36965.
26. MacRae, I.J. and Doudna, J.A. (2007) Ribonuclease revisited: structural insights into ribonuclease III family enzymes. *Curr Opin Struct Biol*, 17, 138–145.
27. Bernstein, E., Caudy, A.A., Hammond, S.M., and Hannon, G.J. (2001) Role for a bidentate ribonuclease in the initiation step of RNA interference. *Nature*, 409, 363–366.
28. Yang, W.J., Yang, D., Na, S., Sandusky, G., Zhang, Q., and Zhao, G. (2004) Dicer is required for embryonic angiogenesis during mouse development. *J Biol Chem.*
29. Kanellopoulou, C., Muljo, S.A., Kung, A.L., Ganesan, S., Drapkin, R., Jenuwein, T., Livingston, D.M., and Rajewsky, K. (2005) Dicer-deficient mouse embryonic stem cells are defective in differentiation and centromeric silencing. *Genes Dev*, 19, 489–501.
30. Karube, Y., Tanaka, H., Osada, H., Tomida, S., Tatematsu, Y., Yanagisawa, K., Yatabe, Y., Takamizawa, J., Miyoshi, S., Mitsudomi, T. (2005) Reduced expression of Dicer associated with poor prognosis in lung cancer patients. *Cancer Sci*, 96, 111–115.
31. Chiosea, S., Jelezcova, E., Chandran, U., Acquafondata, M., McHale, T., Sobol, R.W., and Dhir, R. (2006) Up-regulation of dicer, a component of the MicroRNA machinery, in prostate adenocarcinoma. *Am J Pathol*, 169, 1812–1820.
32. Chiosea, S., Jelezcova, E., Chandran, U., Luo, J., Mantha, G., Sobol, R.W., and Dacic, S. (2007) Overexpression of Dicer in precursor lesions of lung adenocarcinoma. *Cancer Res*, 67, 2345–2350.
33. Cerutti, L., Mian, N., and Bateman, A. (2000) Domains in gene silencing and cell differentiation proteins: the novel PAZ domain and redefinition of the Piwi domain. *Trends Biochem Sci*, 25, 481–482.
34. Yan, K.S., Yan, S., Farooq, A., Han, A., Zeng, L., and Zhou, M.M. (2003) Structure and conserved RNA binding of the PAZ domain. *Nature*, 426, 468–474.
35. Lingel, A., Simon, B., Izaurralde, E., and Sattler, M. (2004) Nucleic acid 3 -end recognition by the Argonaute2 PAZ domain. *Nat Struct Mol Biol*, 11, 576–577.
36. Song, J.J., Smith, S.K., Hannon, G.J., and Joshua-Tor, L. (2004) Crystal structure of Argonaute and its implications for RISC slicer activity. *Science*, 305, 1434–1437.
37. Lingel, A., Simon, B., Izaurralde, E., and Sattler, M. (2003) Structure and nucleic-acid binding of the Drosophila Argonaute 2 PAZ domain. *Nature*, 426, 465–469.
38. Yuan, Y.R., Pei, Y., Ma, J.B., Kuryavyi, V., Zhadina, M., Meister, G., Chen, H.Y., Dauter, Z., Tuschl, T., and Patel, D.J. (2005) Crystal structure of *A. aeolicus argonaute, a site-specific DNA-guided endoribonuclease, provides insights into RISC-mediated mRNA cleavage*. *Mol Cell*, 19, 405–419.
39. Hutvagner, G., and Zamore, P.D. (2002) A microRNA in a multiple-turnover RNAi enzyme complex. *Science*, 297, 2056–2060.
40. Doench, J.G., Petersen, C.P., and Sharp, P.A. (2003) siRNAs can function as miRNAs. *Genes Dev*, 17, 438–442.
41. Doench, J.G. and Sharp, P.A. (2004) Specificity of microRNA target selection in translational repression. *Genes Dev*, 18, 504–511.
42. Liu, J., Carmell, M.A., Rivas, F.V., Marsden, C.G., Thomson, J.M., Song, J.J., Hammond, S.M., Joshua-Tor, L., and Hannon, G.J. (2004) Argonaute2 is the catalytic engine of mammalian RNAi. *Science*, 305, 1437–1441.
43. Morita, S., Horii, T., Kimura, M., Goto, Y., Ochiya, T., and Hatada, I. (2007) One Argonaute family member, Eif2c2 (Ago2), is essential for development and appears not to be involved in DNA methylation. *Genomics*, 89, 687–696.
44. O'Carroll, D., Mecklenbrauker, I., Das, P.P., Santana, A., Koenig, U., Enright, A.J., Miska, E.A., and Tarakhovsky, A. (2007) A Slicer-independent role for Argonaute 2 in hematopoiesis and the microRNA pathway. *Genes Dev*, 21, 1999–2004.
45. Liu, J., Valencia-Sanchez, M.A., Hannon, G.J., and Parker, R. (2005) MicroRNA-dependent localization of targeted mRNAs to mammalian P-bodies. *Nat Cell Biol*, 7, 719–723.
46. Teixeira, D., Sheth, U., Valencia-Sanchez, M.A., Brengues, M., and Parker, R. (2005) Processing bodies require RNA for assembly and contain nontranslating mRNAs. *RNA*, 11, 371–382.
47. Eystathioy, T., Chan, E.K, Tenenbaum, S.A., Keene, J.D., Griffith, K., and Fritzler, M.J.

(2002) A phosphorylated cytoplasmic autoantigen, GW182, associates with a unique population of human mRNAs within novel cytoplasmic speckles. *Mol Biol Cell*, 13, 1338–1351.

48. Eystathioy, T., Jakymiw, A., Chan, E.K., Seraphin, B., Cougot, N., and Fritzler, M.J. (2003) The GW182 protein colocalizes with mRNA degradation associated proteins hDcp1 and hLSm4 in cytoplasmic GW bodies. *RNA*, 9, 1171–1173.
49. Pauley, K.M., Eystathioy, T., Jakymiw, A., Hamel, J.C., Fritzler, M.J., and Chan, E.K. (2006) Formation of GW bodies is a consequence of microRNA genesis. *EMBO Rep*, 7, 904–910.
50. Eystathioy, T., Chan, E.K., Takeuchi, K., Mahler, M., Luft, L.M., Zochodne, D.W., and Fritzler, M.J. (2003) Clinical and serological associations of autoantibodies to GW bodies and a novel cytoplasmic autoantigen GW182. *J Mol Med*, 81, 811–818.
51. Jakymiw, A., Ikeda, K., Fritzler, M.J., Reeves, W.H., Satoh, M., and Chan, E.K. (2006) Autoimmune targeting of key components of RNA interference. *Arthritis Res Ther*, 8, R87.
52. Shiohama, A., Sasaki, T., Noda, S., Minoshima, S., and Shimizu, N. (2003) Molecular cloning and expression analysis of a novel gene DGCR8 located in the DiGeorge syndrome chromosomal region. *Biochem Biophys Res Commun*, 304, 184–190.
53. Levy, A., Demczuk, S., Aurias, A., Depetris, D., Mattei, M.G., and Philip, N. (1995) Interstitial 22q11 microdeletion excluding the ADU breakpoint in a patient with DiGeorge syndrome. *Hum Mol Genet*, 4, 2417–2419.
54. Kurahashi, H., Nakayama, T., Osugi, Y., Tsuda, E., Masuno, M., Imaizumi, K., Kamiya, T., Sano, T., Okada, S., and Nishisho, I. (1996) Deletion mapping of 22q11 in CATCH22 syndrome: identification of a second critical region. *Am J Hum Genet*, 58, 1377–1381.
55. Lindsay, E.A., Halford, S., Wadey, R., Scambler, P.J., and Baldini, A. (1993) Molecular cytogenetic characterization of the DiGeorge syndrome region using fluorescence in situ hybridization. *Genomics*, 17, 403–407.
56. Gong, W., Emanuel, B.S., Collins, J., Kim, D.H., Wang, Z., Chen, F., Zhang, G., Roe, B., and Budarf, M.L. (1996) A transcription map of the DiGeorge and velo-cardio-facial syndrome minimal critical region on 22q11. *Hum Mol Genet*, 5, 789–800.
57. Lindsay, E.A., Rizzu, P., Antonacci, R., Jurecic, V., Delmas-Mata, J., Lee, C.C., Kim, U.J., Scambler, P.J., and Baldini, A. (1996) A transcription map in the CATCH22 critical region: identification, mapping, and ordering of four novel transcripts expressed in heart. *Genomics*, 32, 104–112.
58. Yamagishi, H. and Srivastava, D. (2003) Unraveling the genetic and developmental mysteries of 22q11 deletion syndrome. *Trends Mol Med*, 9, 383–389.
59. Wang, Y., Medvid, R., Melton, C., Jaenisch, R., and Blelloch, R. (2007) DGCR8 is essential for microRNA biogenesis and silencing of embryonic stem cell self-renewal. *Nat Genet*, 39, 380–385.
60. Brownawell, A.M. and Macara, I.G. (2002) Exportin-5, a novel karyopherin, mediates nuclear export of double-stranded RNA binding proteins. *J Cell Biol*, 156, 53–64.
61. Lund, E., Guttinger, S., Calado, A., Dahlberg, J.E., and Kutay, U. (2004) Nuclear export of microRNA precursors. *Science*, 303, 95–98.
62. Yi, R., Doehle, B.P., Qin, Y., Macara, I.G., and Cullen, B.R. (2005) Overexpression of Exportin 5 enhances RNA interference mediated by short hairpin RNAs and microRNAs. *Rna*, 11, 220–226.
63. Gwizdek, C., Ossareh-Nazari, B., Brownawell, A.M., Doglio, A., Bertrand, E., Macara, I.G., and Dargemont, C. (2003) Exportin-5 mediates nuclear export of minihelix-containing RNAs. *J Biol Chem*, 278, 5505–5508.
64. Aparicio, O., Razquin, N., Zaratiegui, M., Narvaiza, I., and Fortes, P. (2006) Adenovirus virus-associated RNA is processed to functional interfering RNAs involved in virus production. *J Virol*, 80, 1376–1384.
65. Lu, S. and Cullen, B.R. (2004) Adenovirus VA1 noncoding RNA can inhibit small interfering RNA and MicroRNA biogenesis. *J Virol*, 78, 12868–12876.
66. Gatignol, A., Buckler-White, A., Berkhout, B., and Jeang, K.T. (1991) Characterization of a human TAR RNA-binding protein that activates the HIV-1 LTR. *Science*, 251, 1597–1600.
67. Donzeau, M., Winnacker, E.L., and Meisterernst, M. (1997) Specific repression of Tax trans-activation by TAR RNA-binding protein TRBP. *J Virol*, 71, 2628–2635.
68. Daher, A., Longuet, M., Dorin, D., Bois, F., Segeral, E., Bannwarth, S., Battisti, P.L., Purcell, D.F., Benarous, R., Vaquero, C. (2001) Two dimerization domains in

the trans-activation response RNA-binding protein (TRBP) individually reverse the protein kinase R inhibition of HIV-1 long terminal repeat expression. *J Biol Chem*, 276, 33899–33905.

69. Benkirane, M., Neuveut, C., Chun, R.F., Smith, S.M., Samuel, C.E., Gatignol, A., and Jeang, K.T. (1997) Oncogenic potential of TAR RNA binding protein TRBP and its regulatory interaction with RNA-dependent protein kinase PKR. *EMBO J*, 16, 611–624.
70. Lee, J.Y., Kim, H., Ryu, C.H., Kim, J.Y., Choi, B.H., Lim, Y., Huh, P.W., Kim, Y.H., Lee, K.H., Jun, T.Y. (2004) Merlin, a tumor suppressor, interacts with transactivation-responsive RNA-binding protein and inhibits its oncogenic activity. *J Biol Chem*, 279, 30265–30273.
71. Duarte, M., Graham, K., Daher, A., Battisti, P.L., Bannwarth, S., Segeral, E., Jeang, K.T., and Gatignol, A. (2000) Characterization of TRBP1 and TRBP2. Stable stem-loop structure at the 5′ end of TRBP2 mRNA resembles HIV-1 TAR and is not found in its processed pseudogene. *J Biomed Sci*, 7, 494–506.
72. Gatignol, A., Duarte, M., Daviet, L., Chang, Y.N., and Jeang, K.T. (1996) Sequential steps in Tat trans-activation of HIV-1 mediated through cellular DNA, RNA, and protein binding factors. *Gene Expr*, 5, 217–228.
73. Provost, P., Barat, C., Plante, I., and Tremblay, M.J. (2006) HIV-l and the microRNA-guided silencing pathway: An intricate and multifaceted encounter. *Virus Res*, 121, 107–115.
74. Lee, Y., Hur, I., Park, S.Y., Kim, Y.K., Suh, M.R., and Kim, V.N. (2006) The role of PACT in the RNA silencing pathway. *EMBO J*, 25, 522–532.
75. Kok, K.H., Ng, M.H., Ching, Y.P., and Jin, D.Y. (2007) Human TRBP and PACT Directly Interact with Each Other and Associate with Dicer to Facilitate the Production of Small Interfering RNA. *J Biol Chem*, 282, 17649–17657.
76. Patel, R.C. and Sen, G.C. (1998) PACT, a protein activator of the interferon-induced protein kinase, PKR. *EMBO J*, 17, 4379–4390.
77. Peters, G.A., Hartmann, R., Qin, J., and Sen, G.C. (2001) Modular structure of PACT: distinct domains for binding and activating PKR. *Mol Cell Biol*, 21, 1908–1920.
78. Gupta, V., Huang, X., and Patel, R.C. (2003) The carboxy-terminal, M3 motifs of PACT and TRBP have opposite effects on PKR activity. *Virology*, 315, 283–291.
79. Hitti, E.G., Sallacz, N.B., Schoft, V.K., and Jantsch, M.F. (2004) Oligomerization activity of a double-stranded RNA-binding domain. *FEBS Lett*, 574, 25–30.
80. Taylor, S.S., Haste, N.M., and Ghosh, G. (2005) PKR and eIF2alpha: integration of kinase dimerization, activation, and substrate docking. *Cell*, 122, 823–825.
81. Garcia, M.A., Gil, J., Ventoso, I., Guerra, S., Domingo, E., Rivas, C., and Esteban, M. (2006) Impact of protein kinase PKR in cell biology: from antiviral to antiproliferative action. *Microbiol Mol Biol Rev*, 70, 1032–1060.
82. Thimmappaya, B., Weinberger, C., Schneider, R.J., and Shenk, T. (1982) Adenovirus VAI RNA is required for efficient translation of viral mRNAs at late times after infection. *Cell*, 31, 543–551.
83. O'Donnell, W.T. and Warren, S.T. (2002) A decade of molecular studies of fragile X syndrome. *Annu Rev Neurosci*, 25, 315–338.
84. Bardoni, B. and Mandel, J.L. (2002) Advances in understanding of fragile X pathogenesis and FMRP function, and in identification of X linked mental retardation genes. *Curr Opin Genet Dev*, 12, 284–293.
85. Bardoni, B., Schenck, A., and Mandel, J.L. (2001) The Fragile X mental retardation protein. *Brain Res Bull*, 56, 375–382.
86. Khandjian, E.W. (1999) Biology of the fragile X mental retardation protein, an RNA-binding protein. *Biochem Cell Biol*, 77, 331–342.
87. Jin, P., Zarnescu, D.C., Ceman, S., Nakamoto, M., Mowrey, J., Jongens, T.A., Nelson, D.L., Moses, K., and Warren, S.T. (2004) Biochemical and genetic interaction between the fragile X mental retardation protein and the microRNA pathway. *Nat Neurosci*, 7, 113–117.
88. Siomi, H., Siomi, M.C., Nussbaum, R.L., and Dreyfuss, G. (1993) The protein product of the fragile X gene, FMR1, has characteristics of an RNA-binding protein. *Cell*, 74, 291–298.
89. Siomi, H., Choi, M., Siomi, M.C., Nussbaum, R.L., and Dreyfuss, G. (1994) Essential role for KH domains in RNA binding: impaired RNA binding by a mutation in the KH domain of FMR1 that causes fragile X syndrome. *Cell*, 77, 33–39.
90. Laggerbauer, B., Ostareck, D., Keidel, E.M., Ostareck-Lederer, A., and Fischer, U. (2001) Evidence that fragile X mental

retardation protein is a negative regulator of translation. *Hum Mol Genet*, 10, 329–338.

91. Li, Z., Zhang, Y., Ku, L., Wilkinson, K.D., Warren, S.T., and Feng, Y. (2001) The fragile X mental retardation protein inhibits its translation via interacting with mRNA. *Nucleic Acids Res*, 29, 2276–2283.
92. Mazroui, R., Huot, M.E., Tremblay, S., Filion, C., Labelle, Y., and Khandjian, E.W. (2002) Trapping of messenger RNA by Fragile X Mental Retardation protein into cytoplasmic granules induces translation repression. *Hum Mol Genet*, 11, 3007–3017.
93. Costa, A., Wang, Y., Dockendorff, T.C., Erdjument-Bromage, H., Tempst, P., Schedl, P., and Jongens, T.A. (2005) The Drosophila fragile X protein functions as a negative regulator in the orb autoregulatory pathway. *Dev Cell*, 8, 331–342.
94. Zhang, Y.Q., Bailey, A.M., Matthies, H.J., Renden, R.B., Smith, M.A., Speese, S.D., Rubin, G.M., and Broadie, K. (2001) Drosophila fragile X-related gene regulates the MAP1B homolog Futsch to control synaptic structure and function. *Cell*, 107, 591–603.
95. Plante, I., Davidovic, L., Ouellet, D.L., Gobeil, L.A., Tremblay, S., Khandjian, E.W., and Provost, P. (2006) Dicer-Derived MicroRNAs Are Utilized by the Fragile X Mental Retardation Protein for Assembly on Target RNAs. *J Biomed Biotechnol*, 2006, 64347.
96. Plante, I. and Provost, P. (2006) Hypothesis: A Role for Fragile X Mental Retardation Protein in Mediating and Relieving MicroRNA-Guided Translational Repression? *J Biomed Biotechnol*, 2006, 16806.

Chapter 19

Intron-Mediated RNA Interference and microRNA Biogenesis

Shao-Yao Ying and Shi-Lung Lin

Abstract

Nearly 97% of the human genome is non-coding DNA, and introns occupy most of it around the gene-coding regions. Numerous intronic sequences have been recently found to encode microRNAs, which are responsible for RNA-mediated gene silencing through RNA interference (RNAi)-like pathways. microRNAs (miRNAs), small single-stranded regulatory RNAs capable of interfering with intracellular messenger RNAs (mRNAs) that contain either complete or partial complementarity, are useful for the design of new therapies against cancer polymorphism and viral mutation. This flexible characteristic is different from double-stranded siRNAs (small interfering RNAs) because a much more rigid complementarity is required for siRNA-induced RNAi gene silencing. miRNAs were firstly discovered in *Caenorhabditis elegans* as native RNA fragments that modulate a wide range of genetic regulatory pathways during embryonic development. Currently, varieties of miRNAs are widely reported in plants, animals and even microbes. Intronic microRNA is a new class of miRNAs derived from the processing of gene introns. The intronic miRNAs differ uniquely from previously described intergenic miRNAs in the requirement of type II RNA polymerases (Pol-II) and spliceosomal components for their biogenesis. Several kinds of intronic miRNAs have been identified in *C. elegans*, mouse and human cells; however, neither function nor application has been reported. Here, we show for the first time that intron-derived miRNAs are able to induce RNA interference in not only human and mouse cells but also zebrafishes, chicken embryos and adult mice, demonstrating the evolutionary preservation of the intron-mediated gene silencing through miRNA functionality in cell and in vivo. These findings suggest an intracellular miRNA-mediated gene regulatory system, fine-tuning the degradation of protein-coding messenger RNAs.

Key words: RNA interference, intronic microRNA, microRNA, RNA splicing, nonsense-mediated RNA decay, RNA-induced gene silencing complex, RNA-induced transcriptional silencing.

M. Sioud (ed.), *Methods in Molecular Biology, siRNA and miRNA Gene Silencing, vol. 487*

DOI: 10.1007/978-1-60327-547-7_19

1. Introduction

The first microRNA (miRNA) molecules, *lin-4* and *let-7*, were identified in 1993 *(1)*, since then the advance of small RNA research has gradually progressed in identifying more miRNA identities and understanding their biogenesis, functionality and target gene regulation. Most of these early miRNAs were located in the non-coding regions between genes and transcribed by unidentified promoters; these are intergenic miRNA. All miRNAs studied at this stage were recognized as the intergenic miRNA until 2003, when Ambros et al. *(1)* discovered some tiny non-coding RNA derived from the intron regions of gene transcripts. In the meantime, Lin et al. *(2)* proved the biogenesis and gene silencing mechanism of these intron-derived miRNAs, providing the first functional evidence for a new miRNA category, intronic miRNA. As shown in **Table 19.1**, several intronic miRNA molecules have been identified in *C. elegans*, mouse and human genomes *(1, 2, 3)* and some of their functions have been related to RNA interference (RNAi).

Introns occupy the largest proportion of non-coding sequences in the protein-coding DNA of a genome. The transcription of the genomic protein-coding DNA generates precursor messenger RNA (pre-mRNA), which contains four major parts including 5′-untranslational region (UTR), protein-coding exon, non-coding intron and 3′-UTR. In broad definition, both 5′- and 3′-UTR can be seen as a kind of intron extension; however, their processing during mRNA translation is different from the intron located between two protein-coding exons, or termed the in-frame intron. The in-frame intron can be sized up to several ten kilobase nucleotides and was thought to be a huge genetic waste in gene transcripts. Recently, this stereotype mis-understanding was changed after the finding of the intronic miRNA. miRNA is usually sized about 18–27 oligonucleotides capable of either directly degrading its intracellular messenger RNA (mRNA) target or suppressing the protein translation of its targeted mRNA, depending on the complementarity between the miRNA and its target. In this way, the intronic miRNA is similar to previously described intergenic miRNAs structurally and functionally, but slightly differs from them in its unique requirement of Pol-II and RNA splicing components for biogenesis *(2,4,5)*. Approximately 10–30% of a spliced intron are exported into the cytoplasm with a moderate half-life *(6)*.

RNA interference (RNAi) is a post-transcriptional gene silencing mechanism in eukaryotes, which can be triggered by small RNA molecules such as microRNA (miRNA) and small interfering RNA (siRNA). These small RNA molecules usually function as a gene silencer, interfering with the intracellular expression of

Table 19.1
Currently studied intronic microRNA (Id-miRNA)

miRNA	Species	Host gene (Intron) [#]	Target gene(s)
miR-2a, -b2	Worm	Spi	
miR-7b	Mammal	Pituitary gland specific factor 1A (2) [NM174947]	Paired mesoderm homeobox protein 2b; HLHm5
miR-10b	Mammal	Homeobox protein HOX-4 *(4)*	
miR-11	*Drosophila*	E2F	
miR-13b2	*Drosophila*	CG7033	
miR-15b, -16-2	Mammal	Chromosome-associated polypeptide C	
miR-17-92	Human	C13orf25	
miR-25, -93, -106b	Mammal	CDC47 homolog *(13)*	
miR-26a1, -26a2, -26b	Vertebrate	Nuclear LIM interactor-interacting factor 1, 2, 3	
miR-28	Human	LIM domain-containing preferred translocation partner in lipoma [NM005578]	
miR-30c1, -30e	Mammal	Nuclear transcription factor Y subunit γ *(5)*	Transcription factor HES-1; PAI-1 mRNA-binding protein
miR-33	Vertebrate	Sterol regulatory element binding protein-2 *(15)*	RNA-dependent helicase p68; NAG14 protein
miR-101b	Human	RNA 3′-terminal phospate cyclase-like protein *(8)*	
miR-103, -107	Human	Pantothenate kinase 1, 2, 3	
miR-105-1, -105-2, -224	Mammal	γ-Aminobutyric acid receptor α-3 subunit precursor, epsilon subunit precursor	
miR-126, -126*	Mammal	EGF-like, Notch4-like, NEU1 protein *(6)* [NM178444]	
miR-128b	Mammal	cAMP-regulated phospho-protein 21 *(11)*	
miR-139	Mammal	cGMP-dependent 3′,5′-cyclic phosphodiesterase *(2)*	
miR-140	Human	NEDD4-like ubiquitin-protein ligase WWP2 *(15)*	
miR-148b	Mammal	Coatomer ζ-1 subunit	
miR-151	Mammal		
miR-152	Human	Coatomer ζ-2 subunit	N-myc proto-oncogene protein; noggin precursor
miR-153-1, -153-2	Human	Protein-tyrosine phosphatase N precursors	
miR-208	Mammal	Myosin heavy chain, cardiac muscle α isoform *(28)*	
miR-218-1, -218-2	Human	Slit homolog proteins [NM003062]	

genes either completely or partially complementary to the small RNAs. In principle, siRNAs are double-stranded RNAs capable of degrading target gene transcripts with almost perfect complementarity *(7,8)*. Unlike the stringent complementarity of siRNAs to their RNA targets, miRNAs are single-stranded and able to pair with target RNAs that have partial complementarity to the miRNAs *(9,10)*. Numerous natural miRNAs have been found to be derived from hairpin-like RNA precursors in almost all eukaryotes, including yeast (*Schizosaccharomyces pombe*), plant (*Arabidopsis spp.*), nematode (*Caenorhabditis elegans*), fly (*Drosophila melanogaster*), mouse and human, involving defense against viral infections and regulation of certain gene expressions during development *(11–21)*. In contrast, natural siRNAs were abundantly discovered in plants and low-level animals (worms and flies), but rarely in mammals *(10)*. Because of the widespread detection of miRNAs in eukaryotes, these small RNAs have recently been used to design novel therapeutics against cancers and viral infections *(22,23)*. In fact, gene silencing mechanisms involving miRNAs have been proposed to be an intracellular defense system for eliminating undesired transgenes and foreign RNAs, such as viral infections and retrotransposon activities *(22,24)*.

2. Biogenesis and Definition

The definition of intronic miRNA is based on two factors; first, they must share the same promoter with their encoded genes, and second, they are spliced out of the transcript of their encoded genes and further processed into mature miRNAs **(Fig. 19.1)**. Although some of currently identified miRNAs are encoded in the genomic intron region of a gene but in the opposite orientation to the gene transcript, those miRNAs are not intronic miRNAs because they neither share the same promoter with the gene nor need to be released from the gene transcript by RNA splicing. For the transcription of those miRNAs, their promoters are located in the antisense direction to the gene, likely using the gene transcript as a potential target for the antisense miRNAs. A good example is *let-7c*, which was found to be an intergenic miRNA located in the antisense region of a gene intron. Current computer programs for miRNA prediction cannot distinguish the intronic miRNAs from the intergenic miRNAs. Because intronic miRNAs are encoded in the gene transcript precursors (pre-mRNAs) and share the same promoter with the encoded gene transcripts, the miRNA prediction programs tend to classify the intronic miRNAs as same as the intergenic miRNAs located in the exonic regions of genes. According to the different biogenic mechanisms between the

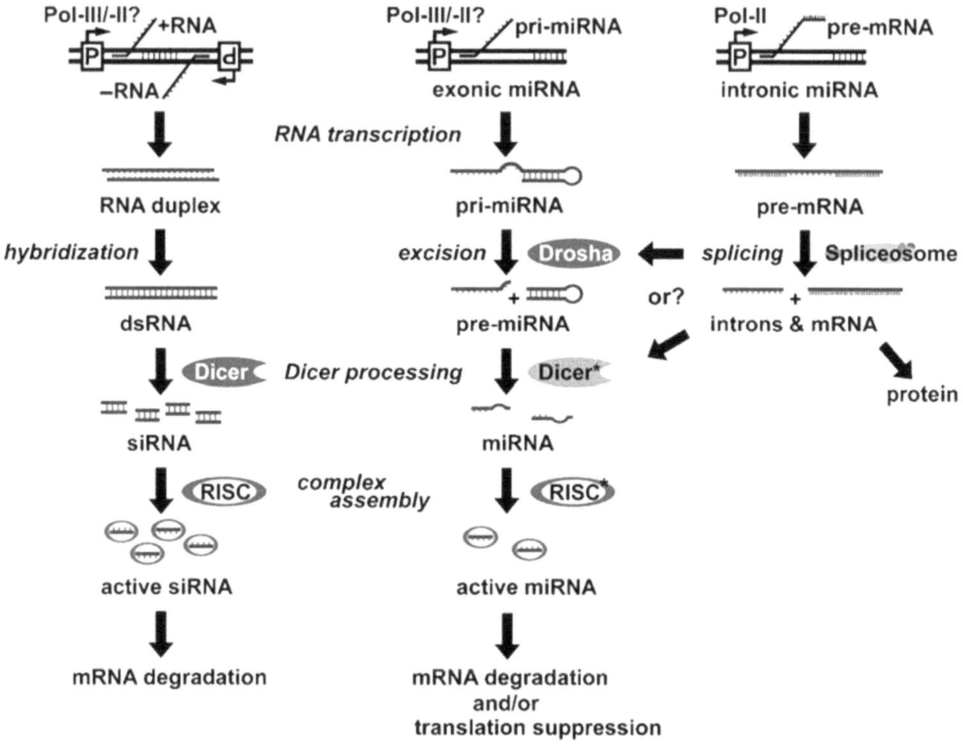

Fig. 19.1. Comparison of biogenesis and RNAi mechanisms among siRNA, intergenic (exonic) miRNA and intronic miRNA. siRNA is likely formed by two perfectly complementary RNAs transcribed from two different promoters (remains to be determined) and further processing into 19–22 bp duplexes by the RNase III-familial endonuclease, Dicer. The biogenesis of intergenic miRNAs, e.g., *lin*-4 and *let*-7, involves a long transcript precursor (pri-miRNA), which is probably generated by Pol-II or Pol-III RNA promoters, while intronic miRNAs are transcribed by the Pol-II promoters of its encoded genes and co-expressed in the intron regions of the gene transcripts (pre-mRNA). After RNA splicing and further processing, the spliced intron may function as a pri-miRNA for intronic miRNA generation. In the nucleus, the pri-miRNA is excised by Drosha RNase to form a hairpin-like pre-miRNA template and then exported to the cytoplasm for further processing by Dicer* to form mature miRNAs. The Dicers for siRNA and miRNA pathways are different. All three small regulatory RNAs are finally incorporated into an RNA-induced silencing complex (RISC), which contains either strand of siRNA or the single-strand of miRNA. The effect of miRNA is considered to be more specific and less adverse than that of siRNA because only one strand is involved. On the other hand, siRNAs primarily trigger mRNA degradation, whereas miRNAs can induce either mRNA degradation or suppression of protein synthesis depending on the sequence complementarity to the target gene transcripts.

intronic and exonic miRNAs, these two types of miRNAs may function as different gene regulators in the adjustment of cellular physiology. Thus, a miRNA-prediction program including a database of non-coding sequences located in the protein-coding pre-mRNA regions is urgently needed for thoroughly screening and understanding the distribution and variety of hairpin-like intronic miRNAs in the genomes.

Intronic miRNA is a new class of small regulatory RNAs derived from the non-coding DNA regions of a gene, such as intron, 5′- and 3′-untranslated region (UTR). Many introns and UTRs have been identified to contain tri-or tetra-nucleotide repeat expansions, capable of being transcribed and processed into repeat-associated miRNAs *(25,26)*. The biogenesis of intronic miRNA in eukaryotes involves five steps. First, miRNA is generated as a part of a long primary precursor miRNA (pri-miRNA) located in the intron or UTR of a gene transcript, by type II RNA polymerases (Pol-II) *(2)*. Second, after intron splicing, the long pri-miRNA is excised by spliceosomal components and/or further processed by *Drosha*-like RNaseIII endonucleases/microprocessors to form precursor miRNA (pre-miRNA) *(2,27)*. Third, the pre-miRNA is exported out of the nucleus probably by Ran-GTP and a receptor of Exportins *(28,29)*. Fourth, once in the cytoplasm, *Dicer*-like nucleases cleave the pre-miRNA to form mature miRNA. Lastly, the mature miRNA is assembled into a ribonuclear particle (RNP) to form a RNA-induced silencing complex (RISC) or RNA-induced transcriptional silencing (RITS) complex for executing the RNA interference (RNAi)-related gene silencing mechanisms *(27,30)*. Although the in vitro model of siRNA-associated RISC assembly has been reported, the link between the final miRNA maturation and RISC assembly remains to be determined. The characteristics of Dicer and RISC have been reported to be distinct between the siRNA and miRNA mechanisms *(31,32)*. In zebrafish, we have recently observed that the stem-loop structure of pre-miRNAs is involved in the strand selection for mature miRNA during RISC assembly *(27)*. These findings suggest that the duplex structure of siRNA may be not essential for the assembly of miRNA-associated RISC in vivo. The biogenesis of miRNA and siRNA seems to share a certain similarity; however, the miRNA mechanisms previously proposed were based on the model of siRNA, and thus one must distinguish their individual properties and differences in these two types of RNAs to understand the evolutional and functional relationship of these gene silencing pathways. In addition, the differences may provide insight into the prevalence of native siRNAs in invertebrates but rarely in mammals.

3. Intronic micro-RNA and Disease

The majority of human gene transcripts contains introns, which vary from species to species, and changes in these non-protein-coding sequences are frequently observed in clinical and circumstantial malfunction. Numerous introns encode miRNAs, which are involved in RNAi-related chromatin silencing mechanisms. Over 90 intronic miRNAs have been identified using bioinformatic approaches to date, but the function of the vast majority of these molecules remains to be determined *(3)*. According to the strictly expressive correlation between intronic miRNAs and their encoded genes, one may speculate that the levels of condition-specific, time-specific and tissue-specific miRNA expression are determined by the intracellular modulation of the encoded gene. This interpretation accounts for a more accurate genetic expression of various traits and any dysregulation of this miRNA-encoded gene correlation will thus result in genetic diseases. For instance, many introns contain tri-or tetra-nucleotide repeat expansions capable of being transcribed and processed into repeat-associated miRNAs *(25,26)*. They play an important role in the modulation of epigenetic alterations of genes which are involved in several genetic diseases that have been identified to be caused by triplet repeat expansions in specific single genes, collectively termed triplet repeat expansion diseases (TREDs). TREDs include, but not limit, fragile X syndrome (FXS), Huntington's disease (HD), myotonic dystrophy (DM) and a number of spinocerebellar ataxias (SCAs).

Observation of intron-related TREDs starts from monozygotic twins who frequently demonstrate slightly, but definitely distinguishing, disease susceptibility and physiological behavior. A long CCTG expansion in the intron 1 of a zinc finger protein ZNF9 gene has been correlated to type 2 myotonic dystrophy (DM2) in either one of the twins with higher susceptibility *(33)*. Since the repeat expansion motifs often obtain high affinity to certain RNA-binding proteins, the interfering role of intron-derived repeat expansion fragments in DM and HD has been recently found to trigger RNAi-like gene silencing *(26)*; however, the involved mechanism is still unclear. Another more established example is the fragile X syndrome (FXS), which represents about 30% of human inherited mental retardation, affecting approximately one in 2000 males and one in 4000 females. This disease is caused by a dynamic pre-mutation [expansion of microsatellite-like trinucleotide –(cytosine-guanine-guanine)– repeats or r(CGG)] at an inherited fragile site on the long-arm of the X chromosome, where the *FMR1* gene *(34, 35, 36)* is located. Because this pre-mutation is dynamic, it can change in length and hence in severity from generation to generation, from person to person, and even within a given person. Patients with the FXS usually have an increased number of r(CGG) > 230 copies in the 5′-UTR of the *FMR1* gene, and this CpG-rich r(CGG) expansion region is often heavily methylated *(37)*. Such r(CGG) expansion and methylation leads to physical, neurocognitive and emotional characteristics linked to the inactivation of the *FMR1* gene and the deficiency of its protein product *(30)*. Two theories have been proposed to explain this *FMR1* methylation mechanism in FXS. First, Handa et al. *(25)* suggested that non-coding

RNA transcripts transcribed from the *FMR1* 5′-UTR r(CGG) expansion can fold into RNA hairpins and are further processed by *Dicer* RNases to form microRNA (miRNA)-like molecules directed against the *FMR1* expression. Second, Jin et al. *(38)* proposed that miRNA-mediated gene methylation occurs in the CpG regions of the *FMR1* r(CGG) expansion, which are targeted by hairpin RNAs derived from the 3′-UTR of the *FMR1* expanded allele transcript. The *Dicer*-processed hairpin RNA may trigger the formation of an RNA-induced initiator of transcriptional silencing (RITS) on the homologous r(CGG) sequences and leads to methylated heterochromatin repression of the *FMR1* locus in 99% of FXS patients.

Using a fish-compatible retrovector, *pGABAR2-rT*, to deliver an intronic transgene, *SpRNAi-rGFP*, containing the *fmr1* 5′-UTR r(CGG) expansion in zebrafish, Lin et al. *(30)* have shown that a special type of intronic miRNA, the repeat-associated miRNA (ramRNA), is highly produced in the diseased neurons with fragile X syndrome. One of the most abundant ramRNA species is located in the 25–45 nucleotide (nt) region of the fish *fmr1* 5′-UTR r(CGG) expansion (accession number NM152963), capable of causing the r(CGG)-methylation and inactivation of the *FMR1* gene. This ramRNA is specially characterized by its unique pre-miRNA structure consisting of (a) multiple loops and short stems in a relatively long hairpin region, (b) a nuclear entry signal (NES) motif to allow the re-entry of the mature ramRNA into the cell nucleus, and (c) a gene silencing motif to recruit the DNA methylation machinery. Recent FXS studies using this ramRNA-induced disease model have found that the formation of a synaptic connection was markedly reduced among the dendrites of the *FMR1*-deficient neurons, similar to the diseased hippocampal neurons in human FXS. *FMR1* deficiency often caused synapse deformity in the neurons essential for cognition and memory activities, damaging the activity-dependent synaptic neuron plasticity. Furthermore, the metabotropic glutamate receptor (mGluR)-activated long-term depression (LTD) was augmented after *FMR1* inactivation, suggesting that exaggerated LTD may be responsible for aspects of abnormal neuronal responses in FXS, such as autism. These findings support a novel disease model in which mature ramRNAs originating from the triplet repeat expansion of a gene can reversely bind back to the corresponding triplet repeat regions of the gene. More triplet repeats in the gene generate more mature ramRNAs. With accumulated ramRNAs binding to their targeted gene, DNA methylation then takes place in the triplet repeat regions, consequently inactivate the targeted gene expression. An animal model like this would be most suitable for studying TREDs involving epigenetic alterations.

4. Man-Made Intronic microRNA

To understand the disease caused by dysregulation of miRNAs, an artificial expression system is needed to recapitulate the function and mechanism of the miRNA in vitro and in vivo. The same strategy may be used to design and develop therapies for the disease. Previously, several vector-based RNAi expression systems have been developed, using type-III RNA polymerase (Pol-III)-directed transcription activities, to generate more stable RNAi efficacy in vitro as well as in vivo *(39, 40, 41, 42, 43)*. Although these approaches have succeeded in maintaining constant gene silencing efficacy in vivo, their delivery strategies failed to target a specific cell population due to the ubiquitous existence of Pol-III activity in all cell types. Moreover, the requirement of using Pol-III RNA promoters, e.g., U6 and H1, for small RNA expression causes another drawback. Because the read-through side-effect of Pol-III occurs on a short transcription template if without proper termination, large RNA products longer than the desired 18–25 base-pairs (bp) can be synthesized and then cause unexpected interferon cytotoxicity *(44,45)*. Such a problem can also result from the competitive conflict between the Pol-III promoter and another vector promoter (i.e., LTR and CMV promoters). Sledz et al. and us have found that high dosage of siRNAs (e.g., > 250 nM in human T cells) was able to cause strong cytotoxicity similar to that of long double-stranded dsRNAs *(46,47)*. This toxicity is due to the double-stranded structure of siRNAs and dsRNAs, which activates the interferon-mediated non-specific RNA degradation and programmed cell death through the signaling of PKR and 2-5A systems. It is known that interferon-induced protein kinase PKR can trigger cell apoptosis, while activation of the interferon-induced 2′,5′-oligoadenylate synthetase (2-5A) system leads to extensive cleavage of single-stranded RNAs (i.e., mRNAs) *(48)*. High siRNA/shRNA concentrations generated by the Pol-III-directed RNAi systems can also over-saturate the cellular microRNA pathway and thus cause global miRNA inhibition and cell death *(49)*. In contrast, a Pol-II-directed intronic miRNA expression system does not show these problems due to their precise regulation under cellular RNA splicing and nonsense-mediated decay (NMD) mechanisms *(50, 51, 52, 53, 54)*, which degrade excessive RNA accumulation to prevent potential cytotoxicity. For therapeutic purpose in vivo, the Pol-II-directed intronic miRNA expression system is likely a better solution than Pol-III-based siRNA/shRNA expression systems.

The intron-derived miRNA system is activated in a specific cell type under the control of a type-II RNA polymerases (Pol-II)-directed transcriptional machinery. To overcome the Pol-III mediated siRNA side-effects, we have successfully developed a novel Pol-II-based miRNA biogenesis strategy, employing intronic miRNA molecules

(2) to knock down more than 85% of selected oncogene function or viral genome replication *(22,23,47)*. Because of its flexibility in binding with partially complementary mRNA targets, miRNA can serve as an anti-cancer drug or vaccine to achieve a major breakthrough in the treatments of cancer polymorphisms and viral mutations. We are the first research group who discovered the biogenesis of miRNA-like precursors from the 5′-proximal intron regions of gene transcripts (pre-mRNAs) produced by the mammalian Pol-II. Depending on the promoter of the miRNA-encoded gene transcript, intronic miRNA is co-expressed with its encoding gene in the specific cell population, which activates the promoter and expresses the gene. It has been noted that a spliced intron was not completely digested into monoribonucleotides for transcriptional recycling since approximately 10–30% of the intron was found in the cytoplasm with a moderate half-life *(6,55)*. This type of miRNA generation has been found to rely on the coupled interaction of nascent Pol-II-mediated pre-mRNA transcription and intron excision, occurring within certain nuclear regions proximal to genomic perichromatin fibrils *(22,56, 57, 58)*. After Pol-II RNA processing and splicing excision, some of the intron-derived miRNA fragments can form mature miRNAs and effectively silence the target genes through the RNAi mechanism, while the exons of pre-mRNA are ligated together to form a mature mRNA for protein synthesis **(Fig. 19.2a)**. Because miRNAs are single-stranded molecules insensitive to PKR- and 2-5A-induced interferon systems, the utilization of this Pol-II-mediated miRNA generation can be safe in vitro and in vivo without the cytotoxic effects of dsRNAs and siRNAs. These findings indicate new functions for mammalian introns in intracellular miRNA generation and gene silencing, which can be used as a tool for analysis of gene functions and development of gene-specific therapeutics against cancers and viral infections.

Using artificial introns carrying hairpin-like miRNA precursors (pre-miRNA), we have successfully generated mature miRNA molecules with full capacity in triggering RNAi-like gene silencing in human prostate cancer LNCaP, human cervical cancer HeLa and rat neuronal stem HCN-A94-2 cells *(2,59,60)* as well as in zebrafish, chicken and mouse in vivo *(30,61)*. As shown in **Fig. 19.2b,** the artificial intron (SpRNAi) was co-transcribed within a precursor messenger RNA (pre-mRNA) by Pol-II and cleaved out of the pre-mRNA by RNA splicing. Then, the spliced intron containing a pre-miRNA structure was further processed into mature miRNAs capable of triggering RNAi-related gene silencing effects. Based on this artificial miRNA model, we have tested various pre-miRNA constructs, and observed that the production of intron-derived miRNA fragments was originated from the 5′-proximity of the intron sequence between the 5′-splice site and the branching point. These miRNAs were able to trigger strong

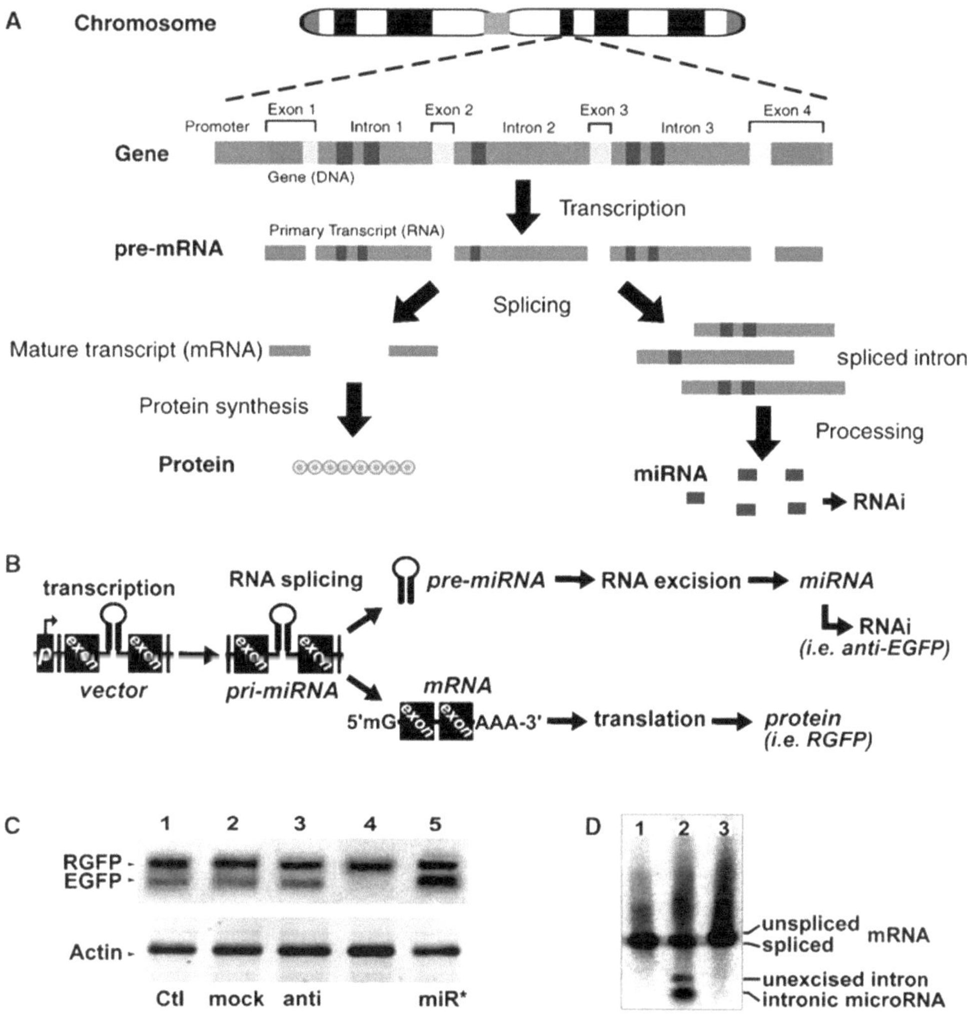

Fig. 19.2. Biogenesis and function of intronic miRNAs. **(a)** The native intronic miRNA is co-transcribed with a precursor messenger RNA (pre-mRNA) by Pol-II and cleaved out of the pre-mRNA by an RNA splicing machinery, spliceosome. The spliced intron with hairpin-like secondary structures is further processed into mature miRNAs capable of triggering RNAi effects, while the ligated exons become a mature messenger RNA (mRNA) for protein synthesis. **(b)** We designed an artificial intron containing pre-miRNA, namely SpRNAi, mimicking the biogenesis processes of the native intronic miRNAs. **(c)** When a designed miR-EGFP(280–302)–stemloop RNA construct was tested in EGFP-expressing Tg(UAS:gfp) zebrafishes, we detected a strong RNAi effect only on the target EGFP (lane 4). No detectable gene silencing effect was observed in other lanes from left to right: 1, blank vector control (Ctl); 2, miRNA–stemloop targeting HIV-p24 (mock); 3, miRNA without stemloop (anti); and 5, stemloop–miRNA* complementary to the miR-EGFP(280–302) sequence (miR*). The off-target genes such as vector RGFP and fish actin were not affected, indicating the high target specificity of miRNA-mediated gene silencing. **(d)** Three different miR-EGFP(280–302) expression systems were tested for miRNA biogenesis from left to right: 1, vector expressing intron-free RGFP, no pre-miRNA insert; 2, vector expressing RGFP with an intronic 5′-miRNA-stemloop-miRNA*-3′ insert; and 3, vector similar to the 2 construct but with a defected 5′-splice site in the intron. In Northern bolt analysis probing the miR-EGFP(280–302) sequence, the mature miRNA was released only from the spliced intron that resulted from the vector 2 construct in the cell cytoplasm.

suppression of genes possessing over 70% of complementarity to the miRNA sequences, whereas non-homologous miRNAs, i.e., empty intron without the pre-miRNA insert, intron with an off-target miRNA insert (negative control), and splicing-defective intron showed no silencing effects on the targeted gene. The same results can also be reproduced in the zebrafish directed against target EGFP expression **(Fig. 19.2c)**, indicating the consistent preservation of the intronic miRNA biogenesis system in vertebrates. Further, no effect was detected on off-target genes, such as RGFP and β-actin, suggesting the high specificity of miRNA-directed RNA interference (RNAi). We have confirmed the identity of the intron-derived miRNAs, which were sized about 18–27 base nt, approximately similar to the newly identified Piwi-interacting RNAs. Moreover, the intronic small RNAs isolated by guanidinium chloride ultracentrifugation can elicit strong, but short-term gene silencing effects on the homologous genes in the transfected cells, indicating their temporary RNAi effects. However, the long-term (>1 month) gene silencing effect that we observed in vivo occurs only in nuclear transfection of the Pol-II-mediated intronic miRNA system by retrovectors.

The components of the Pol-II-mediated SpRNAi system include several consensus nucleotide elements consisting of a 5′-splice site, a branch-point domain, a poly-pyrimidine tract and a 3′-splice site (**Fig. 19.3**). Additionally, a pre-miRNA insert sequence is placed within the artificial intron between the 5′-splice site and the branch-point domain. This portion of the intron would normally form a lariat structure during RNA splicing and processing. We currently know that spliceosomal U2 and U6 snRNPs, both helicases, may be involved in the unwinding and excision of the lariat RNA fragment into pre-miRNA; however, the detailed processing remains to be elucidated. Further, the SpRNAi contains a translation stop codon domain (T codon) in its 3′-proximal region to facilitate the accuracy of RNA splicing, which if present in a cytoplasmic mRNA,

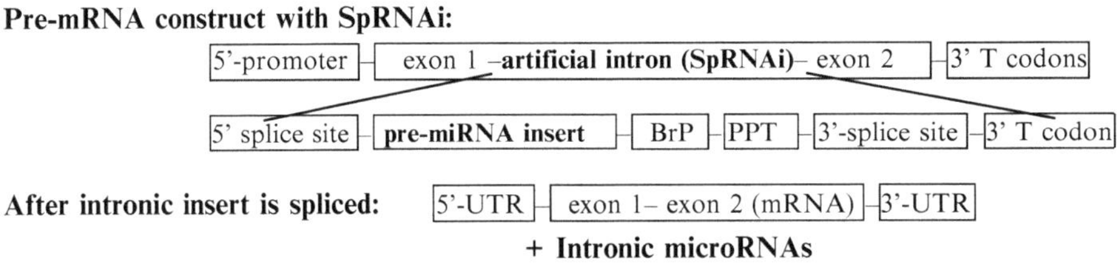

Fig. 19.3. Schematic construct of the artificial SpRNAi intron in a recombinant gene SpRNAi-RGFP for intracellular expression and processing. The components of the Pol-II-mediated SpRNAi system include several consensus nucleotide elements consisting of a 5′-splice site, a branch-point domain (BrP), a poly-pyrimidine tract (PPT), a 3′-splice site and a pre-miRNA insert located between the 5′-splice site and the BrP domain. The expression of the recombinant gene is under the regulation of either a mammalian Pol-II RNA promoter or a compatible viral promoter for cell-type-specific effectiveness. Mature miRNAs are released from the intron by RNA splicing and further Dicer processing.

would signal the diversion of a splicing-defective pre-mRNA to the nonsense-mediated decay (NMD) pathway and thus cause the elimination of any unspliced pre-mRNA in the cell. For intracellular expression of the SpRNAi, we need to insert the SpRNAi construct into the *Dra*II cleavage site of a red fluorescent membrane protein (RGFP) gene from mutated chromoproteins of coral reef *Heteractis crispa*. The cleavage of RGFP at its 208th nt site by the restriction enzyme *Dra*II generates an AG–GN nucleotide break with three recessing nucleotides in each end, which will form the 5′ and 3′ splice site, respectively after SpRNAi insertion. Because this intronic insertion disrupts the expression of functional RGFP, it becomes possible to determine the occurrence of intron splicing and RGFP-mRNA maturation through the appearance of red fluorescent emission around the membrane surface of the transfected cells. The RGFP also provides multiple exonic splicing enhancers (ESEs) to increase RNA splicing efficiency.

5. Strand-Specific Gene Silencing in Zebrafish

The foregoing establishes the fact that intronic miRNAs can be used as an effective strategy to silence specific target gene in vivo *(27)*. We firstly tried to resolve the structural design of pre-miRNA inserts for the best gene silencing effect and found out that a strong structural bias exists in the selection of a mature miRNA strand during assembly of the RNAi effector, RNA-induced gene silencing complex (RISC). RISC is a protein-RNA complex that directs either target gene transcript degradation or translational repression through the RNAi mechanism. Formation of siRNA duplexes has been reported to play a key role in assembly of the siRNA-associated RISC. The two strands of the siRNA duplex are functionally asymmetric, but assembly into the RISC complex is preferential for only one strand. Such preference is determined by the thermodynamic stability of each 5′-end base-pairing in the strand. Based on this siRNA model, the formation of miRNA and its complementary miRNA (miRNA*) duplexes was thought to be an essential step for the assembly of miRNA-associated RISC. If this were true, no functional bias would be observed in the stemloop of a pre-miRNA. Nevertheless, we observed that the stemloop of the intronic pre-miRNA was involved in the strand selection of a mature miRNA for RISC assembly in zebrafish. In these experiments, we constructed miRNA-expressing *SpRNAi-rGFP* vectors as previously described *(2)* and two symmetric pre-miRNAs, miRNA-stemloop-miRNA* [**1**] and miRNA*-stemloop-miRNA [**2**], were synthesized and inserted into the vectors, respectively. Both pre-miRNAs contained the same double-stranded stem arm region, which was directed against the EGFP nts 280–302 sequence. Because the intronic insert region of the SpRNAi-RGFP recombined gene is flanked with a *Pvu*I and an *Mlu*I restriction site at the 5′- and 3′-ends, respectively, the primary insert can be easily removed and replaced by various gene-specific inserts (e.g., anti-EGFP) possessing cohesive ends. By changing the pre-miRNA inserts directed against different gene transcripts, this intronic miRNA generation system provides a valuable tool for genetic and miRNA-associated research in vivo.

To determine the structural preference of the designed pre-miRNAs, we have isolated the zebrafish small RNAs by mirVana miRNA isolation columns (Ambion, Austin, TX) and then precipitated all potential miRNAs complementary to the target EGFP region by latex beads containing the target RNA sequence. One full-length miRNA identity, miR-EGFP(280–302), was verified to be active in the transfections of the 5′-miRNA-stemloop-miRNA*-3′ construct, as shown in the **Fig. 19.4a** (gray-shading sequences). Since the mature miRNA was detected only in the zebrafish transfected by the 5′-miRNA-stemloop-miRNA*-3′ construct, the miRNA-associated RISC tends to preferably interact with the construct [**2**] rather than the [**1**] pre-miRNA. The green fluorescent protein EGFP expression was constitutively driven by the β-actin promoter located in almost all cell types of the zebrafish, while **Fig. 19.4b** shows that transfection of the SpRNAi-RGFP vector into the Tg(UAS:gfp) zebrafish co-expressed the red fluorescent protein RGFP, serving as a positive indicator for the miRNA generation in the transfected cells. This approach has been successfully used in several mouse and human cell lines to show RNAi effects *(59,60)*. We applied the liposome-capsulated vector (total 60 μg) to fishes and found that the vector easily penetrated almost all tissues of two-week-old zebrafish larvae within 24 h, reaching fully systemic delivery of the miRNA effect. The indicator RGFP was detected in both of the fishes transfected

Fig. 19.4 (continued) pre-miRNA insert resulted in no gene silencing significance. **(d)–(g)** Silencing of endogenous β-catenin and noggin genes in chicken embryos. **(d)** The pre-miRNA construct and fast green dye mixtures were injected into the ventral side of chicken embryos near the liver primordia below the heart. **(e)** Northern blot analysis of extracted RNAs from chicken embryonic livers with anti-β-catenin miRNA transfections (lanes 4–6) in comparison with wild types (lanes 1–3) showed a more than 98% silencing effect on β-catenin mRNA expression, while the house-keeping gene, GAPDH, was not affected. **(f)** Liver formation of the β-catenin knockouts was significantly hindered (upper right 2 panels). Microscopic examination revealed a loose structure of hepatocytes, indicating the loss of cell-cell adhesion due to breaks in adherin junctions formed between β-catenin and cell membrane E-cadherin in early liver development. In severely affected regions, feather growth in the skin close to the injection area was also inhibited (lower right 2 panels). Immunohistochemistry staining of β-catenin protein expression (brown) showed a significant decrease in the feather follicle sheaths. **(g)** The lower beak development was increased by the mandibular injection of the anti-noggin pre-miRNA construct (down panel) in comparison to the wild type (up panel). Right panels showed bone (alizarin red) and cartilage (alcian blue) staining to demonstrate the out growth of bone tissues in the lower beak of the noggin knockout. Northern blot analysis (small windows) confirmed a ~60% decrease of noggin mRNA expression in the lower beak area.

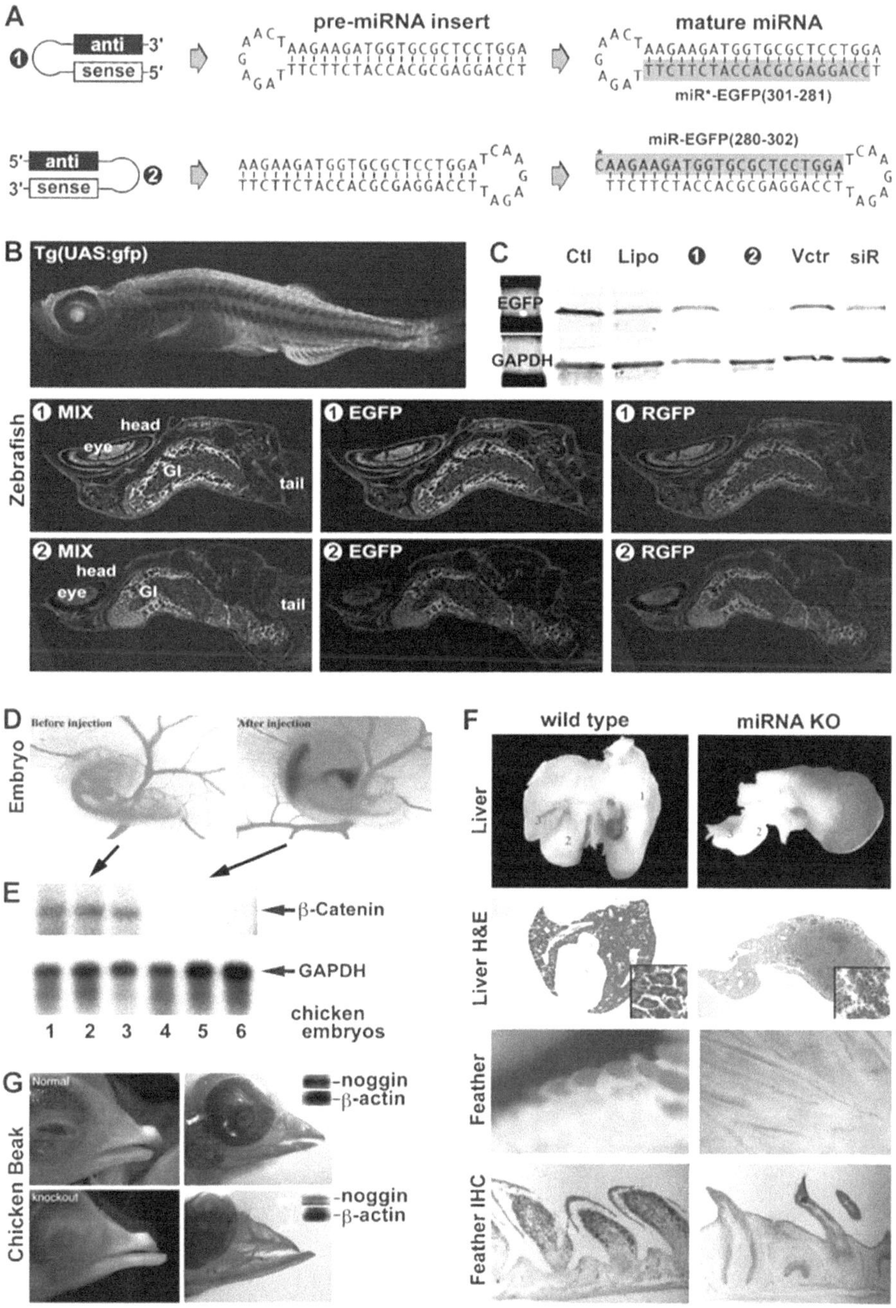

Fig. 19.4. Intronic miRNA-mediated gene silencing effects in vivo. **(a)–(c)** Different preferences of RISC assembly were observed by transfection of 5′-miRNA*-stemloop-miRNA-3′ **[1]** and 5′-miRNA-stemloop-miRNA*-3′ **[2]** pre-miRNA structures in zebrafish, respectively. **(a)** One mature miRNA, namely miR-EGFP(280/302), was detected in the **[2]**-transfected zebrafishes, whereas the **[1]**-transfection produced another kind of miRNA, miR*-EGFP(301–281), which was partially complementary to the miR-EGFP(280/302). **(b)** The RNAi effect was only observed in the transfection of the **[2]** pre-miRNA, showing less EGFP (green) expression in the transfectant **[2]** than **[1]**, while the miRNA indicator RGFP (red) was evenly present in all vector transfections. **(c)** Western blot analysis of the EGFP protein levels confirmed the specific silencing result of **(b)**. No detectable gene silencing was observed in fishes without (Ctl) and with liposome only (Lipo) treatments. The transfection of either a U6-driven siRNA vector (siR) or an empty vector (Vctr) without the designed

by either 5′-miRNA*-stemloop-miRNA-3′ or 5′-miRNA-stemloop-miRNA*-3′ pre-miRNA, whereas the silencing of target EGFP expression (green) was observed only in the fish transfected by the 5′-miRNA-stemloop-miRNA*-3′ pre-miRNA (**Fig. 19.4b** and **c**). The suppression level in the gastrointestinal (GI) tract was found to be less effective, probably due to the high RNase activity in this region. Because thermostability in the 5′ end of the siRNA duplexes resulting from both of the designed pre-miRNAs is the same, we suggest that the stemloop of pre-miRNA is involved in strand selection of mature miRNA during RISC assembly. Given that the cleavage site of Dicer in the stemarm determines the strand selection of mature miRNA *(62)*, the stemloop may function as a determinant for the recognition of a special cleavage site. Therefore, the different stemloop structures among various species may also provide a clue for the prevalence of native siRNAs in invertebrates but rarely in mammals.

6. Intronic piRNA-Mediated RNA Interference in Chicken

The in vivo model of chicken embryos has been widely utilized in research in developmental biology, signal transduction and flu vaccine development. We have successfully tested the feasibility of localized gene silencing in vivo using the intronic miRNA approach and also discovered that the interaction between pre-mRNA and perichromatin DNA may be essential for the intronic miRNA biogenesis. As an example, the β-catenin gene was selected because its products play a critical role in the biological development and ontogenesis *(63)*. The β-catenin is known to be involved in the growth control of skin and liver tissues in chicken embryos. The loss-of-function of β-catenin is lethal in transgenic animals. As shown in **Fig. 19.4d, e, f g**, experimental results demonstrated that the miRNAs derived from a long RNA–DNA hybrid construct (≥150 bp) were capable of inhibiting β-catenin gene expression in the liver and skin of developing chicken embryos. This mimics the mechanism by which interaction between the intronic miRNA precursor and genomic DNA may account for a part of its specific gene silencing effect *(22,23,51)*. We have demonstrated that the [P^{32}]-labeled DNA component of a long RNA–DNA duplex construct in cell nuclear lysates was intact during the effective period of miRNA-induced RNA interference (RNAi) phenomena, while the labeled RNA part was replaced by cold homologues and excised into small 18~27-nt RNA fragments in a 3-day incubation period *(22)*. Since the intronic miRNA generation relies on a coupled interaction of nascent Pol-II-directed pre-mRNA transcription and intron excision occurring proximal to genomic perichromatin fibrils, the above observation indicates that the pre-mRNA–perichromatin DNA interaction may facilitate new intronic miRNA generation by the Pol-II RNA transcription and excision for long-term gene silencing. Thus, Pol-II may actually function as an RNA-dependent RNA polymerase (RdRp) for producing more small miRNAs.

Recent studies have shown that the pre-mRNA–perichromatin DNA interaction results in the generation of Piwi-interacting RNAs (piRNA), which are similar to intronic miRNAs but distinct from other small double-stranded siRNAs and shRNAs by their relatively larger size (approximately 26–31 nucleotides), single-strandedness and strand-specificity as well as by the clustered arrangement of their origins *(58)*. The piRNA class of small RNAs is likely transcribed by an intracellular RNA polymerase, similar to RdRp, from the pre-mRNA–perichromatin DNA duplex region of a replicating cell genome during mitosis or meiosis. Mammalian type-II RNA polymerases (Pol-II) have been observed to possess the RdRp-like activities *(4,64,65)*. Nuclear transfection of long DNA-RNA duplex templates has also been shown to trigger piRNA-like gene silencing effects against viral infection and retrotransposon activity *(22)*. In *Drosophila* and zebrafish, Piwi proteins are recently found to be directly implicated in piRNA biogenesis to maintain transposon silencing in the germline genome *(24,66)*. This function may be conserved in mice as loss of Miwi2, a mouse Piwi homolog, leads to germline stem cell and meiotic defects correlated with increased transposon activity *(67)*. Because the RNAi effector of the piRNA-mediated gene silencing requires Piwi proteins rather than siRNA/shRNA-associated Dicer RNases, this suggests that the piRNA-mediated RNAi mechanism is slightly different from the siRNA/shRNA-mediated RNAi pathway.

In an effort to examine the pre-mRNA–perichromatin DNA interaction theory, we tested intracellular transfection of a long RNA–DNA hybrid construct containing a hairpin anti-β-catenin intronic pre-miRNA, which was directed against the central region of the β-catenin coding sequence (aa 306–644) with perfect complementarity. A perfectly complementary miRNA theoretically directs target mRNA degradation more efficient than translational repression. Using embryonic day 3 chicken embryos, a dose of 25 nM of the pre-miRNA construct was injected into the ventral body cavity, which is close to where the liver primordia would form (**Fig. 19.4d**). For efficient delivery into target tissues, the pre-miRNA construct was mixed with a liposomal transfection reagent (Roche Biomedicals, Indianapolis, IN). A 10% (v/v) fast green solution was concurrently added during the injection as a dye indicator. The mixtures were injected into the ventral side near the liver primordia below the heart using heat pulled capillary needles. After injection, the embryonic eggs were sealed with sterilized scotch tapes and incubated in a humidified incubator at 39–40°C until day 12 when the embryos were examined and photographed under a dissection microscope. Several malforma tions were observed, while the embryos still survived and there was no visible overt toxicity or overall perturbation of embryo development. The liver was the closest organ to the injection site and thus was most dramatically affected in its phenotypes. Other regions, particularly the skin close to the injection site, were also affected by the diffused miRNA effects. As shown in **Fig. 19.4e**, Northern blot analysis detecting the target β-catenin mRNA expression in the dissected livers showed that β-catenin expression in the wild-type livers remained normal (lanes 1–3), whereas those of the miRNA-treated samples was decreased dramatically (lanes 4–6). The miRNA silencing effect degraded more than 98% of β-catenin mRNA expression in embryonic chicken, but has no influence in the house-keeping gene GAPDH expression, indicating its high target specificity and very limited interferon-related cytotoxicity in vivo.

After 10 days of primordial injection with the anti-β-catenin pre-miRNA template, the embryonic chicken livers showed an enlarged and engorged first lobe, but the size of the second and third lobes of the livers were dramatically decreased **(Fig. 19.4f)**. Histological sections of normal livers showed hepatic cords and sinusoidal space with few blood cells. In the anti-β-catenin miRNA-treated embryos, the general architecture of the hepatic cells in lobes 2 and 3 remained unchanged; however, there were islands of abnormal regions in lobe 1. The endothelium development appeared to be defective and blood leaked outside of the blood vessels. Abnormal types of hematopoietic cells were also observed between the space of hepatocytes, particularly dominated by a population of small cells with round nuclei and scanty cytoplasm. In severely affected regions, hepatocytes were disrupted (**Fig. 19.4f,** small windows) and the diffused miRNA effect further inhibited the feather growth in the skin area close to the injection site. The results discussed above showed that the anti-β-catenin miRNA was very effective in knocking out the targeted gene expression at a very low dose of 25 nM and was effect over a long period of time (≥10 days). Further, the miRNA gene silencing effect appeared to be very specific as off-targeted organs appear to be normal, indicating that the small single-strand composition of miRNA herein possessed no overt toxicity. In another attempt to silencing noggin expression in the mandible beak area using the same approach **(Fig. 19.4g)**, it was observed that an enlarged lower beak morphology is reminiscent of those of BMP4-overexpressing chicken embryos reported previously *(68,69)*. Skeleton staining showed the outgrowth of bone and cartilage tissues in the injected mandible area (**Fig. 19.4g**, right panels) and Northern blot analysis further confirmed that about

60% of noggin mRNA expression was knocked out in this region (small windows). Since bone morphogenetic protein 4 (BMP4), a member of transforming growth factor-β (TGF-β) superfamily, is known to promote bone development and that noggin is an antagonist of BMP2/4/7 genes, it is not surprising to find out that our miRNA-mediated noggin knockouts created a morphological change, resembling the BMP4-overexpression results as previous reported in chicken and other avian models. Thus, the gene silencing in chicken by the pre-miRNA transfection presents a great potential of localized transgene-like approach in creating animal models for developmental biology research.

7. Localized RNA Interference Effects on Mouse Skin

To evaluate the efficacy and safety of intronic miRNA in animals, we have tested the vector-based intronic miRNA transfection in mice as previously described *(61)*. As shown in **Fig. 19.5**, patched albino (white) skins of melanin-knockout mice (W-9 black) were created by a succession of intracutaneous (i.c.) transduction of anti-*tyrosinase* (*Tyr*) pre-miRNA construct (50 μg) for 4 days (total 200 μg). Tyr, a type-I membrane protein and copper-containing enzyme, catalyzes the critical and rate-limiting step of tyrosine hydroxylation in the biosynthesis of melanin (black pigment) in skin and hair. After 13-day incubation, the expression of melanin has been blocked in the miRNA transfections due to a significant loss of its intermediates resulting from the anti-*Tyr* miRNA-triggered gene silencing effect. Contrarily, the blank control and the Pol-III (U6)-directed siRNA transfections presented normal black skin color under the same dosage. Northern blot analysis using RNA–PCR-amplified mRNAs from hair follicles showed a 76.1 ± 5.3% reduction of *Tyr* expression 2-day after the miRNA transfection, consistent with the immunohistochemical staining results from the same skin area, whereas mild, non-specific degradation of common gene transcripts was detected in the siRNA-transfected skins (seen from smearing patterns of both house-keeping control *GAPDH* and targeted *Tyr* mRNAs). Given that Grimm et al. *(49)* have recently reported that high siRNA/shRNA concentrations generated by the Pol-III-directed RNAi systems could over-saturate the cellular microRNA pathway and caused global miRNA dysregulation, the siRNA pathway may be incompatible with the native miRNA pathway in some tissues of mammals. Therefore, these findings have shown that the utilization of intronic miRNA expression systems provides a powerful new approach for transgenic animal generation and in vivo gene therapy. It was noted that non-targeted skin hair appears to be

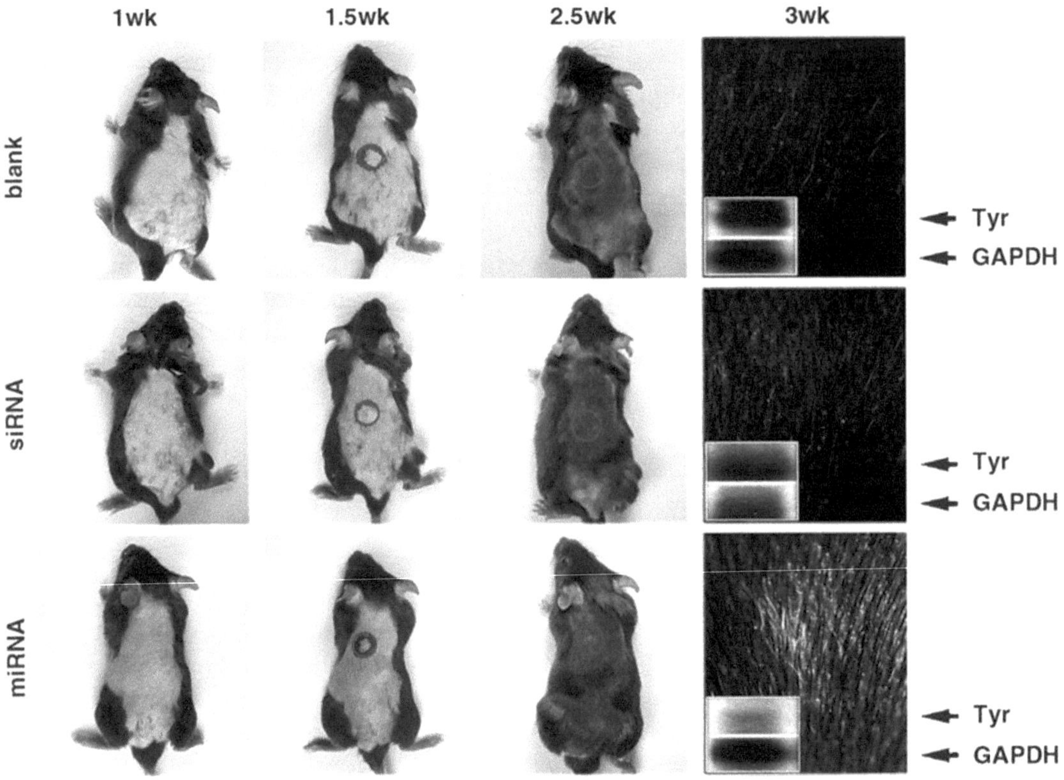

Fig. 19.5. In vivo effects of anti-tyrosinase miRNA on the mouse pigment production of local skins. Transfection of the miRNA induced strong gene silencing of *tyrosinase* (*Tyr*) mRNA expression but not house-keeping *GAPDH*, whereas that of U6-directed siRNA triggered mild non-specific RNA degradation of both *Tyr* and *GAPDH* gene transcripts. Since Tyr is an essential enzyme for black pigment melanin production, the success of gene silencing can be observed by a significant loss of the black color in mouse hairs. The red circles indicate the location of i.c. injections. Northern blot analysis of *Tyr* mRNA expression in local hair follicles confirmed the effectiveness and specificity of the miRNA-mediated gene silencing effect (small windows).

normal after miRNA transfection. This underscores the fact that the intronic miRNA is safe and effective in vivo. The results also indicated that the miRNA-mediated gene silencing effect is stable and efficient in knocking down the targeted gene expression over a relatively long period of time since the hair re-growth takes at least 10 days. Taken together, the intronic miRNA-mediated transgene approach may offer relatively safe, effective and long-term gene manipulation in animals, preventing the non-specific lethal effects of the conventional transgenic methods.

More recent advances of the utilization of intronic miRNA expression systems have been reported in mice. Chung et al. *(53)* have succeeded in expression of a cluster of polycistronic miRNAs using the Pol-II-mediated intronic miRNA expression system. A polycistronic miRNA cluster can be processed into multiple miRNAs via the cellular miRNA pathway. This new RNAi approach has a few advantages over the conventional Pol-III-mediated shRNA expression systems. First, Pol-II expression is tissue-specific, whereas Pol-III expression cannot. Second, Pol-II expression is compatible with the native microRNA pathway, while Grimm et al. *(49)* have reported some incompatibility between the Pol-III-mediated shRNA and Pol-II-mediated native miRNA pathways. Third, excessive RNA accumulation and toxicity can be prevented by the NMD mechanism of a cellular Pol-II-mediated intron expression system, but not a Pol-III exon-like expression system *(70)*. Lastly, one Pol-II is able to express a large size cluster (>10 kb) of polycistronic shRNAs, which can be further excised into multiple shRNAs via the native miRNA pathway, preventing the promoter conflict that often occurs in a multiple promoter vector system. For example, in many commercial U6-mediated shRNA expression systems, a self-inactivated vector promoter is often used to increase the U6 promoter activity.

8. Development of MicroRNA/piRNA-Based Gene Therapy

The following experimentations have shown the preliminary success of silencing exogenous retrovirus replication in an ex vivo cell model of patient-extracted $CD4^+$ T lymphocytes. The specific anti-HIV *SpRNAi-rGFP* vectors were designed to target the *gag-pol* region of about the nts +2113 to +2450 of HIV-1 genome. This region is relatively conserved and can serve as a good target for anti-HIV treatment *(71)*. The viral genes located in this target region include 3′-proximal $Pr55^{gag}$ polyprotein (i.e., matrix p17 + capsid p24 + nucleocapsid p7) and 5′-proximal $p66/p51^{pol}$ polyprotein (i.e., protease p10 + reverse transcriptase); all these components have critical roles in viral replication and infectivity. During the early infection phase, the viral reverse transcriptase transcribes the HIV RNA genome into a double-stranded cDNA sequence, which forms a pre-integration complex with the matrix, integrase and viral protein R (Vpr). This complex is then transferred to the cell nucleus and integrated into the host chromosome, consequently establishing the HIV provirus. We hypothesized that although HIV carries few reverse transcriptase and matrix proteins during its first entry into host cells, the co-suppression of $Pr55^{gag}$ and $p66/p51^{pol}$ gene expressions by miRNAs is expected to eliminate the production of infectious viral particles in the late infection phase. Silencing $Pr55^{gag}$ may prevent the assembly of intact viral particles due to the lack of matrix and capsid proteins, while suppression of protease in $p66/p51^{pol}$ can inhibit the maturation of several viral proteins. HIV expresses about nine viral gene transcripts, which encode at least 15 various proteins; thus, the separation of a polyprotein into individual functional proteins requires the viral protease activity. As shown in **Fig. 19.6**, this therapeutic approach has been reported to be feasible *(22,47)*.

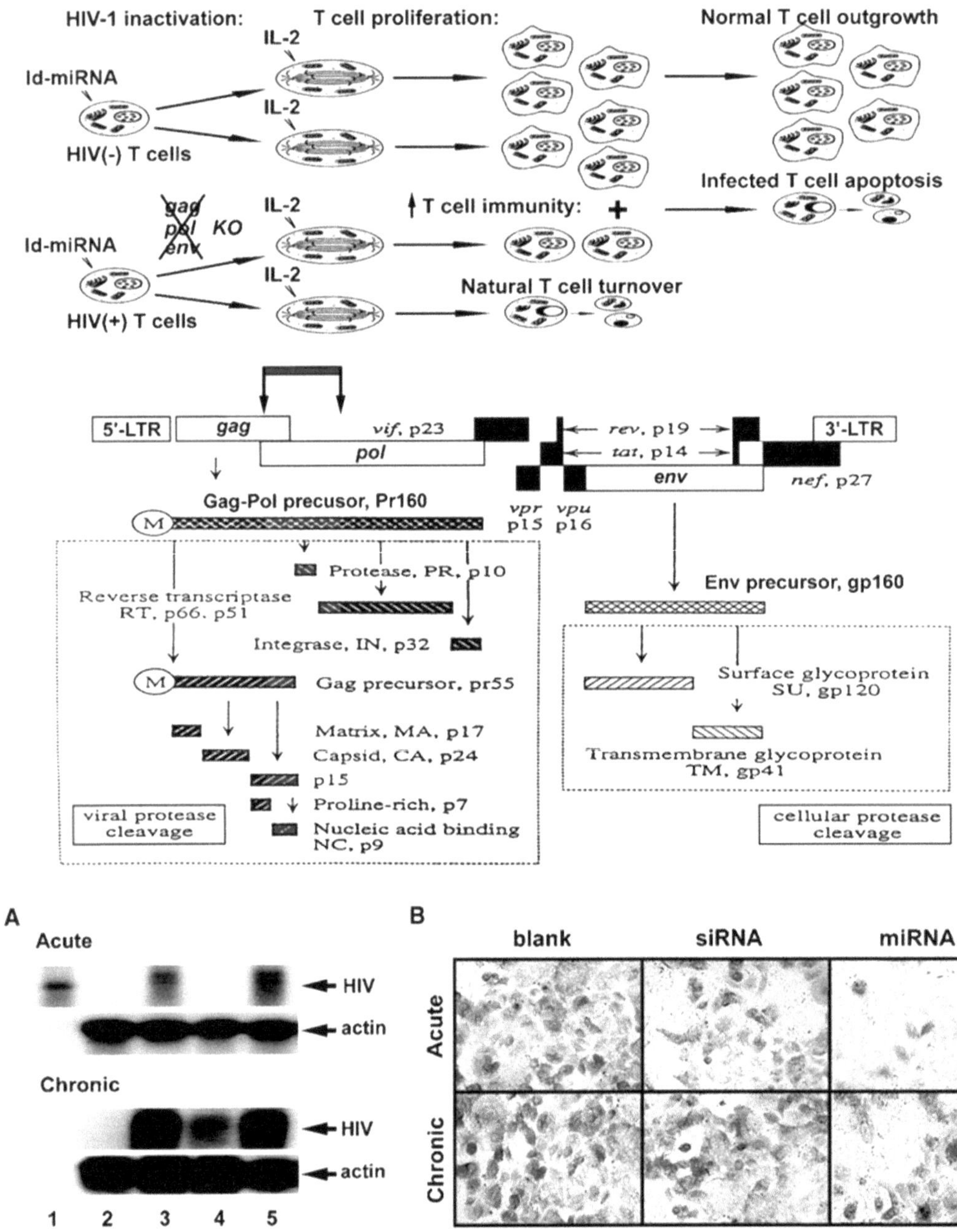

Fig. 19.6. Silencing of HIV-1 genome replication using anti-*gag/pro/pol* miRNA transfections into CD4+ T lymphocytes isolated from the acute and chronic phases of AIDS infections. **(a)** Northern blot analysis showed about 98 and 70% decreases of HIV genome in the acute and chronic infections after miRNA treatments (lanes 4), respectively. No effect was detected in the T cells transfected by miRNA* targeting the same *gag/pro/pol* region of the viral genome (lane 5). The size of pure HIV-1 provirus was measured about 9,700 nucleotide bases (lanes 1). RNA extracts from normal non-infected CD4+ Th lymphocytes were used as a negative control (lanes 2), whereas those from HIV-infected T cells were used as a positive control (lanes 3). **(b)** Immunostaining of HIV p24 marker confirmed the results of **(a)**. Since the ex vivo HIV-silenced T lymphocytes were resistant to any further infection by the same strains of HIV, they may be transfused back to the donor patient for eliminating HIV-infected cells.

The anti-HIV *SpRNAi-rGFP* vectors were tested in the $CD4^+$ T lymphocyte cells from HAART-treated, HIV-seropositive patients. Because only partial complementarity between miRNA and its target RNA is needed to trigger the gene silencing effect, it would be very advantageous to overcome the daunting challenge of high HIV mutations, which frequently generate new drug resistance to current small molecule drugs. Northern blot analysis in **Fig. 19.6a** demonstrated the ex vivo gene silencing effect of anti-HIV miRNA transfections ($n = 3$ for each set) on HIV-1 replication in $CD4^+$ T lymphocytes from both acute and chronic phase AIDS patients. In the acute phase (≤1 month), the 50 nM miRNA vector transfection degraded an average of 99.8% viral RNA genome (lane 4), whereas the same treatment knocked down only an average of 71.4 ± 12.8% viral genome replication in the chronic phase (about 2-year infection). Immunocytochemical staining of HIV p24 marker protein confirmed the results of Northern blot analysis (**Fig. 19.6b**). Sequencing analysis has revealed that at least two HIV-1b mutants in the acute phase and seven HIV-1b mutants in the chronic phase were found within the targeted HIV genome domain. It is likely that the higher genome complexity of HIV mutations in chronic infections is able to counteract the miRNA-mediated silencing efficacy. Transfection of 50 nM miRNA* vector homologous to the HIV-1 genome failed to induce any RNAi effect on viral genome, indicating the specificity of the miRNA effect (**Fig. 19.6b**, lane 5). Expression of cellular house-keeping gene, β-actin, was at a normal level and showed no interferon-induced non-specific RNA degradation. These results suggest that the designed anti-HIV *SpRNAi-rGFP* vector is highly specific and efficient in suppressing HIV-1 replication in the early infections. In conjunction with an intermittent interleukin-2 therapy *(47)*, we may stimulate the growth of non-infected $CD4^+$ T lymphocytes to eliminate the HIV-infected cells.

9. Conclusions

The consistent evidence of miRNA-induced gene silencing effects in zebrafish, chicken embryos, mouse stem cells and human diseases demonstrates the preservation of an ancient intron-mediated gene regulation system in eukaryotes. In these animal models, the intron-derived miRNAs determine the activation of RNAi-like gene silencing pathways. We herein provide the first time evidence for the biogenesis and function of intronic miRNAs in vivo. Given that natural evolution gives rise to more complexity and more variety of introns in higher animal and plant species for coordinating their vast gene expression volumes and interactions, dysregulation of these miRNAs due to intronic expansion or deletion will likely cause genetic diseases, such as myotonic dystrophy and fragile X mental retardation. Thus, gene expression produces not only gene transcripts for its own protein synthesis but also intronic miRNAs, capable of interfering with other gene expressions. Based on this concept, the expression of a gene results in gain-of-function of the gene and also loss-of-function of some other genes, which contain complementarity to the mature intronic miRNAs. An array of genes can swiftly and accurately coordinate their expression patterns with each other through the mediation of their intronic miRNAs, bypassing the time-consuming translation processes under quickly changing environments. Conceivably, intron-mediated gene regulation may be as important as the mechanisms by which transcription factors regulate the gene expression. It is likely that intronic miRNA is able to trigger cell transitions quickly in response to external stimuli without the tedious protein synthesis. Undesired gene products are reduced by both transcriptional inhibition and/or translational suppression via miRNA regulation. This could enable a rapid switch to a new gene expression pattern without the need to produce various transcription factors. This regulatory property of miRNAs may serve as one of the most ancient gene modulation systems before the emergence of proteins. According to the variety of microRNAs and the complexity of genomic introns, a thorough investigation of miRNA variants in the human genome will markedly improve the understanding of genetic diseases and also the design of miRNA-based drugs. Learning how to exploit such a novel gene regulation system in future therapy will be a forthcoming challenge.

References

1. Ambros, V., Lee, R.C., Lavanway, A., Williams, P.T., and Jewell, D. (2003). MicroRNAs and other tiny endogenous RNAs in C. elegans. *Curr Biol.* 13, 807–818.
2. Lin, S.L., Chang, D., Wu, D.Y., and Ying, S.Y. (2003). A novel RNA splicing-mediated gene silencing mechanism potential for genome evolution. *Biochem Biophys Res Commun.* 310, 754–760.
3. Rodriguez, A., Griffiths-Jones, S., Ashurst, J.L., and Bradley, A. (2004). Identification of mammalian microRNA host genes and transcription units. *Genome Res.* 14, 1902–1910.
4. Lin, S.L., Chuong, C.M., and Ying, S.Y. (2001). A Novel mRNA-cDNA interference phenomenon for silencing bcl-2 expression in human LNCaP cells. *Biochem. Biophys. Res. Commun.* 281, 639–644.
5. Ying, S.Y. and Lin, S.L. (2004). Intron-derived microRNAs–fine tuning of gene functions. *Gene* 342, 25–28.
6. Clement, J.Q., Qian, L., Kaplinsky, N., and Wilkinson, M.F. (1999). The stability and fate of a spliced intron from vertebrate cells. *RNA* 5, 206–220.
7. Parrish, S., Fleenor, J., Xu, S., Mello, C., and Fire, A. (2000). Functional anatomy of a dsRNA trigger: Differential requirement for the two trigger strands in RNA interference. *Mol. Cell* 6, 1077–1087.
8. Holen, T., Amarzguioui, M., Wiiger, M.T., Babaie, E., and Prydz, H. (2002). Positional effects of short interfering RNAs targeting the human coagulation trigger Tissue Factor. *Nucleic Acids Res.* 30, 1757–1766.

9. Hutvagner, G. and Zamore, P.D. (2002). A microRNA in a multiple-turnover RNAi enzyme complex. *Science* 297, 2056–2060.
10. Zeng, Y., Yi, R., and Cullen, B.R. (2003). MicroRNAs and small interfering RNAs can inhibit mRNA expression by similar mechanisms. *Proc. Natl. Acad. Sci. USA* 100, 9779–9784.
11. Hall, I.M., Shankaranarayana, G.D., Noma, K., Ayoub, N., Cohen, A., and Grewal, S.I. (2002). Establishment and maintenance of a heterochromatin domain. *Science* 297, 2232–2237.
12. Llave, C., Xie, Z., Kasschau, K.D., and Carrington, J.C. (2002). Cleavage of Scarecrow-like mRNA targets directed by a class of Arabidopsis miRNA. *Science* 297, 2053–2056.
13. Rhoades, M.W., Reinhart, B.J., Lim, L.P., Burge, C.B., Bartel, B., and Bartel, D.P. (2002). Prediction of plant microRNA targets. *Cell* 110, 513–520.
14. Lee, R.C., Feibaum, R.L., and Ambros, V. (1993). The C. elegans heterochromic gene lin-4 encodes small RNAs with antisense complementarity to lin-14. *Cell* 75, 843–854.
15. Reinhart, B.J., Slack, F.J., Basson, M., Pasquinelli, A.E., Bettinger, J.C., Rougvie, A.E., Horvitz, H.R., and Ruvkun, G. (2000). The 21-nucleotide let-7 RNA regulates developmental timing in Caenorhabditis elegans. *Nature* 403, 901–906.
16. Lau, N.C., Lim, L.P., Weinstein, E.G., and Bartel, D.P. (2001). An abundant class of tiny RNAs with probable regulatory roles in Caenorhabditis elegans. *Science* 294, 858–862.
17. Brennecke, J., Hipfner, D.R., Stark, A., Russell, R.B., and Cohen, S.M. (2003). Bantam encodes a developmentally regulated microRNA that controls cell proliferation and regulates the proapoptotic gene hid in Drosophila. *Cell* 113, 25–36.
18. Xu, P., Vernooy, S.Y., Guo, M., and Hay, B.A. (2003). The Drosophila microRNA Mir-14 suppresses cell death and is required for normal fat metabolism. *Curr Biol.* 13, 790–795.
19. Lagos-Quintana, M., Rauhut, R., Meyer, J., Borkhardt, A., and Tuschl, T. (2003). New microRNAs from mouse and human. *RNA* 9, 175–179.
20. Mourelatos, Z., Dostie, J., Paushkin, S., Sharma, A., Charroux, B., Abel, L., Rappsilber, J., Mann, M., and Dreyfuss, G. (2002). miRNPs: a novel class of ribonucleoproteins containing numerous micro-RNAs. *Genes Dev.* 16, 720–728.
21. Zeng, Y., Wagner, E.J., and Cullen, B.R. (2002). Both natural and designed micro RNAs can inhibit the expression of cognate mRNAs when expressed in human cells. *Mol Cell* 9, 1327–1333.
22. Lin, S.L., Chuong, C.M., and Ying, S.Y. (2001). D-RNAi (messenger RNA-antisense DNA interference) as a novel defense system against cancer and viral infections. *Curr. Cancer Drug Targets* 1, 241–247.
23. Lin, S.L. and Ying, S.Y. (2005). D-RNAi-Based Therapeutics. In Los, M. and Gibson, S. (Eds.), Apoptotic Pathways as Targets for Novel Therapies in Cancer & Other Diseases. Kluwer press, New York, pp. 275–296.
24. O'Donnell, K.A. and Boeke, J.D. (2007). Mighty Piwis defend the germline against genome intruders. *Cell* 129, 37–44.
25. Handa, V., Saha, T., and Usdin, K. (2003). The fragile X syndrome repeats form RNA hairpins that do not activate the interferon-inducible protein kinase, PKR, but are cut by Dicer. *Nucleic Acids Res.* 31, 6243–6248.
26. Krol, J., Fiszer, A., Mykowska, A., Sobczak, K., de Mezer, M., and Krzyzosiak, W.J. (2007). Ribonuclease dicer cleaves triplet repeat hairpins into shorter repeats that silence specific targets. *Mol Cell* 25, 575–586.
27. Lin, S.L., Chang, D.C., and Ying, S.Y. (2005). Asymmetry of intronic pre-microRNA structures in functional RISC assembly. *Gene* 356, 32–38.
28. Yi, R., Qin, Y., Macara, I.G., and Cullen, B.R. (2003). Exportin-5 mediates the nuclear export of pre-microRNAs and short hairpin RNAs. *Genes Dev.* 17, 3011–3016.
29. Lund, E., Guttinger, S., Calado, A., Dahlberg, J.E., and Kutay, U. (2004). Nuclear export of microRNA precursors. *Science* 303, 95–98.
30. Lin, S.L., Chang, S.J.E., and Ying, S.Y. (2006) First in vivo evidence of micro-RNA-induced fragile X mental retardation syndrome. *Mol Psychiatry* 11, 616–617.
31. Lee, Y.S., Nakahara, K., Pham, J.W., Kim, K., He, Z., Sontheimer, E.J., and Carthew, R.W. (2004). Distinct roles for Drosophila Dicer-1 and Dicer-2 in the siRNA/miRNA silencing pathways. *Cell* 117, 69–81.
32. Tang, G. (2005). siRNA and miRNA: an insight into RISCs. *Trends Biochem. Sci.* 30, 106–114.
33. Liquori, C.L., Ricker, K., Moseley, M.L., Jacobsen, J.F., Kress, W., Naylor, S.L., Day, J.W., and Ranum, L.P.W. (2001). Myotinic dystrophy type 2 caused by a CCTG expansion in intron 1 of ZNF9. *Science* 293, 864–867.
34. Fu, Y.H., Kuhl, D.P.A., Pizzuti, A., Pieretti, M., Sutcliffe, J.S., Richards, S., (1991).

Variation of the CGG repeat X site results in genetic instability. Resolution of Sherman paradox. *Cell* 67, 1047–1058.

35. Hagerman, R.J., Staley, L.W., O'Conner, R., Lugenbeel, K., Nelson, D., McLean, S.D., and Taylor, A. (1996). Learning-disabled males with a fragile X CGG expansion in the upper premutation size range. *Pediatrics* 97, 122–126.
36. Jin, P., Zarnescu, D.C., Zhang, F., Pearson, C.E., Lucchesi, J.C., Moses, K., and Warren, S.T. (2003). RNA-mediated neurodegeneration caused by the fragile X premutation rCGG repeats in Drosophila. *Neuron* 39, 739–747.
37. Genc, B., Muller-Hartmann, H., Zeschnigk, M., Deissler, H., Schmitz, B., Majewski, F., von Gontard, A., and Doerfler, W. (2000). Methylation mosaicism of 5′-(CGG)(n)-3′ repeats in fragile X, premutation and normal individuals. *Nucleic Acids Res.* 28, 2141–2152.
38. Jin, P., Alisch, R.S., and Warren, S.T. (2004). RNA and microRNAs in fragile X mental retardation. *Nat Cell Biol.* 6, 1048–1053.
39. Miyagishi, M. and Taira, K. (2002). U6 promoter-driven siRNAs with four uridine 3′ overhangs efficiently suppress targeted gene expression in mammalian cells. *Nat Biotechnol.* 20, 497–500.
40. Lee, N.S., Dohjima, T., Bauer, G., Li, H., Li, M.J., Ehsani, A., Salvaterra, P., and Rossi, J. (2002). Expression of small interfering RNAs targeted against HIV-1 rev transcripts in human cells. *Nat Biotechnol.* 20, 500–505.
41. Paul, C.P., Good, P.D., Winer, I., and Engelke, D.R. (2002). Effective expression of small interfering RNA in human cells. *Nat Biotechnol.* 20, 505–508.
42. Xia, H., Mao, Q., Paulson, H.L., and Davidson, B.L. (2002). siRNA-mediated gene silencing in vitro and in vivo. *Nat Biotechnol.* 20, 1006–1010.
43. McCaffrey, A.P., Meuse, L., Pham, T.T., Conklin, D.S., Hannon, G.J., and Kay, M.A. (2002). RNA interference in adult mice. *Nature* 418, 38–39.
44. Gunnery, S., Ma, Y., and Mathews, M.B. (1999). Termination sequence requirements vary among genes transcribed by RNA polymerase III. *J Mol Biol.* 286, 745–757.
45. Schramm, L. and Hernandez, N. (2002). Recruitment of RNA polymerase III to its target promoters. *Genes Dev.* 16, 2593–2620.
46. Sledz, C.A., Holko, M., de Veer, M.J., Silverman, R.H., and Williams, B.R. (2003). Activation of the interferon system by short-interfering RNAs. *Nat Cell Biol.* 5, 834–839.
47. Lin, S.L. and Ying, S.Y. (2004). Combinational therapy for HIV-1 eradication and vaccination. *Int. J. Oncol.* 24, 81–88.
48. Stark, G.R., Kerr, I.M., Williams, B.R., Silverman, R.H., and Schreiber, R.D. (1998). How cells respond to interferons. *Annu. Rev. Biochem.* 67, 227–264.
49. Grimm, D., Streetz, K.L., Jopling, C.L., Storm, T.A., Pandey, K., Davis, C.R., Marion, P., Salazar, F., and Kay, M.A. (2006). Fatality in mice due to oversaturation of cellular microRNA/short hairpin RNA pathways. *Nature* 441, 537–541.
50. Lin, S.L. and Ying, S.Y. (2004). New drug design for gene therapy–Taking advantage of introns. *Lett Drug Des. Discov.* 1, 256–262.
51. Lin, S.L. and Ying, S.Y. (2004) Novel RNAi therapy–Intron-derived microRNA drugs. *Drug Des. Rev.* 1, 247–255.
52. Zhou, H.S.L., Xia, X.G., and Xu, Z. (2005). An RNA polymerase II construct synthesizes short hairpin RNA with a quantitative indicator and mediates high efficient RNAi. *Nucleic Acid Res.* 33, e62.
53. Chung, K.H., Hart, C.C., Al-Bassam, S., Avery, A., Taylor, J., Patel, P.D., Vojtek, A.B., and Turner, D.L. (2006). Polycistronic RNA polymerase II expression vectors for RNA interference based on BIC/miR-155. *Nucleic Acid Res.* 34, e53.
54. Xia, X.G., Zhou, H., Samper, E., Melov, S., and Xu, Z. (2006). Pol II-expressed shRNA knocks down Sod2 gene expression and causes phenotypes of the gene knockout in mice. *PLoS Genet.* 2, e10.
55. Nott, A., Meislin, S.H., and Moore, M.J. (2003). A quantitative analysis of intron effects on mammalian gene expression. *RNA* 9, 607–617.
56. Zhang, G., Taneja, K.L., Singer, R.H., and Green, M.R. (1994). Localization of pre-mRNA splicing in mammalian nuclei. *Nature* 372, 809–812.
57. Ghosh, S. and Garcia-Blanco, M.A. (2000). Coupled in vitro synthesis and splicing of RNA polymerase II transcripts. *RNA* 6, 1325–1334.
58. Bateman, J.R. and Wu, C.T. (2007). DNA replication and models for the origin of piRNAs. *Bioessays* 29, 382–385.
59. Lin, S.L., Chang, D.C., and Ying, S.Y. (2007). Hyaluronan stimulates transformation of androgen-independent prostate cancer. *Carcinogenesis* 28, 310–320.

60. Lin, S.L., and Ying, S.Y. (2006). Gene silencing in vitro and in vivo using intronic microRNAs. *Methods Mol Biol.* 342, 295–312.

61. Lin, S.L., Chang, S.J.E., and Ying, S.Y. (2006). Transgene-like animal model using intronic microRNAs. *Methods Mol Biol.* 342, 321–334.

62. Lee, Y., Ahn, C., Han, J., Choi, H., Kim, J., Yim, J., Lee, J., Provost, P., Radmark, O., Kim, S., and Kim, V.N. (2003). The nuclear RNase III Drosha initiates microRNA processing. *Nature* 425, 415–419.

63. Butz, S. and Larue, L. (1995). Expression of catenins during mouse embryonic development and in adult tissues. *Cell Adhes Commun.* 3, 337–352.

64. Filipovska, J. and Konarska, M.M. (2000). Specific HDV RNA-templated transcription by pol II in vitro. *RNA* 6, 41–54.

65. Modahl, L.E., Macnaughton, T.B., Zhu, N., Johnson, D.L., and Lai, M.M. (2000). RNA-Dependent replication and transcription of hepatitis delta virus RNA involve distinct cellular RNA polymerases. *Mol Cell Biol.* 20, 6030–6039.

66. Vagin, V.V., Sigova, A., Li, C., Seitz, H., Gvozdev, V., and Zamore, P.D. (2006). A distinct small RNA pathway silences selfish genetic elements in the germline. *Science* 313, 320–324.

67. Carmell, M.A., Girard, A., van de Kant, H.J., Bourc'his, D., Bestor, T.H., de Rooij, D.C., and Hamnon, G.J. (2007). MIWI2 is essential for spermatogenesis and repression of transposons in the mouse male germline. *Dev. Cell* 12, 503–514.

68. Abzhanov, A., Protas, M., Grant, B.R., Grant, P.R., and Tabin, C.J. (2004). Bmp4 and morphological variation of beaks in Darwin's finches. *Science* 305, 1462–1465.

69. Wu, P., Jiang, T.X., Suksaweang, S., Widelitz, R.B., and Chuong, C.M. (2004). Molecular shaping of the beak. *Science* 305, 1465–1466.

70. Sheth, U. and Parker, R. (2006). Targeting of aberrant mRNAs to cytoplasmic processing bodies. *Cell* 125, 1095–1109.

71. Kovacs, J.A., Vogel, S., Albert, J.M., Falloon, J., Davey, R.T. Jr., Walker, R.E., Polis, M.A., Spooner, K., Metcalf, J.A., Baseler, M., Fyfe, G., and Lane, H.C. (1996). Controlled trial of interleukin-2 infusions in patients infected with the human immunodeficiency virus. *N Engl. J. Med.* 335, 1350–1356.

Chapter 20

Emergence of a Complex Relationship Between HIV-1 and the microRNA Pathway

Dominique L. Ouellet, Isabelle Plante, Corinne Barat, Michel J. Tremblay, and Patrick Provost

Abstract

Recent experimental evidences support the existence of an increasingly complex and multifaceted interaction between viruses and the microRNA-guided RNA silencing machinery of human cells. The discovery of small interfering RNAs (siRNAs), which are designed to mediate cleavage of specific messenger RNAs (mRNAs), prompted virologists to establish therapeutic strategies based on siRNAs with the aim to suppress replication of several viruses, including human immunodeficiency virus type 1 (HIV-1). It has been appreciated only recently that viral RNAs can also be processed endogenously by the microRNA-generating enzyme Dicer or recognized by cellular miRNAs, in processes that could be viewed as an adapted antiviral defense mechanism. Known to repress mRNA translation through recognition of specific binding sites usually located in their 3′ untranslated region, miRNAs of host or viral origin may exert regulatory effects towards host and/or viral genes and influence viral replication and/or the host response to viral infection. This article summarizes our current state of knowledge on the relationship between HIV-1 and miRNA-guided RNA silencing, and discusses the different aspects of their interaction.

Key words: HIV-1, RNA silencing, microRNA, small interfering RNA, gene expression.

1. Biology of microRNAs

1.1. microRNAs as Key Regulators of Gene Expression

microRNAs (miRNAs) are short ~21 to 24-nucleotide (nt) RNA species expressed in most eukaryotes and are known as key regulators of gene expression that act through imperfect base pairing with their target messenger RNA (mRNA) *(1,2)*. According to the latest update of miRBase (release 11.0, April 2008), the repository of miRNA data on the Web, more than 678 human

M. Sioud (ed.), *Methods in Molecular Biology, siRNA and miRNA Gene Silencing, vol. 487*

DOI: 10.1007/978-1-60327-547-7_20

miRNA sequences have been identified so far, among a total of 6396 entries *(3,4)*. A recent study suggested that miRNAs may regulate up to 92% of the genes in humans *(5)*!

The panoply of small gene regulatory RNAs has recently gained in complexity with the discovery in eukaryotic organisms of additional classes distinct from miRNAs, such as repeat-associated small interfering RNAs (rasiRNAs) *(6)*, tiny noncoding RNAs (tncRNAs) *(7)* and Piwi-interacting RNAs (piRNAs) *(6)*.

1.2. The Endogenous microRNA-Based RNA Silencing Machinery

Encoded by the genome of most eukaryotes examined so far, miRNA genes are transcribed by RNA polymerase (pol) II into stem-loop structured primary miRNAs (pri-miRNAs). Harboring a 5′m7G cap and a 3′ poly(A) tail *(8,9)*, these pri-miRNAs are then trimmed into ~60–70-nt miRNA precursors (pre-miRNAs) (*see* **Fig. 20.1a**) by the nuclear ribonuclease (RNase) III Drosha *(10)*, acting in concert with the DiGeorge syndrome critical region 8 (DGCR8) protein within the microprocessor complex *(11–14)*. The pre-miRNAs are subsequently exported to the cytoplasm via Exportin-5 *(15–18)* and the base of their stem recognized by the PAZ domain of Dicer *(19)*. Acting as an intramolecular dimer, the RNase IIIa and IIIb domains cleave the stem at the base of the loop to generate miRNA:miRNA* duplexes *(19, 20, 21, 22)*. Dicer was recently shown to operate with the transactivating response RNA-binding protein (TRBP) *(23)* within a pre-miRNA processing complex *(24,25)*. Following a strand selection and separation step, which is based on the thermodynamic stability of the RNA duplex *(26)*, the miRNA strand (~21 to 24-nt) with the least stable 5′ end pairing (called the guide strand) is incorporated into effector miRNA-containing ribonucleoprotein (miRNP) complexes, containing Argonaute 2 (Ago2), TRBP and Dicer *(25)*, and guiding them towards specific messenger RNAs (mRNAs). The opposite miRNA* strand (also called passenger strand) is encountered much less frequently and is presumably degraded *(27)*. miRNA assembly on specific mRNA sequences may be facilitated by the fragile X mental retardation protein, which can accept and use miRNAs derived from Dicer *(28)*. The targeted mRNA will be primarily subjected to translational repression, although mRNAs containing partial miRNA complementary sites may also be targeted for degradation in vivo *(29)*. These regulatory events may occur at specific cytoplasmic foci referred to as processing bodies (P-bodies) *(30,31)*, or GW182-containing bodies (GW-bodies) *(32)*, which are formed as a consequence of the presence of miRNAs *(33)*. P-bodies are enriched in proteins involved in RNA-mediated gene silencing, such as Ago2 *(30)*, mRNA degradation *(34)* and nonsense-mediated mRNA decay *(35,36)*

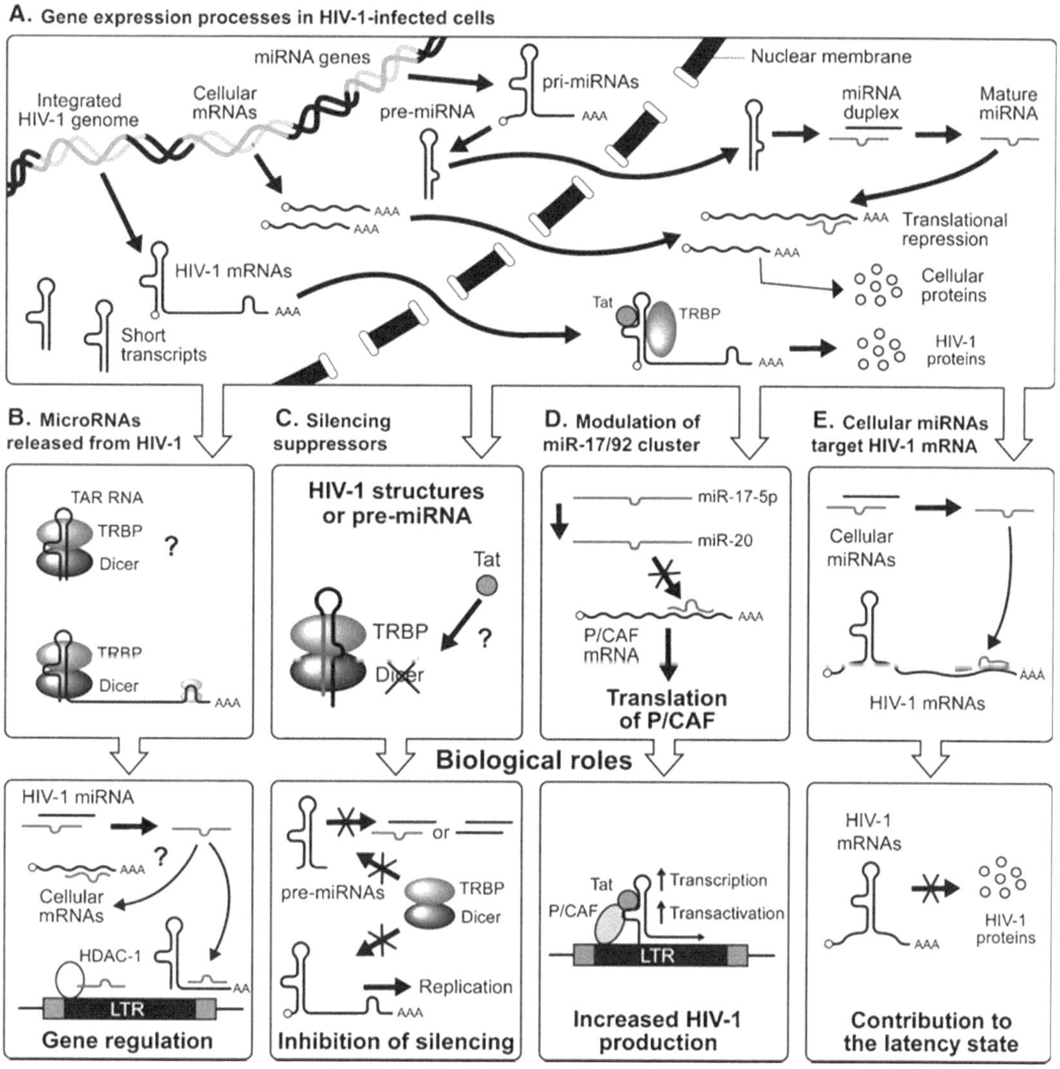

Fig. 20.1. Schematic representation of the intricate relationship between the miRNA silencing pathway and HIV-1. (**a**) In HIV-1-infected cells, full-length and short viral RNA transcripts are produced from the integrated HIV-1 genome. In these cells, like in normal non-infected cells, miRNA genes are transcribed into primary miRNAs (pri-miRNAs), which are trimmed into miRNA precursors (pre-miRNAs) in the nucleus. Pre-miRNAs are then exported to the cytoplasm where they are cleaved by Dicer to generate miRNA:miRNA* duplexes. Following a strand selection and separation step, the mature miRNA is incorporated into effector complexes to mediate recognition and translational repression of specific cellular mRNAs, as reviewed recently *(107)*. (**b**) Secondary structures in HIV-1 mRNAs may themselves be recognized and processed by Dicer into viral miRNAs or siRNAs, which could act on gene regulation in the cell, i.e., by restricting LTR-driven transcription, recruit HDAC-1 to the LTR or repress cellular mRNAs. (**c**) TAR recognition by the Dicer•TRBP complex could be hampered by the viral transactivating protein Tat, which has been suggested to inhibit Dicer activity and could lead to suppression of silencing. This inhibitory effect of Tat remains unclear. (**d**) In infected cells, HIV-1 could alter host gene expression through the modulation of miRNA expression. For instance, downregulation of the miR-17/92 cluster, via miR-17-5p and miR-20, has been reported to increase expression of a Tat cofactor, the P/CAF protein. This regulation could lead to an enhanced transactivation of the TAR element and could contribute to activating latent reservoirs. (**e**) A 1.2-kb fragment present in the 3′UTR of almost all HIV 1 mRNAs can be recognized by cellular miRNAs with a negative impact on viral protein production in $CD4^+$ T cells, following a process that could contribute to keep the virus in its latency phase.

1.3. Biological Roles of microRNAs

miRNAs have been shown to control various processes, such as cell proliferation and apoptosis in flies, and hematopoietic cell differentiation in mice *(37)*. Their biological role is linked to their ability to initially repress translation of specific mRNAs, although a case of enhancement of translation mediated by miRNAs/miRNPs during the cell cycle was recently reported *(38)*. The action of miRNAs is accomplished through recognition of specific miRNA binding sites usually located in the mRNA 3′ untranslated region (UTR), thereby inhibiting translation initiation *(39)*. Because recognition by miRNAs is mainly based on imperfect sequence complementarity, the identification of their physiological mRNA targets is difficult to predict and is rather arduous. Characterization of a few experimentally validated miRNA:mRNA interactions (e.g., let-7 and lin-41) *(40)* allowed to establish a context in which this interaction is favored. For example, the critical miRNA:mRNA pairing region, referred to as the "miRNA seed," involves nucleotides 2–8 of the miRNA in the 5′ to 3′ orientation. Although it appears to be less important, pairing of the miRNA 3′ region may compensate a weaker binding of the 5′ region *(40)*. A better understanding of mRNA recognition by miRNAs helped develop bioinformatic approaches that have proven to be instrumental for identifying potential miRNA targets and initiating characterization of miRNA function.

1.4. A Role for Small RNAs in Antiviral Host Defenses

In addition to fulfilling important gene regulatory functions in their eukaryotic hosts, small RNAs may also help defend against invasion of the host genome by RNAs of foreign origin, such as viruses. Initial evidences for such a role came from observations made by plant biologists. Indeed, while investigating the natural antiviral defense mechanism known as posttranscriptional gene silencing (PTGS), Hamilton and Baulcombe detected the presence of antisense viral RNA of ~25-nt in virus-infected plants by Northern blot *(41)*. The authors noted that these small RNAs, which were later found to originate from viral double-stranded RNA (dsRNA) processing by Dicer or DICER-LIKE 1 (DCL1 in *Arabidopsis*) *(42)*, were long enough to convey sequence specificity and suggested their probable role in limiting virus infection in plants. The antiviral function of small RNAs and their biosynthetic machinery in plants has recently been extended to insects, nematodes *(43,44)* and mammals.

2. Biology of HIV-1

2.1. HIV-1 Life Cycle

HIV-1 is an enveloped virus that binds cell receptor CD4 and most common coreceptors C-X-C motif receptor 4 (CXCR4) in lymphocytes or C-C motif receptor 5 (CCR5) in macrophages. For more details about HIV-1 life cycle, please refer to recent reviews

(45, 46, 47). Viral gp120 is responsible for binding, whereas gp41 is essential for fusion of the viral particle to the cellular membrane. The genome of HIV-1 is composed of two identical single-stranded 9.6 kb RNA molecules. After cell entry and uncoating, the viral genetic material is reverse transcribed into cDNA by the HIV-1 viral reverse transcriptase (RT) enzyme and integrated as double-strand DNA into the host genome before directing viral gene expression. HIV-1 relies heavily on the cellular transcription and translation machineries for the synthesis of viral genomic RNA and proteins. As partly illustrated in **Fig. 20.1a**, the full-length RNA is transcribed by RNA pol II and serves both as genomic RNA and as a template for expression of the structural proteins Gag and Gag/Pol. The singly spliced 4 kb mRNA encodes Vif, Vpr, Vpu and Env and the fully spliced 2 kb mRNAs encode Tat, Rev and Nef. Some HIV-1 RNA transcripts are produced from the DNA bipartite element, known as the inducer of shorts transcripts (IST), located downstream of the start site of transcription into the long terminal repeat *(48,49)*. The subpopulation of non-polyadenylated short transcripts have heterogenous 3′ nucleotide ends situated around position +60 and contain the Trans-Activation Responsive (TAR) element *(49)*. All these RNA species adopt complex and dynamic secondary structures that are reminiscent of pre-miRNAs and thus could potentially be targeted by the host miRNA-guided RNA silencing machinery. Although the high variability of the HIV-1 genome gives rise to a multitude of RNA folding possibilities, a number of structures are very well conserved because of their essential function in the virus life cycle. Among these are the dimerization sites TAR region and the Rev-Responsive Element (RRE). These elements all have in common dsRNA structures (dimerization or stem-loop), which may potentially be processed into miRNAs. Whether these structures are recognized and processed by the Drosha•DGCR8 or Dicer•TRBP complexes remains to be determined.

2.2. HIV-1 Dimerization Initiation Site

In virus particles, the genome consists of two identical molecules of RNA that are non-covalently linked near their 5′ ends. The dimerization process involves a series of conformational changes of the untranslated leader region in which a first structure referred as the kissing-loop complex is rearranged into a more extended molecular duplex *(50)* that can be targeted by synthetic molecules for potential inhibition of the dimerization initiation site (DIS). However, these conformations possess a number of dimer RNA molecules and stem-loop structures that could potentially be recognized by Drosha and/or Dicer, whereby the latter preferentially cleaves dsRNAs at their termini *(22)*, to generate miRNAs.

2.3. HIV-1 TAR

The TAR region is a 59 nt stem bulge loop structure located at the 5′ end of all spliced and unspliced HIV-1 transcripts found in the nucleus and cytoplasm, which is essential for efficient viral transcription. TAR is a short leader RNA structure targeted by the viral transactivating protein Tat that is known to act at the RNA level to enhance virus gene expression by more than 100-fold *(51)*. Upon binding to the trinucleotide pyrimidine-rich bulge of TAR, Tat recruits the positive transcription-elongation factor b (P-TEFb), a complex made of cyclin T1 and the cyclin dependent kinase 9 (CDK9), to the initiation complex. The CDK9 then phosphorylates the C-terminal domain (CTD) of the RNA polymerase II, which promotes the formation of an efficient elongation transcription complex *(52)*. Apart from this nuclear-based function, TAR is also important after RNA export to the cytoplasm since it inhibits translation by two mechanisms, i.e., through a direct block of translation initiation by its secondary structure and by activation of the dsRNA binding protein kinase R (PKR) which, in turn, phosphorylates the eukaryotic initiation factor 2 alpha (eIF2a), leading to an arrest of translation initiation. Both of these negative effects are alleviated by TRBP, which inhibits PKR *(53)* and releases the translational block due to the TAR structure *(54)*. (For more details about HIV-1 TAR RNA and Tat protein, please refer to these reviews. *(52,55,56)*)

2.4. HIV-1 RRE

The RRE domain is a large RNA structure present in all 9 kb and 4 kb RNAs located within the Env intron. Through its interaction with RRE, Rev protein is responsible for the nuclear export of these unspliced or singly spliced RNAs. In the absence of Rev, these RNAs are sequestered in the nucleus and only the multiply spliced 2 kb RNA encoding the regulatory proteins Tat, Rev and Nef are exported in the cytoplasm and translated. The interaction between Rev and RRE promotes the transition between this early phase of the viral life cycle to the late phase where structural proteins are produced *(57)*. The RRE is a 351 nt complex structure that comprises five stem-loop structures on which Rev assembles as a multimeric complex *(58,59)*. This structure may resemble pri-miRNAs, which are often composed of multiple stem-loop structures, and represent a very good candidate for the source of viral miRNA. Although it is interesting to note that, like TAR, RRE interacts with TRBP *(53)*, the latter does not appear to influence the effect of Rev on RRE-containing sequence.

3. Relationship Between HIV-1 and RNA Silencing

3.1. Small RNAs Directed Against HIV-1

Small interfering RNAs (siRNAs) are synthetic 21-nt RNA duplexes that have been designed to mimic the endogenous miRNAs or Dicer-generated siRNAs. Their efficiency in downreg-

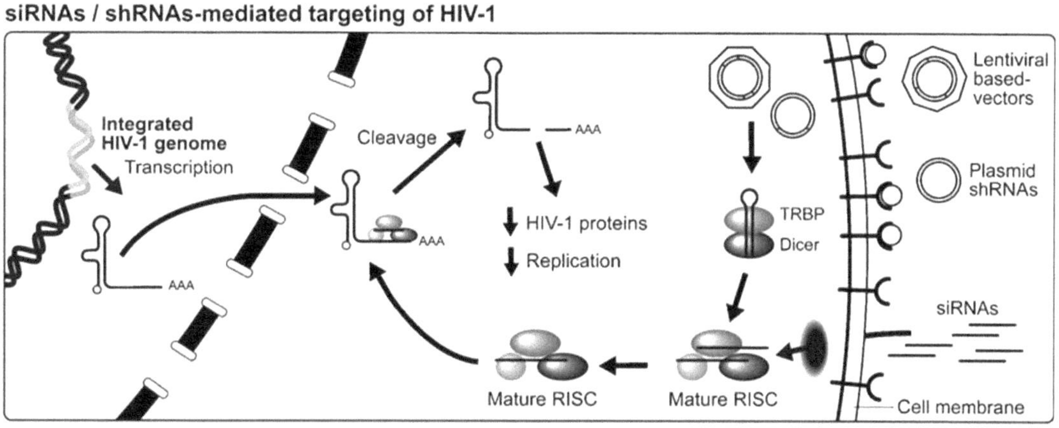

Fig. 20.2. siRNAs and shRNAs expressed from viral or mammalian expression vectors can mediate cleavage of HIV-1 RNAs and inhibit viral protein synthesis.

ulating expression of specific genes in cultured mammalian cells *(60)* established the basis for the use of siRNAs or RNA interference (RNAi) technology in therapeutics. This strategy was exploited successfully to inhibit the replication of several viruses, including HIV-1 (*see* **Fig. 20.2**, and Chap. 16). Approaches based on siRNA targeting of host genes have been also used to restrict HIV-1 production *(61)*. This anti-HIV-1 therapy is one of the RNA-based strategies that include antisenses, ribozymes and aptamers. Some of them are currently being tested in clinical trials *(62)*.

Short hairpin RNA (shRNA) precursors are also used to trigger RNAi against HIV-1. shRNAs are produced either from mammalian expression vectors or viral vectors bearing H1, U6 or 7SK promoters for expression *(63)* (*see* **Fig. 20.2**). The gene inhibitory potency of shRNAs, which requires prior processing into effector siRNAs by Dicer, is superior to siRNAs themselves, presumably because they enter the miRNA pathway upstream to siRNAs. Such approaches may be combined, for example, with protein-based anti-HIV-1 agents, for increased therapeutic efficiency *(64)*.

Although RNAi-based antiviral therapies are promising, HIV-1 has been shown to escape RNAi induced by a specific siRNA. In these cases, the emergence of mutants was observed, either showing nucleotide substitutions or deletions within the targeted sequence *(65)*, or evolving an alternative structure in its RNA genome occluding the siRNA binding site *(66)*. A single substitution in the targeted sequence is sometimes sufficient to abolish the antiviral activity of siRNAs *(67)*. Such problems may be circumvented by targeting the most conserved sequences at multiple locations in the HIV-1 genome *(68)*.

3.2. Viral microRNAs

The observed sequence-specific HIV-1 RNA degradation induced by siRNAs implies that certain HIV-1 RNA sequences are accessible to the RNA silencing machinery in vivo. Likely to be applicable to other viruses, this concept is supported by the demonstrated improvement of RNA-induced silencing complex (RISC)-mediated target RNA cleavage when the target site access is increased *(69)*. Accessibility to the viral RNA is mandatory for an antiviral function of the miRNA-guided RNA silencing pathway, a research theme that has attracted the interest of several laboratories. A research group led by Thomas Tuschl was the first to investigate the role of RNA silencing in human cells infected with viruses. Regarding the small RNA profile of a Burkitt's lymphoma cell line latently infected with Epstein-Barr virus (EBV), they found that this large DNA virus expresses several miRNAs *(70)*. Bioinformatic analysis of the genomic sequences flanking the cloned RNAs, which were detectable by Northern blot, unveiled fold-back structures characteristic of miRNA precursors. As for plant virus-derived siRNAs *(71)*, EBV miRNAs may originate predominantly from Dicer processing of highly structured single-stranded RNA. EBV is now known to express in latently infected cells at least 17 distinct miRNAs that are originating from two clusters located in the introns of the viral BART and adjacent BHRF1 genes. Differential regulation of EBV miRNA expression is observed and implies distinct roles during infection of different human tissues *(72)*. Recently, EBV LMP1 was discovered as a target for miRNAs derived from BART cluster 1. This protein is implicated in the activation of cell signaling and gene expression in infected cells *(73)*.

Using similar approaches, investigation of several other viruses identified miRNAs encoded in the Kaposi's sarcoma-associated herpesvirus (KSHV or HHV8), mouse gammaherpesvirus 68 and human cytomegalovirus (also called HHV5) *(74)*. However, viral miRNAs derived from HIV-1 were neither predicted (using an algorithm identifying genomic regions that may assume a secondary structure similar to that of pri- or pre-miRNAs) nor found among 260 cloned miRNA sequences derived from HeLa cells stably expressing CD4 and CXCR4, and infected by HIV-1, isolate Bru (LAV-1) *(74)*. These findings suggested that HIV-1 may effectively hide its highly structured RNA from RNase III cleavage.

3.3. HIV-1-Derived microRNAs

However, this assertion is being challenged, as concurrent studies about HIV-1 miRNAs have been reported. Using a computational method designed to uncover well-ordered folding patterns in nucleotide sequences, five candidate pre-miRNAs encoded by different regions of the HIV-1 genome were flagged *(75)*. Omoto and colleagues *(76)* reported a miRNA (miR-N367) derived from the *nef* region, an accessory gene partially overlapping with the 3′ long terminal repeat (LTR) (*see* **Fig. 20.1b**). This HIV-1 miRNA could be detected by Northern blot analysis in MT-4 T cells persistently infected with HIV-1 IIIB and cloned from a ~25-nt RNA sub-population. Overexpression of miR-N367, which shows perfect complementarity with *nef*, seemed to suppress HIV-1 LTR-driven transcription in reporter gene assays *(77)*, suggesting that this *nef*-derived miRNA could act as a negative regulator of HIV-1 transcription. The biogenesis and action of this particular miRNA require further investigation.

Another study reported that the HIV-1 RNA genome also encodes an siRNA derived from the *env* gene *(78)*. The authors observed that two RNA strands forming a perfect 19-bp duplex, and joined by an extended 198-nt loop, could be converted into siRNAs upon incubation with recombinant Dicer in vitro. A probe specific for the viral siRNA detected a ~24-nt signal not seen in mock-infected cells by Northern blot analysis *(78)*. Overexpression of this viral siRNA effectively reduced Env mRNA levels and viral replication, whereas its neutralization with complementary 2′-*O*-methyl oligonucleotides led to a dose-dependent increase in HIV-1 replication in human cells *(78)*. These results suggest that an HIV-1-derived siRNA can reduce virus production (*see* **Fig. 20.1b**).

Another natural HIV-1 RNA structure, the TAR element, was recently reported to be cleaved by Dicer to generate a miRNA that has been suggested to recruit the histone deacetylase HDAC-1 to the HIV-1 LTR promoter to silence transcription by chromatin remodeling *(79)* (*see* **Fig. 20.1b**), a concept that has been proposed previously *(80)*. The authors hypothesize that this sequence of events may suppress transcription of viral as well as cellular genes, thereby influencing particular steps of HIV-1 pathogenesis, such as latency.

Whether or not HIV-1 miRNAs are effectively produced in infected cells and fulfill important biological roles warrants further experimental validation and confirmation. A recent study by Lin and Cullen *(81)* is challenging the existence of miRNAs derived from primate retroviruses, such as HIV-1 and human T cell leukemia/lymphoma virus type 1 (HTLV-1), and is questioning the suppressive properties of HIV-1 Tat on RNA silencing that has been reported by Bennasser and colleagues *(78)* (*see* **Fig. 20.1c**). It may be that the identification of some miRNAs are restricted to specific viral strains or that miRNAs may escape detection by standard small RNA cloning strategies, since methylation of the 2′ hydroxyl of the terminal ribose significantly reduces the cloning efficiency of silencing-associated small RNAs *(82)*. This would explain some of the discrepancies observed between laboratories using different techniques to identify viral miRNAs.

3.4. Biosynthetic Mechanism of HIV-1 microRNAs

The controversy surrounding the existence of miRNAs derived from HIV-1 may be related to their levels of expression that may be barely detectable using the techniques currently available. If their existence is proven unequivocally, their biosynthesis would merit due mechanistic considerations. For instance, are these viral miRNAs processed by the successive action of Drosha and Dicer like cellular miRNAs, by Dicer only or, as in plants, by an RNAi pathway adapted to viruses? How does the cleavage of an RNA substrate flanked by genomic single-stranded RNA sequences occur, knowing the preference of Dicer for RNA duplexes bearing terminal 2-nt 3′overhangs *(22)*? Is this processing linked in any way to the infectious state, considering that the latency period of HIV-1 is associated with the expression of aborted mRNA transcripts? How does the presence of an exceedingly expanded loop in *env* siRNA precursor influence its processing? Is the expression level of viral miRNAs related to the relatively inefficient processing of HIV-1 dsRNA substrates by RNases III and/or to the limited access to a structure embedded within the HIV-1 RNA? Although regions of the HIV-1 genome show structures relatively close to that of pri- or pre-miRNAs, the fact that they are decorated with many cellular and viral proteins may also affect their recognition and processing by Drosha and/or Dicer.

3.5. Biological Significance of HIV-1 microRNAs

The possibility that HIV-1 miRNAs exert significant biological roles and directly influence viral pathogenesis and persistence in human cells is appealing. The results of recent studies have suggested a role for HIV-1 miRNAs in transcriptional repression induced either by miR-N367 *(76)* or TAR miRNA binding to the LTR-driven promoter *(79)*. Cellular mRNAs that could potentially be regulated by these viral miRNAs have been tentatively identified *(75)*. Investigation of the human and viral genes, as well as the processes possibly regulated by HIV-1 miRNAs, which have been the subject of speculations, awaits their prior experimental confirmation and validation.

3.6. Cellular microRNAs and HIV-1 Infection

A few years ago, candidate HIV-1 genes that could be controlled by host miRNAs have been predicted in view of thermodynamically favorable miRNA:target pairing *(83)*. In addition, changes in miRNA expression profiles, i.e., downregulation of a large pool of miRNAs, have been observed in human HeLa cells transfected with the infectious molecular clone pNL4-3 *(84)*. A more recent study explored the importance of the miRNA pathway in the control of HIV-1 replication *(85)*. Using siRNAs against Drosha and Dicer in peripheral blood mononuclear cells (PBMCs) from HIV-1-infected patients, Triboulet et al. *(85)* noticed a faster virus replication kinetic in Drosha- or Dicer-depleted cells, as compared to cells treated with a control siRNA. The authors also confirmed in latently infected U1 cells that both Drosha and Dicer contribute to the suppression of HIV-1 replication.

HIV-1 infection was also associated with either up- or down-regulation of specific miRNA clusters. For instance, the miR-17/92 cluster, which encode for seven miRNAs, among which miR-17-5p and miR-20 may target the histone acetyltransferase and HIV-1 Tat cofactor p300/CBP-associated factor (PCAF), was substantially decreased (*see* **Fig. 20.1d**). The authors proposed that this gene regulatory axis may help understand how latent virus reservoirs could be activated.

A more recent study explored the possible contribution of cellular miRNAs to HIV-1 infection. Huang et al. *(86)* showed that the 3′ UTR of almost all HIV-1 mRNA produced during latency in resting primary $CD4^+$ T lymphocytes contain a 1.2-kb fragment that can be recognized by cellular miRNAs, with a negative impact on viral protein production (*see* **Fig. 20.1e**). Combined with the relatively inefficient synthesis of Tat and Rev, miRNAs harbored by resting $CD4^+$ T cells may participate in post-transcriptional regulation of HIV-1 mRNA and contribute to keep the virus in its latency phase, as observed in patients with suppressive highly active antiretroviral therapy (HAART) *(86)*. These new elements contribute to our understanding of the molecular basis of viral latency and help us design therapeutic strategies aimed at purging HIV-1-infected patients of the quiescent virus.

3.7. HIV-1, RNA Silencing, and RNA Editing

The susceptibility of viral RNAs to RNases III may also be modified by structural changes produced by adenosine deaminases that act on RNA (ADARs) *(87,88)*. The predominant form of RNA editing in human consists in the specific conversion of adenosine (A) into inosine (I) within largely double-stranded cellular and viral RNAs [reviewed in Ref. *(89)*]. A-to-I RNA editing may thus alter base pairing of a dsRNA substrate and reduce its susceptibility to Dicer cleavage, preventing it from initiating RNAi *(87)*. Several viral genomes or transcripts show sequence changes consistent with such modification, including HIV-1. Indeed, TAR was previously reported to be a substrate for ADAR in *Xenopus* oocytes and edited in a process dependent on Tat *(90)*.Whether editing of viral RNAs may lead to viral persistence, as speculated previously *(89)*, remains to be confirmed.

Another group of deaminases from the APOBEC3 family has been reported to counteract HIV-1 replication. For example, APOBEC3G is incorporated into HIV-1 particles and acts to restrict HIV-1 replication in infected cells by deaminating dC to dU in the first (minus)-strand cDNA replication intermediate during the viral reverse transcription process, *(91)* which is correlated with a G-to-A modification of the second (positive)-strand. Despite its action on cDNA, APOBEC3 family members may induce mutations in the HIV-1 genomic DNA prior to its integration and thereby contribute to a pool of mutant RNA transcripts *(92)*. The viral accessory protein Vif is known to be an inhibitor of APOBEC3 *(93)*. The effect of these editing events on the generation of mutations in dsRNA structures remains to be elucidated.

3.8. Interaction between HIV-1 Tat and the microRNA-Based RNA Silencing Machinery

In addition to protein-RNA and RNA-RNA interactions, recent studies have revealed an intriguing link between protein components involved in HIV-1 pathogenesis and RNA silencing, such as the virally encoded Tat protein and the cellular TRBP.

Overexpression of Tat in mammalian cells was shown to attenuate silencing of reporter genes induced by short hairpin RNAs (shRNAs), but not siRNA *(78)*. Knowing that the former elicits RNAi upon Dicer processing, the authors investigated and determined that Tat could inhibit Dicer activity in vitro (*see* **Fig. 20.1b**). However, prior to qualifying HIV-1 Tat as a proven inhibitor of Dicer function, it would be prudent (i) to determine if the Dicer inhibitory effect of Tat can be extended in vivo and occurs at physiological expression levels, (ii) to confirm that the observed inhibitory effects of Tat are specific and not due to random binding to dsRNAs, (iii) to verify if RNAi proceeds normally in the context of HIV-1 infection, and (iv) to assess whether Dicer function is indeed inhibited by HIV-1 in infected cells. Whether HIV-1 Tat suppresses RNA silencing remains controversial. *(81)*.

3.9. TRBP and PACT Function in RNA Silencing

HIV-1 TAR RNA-binding protein (TRBP) was originally discovered as a cellular protein that cooperates synergistically with viral Tat function and enhances transactivation of the HIV-1 5′ LTR (*see* **Fig. 20.1a**) *(23)*. TRBP is also known to inhibit the interferon (IFN)-induced dsRNA-regulated protein kinase R (PKR) *(94)*, and to be involved in miRNA-guided RNA silencing, more specifically, in assisting Dicer function within a pre-miRNA processing complex (*see* **chapter 18, section 3.3**) *(24,25)*. Immunoprecipitation approaches identified TRBP as a Dicer-interacting protein *(24,25)*. The Dicer-binding region on TRBP could be delineated to its third C-terminal dsRNA-binding domain (dsRBD) and depletion of TRBP was found to negatively affect pre-miRNA processing using cell extracts in vitro *(24)*.

Similar to TRBP, PKR-activating protein (PACT) *(95)* has been found to interact with the N-terminal domain of Dicer via its third dsRBD *(96)*. In fact, PACT can bind directly to TRBP and form a ternary complex with Dicer and TRBP to facilitate the production of siRNAs by Dicer. Knockdown of both TRBP and PACT in cultured mammalian cells led to a significant inhibition of gene silencing mediated by shRNAs, but not by siRNAs, suggesting that TRBP and PACT function primarily at the step of siRNA production *(97)*. Despite exerting opposite effects on PKR, PACT and TRBP may thus play a similar, possibly redundant, role in miRNA biogenesis and function. The exact role of PACT in HIV-1 pathogenesis and RNA silencing remains to be clearly defined.

3.10. A Dual Role for TRBP – Implications for HIV-1

TRBP may thus exert a dual role in HIV-1 pathogenesis and RNA silencing, as recently discussed *(98)*. The requirement of TRBP to achieve a higher virus production may have forced the virus to evolve under selective pressure from the RNA silencing machinery. There is a possibility that TAR and RRE RNA structures could also compete with Dicer for TRBP binding, and thus inhibit RNA silencing *(98)*. The delicate balance thereby created may have conferred to HIV-1 the ability to replicate in TRBP-expressing cells and be responsible, to some extent, for the low virus load and persistence in HIV-1-infected individuals. Pharmacological interventions aimed at dissociating TRBP functions may represent a relevant therapeutic area to combat the HIV-1 pandemic.

3.11. Perspectives for HIV-1

A number of studies published recently have provided key insights into the increasingly complex interaction between HIV-1, miRNAs and host RNA silencing machineries. It has been known that HIV-1 induces drastic changes in gene expression programming of infected cells. With the recent idea that HIV-1 may encode miRNAs, the identification and validation of the complete HIV-1 miRNA array as well as their cellular and viral mRNA targets, which pose a considerable challenge, may significantly improve our understanding of HIV-1 pathogenesis. In particular, it may help determine to what extent the perturbed gene expression profiles in HIV-1-infected cells *(99, 100, 101)* can be related to virus-derived miRNAs and how it ultimately influences viral replication, latency as well as the efficiency of host defenses. This raises the attractive hypothesis that HIV-1 replication may result, at least in part, from a delicate balance between the structural requirements to support HIV-1 replication versus the potential beneficial role of HIV-1 miRNAs in conferring an advantage to the virus and/or thwarting host defenses.

3.12. Applicability to Other Viruses

The various aspects of the interaction between HIV-1 and the RNA silencing machinery may also be applicable to other viruses of global importance for human health. In turn, mechanisms described for other viruses may ultimately be transposed to HIV-1. A few examples that may be relevant include (i) herpesvirus Kaposi's sarcoma-associated herpesvirus (KSHV), which encodes as much as 11 distinct miRNAs that may play critical roles in the establishment and/or maintenance of KHSV latent infection *(102)*, (ii) EBV, which expresses miRNAs from its BART and BHRF transcripts that either target the viral protein LMP1 *(73)* or characterize type III latency in infected B lymphocytes *(103)*, (iii) simian virus 40 (SV40), which encodes miRNAs that regulate viral gene expression and reduce susceptibility to cytotoxic T cells *(104)*, (iv) a cellular miRNA, miR-32, that was recently shown to restrict the accumulation of the retrovirus primate foamy virus type 1 (PFV-1) *(105)*, and, in contrast, (v) an abundant miRNA specifically expressed in the human liver, miR-122, that has been shown to assist hepatitis C virus (HCV) replication through a genetic interaction with the 5′ noncoding region of the viral genome *(106)*.

4. Conclusion

Further investigation on the relationship between HIV-1 and RNA silencing pathways may unveil key aspects of host-pathogen interactions, provide new insights into the persistence of the virus in infected patients and offer novel basis for anti-HIV-1 therapies.

5. Notes

During the preparation of this manuscript, Ouellet et al. *(108)* reported the identification of two miRNAs originating from the HIV-1 TAR element, namely miR-TAR-5p and miR-TAR-3p. The functional implication of these miRNAs in HIV-1 pathogenesis remains to be elucidated.

Acknowledgments

We express our gratitude to Gilles Chabot for the graphic illustrations. P. P. is a New Investigator of the Canadian Institutes of Health Research (CIHR) and Junior 2 Scholar from the Fonds de la Recherche en Santé du Québec. M.J.T. is the recipient of the Canada Research Chair in Human Immuno-Retrovirology (senior level). This work was financially supported by grant HOP-83069 from Health Canada/CIHR (P.P. and M.J.T.).

References

1. Zamore, P.D. and Haley, B. (2005) Ribo-gnome: the big world of small RNAs. *Science*, 309, 1519–1524.
2. Kim, V.N. (2005) MicroRNA biogenesis: coordinated cropping and dicing. *Nat Rev Mol Cell Biol*, 6, 376–385.
3. Griffiths-Jones, S. (2004) The microRNA Registry. *Nucleic Acids Res*, 32, D109–111.
4. Griffiths-Jones, S., Grocock, R.J., van Dongen, S., Bateman, A., and Enright, A.J. (2006) miRBase: microRNA sequences, targets and gene nomenclature. *Nucleic Acids Res*, 34, D140–144.
5. Miranda, K.C., Huynh, T., Tay, Y., Ang, Y.S., Tam, W.L., Thomson, A.M., Lim, B., and Rigoutsos, I. (2006) A pattern-based method for the identification of MicroRNA binding sites and their corresponding heteroduplexes. *Cell*, 126, 1203–1217.
6. Aravin, A.A., Lagos-Quintana, M., Yalcin, A., Zavolan, M., Marks, D., Snyder, B., Gaasterland, T., Meyer, J., and Tuschl, T. (2003) The small RNA profile during Drosophila melanogaster development. *Dev Cell*, 5, 337–350.

7. Ambros, V., Lee, R.C., Lavanway, A., Williams, P.T., and Jewell, D. (2003) MicroRNAs and other tiny endogenous RNAs in C. elegans. *Curr Biol*, 13, 807–818.
8. Cai, X., Hagedorn, C.H., and Cullen, B.R. (2004) Human microRNAs are processed from capped, polyadenylated transcripts that can also function as mRNAs. *RNA*, 10, 1957–1966.
9. Lee, Y., Kim, M., Han, J., Yeom, K.H., Lee, S., Baek, S.H., and Kim, V.N. (2004) MicroRNA genes are transcribed by RNA polymerase II. *Embo J*, 23, 4051–4060.
10. Lee, Y., Ahn, C., Han, J., Choi, H., Kim, J., Yim, J., Lee, J., Provost, P., Radmark, O., Kim, S. et-al. (2003) The nuclear RNase III Drosha initiates microRNA processing. *Nature*, 425, 415–419.
11. Denli, A.M., Tops, B.B., Plasterk, R.H., Ketting, R.F., and Hannon, G.J. (2004) Processing of primary microRNAs by the Microprocessor complex. *Nature*, 432, 231–235.
12. Gregory, R.I., Yan, K.P., Amuthan, G., Chendrimada, T., Doratotaj, B., Cooch, N., and Shiekhattar, R. (2004) The Microprocessor complex mediates the genesis of microRNAs. *Nature*, 432, 235–240.
13. Han, J., Lee, Y., Yeom, K.H., Kim, Y.K., Jin, H., and Kim, V.N. (2004) The Drosha-DGCR8 complex in primary microRNA processing. *Genes Dev*, 18, 3016–3027.
14. Landthaler, M., Yalcin, A., and Tuschl, T. (2004) The human DiGeorge syndrome critical region gene 8 and Its D. melanogaster homolog are required for miRNA biogenesis. *Curr Biol*, 14, 2162–2167.
15. Bohnsack, M.T., Czaplinski, K., and Gorlich, D. (2004) Exportin 5 is a RanGTP-dependent dsRNA-binding protein that mediates nuclear export of pre-miRNAs. *RNA*, 10, 185–191.
16. Brownawell, A.M. and Macara, I.G. (2002) Exportin-5, a novel karyopherin, mediates nuclear export of double-stranded RNA binding proteins. *J Cell Biol*, 156, 53–64.
17. Lund, E., Guttinger, S., Calado, A., Dahlberg, J.E., and Kutay, U. (2004) Nuclear export of microRNA precursors. *Science*, 303, 95–98.
18. Yi, R., Qin, Y., Macara, I.G., and Cullen, B.R. (2003) Exportin-5 mediates the nuclear export of pre-microRNAs and short hairpin RNAs. *Genes Dev*, 17, 3011–3016.
19. Zhang, H., Kolb, F.A., Jaskiewicz, L., Westhof, E., and Filipowicz, W. (2004) Single processing center models for human Dicer and bacterial RNase III. *Cell*, 118, 57–68.
20. Bernstein, E., Caudy, A.A., Hammond, S.M., and Hannon, G.J. (2001) Role for a bidentate ribonuclease in the initiation step of RNA interference. *Nature*, 409, 363–366.
21. Provost, P., Dishart, D., Doucet, J., Frendewey, D., Samuelsson, B., and Radmark, O. (2002) Ribonuclease activity and RNA binding of recombinant human Dicer. *Embo J*, 21, 5864–5874.
22. Zhang, H., Kolb, F.A., Brondani, V., Billy, E., and Filipowicz, W. (2002) Human Dicer preferentially cleaves dsRNAs at their termini without a requirement for ATP. *Embo J*, 21, 5875–5885.
23. Gatignol, A., Buckler-White, A., Berkhout, B., and Jeang, K.T. (1991) Characterization of a human TAR RNA-binding protein that activates the HIV-1 LTR. *Science*, 251, 1597–1600.
24. Haase, A.D., Jaskiewicz, L., Zhang, H., Laine, S., Sack, R., Gatignol, A., and Filipowicz, W. (2005) TRBP, a regulator of cellular PKR and HIV-1 virus expression, interacts with Dicer and functions in RNA silencing. *EMBO Rep*, 6, 961–967.
25. Chendrimada, T.P., Gregory, R.I., Kumaraswamy, E., Norman, J., Cooch, N., Nishikura, K., and Shiekhattar, R. (2005) TRBP recruits the Dicer complex to Ago2 for microRNA processing and gene silencing. *Nature*, 436, 740–744.
26. Schwarz, D.S., Hutvagner, G., Du, T., Xu, Z., Aronin, N., and Zamore, P.D. (2003) Asymmetry in the assembly of the RNAi enzyme complex. *Cell*, 115, 199–208.
27. Matranga, C., Tomari, Y., Shin, C., Bartel, D.P., and Zamore, P.D. (2005) Passenger-Strand Cleavage Facilitates Assembly of siRNA into Ago2-Containing RNAi Enzyme Complexes. *Cell*, 123, 607–620.
28. Plante, I., Davidovic, L., Ouellet, D.L., Gobeil, L.A., Tremblay, S., Khandjian, E.W., and Provost, P. (2006) Dicer-Derived MicroRNAs Are Utilized by the Fragile X Mental Retardation Protein for Assembly on Target RNAs. *J Biomed Biotechnol*, 2006, 64347.
29. Bagga, S., Bracht, J., Hunter, S., Massirer, K., Holtz, J., Eachus, R., and Pasquinelli, A.E. (2005) Regulation by let-7 and lin-4 miRNAs results in target mRNA degradation. *Cell*, 122, 553–563.
30. Liu, J., Valencia-Sanchez, M.A., Hannon, G.J., and Parker, R. (2005) MicroRNA-dependent localization of targeted mRNAs to mammalian P bodies. *Nat Cell Biol*, 7, 719–723.
31. Teixeira, D., Sheth, U., Valencia-Sanchez, M.A., Brengues, M., and Parker, R. (2005)

Processing bodies require RNA for assembly and contain nontranslating mRNAs. *RNA*, 11, 371–382.
32. Eystathioy, T., Chan, E.K., Tenenbaum, S.A., Keene, J.D., Griffith, K., and Fritzler, M.J. (2002) A phosphorylated cytoplasmic autoantigen, GW182, associates with a unique population of human mRNAs within novel cytoplasmic speckles. *Mol Biol Cell*, 13, 1338–1351.
33. Eulalio, A., Behm-Ansmant, I., Schweizer, D., and Izaurralde, E. (2007) P-body formation is a consequence, not the cause, of RNA-mediated gene silencing. *Mol Cell Biol*, 27, 3970–3981.
34. Eystathioy, T., Jakymiw, A., Chan, E.K., Seraphin, B., Cougot, N., and Fritzler, M.J. (2003) The GW182 protein colocalizes with mRNA degradation associated proteins hDcp1 and hLSm4 in cytoplasmic GW bodies. *RNA*, 9, 1171–1173.
35. Fukuhara, N., Ebert, J., Unterholzner, L., Lindner, D., Izaurralde, E., and Conti, E. (2005) SMG7 is a 14-3-3-like adaptor in the nonsense-mediated mRNA decay pathway. *Mol Cell*, 17, 537–547.
36. Unterholzner, L. and Izaurralde, E. (2004) SMG7 acts as a molecular link between mRNA surveillance and mRNA decay. *Mol Cell*, 16, 587–596.
37. Bartel, D.P. (2004) MicroRNAs: genomics, biogenesis, mechanism, and function. *Cell*, 116, 281–297.
38. Vasudevan, S., Tong, Y., and Steitz, J.A. (2007) Switching from repression to activation: microRNAs can up-regulate translation. *Science*.
39. Pillai, R.S., Bhattacharyya, S.N., Artus, C.G., Zoller, T., Cougot, N., Basyuk, E., Bertrand, E., and Filipowicz, W. (2005) Inhibition of translational initiation by Let-7 MicroRNA in human cells. *Science*, 309, 1573–1576.
40. Vella, M.C., Reinert, K., and Slack, F.J. (2004) Architecture of a validated microRNA:target interaction. *Chem Biol*, 11, 1619–1623.
41. Hamilton, A.J. and Baulcombe, D.C. (1999) A species of small antisense RNA in posttranscriptional gene silencing in plants. *Science*, 286, 950–952.
42. Reinhart, B.J., Weinstein, E.G., Rhoades, M.W., Bartel, B., and Bartel, D.P. (2002) MicroRNAs in plants. *Genes Dev*, 16, 1616–1626.
43. Lu, R., Maduro, M., Li, F., Li, H.W., Broitman-Maduro, G., Li, W.X., and Ding, S.W. (2005) Animal virus replication and RNAi-mediated antiviral silencing in Caenorhabditis elegans. *Nature*, 436, 1040–1043.
44. Wilkins, C., Dishongh, R., Moore, S.C., Whitt, M.A., Chow, M., and Machaca, K. (2005) RNA interference is an antiviral defence mechanism in Caenorhabditis elegans. *Nature*, 436, 1044–1047.
45. Han, Y., Wind-Rotolo, M., Yang, H.C., Siliciano, J.D., and Siliciano, R.F. (2007) Experimental approaches to the study of HIV-1 latency. *Nat Rev Microbiol*, 5, 95–106.
46. Levy, J.A. (2006) HIV pathogenesis: knowledge gained after two decades of research. *Adv Dent Res*, 19, 10–16.
47. Suzuki, Y. and Craigie, R. (2007) The road to chromatin - nuclear entry of retroviruses. *Nat Rev Microbiol*, 5, 187–196.
48. Sheldon, M., Ratnasabapathy, R., and Hernandez, N. (1993) Characterization of the inducer of short transcripts, a human immunodeficiency virus type 1 transcriptional element that activates the synthesis of short RNAs. *Mol Cell Biol*, 13, 1251–1263.
49. Ratnasabapathy, R., Sheldon, M., Johal, L., and Hernandez, N. (1990) The HIV-1 long terminal repeat contains an unusual element that induces the synthesis of short RNAs from various mRNA and snRNA promoters. *Genes Dev*, 4, 2061–2074.
50. Huthoff, H. and Berkhout, B. (2002) Multiple secondary structure rearrangements during HIV-1 RNA dimerization. *Biochemistry*, 41, 10439–10445.
51. Gatignol, A. and Jeang, K.T. (2000) Tat as a transcriptional activator and a potential therapeutic target for HIV-1. *Adv Pharmacol*, 48, 209–227.
52. Bannwarth, S. and Gatignol, A. (2005) HIV-1 TAR RNA: the target of molecular interactions between the virus and its host. *Curr HIV Res*, 3, 61–71.
53. Park, H., Davies, M.V., Langland, J.O., Chang, H.W., Nam, Y.S., Tartaglia, J., Paoletti, E., Jacobs, B.L., Kaufman, R.J., and Venkatesan, S. (1994) TAR RNA-binding protein is an inhibitor of the interferon-induced protein kinase PKR. *Proc Natl Acad Sci U S A*, 91, 4713–4717.
54. Dorin, D., Bonnet, M.C., Bannwarth, S., Gatignol, A., Meurs, E.F., and Vaquero, C. (2003) The TAR RNA-binding protein, TRBP, stimulates the expression of TAR-containing RNAs in vitro and in vivo independently of its ability to inhibit the dsRNA-dependent kinase PKR. *J Biol Chem*, 278, 4440–4448.
55. Nekhai, S. and Jeang, K.T. (2006) Transcriptional and post-transcriptional regulation of

HIV-1 gene expression: role of cellular factors for Tat and Rev. *Future Microbiol*, 1, 417–426.
56. Stevens, M., De Clercq, E., and Balzarini, J. (2006) The regulation of HIV-1 transcription: molecular targets for chemotherapeutic intervention. *Med Res Rev*, 26, 595–625.
57. Pollard, V.W. and Malim, M.H. (1998) The HIV-1 Rev protein. *Annu Rev Microbiol*, 52, 491–532.
58. Lesnik, E.A., Sampath, R., and Ecker, D.J. (2002) Rev response elements (RRE) in lentiviruses: an RNAMotif algorithm-based strategy for RRE prediction. *Med Res Rev*, 22, 617–636.
59. Malim, M.H., Bohnlein, S., Fenrick, R., Le, S.Y., Maizel, J.V., and Cullen, B.R. (1989) Functional comparison of the Rev transactivators encoded by different primate immunodeficiency virus species. *Proc Natl Acad Sci U S A*, 86, 8222–8226.
60. Elbashir, S.M., Harborth, J., Lendeckel, W., Yalcin, A., Weber, K., and Tuschl, T. (2001) Duplexes of 21-nucleotide RNAs mediate RNA interference in cultured mammalian cells. *Nature*, 411, 494–498.
61. Martinez, M.A., Gutierrez, A., Armand-Ugon, M., Blanco, J., Parera, M., Gomez, J., Clotet, B., and Este, J.A. (2002) Suppression of chemokine receptor expression by RNA interference allows for inhibition of HIV-1 replication. *AIDS*, 16, 2385–2390.
62. Rossi, J.J., June, C.H., and Kohn, D.B. (2007) Genetic therapies against HIV. *Nat Biotechnol*, 25, 1444–1454.
63. Amarzguioui, M., Rossi, J.J., and Kim, D. (2005) Approaches for chemically synthesized siRNA and vector-mediated RNAi. *FEBS Lett*, 579, 5974–5981.
64. Li, M.J., Kim, J., Li, S., Zaia, J., Yee, J.K., Anderson, J., Akkina, R., and Rossi, J.J. (2005) Long-term inhibition of HIV-1 infection in primary hematopoietic cells by lentiviral vector delivery of a triple combination of anti-HIV shRNA, anti-CCR5 ribozyme, and a nucleolar-localizing TAR decoy. *Mol Ther*, 12, 900–909.
65. Das, A.T., Brummelkamp, T.R., Westerhout, E.M., Vink, M., Madiredjo, M., Bernards, R., and Berkhout, B. (2004) Human immunodeficiency virus type 1 escapes from RNA interference-mediated inhibition. *J Virol*, 78, 2601–2605.
66. Westerhout, E.M., Ooms, M., Vink, M., Das, A.T., and Berkhout, B. (2005) HIV-1 can escape from RNA interference by evolving an alternative structure in its RNA genome. *Nucleic Acids Res*, 33, 796–804.
67. Sabariegos, R., Gimenez-Barcons, M., Tapia, N., Clotet, B., and Martinez, M.A. (2006) Sequence Homology Required by Human Immunodeficiency Virus Type 1 To Escape from Short Interfering RNAs. *J Virol*, 80, 571–577.
68. Leonard, J.N. and Schaffer, D.V. (2005) Computational design of antiviral RNA interference strategies that resist human immunodeficiency virus escape. *J Virol*, 79, 1645–1654.
69. Brown, K.M., Chu, C.Y., and Rana, T.M. (2005) Target accessibility dictates the potency of human RISC. *Nat Struct Mol Biol*, 12, 469–470.
70. Pfeffer, S., Zavolan, M., Grasser, F.A., Chien, M., Russo, J.J., Ju, J., John, B., Enright, A.J., Marks, D., Sander, C. et-al. (2004) Identification of virus-encoded microRNAs. *Science*, 304, 734–736.
71. Molnar, A., Csorba, T., Lakatos, L., Varallyay, E., Lacomme, C., and Burgyan, J. (2005) Plant virus-derived small interfering RNAs originate predominantly from highly structured single-stranded viral RNAs. *J Virol*, 79, 7812–7818.
72. Cai, X., Schafer, A., Lu, S., Bilello, J.P., Desrosiers, R.C., Edwards, R., Raab-Traub, N., and Cullen, B.R. (2006) Epstein-Barr virus microRNAs are evolutionarily conserved and differentially expressed. *PLoS Pathog*, 2, e23.
73. Lo, A.K., To, K.F., Lo, K.W., Lung, R.W., Hui, J.W., Liao, G., and Hayward, S.D. (2007) Modulation of LMP1 protein expression by EBV-encoded microRNAs. *Proc Natl Acad Sci U S A*, 104, 16164–16169.
74. Pfeffer, S., Sewer, A., Lagos-Quintana, M., Sheridan, R., Sander, C., Grasser, F.A., van Dyk, L.F., Ho, C.K., Shuman, S., Chien, M. et-al. (2005) Identification of microRNAs of the herpesvirus family. *Nat Methods*, 2, 269–276.
75. Bennasser, Y., Le, S.Y., Yeung, M.L., and Jeang, K.T. (2004) HIV-1 encoded candidate micro-RNAs and their cellular targets. *Retrovirology*, 1, 43.
76. Omoto, S., Ito, M., Tsutsumi, Y., Ichikawa, Y., Okuyama, H., Brisibe, E.A., Saksena, N.K., and Fujii, Y.R. (2004) HIV-1 nef suppression by virally encoded microRNA. *Retrovirology*, 1, 44.
77. Omoto, S. and Fujii, Y.R. (2005) Regulation of human immunodeficiency virus 1 transcription by nef microRNA. *J Gen Virol*, 86, 751–755.
78. Bennasser, Y., Le, S.Y., Benkirane, M., and Jeang, K.T. (2005) Evidence that HIV-1

encodes an siRNA and a suppressor of RNA silencing. *Immunity*, 22, 607–619.
79. Klase, Z., Kale, P., Winograd, R., Gupta, M.V., Heydarian, M., Berro, R., McCaffrey, T., and Kashanchi, F. (2007) HIV-1 TAR element is processed by Dicer to yield a viral micro-RNA involved in chromatin remodeling of the viral LTR. *BMC Mol Biol*, 8, 63.
80. Weinberg, M.S. and Morris, K.V. (2006) Are viral-encoded microRNAs mediating latent HIV-1 infection? *DNA Cell Biol*, 25, 223–231.
81. Lin, J. and Cullen, B.R. (2007) Analysis of the interaction of primate retroviruses with the human RNA interference machinery. *J Virol*, 81, 12218–12226.
82. Ebhardt, H.A., Thi, E.P., Wang, M.B., and Unrau, P.J. (2005) Extensive 3¢ modification of plant small RNAs is modulated by helper component-proteinase expression. *Proc Natl Acad Sci U S A*, 102, 13398–13403.
83. Hariharan, M., Scaria, V., Pillai, B., and Brahmachari, S.K. (2005) Targets for human encoded microRNAs in HIV genes. *Biochem Biophys Res Commun*, 337, 1214–1218.
84. Yeung, M.L., Bennasser, Y., Myers, T.G., Jiang, G., Benkirane, M., and Jeang, K.T. (2005) Changes in microRNA expression profiles in HIV-1-transfected human cells. *Retrovirology*, 2, 81.
85. Triboulet, R., Mari, B., Lin, Y.L., Chable-Bessia, C., Bennasser, Y., Lebrigand, K., Cardinaud, B., Maurin, T., Barbry, P., Baillat, V. et-al. (2007) Suppression of microRNA-silencing pathway by HIV-1 during virus replication. *Science*, 315, 1579–1582.
86. Huang, J., Wang, F., Argyris, E., Chen, K., Liang, Z., Tian, H., Huang, W., Squires, K., Verlinghieri, G., and Zhang, H. (2007) Cellular microRNAs contribute to HIV-1 latency in resting primary CD4+ T lymphocytes. *Nat Med*, 13, 1241–1247.
87. Knight, S.W. and Bass, B.L. (2002) The role of RNA editing by ADARs in RNAi. *Mol Cell*, 10, 809–817.
88. Tonkin, L.A. and Bass, B.L. (2003) Mutations in RNAi rescue aberrant chemotaxis of ADAR mutants. *Science*, 302, 1725.
89. Bass, B.L. (2002) RNA editing by adenosine deaminases that act on RNA. *Annu Rev Biochem*, 71, 817–846.
90. Sharmeen, L., Bass, B., Sonenberg, N., Weintraub, H., and Groudine, M. (1991) Tat-dependent adenosine-to-inosine modification of wild-type transactivation response RNA. *Proc Natl Acad Sci U S A*, 88, 8096–8100.
91. Chelico, L., Pham, P., Calabrese, P., and Goodman, M.F. (2006) APOBEC3G DNA deaminase acts processively 3¢ --> 5¢ on single-stranded DNA. *Nat Struct Mol Biol*, 13, 392–399.
92. Mbisa, J.L., Barr, R., Thomas, J.A., Vandegraaff, N., Dorweiler, I.J., Svarovskaia, E.S., Brown, W.L., Mansky, L.M., Gorelick, R.J., Harris, R.S. et-al. (2007) Human immunodeficiency virus type 1 cDNAs produced in the presence of APOBEC3G exhibit defects in plus-strand DNA transfer and integration. *J Virol*, 81, 7099–7110.
93. Kao, S., Khan, M.A., Miyagi, E., Plishka, R., Buckler-White, A., and Strebel, K. (2003) The human immunodeficiency virus type 1 Vif protein reduces intracellular expression and inhibits packaging of APOBEC3G (CEM15), a cellular inhibitor of virus infectivity. *J Virol*, 77, 11398–11407.
94. Daher, A., Longuet, M., Dorin, D., Bois, F., Segeral, E., Bannwarth, S., Battisti, P.L., Purcell, D.F., Benarous, R., Vaquero, C. et-al. (2001) Two dimerization domains in the trans-activation response RNA-binding protein (TRBP) individually reverse the protein kinase R inhibition of HIV-1 long terminal repeat expression. *J Biol Chem*, 276, 33899–33905.
95. Patel, R.C. and Sen, G.C. (1998) PACT, a protein activator of the interferon-induced protein kinase, PKR. *Embo J*, 17, 4379–4390.
96. Lee, Y., Hur, I., Park, S.Y., Kim, Y.K., Suh, M.R., and Kim, V.N. (2006) The role of PACT in the RNA silencing pathway. *Embo J*, 25, 522–532.
97. Kok, K.H., Ng, M.H., Ching, Y.P., and Jin, D.Y. (2007) Human TRBP and PACT directly interact with each other and associate with dicer to facilitate the production of small interfering RNA. *J Biol Chem*, 282, 17649–17657.
98. Gatignol, A., Laine, S., and Clerzius, G. (2005) Dual role of TRBP in HIV replication and RNA interference: viral diversion of a cellular pathway or evasion from antiviral immunity? *Retrovirology*, 2, 65.
99. van 't Wout, A.B., Lehrman, G.K., Mikheeva, S.A., O'Keeffe, G.C., Katze, M.G., Bumgarner, R.E., Geiss, G.K., and Mullins, J.I. (2003) Cellular gene expression upon human immunodeficiency virus

type 1 infection of CD4(+)-T-cell lines. *J Virol*, 77, 1392–1402.

100. Chun, T.W., Justement, J.S., Lempicki, R.A., Yang, J., Dennis, G., Jr., Hallahan, C.W., Sanford, C., Pandya, P., Liu, S., McLaughlin, M. et-al. (2003) Gene expression and viral production in latently infected, resting CD4+ T cells in viremic versus aviremic HIV-infected individuals. *Proc Natl Acad Sci U S A*, 100, 1908–1913.
101. Corbeil, J., Sheeter, D., Genini, D., Rought, S., Leoni, L., Du, P., Ferguson, M., Masys, D.R., Welsh, J.B., Fink, J.L. et-al. (2001) Temporal gene regulation during HIV-1 infection of human CD4+ T cells. *Genome Res*, 11, 1198–1204.
102. Cai, X., Lu, S., Zhang, Z., Gonzalez, C.M., Damania, B., and Cullen, B.R. (2005) Kaposi's sarcoma-associated herpesvirus expresses an array of viral microRNAs in latently infected cells. *Proc Natl Acad Sci U S A*, 102, 5570–5575.
103. Xing, L. and Kieff, E. (2007) Epstein-Barr virus BHRF1 micro and stable RNAs during latency III and after induction of replication. *J Virol*, 81, 9967–9975.
104. Sullivan, C.S., Grundhoff, A.T., Tevethia, S., Pipas, J.M., and Ganem, D. (2005) SV40-encoded microRNAs regulate viral gene expression and reduce susceptibility to cytotoxic T cells. *Nature*, 435, 682–686.
105. Lecellier, C.H., Dunoyer, P., Arar, K., Lehmann-Che, J., Eyquem, S., Himber, C., Saib, A., and Voinnet, O. (2005) A cellular microRNA mediates antiviral defense in human cells. *Science*, 308, 557–560.
106. Jopling, C.L., Yi, M., Lancaster, A.M., Lemon, S.M., and Sarnow, P. (2005) Modulation of hepatitis C virus RNA abundance by a liver-specific MicroRNA. *Science*, 309, 1577–1581.
107. Ouellet, D.L., Perron, M.P., Gobeil, L.A., Plante, P., and Provost, P. (2006) MicroRNAs in gene regulation: When the smallest governs it all. *J Biomed Biotechnol*, 2006, 69616.
108. Ouellet, D.L., Plante, I., Landry, P., Barat, C., Janelle, M.E., Flamand, L., Tremblay, M.H., and Provost, P. (2008) Identification of functional microRNAs released through asymmetrical processing of HIV-1 TAR element Nucleic Acids Res, 36, 2353–2365.

Chapter 21

Synthetic microRNA Targeting Glioma-Associated Antigen-1 Protein

Naotake Tsuda, Takahi Mine, Constantin G. Ioannides, and David Z. Chang

Abstract

The transcription factor glioma-associated antigen-1 (Gli-1) mediates activation of the sonic hedgehog (Shh) pathway, a process that precedes the transformation of tissue stem cells into cancerous stem cells and that is involved in early and late epithelial tumorigenesis. Hypothesizing that targeting the 3′-untranslated region (3′-UTR) of Gli-1 mRNA would effectively inhibit epithelial tumor cell proliferation, we evaluated several complementary miRNA molecules for their ability to do so. The synthetic miRNAs and corresponding duplex/small temporal RNAs were introduced as 3-nucleotide (nt) loops into GU-rich portions of the 3′UTR Gli-1 sequence. One particular miRNA (miRNA Gli-1-3548) and its corresponding duplex (Duplex 3548) significantly inhibited proliferation of Gli-1⁺ ovarian (SK-OV-3) and pancreatic (MiaPaCa-2) tumor cells by delaying cell division and activating late apoptosis in MiaPaCa-2 cells. Here, we describe the design of effective miRNA sequences and their applications as anti-gene agents.

Key words: Glioma-associated antigen-1, microRNA, ovarian cancer, pancreatic cancer, apoptosis

1. Introduction

Small RNAs are a novel class of active molecules, which effectively control mRNA translation, mRNA-stability and are abundantly expressed in cells *(1–3)*. To date this class of RNA contains, the short double stranded-RNAs, (siRNA), the single stranded-negative strand microRNAs (miRNA), the Piwi-interacting RNA (piRNA) and the small RNAse L activators *(4–5)*. Most siRNAs originate from the in-frame mRNA-sequence, while miRNAs contain 3′-UTR mRNA sequences and piRNAs are found

M. Sioud (ed.), *Methods in Molecular Biology, siRNA and miRNA Gene Silencing, vol. 487*

DOI: 10.1007/978-1-60327-547-7_21

only in spermatocytes (see Chaps. 18 and 19). RNase L activators are collectively referred to as 2–5*A* ($p_x5'A\ (2'p5'A)_n$; x = 1–3; $n > 2$ *(5)*. Animal miRNAs, siRNAs and piRNAs contain 5′-phosphate and 2′3′-hydroxy-termini. piRNA and rasiRNAs are 2′-*O*-methylated, which protects small RNAs from degradation by 3′-5′ exonucleases. Small molecule activators of RNAse L are cyclical RNA forms *(5)*. In addition to classical small RNAs, a large number of intermediates are present in cells.

Delivery of small RNAs to cells in vitro uses transfection. Novel approaches for RNA delivery are being developed (see Chap. 3). Linking of small RNAs to peptides, proteins, and antibodies to extracellular receptors or particles can ensure targeted delivery. For example, endocytosed proteins via heat-shock protein (HSP) receptors or specific cancer cell receptors can be delivered to the endoplasmic reticulum (ER), and in interfere in situ with protein synthesis and translation *(6, 7)*. Phagocytosed particles by macrophages and dendritic cells are delivered to cytoplasm.

In contrast to siRNAs, miRNAs are genome encoded as long RNA precursors that are processed into 21–22 short RNAs (see Chaps. 18 and 19). These non-coding RNAs have been identified in the genomes of a wide range of multicellular life forms, including plants and animals. The function of miRNAs in vertebrates and mammals is largely unknown, but studies in *Caenorhabditis elegans* and *Drosophila melanogaster* have revealed that miRNAs can bind to target sites in mRNAs with imperfect base pairing and, by unknown mechanisms, significantly reduce translational efficiency *(8, 9)*.

miRNAs are found both in normal tissues and cancer cells. More than half of the human miRNA genes are located at sites known to be involved in cancers, such as fragile sites, minimal regions of loss of heterozygosity, minimal regions of amplification or common breakpoint regions. Such locations suggest that some miRNAs are involved in tumorigenesis *(10)*. miRNA expression profiles of a large number of human tumor samples show that miRNAs are generally, though not always, downregulated in tumors and that such downregulation is often associated with poor prognosis *(11)*. For example, miR-143 and miR-145 expression is downregulated in colorectal neoplasia and miRNA let-7 expression in lung cancer, whereas expression of the precursor of miR-155 is upregulated in pediatric Burkitt lymphoma. These observations, coupled with the observation that naturally occurring miRNAs also regulate stem cell division *(12, 13)*, imply that miRNAs may act as both tumor suppressors and oncogenes.

Although several studies reported on gene silencing by miRNAs, RNA sequences that generate miRNAs inhibiting the expression of regulatory proteins have not yet been described for the glioma-associated antigen 1 (*Gli-1*) gene. Approaches for synthetic miRNA design targeting cancer genes have not been yet established. Here we describe our approaches of designing novel miRNA targeting Gli-1.

We engineered *Gli-1* miRNA to downregulate protein expression. Autonomous activation of the Gli-1/2 pathway provides metastatic tumor cells with a means of efficiently proliferating at a distance from continually Shh-expressing epithelium *(14)*. Because Gli function is a last and essential step of the Shh pathway, its inhibition may inhibit *Shh* signaling at any level *(15, 16)*. Thus, *Gli-1/2* pathway-blocking agents might be used to combat epithelial cancers. For example, cyclopamine and other small molecule inhibitors have been used in animal models to inhibit *Smo* activity in prostate and pancreatic cancer xenografts with good results *(17)*.

We found that our engineered Gli-1 miRNA and Gli-1 small temporal/duplex miRNA (Duplex 3548) (to which we gave the name "butterfly RNA") significantly inhibited the proliferation and division of the tumor cells. We also unexpectedly observed that Duplex 3548 inhibition led to a decrease in the number of tumor cells overexpressing the HER-2 receptor (HER-2^{hi}). Together, our findings suggested that in vitro miRNA inhibition of Gli-1 might be a potentially novel approach to the treatment of pancreatic cancer.

Unlike naturally occurring RNA, our mutant synthetic miRNAs may be able to activate cytokine responses in tumors via RNA receptors *(6)*. A number of cancer cells are sensitive to cytokines, such as IFN-α and TNF-α/β and IFN-γ, which may induce apoptosis in cancer cells. However, systemic administration of cytokines resulted in high toxicity and weak effects because of their very short life. Therefore, if our synthetic miRNAs approach turns out to be clinically feasible, it might help overcome the known shortcomings of systemic cytokine therapy for cancers.

Potential target RNA receptors include the intracellular Toll-like receptors (TLRs) (see chapt. 2). The TLRs are intracellular receptors; some are located in the endoplasmic reticulum. TLR-3 binds dsRNA while TLR-7 and TLR-8 bind single-stranded RNA. Activation of signaling by TLRs results in cytokine production. TLRs, which bind TLR-3, TLR-7 and TLR-8, activate signaling through the adaptor MyD88. The activation of MyD88 enhances TLR expression and production of cytokines which include IFN-α and TNF-α, and in some instances IL-12. These cytokines may promote autoimmunity *(18)*. Small-nuclear RNAs complexed with proteins (RNP) perform similar functions. The reverse phenomenon is the regulation of TLR-expression levels by miRNAs *(19)*.

Foreign RNAs from negative strand viruses such as influenza are well-known activators of cytokine responses in cancer cells and activate TLR7 and -8. The cytoplasmic RNA receptors include retinoic-acid-inducible-gene-I (RIG-1/DDX58) and melanoma differentiation-associated gene (MDA-5/IF1H1) *(20)*. These receptors contain N-terminal caspase activation and recruitment motifs. RIG-1 and MDA-5 relay their signals to interferon regulatory factor-3 (IRF-3) and NF-κB. Recent studies showed that the 5′-triphosphate is essential for RIG-1 recognition of viral RNAs *(21)*. Notably, RNase L is a peculiar RNA-response receptor. It becomes activated by increase in the concentration of RNA and functions to degrade RNA. The small self-RNAs produced by action of RNase L on cellular RNA induce IFN-β expression *(22, 23)*.

The ability of miRNAs to activate IFNs but not IL-6 or IL-1 is important for their anticancer effects. This difference may explain the double-faced "Janus" effects of miRNA described above. Activation of IL-6 became a topic of recent concern in liver and colorectal cancers, because it promotes cancer growth *(24, 25)*. As such it will be important to design small RNAs which can activate more IFN production than IL-6 and IL-1 production. How this can be accomplished is still unknown. Based on the differential cytokine responses to viruses *(26)*, we predict that changes in RNA sequence and structure will be beneficial to learn how to activate anti-cancer responses with small-synthetic RNAs. The methodology for functional small RNA design is presented below.

2. Materials

2.1. Reagents

1. RPMI 1640 medium.
2. Fetal calf serum.
3. Propidium iodide.
4. Ribonuclease A.
5. Lipofectamine-2000.
6. Control siRNA-FITC (Dharmacon, Chicago, IL).
7. MTT cell proliferation assay kit (Molecular Probes, Eugene, OR).
8. TACS Annexin V-FITC apoptosis detection kit (R&D Systems, Minneapolis, MN).

2.2. Buffers

1. Permeabilization buffer: 4X eBioscience permeabilization solution.
2. Phosphate-buffered saline (PBS): 137 mM NaCl, 10 mM Phosphate, 2.7 mM KCl, pH 7.4.

2.3. Cell Culture

MiaPaCa-2 (pancreatic cancer) and SK-OV-3 (ovarian cancer) were obtained from the American Type Culture Collection (ATCC)

(Manassas, VA) and maintained in culture in complete RPMI 1640 medium containing 10% FCS and antibiotics. Cell lines were screened for mycoplasma contamination and were found mycoplasma free. Cells were treated periodically with BM-cyclin (Boehringer) to maintain them mycoplasma free.

2.4. Immunofluorescence

1. Antibodies against Gli-1 and Gli-2 (Santa Cruz Biotechnology, CA). The anti-Gli-1 antibody was raised against a peptide mapping the C-terminus of Gli-1 of human origin, and the antibody against Gli-2 was raised against a peptide mapping near the N-terminus of Gli-2 of human origin.
2. Human IgG (Sigma Chemical Co. St. Louis, MO).
3. Mouse anti-HER-2-PE-conjugated mAb (Sigma chemical Co. St. Louis, MO).
4. Mouse IgG1-PE conjugated.

2.5. microRNAs

1. The used microRNAs are synthesized as a custom order by Dharmacon with 5′-phosphate groups added, and they are purified by high-pressure liquid chromatography (HPLC).
2. microRNAs are dissolved in RNAse-free water at 20 nM, stored in aliquots at –20°C.
3. Lipofectamine2000 transfection reagent (Invitrogen).
4. Negative control siRNA-FITC (fluorescein-labeled luciferase GL2 Duplex) was also obtained from Dharmacon.

3. Methods

3.1. Immunofluorescence

1. Expression of HER-2 protein is determined using the PE-conjugated HER-2-specific mAb clone Neu 24.7 ($IgG1_{\kappa}$). Expression of Gli-1 is detected with mAb listed in Sect. 2.
2. To increase the sensitivity of staining and reduce background, tumor cells are first incubated with 5 µl of a solution of 1 mg/ml of purified human IgG per 10^6 tumor cells, then washed and stained with mouse anti-HER-2-PE-conjugated, mAb or with isotype control mouse IgG1-PE conjugated.
3. These antibodies react with Gli-1 and Gli-2 by immunofluorescence.
4. To detect and quantitate the intracellular Gli-1 and Gli-2 proteins, tumor cells are permeabilized using the eBioscience permeabilizationn buffer. Subsequently, to permeabilization, tumor cells are then incubated for 1 h with human IgG1 as blocking antibody followed by Gli-1- and Gli-2- specific Abs. The incubation was done for 1 h at room temperature.

5. The fluorescence of cells should be determined over a range of Ab concentrations. However, 1.0 μg anti-Gli-1/50,000 cells, and 0.25 μg secondary Ab-FITC conjugated gave the lowest background staining.
6. The sensitivity of the staining can be verified by fluorescent or confocal microscopy.
7. Cells are then analyzed in a Becton Coulter, Excalibur, flow cytometer, and the data are analyzed with a Cell Quest software program.
8. The effects of miRNAs on cell sizes are determined from the forward scatter (FS) geometrical means values (x^2).
9. Cells of FS > 500 are designated as large-size (LS), while cells of FS < 500 are designated as intermediate size (IS).
10. The relative density of the HER-2 and Gli-1 protein in tumor cells is determined from the geometrical mean (y^2) of fluorescence intensity (MFI) of Ab staining.

3.2. Design of miRNAs

1. The sequences of Human HER-2/neu and human Gli-1 cDNA containing 5′- and 3′-UTR are present in NCBI database (accession numbers HER2/neu: M11730//Gli-1: NM 005269).
2. Both UTR sequences are analyzed for the presence of candidate siRNAs using the algorithms listed by Dharmacon on its Web site (www.dharmacon.com).
3. According to the criteria of Dharmacon siDESIGN Center, the following design criteria are used. For more detail, see Chap. 1.
   (a) 30–52% GC Content—add 1 point for satisfying this criterion.
   (b) At least 3 A/Us at positions 15–19 (sense)—add 1 point for each A/U for a total up to 5 points. At least 3 points are required to be scored as positive in the final output.
   (c) Absence of internal repeats—add 1 point for satisfying this criterion.
   (d) A at position 19 (sense)—add 1 point for satisfying this criterion.
   (e) A at position 3 (sense)—add 1 point for satisfying this criterion.
   (f) U at position 10 (sense)—add 1 point for satisfying this criterion.
   (g) No G/C at position 19 (sense)—subtract 1 point for satisfying this criterion.
   (h) No G at position 13 (sense)—subtract 1 point for satisfying this criterion.

All siRNA candidates are evaluated and scored by the SMART selection algorithm and then filtered after a modified BLASTn analysis. The maximum score is 10.

4. Although siRNA binds in the translated region of the gene, microRNA binds in the 3′untranslated regions (3′UTR).
5. Candidate siRNA sequences in 3′ UTR selected by algorithms 1 and 2 are then manipulated to design miRNA.
6. Modification of anti-sense RNA is performed by targeting its central region. We also modified positions 10–11 of the anti-sense to mismatch (if sense is A, it was changed to C or G, if sense is G, it was changed to G or A to avoid G-U wobble) to the identical position of sense, and one base is inserted between position 10 and 11 of anti-sense, which would make a loop toward identical sense sequence (*see* **Note 1**).
7. This inserted base was designed such that it should not bind to either position 10 or 11 of sense strand. For example, in Gli-1 3548miRNA, AU in position 10–11(counting from 5′end) of the anti-sense strand replaced to UCA, CCC, CCA and UCC to create a loop between the anti-sense and sense strands RNA.
8. After defining these candidates, each GC% and ΔG are calculated.
9. Among these candidates, the ones which have lower ΔG than the unmutated sense and antisense strands but still are closer to the ΔG of the unmutated sense and antisense, and 36–50% GC content were selected for experimental testing (*see* **Note 2**).
10. Verification of the specificity for the target sequences of the designed miRNA is performed with the algorithms provided by Drs. Peter Scacheri and Frances Collins, Human Genome Program, National Institute of Health *(31)*.
11. In brief, algorithm 1 identifies the miRNAs which are homologous in the first 5′ nt and the last 3′ nt.
12. Then, if miRNA is 19 nt long, 7 + 7 = 14, 19 − 14 = 5, it follows that only two miRNA differ in the central 5 nt left from Gli-1 3418.
13. Consequently only one miRNA is homologous with Gli-1 3418 (8 + 8 = 16, 19 − 16 = 3).
14. Algorithm 3 searches for matches between the 3′UTRs and the sequence of the 5′ anchor of each miRNAs strand.
15. The miRNA candidates have at least exact match in 7 bases anchor length in the first 5′ nt and the last 3′ nt through algorithm 1, or 12 bases anchor length in the first 5′ nt through

Table 21.1
Selected 3′ UTR mRNA sequences are Gli-1 specific

Start(sense)	Sequence (5′→3′))	Number of mRNA of homologous sequence in the first and last (5–9) nt						
		5	6	7	8	9		
Gli-1 3418	GGAAATGCGATCTGTGATG	333	27	2	1	1		
Gli-1 3414	ATGCGATCTGTGATGGATG	29	3	3	1	1		
Gli-1 3548	CATTATCAAATTTCTCCTC	209	22	2	1	1		
HER-2 4350	AGCCUGGAUGGAUGACACCA	350	25	10	1	0		
		Number of mRNA of homologous sequence in the first (8–14) nt						
		8	9	10	11	12	13	14
Gli-1 3418	GGAAATGCGATCTGTGATG	508	17	3	2	1	1	1
Gli-1 3414	ATGCGATCTGTGATGGATG	49	11	2	1	1	1	1
Gli-1 3548	CATTATCAAATTTCTCCTC	410	126	68	18	7	2	1
HER-2 4350	AGCCUGGAUGGAUGACACCA	723	124	51	11	0	0	0

algorithm 3 were selected as a best candidate miRNAs for in vitro testing (*see* **Note 2**). The results of algorithm were shown in **Table 21.1**.

16. All of these candidates are checked whether they have any conserved sequence in the 5′ends of miRNA (*see* **Note 3**).
17. All miRNA sequences are BLAST searched in the National Center for Biotechnology Information's (NCBI) "search for short nearly exact matches" mode against all human sequences deposited in the GenBank and RefSeq databases and are not found to have significant homology (>17 contiguous nucleotides of identity) to genes other than the targets (*see* **Note 4**).

The design of synthetic miRNA targeting 3′-UTR Gli-1 mRNA are shown in **Table 21.2**.

3.3. dsRNA and miRNA Transfection

1. dsRNA and miRNAs are transfected in SKOV3 and Mia-PaCa-2 cells using Lipofectamine-2000. At the time of transfection, the cells are 50~70% confluent.
2. miRNAs are added to each well in 100 micro of serum-free RPMI medium/well at final concentrations of 83, 167, 250

Table 21.2
Design of synthetic microRNAs targeting 3′UTR Gli-1 mRNA

Gli-1 mRNA	Sequence (5′→3′)	Free energy(kcal/mol) Anti-sense : sense	Design
3414 sense	CATCCATCACAGATCG- CAT-3′		
3414 anti-sense	AUGCGAUCU GU GAUGGAUG	–30.9	3-nt Loop in Position 10
3414 microRNA	AUGCGAUCUAAA- GAUGGAUG	–21.5 (–9.4)	
3418 sense	CATCACAGATCGCATT- TCC		
3418 anti-sense	GGAAAUGCG AU CUGUGAUG	–26.4	3-nt Loop in Position 10
3418 microRNA	GGAAAUGCGUCACU- GUGAUG	–18.9 (–7.5)	
3548 anti-sense miR-361*	CAUUAUCAAA UUUCUCCUC UUAUCA AAUC- UCCAGGGGUAC G		1: Selected the 3'UTR region with best homology with miR-361
3548 sense miR-136** Conserved 8-mer	GAGGAGAAAUUU- GAUAAUG ACUCCAUUUGUUUU- GAUGAUGGA UUUGAUAA		2: The sense strand contains a conserved miR 8-mer.
3548 anti-sense 3548 microRNA	CAUUAUCAA A UUUCUCCUC 5'-CAUUAUCAAUCCU- UCUCCUC-3'	–21.9 –16.0 (–5.9)	3: Loop introduction in the anti-sense strand.
3548 Duplex microRNA	3'-G*UAAUAGUU*AG- GAAGAGGAG-5' 5'-CAU UAUCAAUCCU- UCUCCUC-3'	–27.9 (+6.0)	4: Complementary sequence with microRNA 3548 for the loop in the sense strand.

and 500 nM. The volume of the Lipofectamine 2000 is maintained at a constant 0.0025% of the total volume. The miRNA-cationic liposome complexes are incubated together for 30 min at room temperature prior to adding the tumor cells. Subsequently, the cells are incubated for 4 h in 200 μl serum-free RPMI medium/well.

3. Control transfections are performed in parallel with negative control siRNA-FITC.
4. After 4 h of incubation, 200 μl of RPMI medium containing 20% FCS is added to each well. The final volume is now 400 μl/well.
5. At concentrations below 83 nM, the siRNA-FITC failed to transfect more than 65% of SKOV3 cells.
6. To transfect SKOV3 cells with the same amount of RNA, we used 83 nM of dsRNA and 167 nM of single-stranded miRNAs (*see* **Note 5**). The decreased expression of Gli-1 protein after transfection of Gli-1 miRNA-3418, Gli-1 duplex 3548 or HER-2 miRNA-10GGA is shown in **Fig. 21.1**.

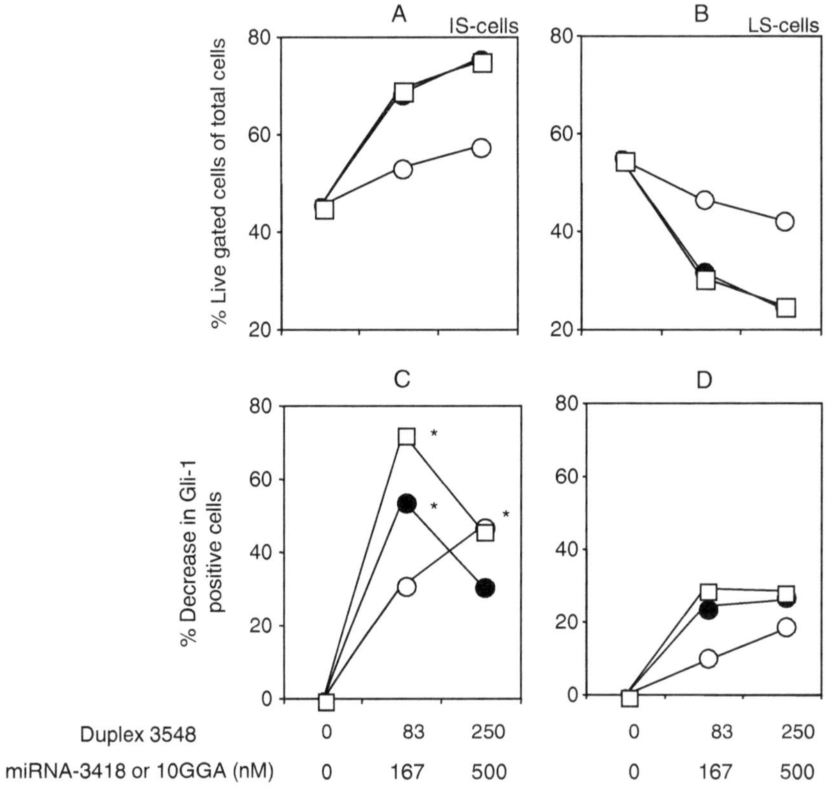

Fig. 21.1. Gli-1 miRNA decreases expression of Gli-1 protein in tumor cells. (**a**) and (**b**) Gli-1 miRNAs increased the number of intermediate-size (IS) tumor cells and decreased the number of large-size (LS) cells. (**c**) and (**d**) Gli-1 miRNAs decreased the number of IS and LS Gli-1+ cells. (○) Gli-1 Duplex 3548, (●) Gli-1 miRNA 3418,. *, $p < 0.05$, at least 50% decrease in cell numbers compared with nontransfected miRNAs.

3.4. Cell Proliferation Assay

1. Prior to transfection with miRNAs, plate the cells for 24 h.
2. After transfection, add 10% FCS and incubate the cells for 48–72 h. Subsequently, detach and determine the cell number.
3. Non-adherent cells are separated from adherent cells by gentle pipetting.
4. Adherent cells are detached by incubation with 10 mM EDTA in phosphate-buffered saline (PBS).
5. Inhibition of tumor cell proliferation is determined using the classic MTT cell proliferation assay kit (Molecular Probes, Eugene, OR).

3.5. Cell Division Assay

1. Tumor cells are resuspended in PBS at a final concentration of 1×10^6 cells/ml.
2. Then, 2 μl of a 5 mM stock CFSE solution is added to each milliliter of cells in solution to a final working concentration of 10 μM. The mixture is then incubated at 37°C for 7 min.
3. Staining is quenched by adding five volumes of ice-cold RPMI medium containing 10% FCS to the cells and incubating the cells on ice for 5 min.
4. The cells are then pelleted by centrifugation.
5. The resulting pellet is washed by being suspended in fresh media. This is done three times.
6. The washed cells are plated onto 24-well plates in 1 ml of RPMI medium containing 10% FCS per well and then cultured for 24 h.
7. The cells are transfected with miRNAs for 48 h, after which they are subjected to flow cytometric analysis.
8. The number of generations and the number of cells in each generation are calculated using Flow-Jo software for Windows (Tree Star, Inc., Ashland, OR).

3.6. Apoptosis Assay

1. To address whether miRNAs induced death by apoptosis, we determined the number of early and late apoptotic MiaPaCa-2 and SKOV-3 cells in response to miRNAs and to duplex-3548.
2. The percentage of early and late apoptotic cells are determined with the TACS Annexin V-FITC apoptosis detection kit.
3. The percentage of apoptotic cells are quantitated by flow cytometry after fluorescent staining with Annexin-FITC and PI. The results of the apoptosis assay are shown in **Fig. 21.2**.

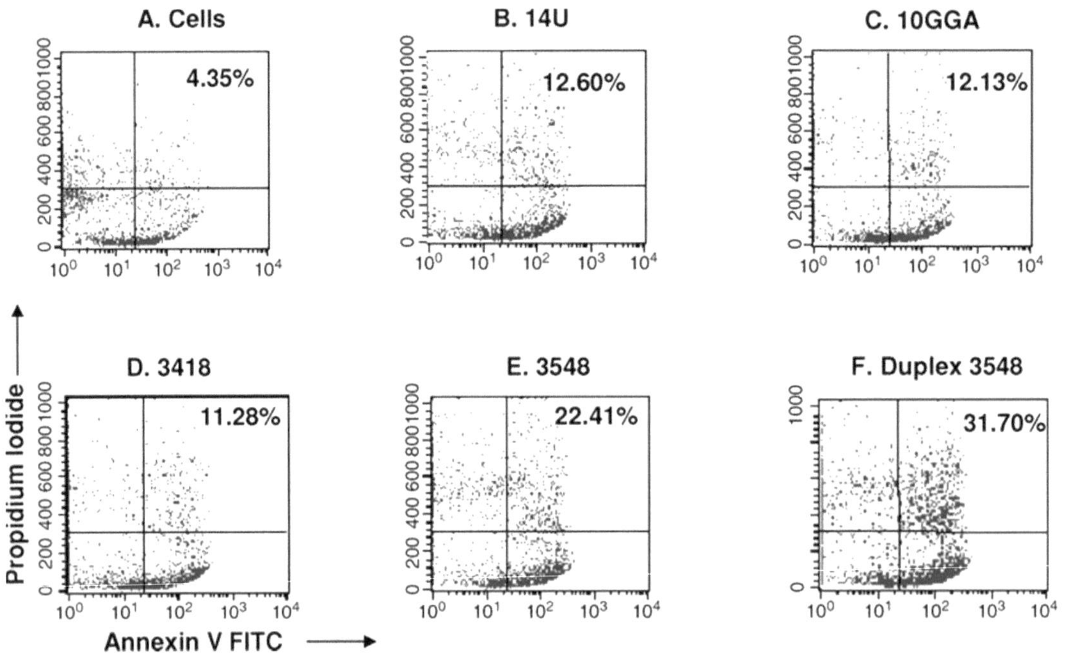

Fig. 21.2. (**a–f**), Gli-1 miRNA 3548 (3548) and Duplex 3548 induced late apoptosis. Cells were stained 48 h after transfection. "Cells," control untransfected MiaPaCa-2 cells. Double-positive cells in the upper right quadrant [PI (Red)+-Annexin-V-FITC(Green)+] represent late apoptotic cells. Cells in the bottom right quadrant [PI(Red)+-AnnexinV-(FITC) +] represent early apoptotic cells.

3.7. Cell Cycle Assay

Cell cycle analysis is performed by flow cytometry after staining with propidium iodide.

1. Cells are harvested, washed twice with 1X PBS, resuspended in 200 μl of 1X PBS, and then fixed with 10 ml of cold 75% ethanol at 4°C for a minimum of 4 h and then washed twice with 1X PBS.
2. Subsequently, cells are resuspended in 500 μl of 1X PBS and stained with 200 μl of propidium iodide (50 μg/ml) and 20 μl of RNase (1 mg/ml) in a 37°C water bath for 15–20 min.
3. Cell cycle analysis is done with a FACStation equipped with CellQuest software.

4. Notes

1. Since in Drosophila embryo extracts the antisense strand of the siRNA sets the ruler for cleavage of target mRNA, at the ninth nucleotide from its paired 5′ end *(33)*, the position of the bulge may be a critical determinant of translational repression activity.

2. The core algorithm predicts a minimum free energy, *G*, as well as minimum free energies for foldings that must contain any particular base pair *(34)*. The ability of a miRNA to translationally repress a target mRNA is largely dictated by the free energy of binding of the first eight nucleotides in the 5′ region of the miRNA. The mutations creating mismatches with the 5′ region of the miRNA inactivated the repression, whereas the order mutations had no effect. The interactions in the 3′ regions were not important, the stability of the miRNA-mRNA interaction in the 5′ region was high. Also G:U wobble in the 5′ region of the miRNA hinders repression despite its favorable contribution to RNA:RNA duplexes *(29,30)*.
3. The 5′ ends of related miRNAs tend to be better conserved than the 3′ end, further supporting the hypothesis that these segments are most critical for mRNA recognition. The critical importance of pairing to segment 2.8 for target identification in silico reflects its importance for target recognition in vivo and speculates that this segment nucleates pairing between miRNAs and miRNAs *(35)*. The 3′UTR analysis yields 106 motifs likely to be involved in post-transcriptional regulation. Nearly one-half are associated with miRNAs, leading to the discovery of many new miRNA genes and their likely target genes *(36)*
4. Studying RNA expression alone may seriously underestimate off-target effects. Once the rules for siRNA and miRNA sequence context are better defined experimentally, improved computational resources will be needed to aid in design of miRNAs, to minimize the potential for off-target interactions.
5. RNAi is often very effective at minimal concentrations, and using the lowest possible concentration of miRNA or siRNA has been suggested to prevent saturation of the RNAi machinery and unwanted side effects.

Acknowledgements

This work has been supported in part by the Topfer Pancreatic Cancer Research Fund (DZC, NT), Grant DOD-01-1-299 (CGI, NT), and grant from the 21st Century COE program from the Kurume University (NT, TM).

References

1. Chu C.Y. and Rana T.M. (2007) Small RNAs: Regulators and guardians of the genome. *J. Cell. Physiol.* 213:412–419.
2. Calin G.A. and Croce C.M. (2007) Chromosomal rearrangements and microRNAs: a new cancer link with clinical implications. *J. Clin. Invest.* 117:2059–2066.

3. Fabbri M., Ivan M., Cimmino A., Negrini M., and Calin G.A. (2007) Regulatory mechanisms of microRNAs involvement in cancer. *Expert Opin. Biol. Ther.* 7:1009–1019.
4. Kirino Y. and Mourelatos Z. (2007) Mouse Piwi-interacting RNAs are 2 -*O*-methylated at their 3 termini. *Nat. Struct. Mol. Biol.* 14:259–260.
5. Thakur C.S., Jha B.K., Dong B., Das Gupta J., Silverman K.M., Mao H., Sawai H., Nakamura A.O., Banerjee A.K., Gudkov A., and Silverman R.H. (2007) Small-molecule activators of RNase L with broad-spectrum antiviral activity. *Proc. Natl. Acad. Sci. U S A* 104(23):9585–9590.
6. Tsuda N., Chang D.Z., Mine T., Efferson C., García-Sastre A., Wang X., Ferrone S., and Ioannides C.G. (2007) Taxol increases the amount and T cell activating ability of self-immune stimulatory multimolecular complexes found in ovarian cancer cells. *Cancer Research* 67(17):8378–8387.
7. Rozema D.B., Lewis D.L., Wakefield D.H., Wong S.C., Klein J.J., Roesch P.L., Bertin S.L., Reppen T.W., Chu Q., Blokhin A.V., Hagstrom J.E., and Wolff J.A. (2007) Dynamic PolyConjugates for targeted in vivo delivery of siRNA to hepatocytes. *Proc. Natl. Acad. Sci. U S A* 104(32):12982–12997.
8. Wightman B., Ha I., and Ruvkun G. (1993) Posttranscriptional regulation of the heterochronic gene lin-14 by lin-4 mediates temporal pattern formation in *C. elegans*. *Cell* 75(5):855–862
9. Olsen P.H. and Ambros V. (1999) The lin-4 regulatory RNA controls developmental timing in *Caenorhabditis elegans* by blocking LIN-14 protein synthesis after the initiation of translation. *Dev. Biol.* 216:671–680.
10. Calin G.A., Sevignani C., Dumitru C.D., Hyslop T., Noch E., Yendamuri S., Shimizu M., Rattan S., Bullrich F., Negrini M., and Croce C.M. (2004) Human microRNA genes are frequently located at fragile sites and genomic regions involved in cancers. *Proc. Natl. Acad. Sci. U S A* 101:2999–3004.
11. Lu J., Getz G., Miska E.A., Alvarez-Saavedra E., Lamb J., Peck D., Sweet-Cordero A., Ebert B.L., Mak R.H., Ferrando A.A., Downing J.R., Jacks T., Horvitz H.R., and Golub T.R. (2005) MicroRNA expression profiles classify human cancers. *Nature* 435: 834–838.
12. Altuvia Y., Landgraf P., Lithwick G., Elefant N., Pfeffer S., Aravin A., Brownstein M.J., Tuschl T., and Margalit H. (2005) Clustering and conservation patterns of human microRNAs. *Nucleic Acids Res.* 33:2697–2706.
13. Hatfield S.D., Shcherbata H.R., Fischer K.A., Nakahara K., Carthew R.W., and Ruohola-Baker H. (2005) Stem cell division is regulated by the microRNA pathway. *Nature* 435:974–978.
14. Sanchez P., Hernández A.M., Stecca B., Kahler A.J., DeGueme A.M., Barrett A., Beyna M., Datta M.W., Datta S., Ruiz I., and Altaba A. (2004) Inhibition of prostate cancer proliferation by interference with SONIC HEDGEHOG-GLI1 signaling. *Proc. Natl. Acad. Sci. U S A* 101: 12561–12566.
15. Thayer S.P., di Magliano M.P., Heiser P.W., Nielsen C.M., Roberts D.J., Lauwers G.Y., Qi Y.P., Gysin S., Fernández-del Castillo C., Yajnik V., Antoniu B., McMahon M., Warshaw A.L., and Hebrok M. (2003) Hedgehog is an early and late mediator of pancreatic cancer tumorigenesis. *Nature* 425:851–856.
16. Frank-Kamenetsky M., Zhang X.M., Bottega S., Guicherit O., Wichterle H., Dudek H., Bumcrot D., Wang F.Y., Jones S., Shulok J., Rubin L.L., and Porter J.A. (2002) Small-molecule modulators of Hedgehog signaling: identification and characterization of Smoothened agonists and antagonists. *J. Biol.* 1:10.
17. Stecca B., Ruiz I., and Altaba A. (2002) The therapeutic potential of modulators of the Hedgehog-Gli signaling pathway. *J. Biol.* 1:9.
18. Groom J.R., Fletcher C.A., Walters S.N., Grey S.T., Watt S.V., Sweet M.J., Smyth M.J., Mackay C.R., and Mackay F. (2007) BAFF and MyD88 signals promote a lupus-like disease independent of T cells. *J. Exp. Med.* 204:1959–1971.
19. Chen X.M., Splinter P.L., O'Hara S.P., and LaRusso N.F. (2007) A cellular miRNA, let-7i, regulates toll-like receptor 4 expression and contributes to cholangiocyte immune responses against Cryptosporidium parvum infection. *J. Biol. Chem.* 282:28929–28938.
20. Kato H., Takeuchi O., Sato S., Yoneyama M., Yamamoto M., Matsui K., Uematsu S., Jung A., Kawai T., Ishii K.J., Yamaguchi O., Otsu K., Tsujimura T., Koh C.S., Reis e Sousa C., Matsuura Y., Fujita T., and Akira S. (2006) Differential roles of MDA5 and RIG-I helicases in the recognition of RNA viruses. *Nature* 441:101–105.
21. Hornung V., Ellegast J., Kim S., Brzózka K., Jung A., Kato H., Poeck H., Akira S., Conzelmann K.K., Schlee M., Endres S., and Hartmann G. (2006) 5 -Triphosphate RNA is the ligand for RIG-I. *Science* 314:994–997.

22. Iordanov M.S., Paranjape J.M., Zhou A., Wong J., Williams B.R., Meurs E.F., Silverman R.H., and Magun B.E. (2000) Activation of p38 mitogen-activated protein kinase and c-Jun NH(2)-terminal kinase by double-stranded RNA and encephalomyocarditis virus: involvement of RNase L, protein kinase R, and alternative pathways. *Mol Cell Biol.* 20:617–627.
23. Malathi K., Dong B., Gale M.Jr., and Silverman R.H. (2007) Small self-RNA generated by RNase L amplifies antiviral innate immunity. *Nature* 448:816–819.
24. Naugler W.E., Sakurai T., Kim S., Maeda S., Kim K., Elsharkawy A.M., and Karin M. (2007) Gender disparity in liver cancer due to sex differences in MyD88-dependent IL-6 production. *Science* 317:121–124.
25. Rakoff-Nahoum S. and Medzhitov R. (2007) Regulation of spontaneous intestinal tumorigenesis through the adaptor protein MyD88. *Science* 317:124–127.
26. Onoguchi K., Yoneyama M., Takemura A., Akira S., Taniguchi T., Namiki H., and Fujita T. (2007) Viral infections activate types I and III interferon genes through a common mechanism. *J. Biol. Chem.* 282:7576–7581.
27. Tsuda N., Kawano K., Efferson C.L., and Ioannides C.G. (2005) Synthetic microRNA and double-stranded RNA targeting the 3 -untranslated region of HER-2/neu mRNA inhibit HER-2 protein expression in ovarian cancer cells. *Int. J. Oncol.* 27:1299–1306.
28. Tsuda N., Ishiyama S., Li Y., Ioannides C.G., Abbruzzese J.L., and Chang D.Z. (2006) Synthetic microRNA designed to target glioma-associated antigen 1 transcription factor inhibits division and induces late apoptosis in pancreatic tumor cells. *Clin .Cancer Res.* 12:6557–6564.
29. Doench J.G., Petersen C.P., and Sharp P.A. (2003) siRNAs can function as miRNAs. *Genes Dev.* 17:438–442.
30. Doench J.G. and Sharp P.A. (2004) Specificity of microRNA target selection in translational repression. *Genes Dev.* 18:504–511.
31. Scacheri P.C., Rozenblatt-Rosen O., Caplen N.J., Wolfsberg T.G., Umayam L., Lee J.C., Hughes C.M., Shanmugam K.S., Bhattacharjee A., Meyerson M., and Collins F.S. (2004) Short interfering RNAs can induce unexpected and divergent changes in the levels of untargeted proteins in mammalian cells. *Proc Natl Acad Sci U S A* 101:1892–1897.
32. Efferson C.L., Tsuda N., Kawano K., Nistal-Villán E., Sellappan S., Yu D., Murray J.L., García-Sastre A., and Ioannides C.G. (2006) Prostate tumor cells infected with a recombinant influenza virus expressing a truncated NS1 protein activate cytolytic CD8+ cells to recognize noninfected tumor cells. *J. Virol.* 80:383–394.
33. Elbashir S.M., Martinez J., Patkaniowska A., Lendeckel W., and Tuschl T. (2001) Functional anatomy of siRNAs for mediating efficient RNAi in Drosophila melanogaster embryo lysate. *EMBO J.* 20:6877–6888.
34. Zuker M. (2003) Mfold web server for nucleic acid folding and hybridization prediction. *Nucleic Acids Res.* 31:3406–3415.
35. Lewis B.P., Shih I.H., Jones-Rhoades M.W., Bartel D.P., and Burge C.B. (2003) Prediction of Mammalian MicroRNA Targets. *Cell* 115:787–798
36. Xie X., Lu J., Kulbokas E.J., Golub T.R., Mootha V., Lindblad-Toh K., Lander E.S., and Kellis M. (2005) Systematic discovery of regulatory motifs in human promoters and 3 UTRs by comparison of several mammals. *Nature* 434:338–345

Chapter 22

Therapeutic Targeting of Gene Expression by siRNAs Directed Against *BCR-ABL* Transcripts in a Patient with Imatinib-Resistant Chronic Myeloid Leukemia

Michael Koldehoff and Ahmet H. Elmaagacli

Abstract

Within the recent years, RNA interference (RNAi) has become an almost-standard method for in vitro knockdown of any target gene of interest. Now, one major focus is to further explore its potential in vivo, including the development of novel therapeutic strategies. From the mechanism, it becomes clear that small interfering RNAs (siRNAs) play a pivotal role in triggering RNAi. Thus, the efficient delivery of target gene-specific siRNAs is one major challenge in the establishment of therapeutic RNAi. Here we show that in vivo application of targeted nonvirally delivered synthetic *bcr-abl* siRNA in a female patient with recurrent Philadelphia chromosome positive chronic myeloid leukemia (CML) resistant to imatinib (Y253F mutation) and chemotherapy after allogeneic hematopoietic stem cell transplantation can silence the expression of *bcr-abl* gene. We found a remarkable inhibition of the overexpressed *bcr-abl* oncogene resulting in increased apoptosis of CML cells. In vivo siRNA application was well tolerated without any clinically adverse events. Our findings imply that the clinical application of synthetic siRNA is feasible, safe and has real potential for genetic-based therapies using synthetic nonviral carriers.

Key words: *bcr-abl*, small interfering RNA, RNAi, CML, oligonucleotherapy, oncogenes.

1. Introduction

Chronic myeloid leukemia (CML) is a relatively well-differentiated myeloproliferative disorder originating from transformed hematopoietic stem cells. The disease arises as a consequence of a rare mutational event resulting in a reciprocal translocation between the long arms of chromosomes 9 and 22. The shortened chromosome 22 formed by this translocation is the Philadelphia (Ph^1) chromosome, named after the city in which it was discovered. CML was the first neoplastic process to be linked to

M. Sioud (ed.), *Methods in Molecular Biology, siRNA and miRNA Gene Silencing, vol. 487*

DOI: 10.1007/978-1-60327-547-7_22

a consistent acquired genetic abnormality and currently it is the best-studied molecular model of leukemia. The translocation creates the chimeric oncogene *bcr-abl* with the protein product BCR-ABL, a tyrosine kinase with constitutive activity. The expression of these consistent molecular changes has been shown to be necessary and sufficient for the transformed phenotype of CML cells. BCR-ABL is responsible for the pathogenesis of CML, demonstrated by the transforming ability of BCR-ABL expression in cell lines and within mice. The natural clinical course in human CML can be divided into three phases. It follows a fairly benign course for several years (chronic phase) before transforming into the more aggressive accelerated phase and life-threatening blast crisis. At the time of diagnosis the majority of patients are in the chronic phase and are asymptomatic. The blast crisis occurs when there is a failure of maturation of the malignant precursors, often accompanied by additional cytogenetic abnormalities and resulting in a disease resembling acute myeloid or lymphoblastic leukemia *(1 – 4)*. Treatment with imatinib mesylate, a selective inhibitor of ABL, the chimeric BCR-ABL fusion protein, platelet-derived growth factor receptor (PDGF-R) α and β, and c-kit has shown remarkable clinical activity with minimal side effects in *bcr-abl*(+), c-kit(+) or PDGF-R(+) leukemias and has resulted in complete cytogenetic remission in a majority of chronic phase CML patients, and in fewer patients with accelerated and blast phase disease. Disease progression remains low but detectable, though this risk may decline over time *(5– 7)*. Although allografting is still considered to be the only potentially curative approach in CML patients, transplantation numbers have dropped significantly in the imatinib era due to transplantation-associated mortality and morbidity. Furthermore, as patients may relapse after allogeneic transplantation or develop imatinib-resistant disease, additional CML therapies are required *(8,9)*.

In 1998 Andrew Fire and Craig Mello discovered, in a series of experiments in *Caenorhabditis elegans*, that injection of sense or antisense RNAs led to negligible decreases of target RNA, whereas introduction of double-stranded RNA (dsRNA) resulted in effective and specific degradation of cytoplasmic mRNA. Furthermore, the aforementioned silencing effects of dsRNA in *C. elegans* were systemic and heritable. An evolutionary conserved cellular mechanism to protect against viral infections, RNA interference (RNAi) inhibits gene expression by degrading mRNA in a sequence-specific post-transcriptional manner, upon introduction of dsRNA. This long dsRNA is cut into 21–23-mer active intermediates, termed small interfering RNAs (siRNAs). siRNAs are the mediators of mRNA degradation in the process of RNAi. Different ways of producing these small interfering molecules, including chemical synthesis, in vitro transcription, or vector-based delivery into mammalian cell lines have been successfully developed over the last few years *(10–13)*. Regardless of the transfection method, the efficiency of the siRNAs is mostly dependent on the successful rational design of the 21-mer sequences *(14*, see Chap. 1). Scherr et al. recently published a study showing that siRNA directed against *bcr-abl* can specifically inhibit *bcr-abl* expression in the $Ph^1(+)$ cell lines and the cells of CML patients *(15*, see Chap. 11). Furthermore, Wohlbold et al. showed that siRNA treatment might sensitize cells to imatinib mesylate contributing to its therapeutic potential *(16)*. We demonstrated that combined transfection with the Wilm's tumor gene (*WT1*) siRNA and *bcr-abl* siRNA in the $Ph^1(+)$ cell lines and the cells of CML patients increased the inhibition of the rate of proliferation and the rate of induced apoptosis compared to transfection with bcr-abl siRNA or WT1 siRNA alone *(17)*. Also we showed that bcr-abl siRNA had antiproliferative and pro-apoptoic effects on $Ph^1(+)$ AML cells in vitro *(18)*. Limitations in the use of oligonucleotides as gene expression inhibitors include their poor stability in biological medium, their weak intracellular penetration, and the poor cytoplasmic delivery when they have to reach their complementary target. It is, however, possible to bypass this problem by using synthetic nonviral carriers, such as cationic liposomes and polymers *(19,20)*. RNAi therapeutics represents a fundamentally new way to treat human disease by addressing targets that are otherwise undruggable with existing medicines. We recently demonstrated the application of targeted, nonviral delivery of *bcr-abl* siRNA as a therapeutic approach in a female CML patient with imatinib-resistant medullar and extramedullar relapse after allogeneic hematopoietic stem cell transplantation (HSCT) and report here in vivo evidence of the efficacy of RNAi-based therapeutics efficacy in this CML disease *(21)*.

2. Methods of Inducing RNAi

One remarkable property of RNAi and cosuppression is that in both processes a signal appears to be generated, which travels through the organism to induce sequence-specific gene silencing at a considerable distance. The discovery that siRNA is the effector mechanism of endogenous RNAi prompted investigation into the utilization of exogenously administered siRNA, or vectors inducing the expression of siRNA, for gene-specific silencing. Genetic manipulation using such a strategy raises several issues that need to be addressed: (1) stability of siRNA; (2) ability to constitutively express the siRNA; (3) possibility of tissue-specific delivery; and (4) finding the best method for identifying effective silencing sites on the mRNA transcript. In order to answer these questions various versions of siRNA have been developed *(22,23)*.

3. RNAi Delivery in Mammalian Cells

The three most important attributes to take into account when designing and selecting siRNA are potency, specificity ,and nuclease stability. As our understanding of the molecular and structural mechanism of RNAi has grown, it is now possible to identify potent, specific, and stable siRNA candidates targeting any gene of interest. RNAi in mammalian cells can be triggered by direct introduction through injection, electroporation, lipid-mediated transfection, nanoparticles, or antibody bound enzymatically generated or chemically synthesized siRNAs, among others. Alternatively, siRNAs or small hairpin RNAs (shRNAs) can be delivered by vector-based intracellular expression. SiRNAs can be synthesized chemically, generated enzymatically through in vitro transcription by T7 phage polymerase, or through endonuclease digestion by recombinant Dicer of in vitro transcribed long dsRNAs. In mammalian cells, direct delivery of siRNAs can only induce transient silencing due to their limited half-life and to their dilution during cell division. Chemical modifications are required to potentiate siRNA nuclease and thermodynamic stability in vivo without compromising their efficacy. Recently, several groups reported different approaches for systemic in vivo delivery of siRNAs. Soutschek et al. described the intravenous injection in mice of chemically modified naked siRNAs coupled to a cholesterol group chemically linked to the terminal hydroxyl group of the sense string to promote entry into the cells *(24)*. In vivo delivery of chemically modified siRNAs encapsulated into liposome particles has been recently reported by Morrisey et al., and Song et al. described an antibody-based delivery system which offered the possibility of systemic, cell-type-specific siRNA delivery *(25,26)*.

4. In Vivo siRNA Delivery Through Dispersion Lipid Solution (DLS) Complexation

Naked siRNAs are degraded in human plasma with a half-life of minutes. To convert siRNAs into optimized drugs, the modification of chemically synthesized siRNAs need to be protected from nuclease digestion and last longer than naked siRNAs, especially when it is exposed to nuclease-rich environments, such as blood to target human leukemias *(23,27)*. Our recently described method for the efficient protection and delivery of siRNAs in vitro and in vivo relies on anionic liposome complexation in leukemic malignancies *(21)*. In brief, the sequence of our siRNA targeted against bcr-abl was AAGCAGAGTTCAAAAGCCCTT (from Qiagen-Xeragon, Germany), as published previously by Scherr and coworkers *(14)*. Chemically synthesized naked siRNA was dissolved in a suspension buffer (100 nM potassium acetate, 30 mM HepesKOH, 2 nm Mg-acetate, pH 7.4), heated at 90°C for 1 min, and then incubated for 1 h at 37°C. Then the solution was diluted 1:10 in a 10% lipid solution (Lipovenös) purchased from Fresenius Kabi (Bad Homburg, Germany) containing, per liter: 100 g soya bean oil, 25 g glycerol, and 6 g phospholipids from egg. The solution was incubated for 15 min at room temperature and mixed multiple times by passing it through a small lumen syringe generating negative pressure using a three-way valve to form the dispersion lipid solution (DLS) and reconstitute siRNA chylomicrons in the lipid solution.

5. Efficiency of siRNA Transfection with DLS Ex Vivo

In vitro transfections with 0.8 μg siRNA were performed in 24-well plates using TransMessenger transfection reagent (Qiagen, Hilden, Germany) or DOTAP Liposomal transfection reagent (Roche Diagnostics, Mannheim, Germany) following the manufacturer's protocol, or using DLS (1×10^5/well) following the above described delivery protocol. Using fluorescence-marked, nonsilencing siRNA (Qiagen) we evaluated the transfection rate of DLS, TransMessenger or DOTAP Liposomal in K562 cells. Twenty-four hours after transfection the number of fluorescently marked cells was evaluated using fluorescence microscopy. 5 × 100 cells were counted per sample. With DLS we found a mean transfection rate of 70% (range 61–78%), with TransMessenger or DOTAP Liposomal we found a mean transfection rate of 50% (range 44–62%) and 59% (range 54–66%), respectively.

Cell proliferation was determined by 5-bromo-2-deoxyuridine (BrdU) incorporation assay and apoptotic cells were determined using the in situ cell Death detection kit, both following the manufacture's instructions from Roche Diagnostics (Mannheim, Germany). Twenty-four hours after transfection with *bcr-abl* siRNA we observed a moderate inhibition of proliferation in the CML cells of the patient prior to siRNA therapy from either spontaneously 36.8 ± 4.6% or nonsilencing siRNA 34.8 ± 3.2% to 14.6 ± 2.1% (reduction of 76%, $p < 0.002$) and a strong induction of apoptosis from spontaneously 8.1 ± 1.7 or nonsilencing siRNA 8.6 ± 2.2% to 16.1 ± 1.2 (induction of 99%, $p < 0.05$) as shown in **Table 22.1**. Transfection with siRNA in K562 cells induced an increased rate of apoptosis from spontaneously 13.0 ± 2.5% or nonsilencing siRNA 13.5 ± 2.7% to 18.5 ± 4.4% (induction of 42%, $p < 0.01$) and an inhibition of proliferation from 48.6 ± 4.2 or nonsilencing siRNA 50.6 ± 3.2% to 28.3 ± 4.9% (reduction of 42%, $p < 0.05$). The effects on *bcr-abl* siRNA amounts were

Table 22.1
Effects of bcr-abl siRNAs on apoptosis and proliferation in patient leukemic cells and K562 cell line

	bcr-abl siRNA	Nonsilencing siRNA	Control (spontaneous)	*P*-value
Effects of bcr-abl siRNA on apoptosis				
CML cells of the patient	16.1 ± 1.2	8.6 ± 2.2	8.1 ± 1.7	<0.05
K562 cells	18.5 ± 4.4	13.5 ± 2.7	13.0 ± 2.5	<0.01
Effects of bcr-abl siRNA on proliferation				
CML cells of the patient	14.6 ± 2.1	34.8 ± 3.2	36.8 ± 4.6	<0.002
K562 cells	28.3 ± 4.9	50.6 ± 3.2	48.6 ± 4.2	<0.05

All experiments were performed at $n = 15$. Data are shown as the mean (with standard deviation) and *P*-values are presented. Controls without siRNA were set to 100%.

Table 22.2
Effect of different siRNA delivery strategies on bcr-abl gene expression

	bcr-abl/gapdh level	*P*-value
Spontaneous (control) without siRNA	147.1 ± 52.7	
DMSO (10%) plus siRNA	111.2 ± 22.3	n.s.
Transmessenger plus siRNA	66.9 ± 12.6	<0.006
DOTAP Liposomal plus siRNA	58.7 ± 6.1	<0.002
DLS plus siRNA	47.0 ± 14.6	<0.002

All experiments were performed at $n = 7$. Data are shown as the mean (with standard deviation) and *P*-values are presented. Controls without siRNA were set to 100%.
Bcr-abl gene expression was measured by real-time PCR and normalized to GAPDH expression

different between primary CML cells and K562 cells in induction of apoptosis ($p < 0.05$) and in inhibition of proliferation ($p < 0.001$) as shown in **Table 22.1**.

To further examine the role and the effects of lipid composition (DLS) on siRNA properties, we compared it to other transfection procedures in the leukemic cell line K562 by measuring bcr-abl expression, which was quantified in relation to the housekeeping gene gapdh by real-time PCR (ratio of bcr-abl/gapdh). Quantitative analyses of bcr-abl mRNA levels from several experiments **(Table 22.2)** revealed a significant reduction of bcr-abl mRNA levels to amounts between 54.5% and 68% (mean) after siRNA transfection compared to untreated or nonsilencing siRNA. These findings suggest that DLS was most effective in generating siRNA mediated silencing of the bcr-abl gene *(21)*.

6. Pharmacokinetics and Theoretical Consideration for Therapeutic siRNA Dosage

For effective drug therapy, it is necessary to deliver therapeutic agents selectively to their target sites, since most drugs are associated with both beneficial effects and unfavourable actions. In general, the lack of selectivity of most conventional drugs is closely related to their pharmacokinetic properties. The in vivo fate of a drug given by a particular administration route is determined by both the physicochemical properties of drug and the anatomical and physiological characteristics of the body; therefore, the biodistribution characteristics of systemically administered siRNAs has attracted much attention *(27)*. Modified siRNAs have an acceptable half-life in vitro and a predictable biodistribution profile similar to that of single-stranded antisense oligonucleotides (ASO). They also have repeatedly been more robust than ASO techniques in terms of consistency of transcript knockdown and threshold concentration. The data from Sewell et al. *(28)* suggests a dosing of 2–4 mg/kg b.w. for ASO. In a phase I and II multicenter study using the bcl-2 ASO oblimersen sodium in patients with advanced chronic lymphocytic leukemia, the drug's maximum-tolerated dose was established at 3 mg/kg b.w. *(29)*. Experimental evidence showed a 100- to 1000-fold increase in the activity of siRNA versus ASO *(30)*; we therefore estimated that the minimal effective siRNA dose administered in vivo should be approximately 10 μg/kg b.w. and we chose this dose for the first application of DLS-formulated *bcr-abl* siRNA *(21)*.

7. Patient's Characteristics and Clinical Course Prior to siRNA Administration

A 47-year-old female patient was transplanted for $Ph^1(+)$ CML (molecular configuration of the *major bcr-abl* mRNA transcript, b3a2 translocation for a p210 oncoprotein, leukocytes of 8.4/nl containing 10% blasts in the bone marrow, a hemoglobin level of 11.7 g/dl and platelets of 136/nl) in accelerated phase with blood stem cells from an HLA partially matched unrelated male donor (HLA-DRB1 allele mismatch) after receiving a myeloablative conditioning therapy with fractionated total body irradiation (10 Gray), cyclophosphamide (120 mg/kg b.w.) and anti-thymocyte globulin (30 mg/kg b.w., Fresenius, Germany). Transplant-associated complications included renal insufficiency with creatinine serum levels of up to 2 mg/ml and acute graft-versus host disease (GVHD) grade II (skin stage 3, gut stage 1, histologically confirmed), which developed into an extensive chronic GVHD with mainly skin involvement. GVHD was initially treated with cyclosporine A (CSA) (3 mg/kg b.w.) that was later changed to tacrolimus (1 mg/day), prednisone (2 mg/kg b.w.), and transiently mycophenolate mofetil (MMF) (2 g/day). Prednisone was continuously tapered to a minimum of 20 mg/day post-transplant. Tacrolimus treatment was ceased subsequent to renal insufficiency and MMF had to be stopped due to intolerance. At day + 155 post-transplant the patient developed a molecular and cytogenetic relapse with repeated positive *major bcr-abl* levels measured by real-time polymerase chain reaction (real-time PCR). In a chromosome assay performed from a bone marrow aspirate at that time, eleven (55%) $Ph^1(+)$ chromosomes of 20 analyzed metaphases were detected. The immunosuppressive therapy with prednisone was reduced to 12.5 mg/day, which caused recurrence of skin GVHD with generalized exanthema (chronic GVHD, extensive disease). Complete blood counts and metabolic profiles defined a complete hematological remission. Chromosomal analysis classified a major cytogenetic response, but the patient remained PCR positive for the *bcr-abl* transcript (range: 0.01–0.3 *bcr-abl/gapdh* quotient). At day + 247 post-transplant the patient relapsed with leukocytes of 5.8/nl containing 1% blasts, a hemoglobin of 10.0 g/dl and platelets of 150/nl. The patient's *bcr-abl* titer increased again to 4.3% (*bcr-abl/gapdh* quotient) with one (20%) $Ph^1(+)$ chromosome of 5 analyzed metaphases in cytogenetic and four $Ph^1(+)$ chromosomes of 200 analyzed interphase nuclei in the fluorescence in situ hybridization (FISH) analysis. The donor cells showed a 99.8% chimerism by the XY-FISH analysis of sex chromosomes. A therapy with imatinib mesylate 400 mg/day was initiated, which had to be reduced to 300 mg/day due to leukocytopenia and thrombocytopenia. The patient again required transient blood (two units/week) and platelet (one unit/week) transfusions. The *bcr-abl* level dropped down to 0.086% at day + 280 post-transplant due to a complete hematological response.

At day + 381 after transplant six subcutaneous nodes were detected at the lower abdomen with a maximum diameter of 12 mm. The hematologic parameters are stable with leukocytes of 3.2/nl, a hemoglobin of 8.7 g/dl and platelets of 36/nl, with a granulocytes of 33.5%. One node was extirpated and histologically examined revealing an extramedullar relapse of CML with a positive *bcr-abl/gapdh* titer of 17.3% as determined by real-time PCR. Moreover, at day + 389 the patient developed bilateral pleural effusions. Diagnostic evaluation of pleural fluid showed

CML cells with a *bcr-abl/gapdh* quotient of 4.4% in real-time PCR and 27% Ph^1(+) chromosomes in FISH analysis. The cytostatic treatment was extended to cytosine arabinoside (ARA-C) (20 mg) administered subcutaneously every five days. At day + 421 post-transplants the *bcr-abl* level in peripheral blood increased to 4.6% accompanied by growth of the extramedullar CML nodes into the abdominal skin under a continued combination therapy with imatinib mesylate and ARA-C (40 mg every five days). This patient, who initially had responded to imatinib, subsequently lost her response in association with the emergence of an imatinib-resistant clone due to both the overexpression and amplification of bcr-abl gene locus or to the development of a bcr-abl gene mutation. Although the patient still received immunosuppressive medication (prednisone 12.5 mg/day) at that time, skin GVHD was active with generalized exanthema (chronic GVHD, extensive disease). SiRNA treatment was started on day + 426 after transplant without discontinuing imatinib mesylate, ARA-C treatment or the immunosuppressive medication.

8. Clinical Performance After Therapeutic siRNA Administration

At day + 426 post-transplant the patient received for the first time, 10 μg/kg b.w. *bcr-abl* siRNA intravenously in a DLS of 20 ml as an accompanying therapy trial (**Fig. 22.1a**). SiRNA was administered over 1–2 min as a bolus injection. In addition, 300 μg siRNA in 2 ml DLS were injected directly into a subcutaneous CML node which was marked as node A. No side effects were observed during the first hours after administration. Two days after siRNA treatment the patient suffered from moderate dizziness (grade 0–1, WHO classification), which resolved completely after three days. Diuretic therapy with hydrochlorothiazid 25 mg/day had to be intensified by adding furosemide 40 mg/day at day + 428. Even though capillary leakage could not be ruled out, we decided to give a single application of prednisone (flat 60 mg, 1 mg prednisone/kg b.w.) on day + 430 in addition to fluid restriction. After the first siRNA administration we found a transient improvement of peripheral blood leucocytes from 9/nl to a minimum of 2.2/nl, with the disappearance of immature cells that defined a complete hematological response, followed by an increase to 3.4/nl before the second siRNA injection. FISH analysis showed an absence of the Ph^1(+) chromosomes of 200 analyzed interphase nuclei. The *bcr-abl* mRNA level decreased from 6.6% to 0.053% nine days after the first siRNA injection. Another two days later the *bcr-abl* expression increased again and reached a *bcr-abl/gapdh* quotient of 16.1% at day + 448. Platelets

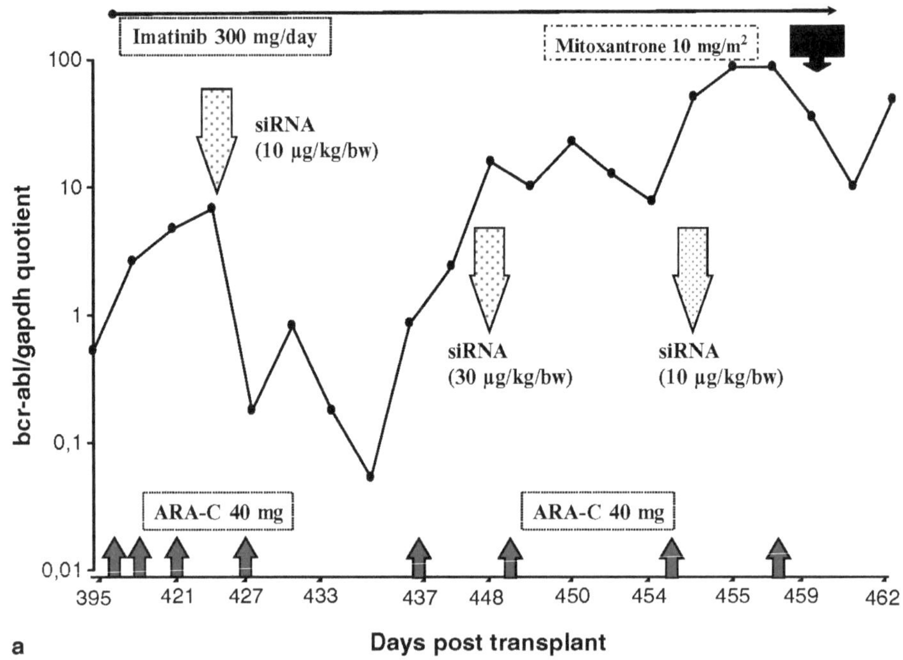

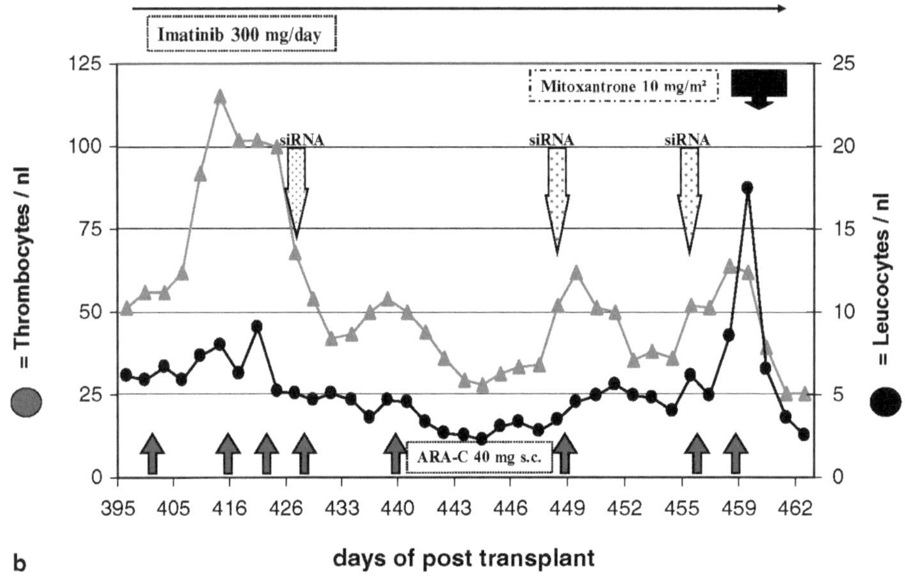

Fig. 22.1. **a** Effect of siRNA application on bcr-abl expression measured by real-time RT PCR. Spotted arrows indicate the time points of siRNA administration with dose in µg per kg/body weight. Grey arrows indicate the time points of cytosine-arabinoside administration (40 mg subcutaneously). Imatinib mesylate was given throughout with 300 mg/day. The black arrow indicates the time point of mitoxantrone application with 10 mg/m^2. After the first siRNA application, the bcr-abl titer fell about a log step. This effect could not be repeated after the second siRNA application even though the siRNA dose was increased. **b** Leukocyte and platelet counts over the time period of siRNA treatment. No considerable hematological toxicities grade ≥ 3–4 WHO classification were seen.

decreased to a minimum of 28/nl on day + 444 followed by an increase to a maximum to 62/nl on day + 449. We repeated the intravenous application of siRNA on day + 447 (21 days later after initial siRNA treatment) with an increased dose 30 µg/kg b.w. in 60 ml DLS (injected over 15 min). After the second siRNA administration with an increased siRNA dose no changes in leukocyte numbers and immature cells were shown and have applied inducing a transient decrease of the *bcr-abl* amount to 7.8% four days later. After the second siRNA administration platelets decreased again to 36/nl on day + 454. At the time of the third siRNA administration on day + 454 with another 10 µg/kg b.w. in 20 ml DLS (injected over 5 min), the *bcr-abl* level had again increased to 50% and dropped only slightly thereafter as shown in Fig. 22.1b. After the third siRNA administration (same dose as the first application) by rapid injection, leukocyte counts decreased for only 2 days to a minimum of 4.9/nl. The platelets increased to a maximum of 64/nl on day + 458 followed by a continuous drop accompanying the progress of the leukemic disease. No considerable hematological toxicities (grade ³ 3–4 WHO classification) between counts before and after siRNA treatment were found, as shown in **Fig. 22.1b**. On day + 455 70% blasts were found in the peripheral blood and a palliative chemotherapy with mitoxantrone (flat 20 mg) was initiated according to the patient's wish. The patient died at day + 473 post-transplant.

9. Course of Extramedullar CML Nodes in the Abdominal Skin and the Malignant Pleural Effusion After siRNA Treatment

At the time of first siRNA administration five nodes were detectable in the abdominal skin. Two of them could be monitored by ultrasound; three others were measurable with a size under 5 mm in diameter. Node A measured 17.8 × 12.2 mm and node B was 14.2 × 7.7 mm. Two ml DLS containing 300 µg siRNA was injected into node A, whereas in node B no siRNA was applied. Both nodes shrank after the first siRNA administration to 2.9 × 1.1 mm (node A) and under 5 mm in diameters (node B).

A malignant pleural effusion was first diagnosed at day + 389 after transplantation with an increasing volume reaching its maximum at day + 430, four days after first siRNA administration. Diagnostic evaluation of pleural fluid showed the detection of CML cells with a *bcr-abl/gapdh* quotient of 4.4% in real-time PCR. The initial diuretic therapy with hydrochlorothiazid 25 mg/day was intensified by adding furosemide 40 mg/day at day + 428. The pleural effusion decreased continuously after day + 430 as demonstrated by ultrasound (right pleura: from 71.6 mm to 4.7 mm and left pleura: from 46.6 mm to 2.7 mm) and a chest x-ray *(21)*.

10. Disease Progression and Resistance to Treatment

Several mechanisms have been proposed to underlie the development of imatinib resistance in CML, including bcr-abl gene mutations, overexpression and amplification of the bcr-abl gene locus, activation of bcr-abl-independent pathway, and increased drug efflux through the multidrug resistance gene *(33)*. To shed further light on the progression before the siRNA treatment, we have analyzed the presence of ABL kinase domain mutations by direct sequencing to assess the extent to which mutations account for resistance. We identified a point mutation with a nucleotide change that resulted in an Y253F mutation in the part of the ATP (phosphate)-binding pocket or P-loop. The presence of bcr-abl positive subclones with point mutations in the ABL kinase domain correlated with acquired resistance to imatinib and especially P-loop mutations are likely to be associated with progression to advanced-phase disease *(34,35)*.

10.1. Data Analysis

The discovery that RNAi could be delivered effectively into mammalian cells has raised the possibility that selective intervention in leukemic cell gene regulation might be feasible in the treatment of leukemia. For effective drug therapy, it is necessary to deliver therapeutic agents selectively to their target sites, since most drugs (Oligonucleotherapy, ASO) are associated with both beneficial effects and unfavorable actions *(36)*. It was demonstrated that the use of siRNA directed against the bcr-abl fusion gene inhibits *bcr-abl* gene expression *(15–18, 37)*. Moreover, *bcr-abl* silencing was accompanied by induction of apoptotic cell death *(15,37)* with apoptosis rates as high as those induced by 1 μM imatinib *(16)*. Unfortunately, the use of oligonucleotides or siRNA requires a sustained delivery system to provide transfection efficiency in mammalian cells. Guo et al. reported efficient gene delivery using anionic liposome-complexed polyplexes containing a constant lipid/oligonucleotide ratio and an increased polycation/oligonucleotide ratio resulting in an increased transfection activity. Anionic lipids such as oleic acid showed significantly less cytotoxicity compared to the commonly used cationic liposomes or polyethylenimine-mediated transfection and several cell lines were transfected with high efficiency *(31)*. Patil and colleagues reported a novel anionic lipoplex DNA delivery system composed of a ternary complex of endogenously occurring nontoxic anionic lipids and physiological Ca^{2+} cations that was characterized by high transfection efficiency and low toxicity *(32)*. Here we describe our experiences with *bcr-abl* siRNA administration to a patient with recurrent $Ph^1(+)$, imatinib-resistant CML after allogeneic HSCT representing an in vivo application of targeted nonvirally delivered synthetic *bcr-abl* siRNA by short volume in rapid injection. Our aim was to evaluate if targeted, nonviral siRNA administration is safe and feasible. To facilitate transfection, we dispensed our siRNA in a DLS with anionic lipoplexes. This anionic lipid delivery system allowed us to overcome the poor stability of siRNA in biological medium resulting in high transfection efficiency. The intravenously administered siRNA DLS was well tolerated. No side effects, apart from mild dizziness (WHO grade 1), were observed, in particular any severe hematotoxicity was absent. Although we used low siRNA doses with initially 10 μg/kg b.w. and 30 μg/kg b.w. at the second application, we observed remarkable antileukemic effects. Assuming an equal body distribution, which might be misleading, this siRNA application (10 μg/kg b.w.) resulted in a tissue concentration of 0.7 nM. Kretschmer-Kazemi Far and colleagues reported on the activity of siRNA compared to antisense oligonucleotides (ASO) in mammalian cells. The biological activity of siRNA seems to be influenced by the local characteristics of the target RNA, including local RNA folding. The concentration dependency of siRNA-mediated suppression indicates a > 1000-fold difference between active siRNA (IC_{50} ~ 0.2–0.5 nM) versus an inactive siRNA (IC_{50} ~ 1 μM), which is consistent with the activity pattern of ASO when relating target suppression to predicted local target accessibility. It remains unclear, however, as to whether such additional substances influence the molecular mechanism of RNAi or whether they catalyze this reaction in an enzyme-like fashion in order to facilitate this process and increase the kinetics without affecting the mechanism *(30)*. Our observations become even more astonishing, when taking into consideration that previous studies on ASO were performed with up to two to three log steps higher oligonucleotide doses than were used by us *(32,33,38)*. The *bcr-abl* level measured from peripheral blood already started to drop one day after siRNA treatment and decreased up to a log step within seven days. The *bcr-abl* level also decreased after the second siRNA injection, though to a lesser extent. After the third siRNA administration no substantial reduction of *bcr-abl* expression was seen suggesting the emergence of siRNA resistance or transfection failure due to induction of serum RNase. In vitro experiments using the peripheral leukemic cells of our patient showed a reduced response rate to *bcr-abl* siRNA transfection in cells which were taken seven days after the last siRNA injection compared to cells taken prior to siRNA treatment *(21)*. A response to siRNA therapy was also suggested by the subsequent regression of cutaneous CML nodes. The malignant pleural effusion decreased in association with siRNA therapy. At the same time since the diuretic therapy was intensified the contribution of siRNA treatment to this improvement remains unclear.

The clinical impact of *bcr-abl* siRNA may, however, very well be based on it functioning as an enhancer of the response to cytostatic agents such as imatinib mesylate or ARA-C. This is supported by recent in vitro experiments they demonstrating that cells that became resistant to imatinib mesylate can be re-sensitized by *bcr-abl* siRNA *(16,39)*. In addition, studies in our laboratory using K562 cells showed that the bcr-abl titer 24 hours after *bcr-abl* siRNA transfection with TransMessengerdecreases by only about 30% (data not shown) indicating that the decrease of *bcr-abl* is not sufficient to explain the antileukemic effects seen in association with the clinical use of *bcr-abl* siRNA. It is also of note that we found comparable effects on the inhibition of proliferation in patient cells and in K562 cells ranging from about 76% in patient cells up to 42% in K562 cells. Apoptosis was induced in up to 99% of patient cells and was less apparent in K562 cells. We speculate that this might indicate that the process of apoptosis is not uniform and shows different patterns of response to chemotherapy. Recent findings by Hornung et al. regarding the stimulation of interferon-aproduction by PDC after siRNA injection into mice have important implications for the assessment of siRNA mediated effects: immunostimulation by siRNA needs to be considered whenever PDC or other toll-like receptor-7 (TLR-7)-expressing cells are present. The administration of non-complexed siRNA in vivo is non-immunostimulatory and siRNA modification with other chemicals is a valuable tool to selectively modify the silencing or the immunostimulatory properties of siRNA for the desired type of application *(40)*. Strategies for the in vivo application of siRNA molecules that include local as well as systemic modes of administration were recently reported by Aigner *(41)*. He reported that in many studies the administration relies on the use of a relatively high amount of siRNAs. Bearing in mind that intracellular immune responses have been shown to be concentration-dependent, this may increase the risk of nonspecific effects in addition to other side effects and cost considerations.

11. Conclusion and Outlook

RNAi has already proven to be a very efficient and specific method for the knockdown of physiologically or pathologically relevant genes of interest. Successful gene therapy by siRNAs depends on the development of efficient delivery systems. Here, we have described recent advances in the clinical application of three doses of targeted, nonvirally delivered siRNA, which provides us with promising initial clinical data of this still experimental therapeutic approach. siRNA administration did not cause adverse events and had some therapeutic value, pursuing further studies and involving additional patients in this treatment. In all probability, one major advantage of formulated, modified, or unmodified siRNAs for gene knockdown will be their "double specificity", i.e., the combination of a high target gene specificity through optimal siRNA sequences and an at least somewhat increased target organ specificity through sophisticated delivery vehicles.

Acknowledgments

The authors would like to thank Katja Ahrens, Melanie Kroll, Silke Gottwald, Ines Riepenhoff, and Christiane Schary for their excellent technical execution of the PCR analyses.

References

1. Goldmann, J.M. and Melo, J.V. (2003) Chronic myeloid leukemia—advances in biology and new approaches to treatment. *N. Engl. J. Med.* 349, 1451–1461.
2. Melo, J.V. (1996) The diversity of BCR-ABL fusion proteins and their relationship to leukemia phenotype. *Blood* 88, 2375–2384.
3. Dean, M., Fojo, T., and Bates, S. (2005) Tumour stem cells and drug resistance. *Nat. Rev. Cancer* 5, 275–284.
4. Elmaagacli, A.H., Freist, A., Hahn, M., et al. (2001) Estimating the relapse stage in chronic myeloid leukaemia patients after allogeneic stem cell transplantation by the amount of bcr-abl fusion transcripts detected using a new real-time polymerase chain reaction method. *Br. J. Haemtol.* 113, 1072–1075.
5. Capdeville, R. and Silberman, S. (2003) Imatinib: a targeted clinical drug development. *Semin. Hematol.* 40, 15–20.
6. Druker, B.J., Guilhot, F., O'Brien, R.A., et al. (2006) Long-term benefits of imatinib (IM) for patients newly diagnosed with chronic myelogenous leukemia in chronic phase (CML-CP): the 5-year update from the IRIS study. *J. Clin. Oncol.* 24, 6506.
7. Hehlmann, R., Berger, U., and Hochhaus, A. (2005) Chronic myeloid leukemia: a model for oncology. *Ann. Hematol.* 84, 487–497.
8. Gratwohl, A., Baldomero, H., Horisberger, B., et al. (2002) Current trends in hematopoietic stem cell transplantation in Europe. *Blood* 100, 2374–2386.
9. Goldman, J. and Gordon, M. (2006) Why do chronic myelogenous leukemia stem cells survive allogeneic stem cell transplantation or imatinib: does it really matter? *Leuk. Lymphoma* 47, 1–7.
10. Fire, A., Xu, S., Montgomery, M.K., et al. (1998) Potent and specific genetic interference by double-stranded RNA in Caenorhabditis elegans. *Nature* 391, 806–811.
11. Elbashir, S.M., Harborth, J., Lendeckel, W., et al. (2001) Duplexes of 21- nucleotide RNAs mediate RNA interference in cultured mammalian cells. *Nature* 411, 494–498.
12. Donze, O. and Picard, D. (2002) RNA interference in mammalian cells using siRNAs synthesized with T7 RNA polymerase. *Nucleic Acids Res.* 30, e46.
13. Mittal, V. (2004) Improving the efficiency of RNA interference in mammals. *Nat. Rev. Genet.* 5, 355–365.
14. Schwarz, D.S., Hutvágner, G., Du, T., et al. (2003) Asymmetry in the assembly of the RNAi enzyme complex. *Cell* 115, 199–208.
15. Scherr, M., Battmer, K., Winkler, T., et al. (2003) Specific inhibition of bcr-abl gene expression by small interfering RNA. *Blood* 101, 1566–1569.
16. Wohlbold, L., van der Kuip, H., Miething, C., et al. (2003) Inhibition of bcr-abl gene expression by small interfering RNA sensitzes for imatinib mesylate (ST571). *Blood* 102, 2236–2239.
17. Elmaagacli, A.H., Koldehoff, M., Peceny, R., et al. (2005) WT1 and BCR-ABL specific

small interfering RNA have additive effects in the induction of apoptosis in leukemic cells. *Haematologica* 90, 326–334.

18. Koldehoff, M., Steckel, N.K., Beelen, D.W., et al. (2006) Synthetics mall interfering RNAs reduce bcr-abl gene expression in leukaemic cells of de novo Philadelphia (+) acute myeloid leukemia. *Clin. Exp. Med.* 6, 45–47.
19. Dass, C.R. (2004) Lipoplex-mediated delivery of nucleic acids: factors affecting in vivo transfection. *J. Mol. Med.* 82, 579–591.
20. Fattal, E., Couvreur, P., and Dubernet, C. (2004) "Smart" delivery of antisense oligonucleotides by anionic pH-sensitive liposomes. *Adv. Drug Delivery Rev.* 56, 931–946.
21. Koldehoff, M., Steckel, N.K., Beelen, D.W., et al. (2007) Therapeutic application of small interfering RNA directed against bcr-abl transcripts to a patient with imatinib-resistant chronic myeloid leukaemia. *Clin. Exp. Med.* 7, 47–55.
22. de Fougerolles, A., Vornlocher, H.P., Maraganore, J., et al. (2007) Interfering with disease: a progress report on siRNA-based therapeutics. *Nat. Rev. Drug Discov.* 6, 443–453.
23. Kawakami, S. and Hashida, M. (2007) Targeted delivery systems of small interfering RNA by systemic administration. *Drug Metab. Pharmacokinet.* 22, 142–151.
24. Soutschek, J., Akinc, A., Bramlage, B., et al. (2004) Therapeutic silencing of an endogenous gene by systemic administration of modified siRNAs. *Nature* 432, 173–178.
25. Morrissey, D.V., Lockridge, J.A., Shaw, L., et al. (2005) Potent and persistent in vivo anti-HBV activity of chemically modified siRNAs. *Nat. Biotechnol.* 23, 1002–1007.
26. Song, E., Zhu, P., Lee, S.-K., et al. (2005) Antibody mediated in vivo delivery of small interfering RNAs via cell-surface receptors. *Nat. Biotechnol.* 23, 709–717.
27. Sezaki, H. and Hashida, M. (1984) Macromolecule-drug conjugates in targeted cancer chemotherapy. *Crit. Rev. Ther. Drug Carrier Syst.* 1, 1–38.
28. Sewell, K.L., Geary, R.S., Baker, B.F., et al.. (2002) Phase I trial of ISIS 104838, a 2¢-methoxyethyl modified antisense oligonucleotide targeting tumor necrosis factor-a. *J. Pharma. Exp. Therap.* 303, 1334
29. O'Brian, S.M., Cunningham, C.C., Golenkov, A.K., et al. (2005) Phase I and II multicenter study of oblimersen sodium, a Bcl-2 antisense oligonucleotide, in patients with advanced chronic lymphocytic leukaemia. *J. Clin. Oncol.* 23, 7697–7702.
30. Kretschmer-Kazemi Far, R. and Sczakiel, G. (2003) The activity of siRNA in mammalian cells is related to structural target accessibility: a comparison with antisense oligonucleotides. *Nucleic Acids Res.* 31, 4417–4424.
31. Guo, W. and Lee, R.J. (2000) Efficient gene delivery using anionic liposome-complexed polyplexes (LPDII). *Biosci. Rep.* 20, 419–432.
32. Patil, S.D., Rhodes, D.G., and Burgess, D.J. (2005) Biophysical characterization of anionic lipoplexes. *BBA-Biomembranes* 1711, 1–11.
33. Advani, R., Peethambaram, P., Lum, B.L., et al. (2004) A phase II trial of aprinocarsen, an antisense oligonucleotide inhibitor of protein kinase C alpha, administered as a 21-day infusion to patients with ovarian carcinoma. *Cancer* 100, 321–6.
34. Marcucci, G., Byrd, J.C., Dai, G., et al. (2003) Phase 1 and pharmacodynamic studies of G3139, a BCL-2 antisense oligonucleotide, in combination with chemotherapy in refractory or relapsed leukemia. *Blood* 101, 425–32.
35. Goldman, J. (2004) Monitoring minimal residual disease in *BCR-ABL*-positive chronic myeloid leukaemia in the imatinib era. *Curr. Opin. Hematol.* 12, 33–39.
36. Branford, S., Rudzki, Z., Walsh, S., et al. (2002) High frequency of point mutation clustered within the adenosine triphosphate-binding region of BCR/ABL in patients with chronic myeloid leukaemia or Ph-positive acute lymphoblastic leukaemia who develop imatinib (STI571) resistance. *Blood* 99, 3472–3475.
37. Branford, S., Rudzki, Z., Walsch, S., et al. (2003) Detection of BCR-ABL mutations in patients with CML treated with imatinib is virtually always accompanied by clinical resistance, and mutations in the ATP phosphate-binding loop (P-loop) are associated with a poor prognosis. *Blood* 102, 276–283.
38. Dias, N. and Stein, C.A. (2002) Antisense Oligonucleotides: Basic concepts and mechanisms. *Mol. Cancer Ther.* 1, 347–355.
39. Wilda, M., Fuchs, U., Wössmann, W., and Borkhardt, A. (2002) Killing of leukemic cells with a BCR/ABL fusion gene by RNA interference. *Oncogene* 21, 5716–5724.
40. Hornung, V., Guenthner-Biller, M., Bourquin, C., et al. (2005) Sequence-specific potent induction of IFN-alpha by short interfering RNA in plasmacytoid dendritic cells through TLR7. *Nat. Med.* 11, 263–270.
41. Aigner, A. (2007) Application of RNA interference: current state and prospects for siRNA-based strategies in vivo. *Appl. Microbiol. Biotechnol.* 76, 9–21.

INDEX

A

B

C

D

P

R

S

T

MIX
Papier aus verantwortungsvollen Quellen
Paper from responsible sources
FSC® C105338

If you have any concerns about our products,
you can contact us on
ProductSafety@springernature.com

In case Publisher is established outside the EU,
the EU authorized representative is:
Springer Nature Customer Service Center GmbH
Europaplatz 3, 69115 Heidelberg, Germany

Printed by Libri Plureos GmbH
in Hamburg, Germany